16 0818474 1

AF616062

THE EVOLVING UNIVERSE

# ASTROPHYSICS AND SPACE SCIENCE LIBRARY

VOLUME 231

# THE EVOLVING UNIVERSE

## SELECTED TOPICS ON LARGE-SCALE STRUCTURE AND ON THE PROPERTIES OF GALAXIES

Edited by

DONALD HAMILTON
*Institute of Astronomy and Astrophysics,*
*University of Munich, Germany*

KLUWER ACADEMIC PUBLISHERS
DORDRECHT / BOSTON / LONDON

Library of Congress Cataloging-in-Publication Data

ISBN 0-7923-5074-X

Published by Kluwer Academic Publishers,
P.O. Box 17, 3300 AA Dordrecht, The Netherlands.

Sold and distributed in North, Central and South America
by Kluwer Academic Publishers,
101 Philip Drive, Norwell, MA 02061, U.S.A.

In all other countries, sold and distributed
by Kluwer Academic Publishers,
P.O. Box 322, 3300 AH Dordrecht, The Netherlands.

*Printed on acid-free paper*



Printed in the Netherlands.

## TABLE OF CONTENTS

**The Las Campanas Redshift Survey**



**The Moderate Redshift Universe**



**Redshift Distortions**

## Future Surveys



## Absorption-Line Based Studies



## Hubble Deep Field Analyses



## Theory and Theoretical Analyses of Observational Data

**Miscellaneous**

# INTRODUCTION AND PREFACE

## 1. The Workshop and this Tome

In the excellent bucolic setting of Schloß Ringberg in Upper Bavaria, over 50 scientists assembled during the week of 23–28 September 1996 to discuss recent results, both theoretical and observational in nature, on the large-scale structure of the Universe. Such a topic is perhaps nowadays far too encompassing, and is essentially all of what we used to call "observational cosmology."

The original philosophy of the organization of this meeting was deliberated aimed at the younger community and their contributions. As a consequence, the content of the presentations was refreshingly new, as it should be. In spite of the deficiences caused by the lack of certain key researchers in this field, for one reason or another, the final result was rewarding to all. Although the conference was held in Fall 1996, the contributions contained herein were submitted as late as Spring 1998, thus the content maintains some degree of trendiness.

Originally the current volume was to be a "proceedings." This refers to the usual archival tome that fills one's shelf and is rarely consulted, except to see the canonical group photo, which by the way, we also have. Nevertheless, I wanted something more than that. Although the field is rapidly changing, with so-called *facts* in a state of constant volubility, now is a good time for reflection prior to the commencement of the Sloan Survey, presumably the definitive large-scale program of low- to moderate-redshift galaxies in our lifetime.

It was decided, therefore, not to limit the space of a contribution but to allow the authors to use as much space as they felt appropriate. Because of this, the quality of the papers is much greater than what would have otherwise been if the normal page restriction had been enforced. At this point this volume became not a proceedings but an edited volume. In other words, what is contained herein are selected topics on large-scale structure and on the properties of galaxies. It is hoped now that such a style will be



*D. Hamilton (ed.), The Evolving Universe*, 1-10.

more enduring and useful to the astronomical community. The centerpiece of this tome is a paper by Hamilton (the European living in America) on linear redshift distortions. Unfortunately, at least one contribution (that of Dalcanton) was not forthcoming. I regret much the lack of this one, as at the Ringberg meeting this contribution was both interesting and substantial.

## 2. Key Issues in Large-Scale Structure

### 2.1. REALLY LARGE-SCALE STRUCTURE

As is almost trite by now, the size of physical structures found by various redshift surveys is on the order of the extent of the survey. In other words, the sizes found continue to increase in size. The size distribution of structures is, as is well known, a measure of the distribution of the original fluctuation spectrum. Naturally, measurements over such a large area are time consuming and are rarely given the telescope time to complete a survey properly.

One structure that is probably the largest suggested to date is that of an extended Corona Borealis Supercluster[1] of galaxies. The classical Corona Borealis supercluster (at about 15hrs, +30°) consists of a collection of dense (mostly X-ray bright) clusters at redshifts about $z$=0.075 (*e.g.*, Abell 2056, 2061, 2065, 2067, 2079, and 2089). In addition, there are several background clusters at $z$=0.11 (Abell 2062, 2069, and 2083) and at $z$=0.23 (Abell 2059 and 2073) that help to give the impression of a more dense region. The traditional lateral dimensions[2] have been given as $\sim$ 100 Mpc by $\sim$ 200-400 Mpc.

Due to their somewhat excessive masses, superclusters would distort the local Hubble flow, and by studying the galaxies on the periphery of these structures an estimate of the total mass density is possible. Such structures are far from virialization (as by the way, are most clusters of galaxies). Dynamically speaking, such structures are quite young in their evolutionary state.

However, the redshift $z \sim 0.08$ is one that should be quite familiar to those who have studied Abell clusters. Examining the redshift distribution of all Abell clusters (see Figure 1) suggests that in fact the redshift interval between $z \sim 0.06$ and $z \sim 0.09$ is one that is *preferred*, *i.e.*, it has a unambiguous, albeit broad, peak. Several famous Abell clusters that are a part of this peak are A2029, A2670, A401, and A104. Selecting clusters that have redshifts between 0.06 and 0.09 and just examining their distribution on the sky in a rather crude manner (Figure 2) presents a rather unusual

[1]Superclusters were originally defined as clusters of clusters of galaxies. Today these can be recognized to be the density enhancements along a filament of galaxies.

[2]The Hubble constant is assumed throughout to be 60 km sec$^{-1}$ Mpc$^{-1}$.

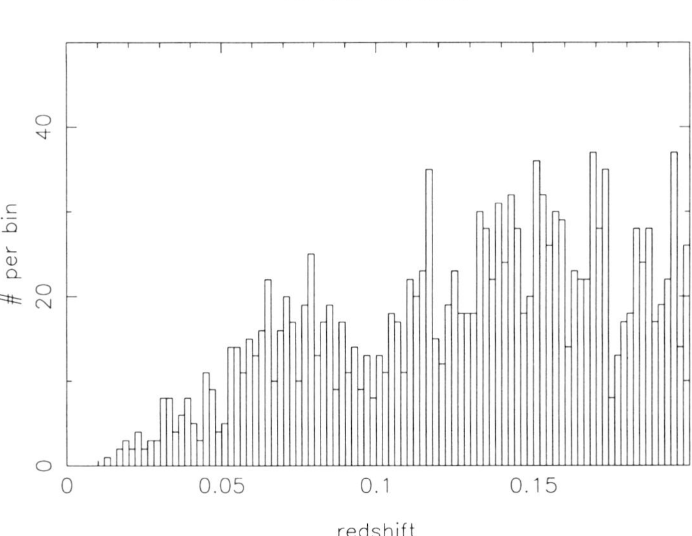


*Figure 1.* Redshift distribution of Abell Clusters, noting the preferred redshift interval $z \sim 0.06$ to $0.09$, which is the same as that of the Corona Borealis Supercluster. The incompleteness of such a diagram begins only at about $z$=0.2, and so the aforementioned peak is unlikely to be due to a deficiency of redshifts just beyond $z \sim 0.08$.

feature – what might be called the *extended* Corona Borealis Supercluster. In the lesser dimensions, this structure is about 150 Mpc deep[3] and $\sim$ 100 Mpc in the shortest dimension perpendicular to the line-of-sight. However, in the longest dimension, the supercluster appears to extend to at least a gigaparsec. The classical structure was mentioned in various reviews by N. Bahcall, but this extended size was not recognized at that time. With these dimensions the supercluster takes a cigar-shaped or triaxial form. Perhaps this supercluster is an example of a Zel'dovich pancake or perhaps just a filament? The mass of this supercluster is $\sim 10^{17-18}$ solar masses, which is orders of magnitude larger than the largest predicted mass from an adiabatic perturbation. Furthermove the probability of finding such a massive structure within a Hubble volume in *any* flavor of CDM is substantially less than one. The characteristic scale of a pancake was derived by Schaeffer and Silk (1985) to be approximately 30 Mpc, somewhat lower by nearly two orders of magnitude than the *CorBor* structure. The structure may be breaking up into clusters as there is the existence of the canonical *CorBor* region plus the covey of clusters located to the south-west (about 12hrs,

[3]The velocity breadth is between 18000 and 27000 km sec$^{-1}$ (based upon results of the Norris Survey at The Hale Telescope).

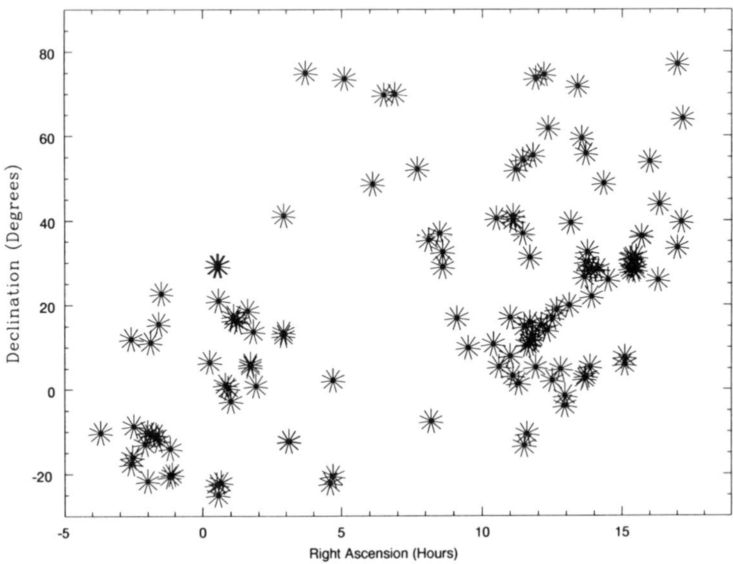


*Figure 2.* Spatial distribution of Abell Clusters with redshifts within that of the Corona Borealis Supercluster, namely 0.060-0.090. The "extended" Corona Borealis supercluster is an essentially linear structure $\sim$ 150 Mpc by 100 Mpc in the shortest dimensions (assuming here $H_o = 60$ km sec$^{-1}$ Mpc$^{-1}$) and 1000 Mpc in the longest direction, thus making it the largest known physical structure. The size of the symbols is approximately that of two Abell diameters. The areas apparently without any Abell clusters are within the zone of avoidance.

$+18°$).

Besides the peak within the redshift distribution of the Abell clusters, the preferred value of $\sim$ 0.075 is also apparent in other surveys. The back side of the Boötes void (located $\sim$ 40°) is at this redshift. There is a significant peak in the redshift distibutions of Broadhurst *et al.* (1990) ($\sim$ 45° removed from *CorBor*); the IRAS north-ecliptic pole region (Ashby *et al.* 1996) and is located at one of the ends of the extended cluster distribution; and also there is a significant *CorBor* peak in the northern fields (especially the 'North 50') of the Las Campanas Redshift Survey (located about at the opposite end of the structure from that of the NEPR field). The Sloan Sky Survey will obviously be able to delineate properly the boundaries of the filament and its contrast relative to the field. It should be noted that the most of the northern sky (the area to be surveyed by Sloan) is dominated relative to that of the southern sky by the presence of *CorBor* and friends (see, Picard, Caltech Ph.D. thesis).

The significance of such a structure within the context of CDM is unclear. Kaiser and Davis (1985) calculated that the *CorBor* structure known at that time[4] was only a $4\sigma$ event at formation. Unfortunately, the possibility of calculating a believable value for the strength of the fluctuation is a nontrivial task (via geometry) given the exceptionally non-spherical form for the cluster. Only through the mass can a qualitative derivation be realized.

## 3. Basic Results at Ringberg

### 3.1. REDSHIFT SURVEYS

It is necessarily the case that the consequences of redshift surveys were presented (and are conveyed herein as well) such as correlation functions and luminosity functions. Important papers were present on the former, but the latter is by far the most extensive discussions present here. In fact a great many emphasize the local Universe, and not to seem too evangelistic, an area sorely neglected by the extragalactic community in their daze caused by higher-redshift fever.

The enthusiasm of the participants for these subjects and others was evident in the endless energy with which they wished to carry on extended discussions – especially during the evening sessions when the so-called "Technical Discussions" occurred. Little airy persiflage was the norm.

This meeting had the particular advantage that several large redshift surveys were finished during the year preceeding that of the Ringberg meeting. These include the Las Campanas Redshift Survey, The Norris Survey (The Hale Telescope), ESP (ESO), the Münster Redshift Project, and recent re-analyses of SSRS2 and CfA2.

Due to the importance of the Las Campanas Redshift Survey, it was given the emphasis that it deserved. In total there were four presentations: the first by Bob Kirshner giving a general review (regretfully not included in these proceedings); one by Huan Lin on the luminosity function and 3D power spectrum results; one by Stephen Landy on the geometry and power spectrum of the LCRS; and finally, one by Douglas Tucker on the autocorrelation function (in $z$-space). This survey will be the largest redshift survey conducted in an 'old-fashioned' style, meaning only a few participants, before the Sloan Digital Sky Survey comes to fruition. It may even have higher number of redshifts per participant than that of the Sloan!

One point was made clear at the workshop: that the observational selection effects (including biases introduced by the design of the instrument) can have significant influence over the result. It can only be emphasized

[4]The size was basically that of the core group of rich clusters of galaxies something around 40 by 40 Mpc. The depth at that time was unknown.

that to correct for such deficiencies that extensive Monte-Carlo simulations be performed.

### 3.2. EVOLUTION OF CLUSTERING

It is expected that clustering of galaxies would evolve over time with galaxies being more clustered locally than at earlier times. It would be then an appropriate goal to establish the strength of the clustering decrease with redshift, and in particular, the relevance of this decrease within a CDM context. The survey that will be able to contribute significantly to this discussion will be the moderate-$z$ CNOC2 survey of field galaxies, and later the SDSS, to some extent.

Nevertheless, surveys that have already been conducted and published can contribute to the understanding, if not quantitatively, at least qualitatively to the determination of the evolution of both the slope of the correlation function and its normalization. The low-redshift correlation data fits of CfA, SSRS, and Stromlo when combined with the moderate-redshift Norris Survey data, and that of the Canada-France-Redshift Survey give a dramatic impression (Figure 3) of a highly evolving proper correlation length. The formalism for this type of evolution is discussed by Le Fèvre (this volume, p 169). Fitting the aforementioned data sets yields $\epsilon = 2.3 \pm 0.1$. This is substantially different than what is expected using linear theory $(\gamma - 1 \sim .7)^5$. Preliminary analysis has revealed that there appears to be no variation of $\gamma$ with redshift, and this is supported by N-body modelling within a CDM context (Davis *et al.* 1985). As a reminder, a random pattern locally is established for $r_o \sim 10$ Mpc.

The interpretation of this result as evidence for clustering evolution should be approached with caution. Any differential selection probability between absorption-line and emission-line galaxies as a function of redshift could mask or mimic such evolution. Ignoring this difficulty, and segregating between these two sets of galaxies, suggests that for the Norris data the absorption-line galaxies show little evidence for clustering evolution whereas the emission-line galaxies show substantial. The former result is in contradistinction with that of the Le Fèvre *et al.* result (see, p167), but the latter is essentially in agreement. At the least, the blue galaxies (here assumed to be emission-line galaxies) have a clustering pattern that evolves significantly with redshift. The clustering evolution of the red galaxies is still uncertain. Furthermore, recent results from Keck suggests that there are significant clustering patterns already established by a redshift of three,

[5]This assumes linear theory, whose applicability in this particular case can be successfully challenged.

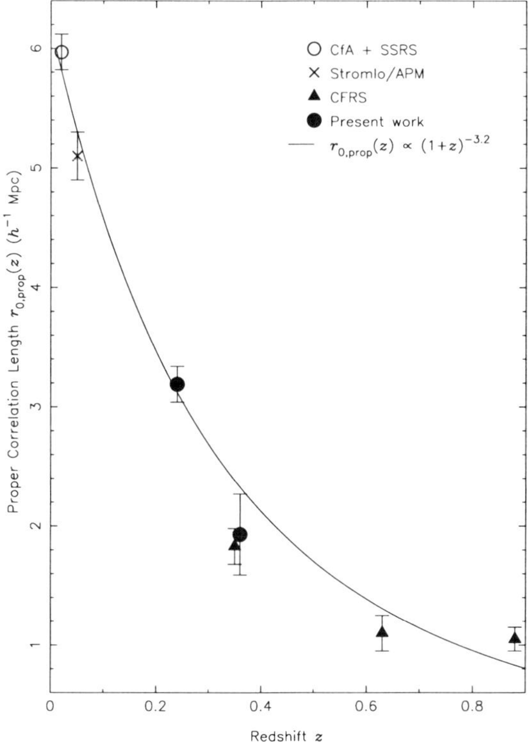


*Figure 3.* Variation of proper correlation length as a function of redshift. Assuming population mix is invariant within these surveys, then what is presented here demonstrates the decrease in clustering with greater look-back times. Segregating galaxies according to color, suggests that blue galaxies are less clustered in the recent past, whereas red galaxies are similarly clustered as locally. The CfA and SSRS contribute to the low-$z$ data. The CFRS concentrates on galaxies with redshifts in excess of 0.5. The Norris Survey from the 200-inch telescope ("Present Work"; presented in Small, Sargent, and Hamilton 1997) fills in between these surveys. Fortunately, the latter two surveys overlap in redshift and their respective data points coincide. These results are obtained by determining $\xi(r)$ from the projected correlation function.

consistent with the low or even no evidence for red-galaxy clustering evolution.

Biasing has now become the savior of cold dark matter theory. By introducing a time-dependent biasing factor, significant clustering at high redshift can be explained. Heretofore, CDM predicted that significant structure formed late. In order to save CDM, the biasing factor is now required to be dependent upon both the epoch in question and upon the environment (local matter density). Remember when CDM was a complete theory with only *one* adjustable parameter – the amplitude of the fluctuation spectrum!

### 3.3. LUMINOSITY FUNCTIONS

Participants of the aforementioned surveys presented their respective analyses for luminosity functions. Marzke reviewed the latest results of the SSRS/CfA/Century and other local galaxy redshift surveys to derive the local luminosity function as well as the local luminosity density. Lin reviewed the LF obtained from the Las Campanas Survey; Small, Sargent, & Hamilton, that of the Norris Survey; Zamorani *et al.*, that of the ESO Slice Project; and da Costa reviewed the status of the local LF, based upon the SSRS2. These analyses usually divide their samples according to either color or emission-line strength (such as that from [OII]3727). The bluer the selection criteria, the steeper the faint-end slope. Biases introduced by object selection criteria or instrumental design must be determined by Monte-Carlo simulations, otherwise comparisons between the various surveys will be meaningless.

One vital resolution of a long-standing problem can be derived from these various results: that the canonical normalization of the luminosity function as given by Loveday *et al.* (1995; the Stromlo/APM Survey) appears to be about a factor of two too low. With the higher normalization, some of the difficulty of modelling galaxy counts at faint magnitudes based upon the local LF has been removed. Furthermore, a steeper faint-end slope than what is normally assumed is also likely and this helps to reduce even further the amount of luminosity evolution required to explain the faint numbers of galaxies.

There now are two unambiguous camps of results. On the one hand, the analyses of the Norris Survey (Small, Sargent, & Hamilton), ESP (Zamorani *et al.*), SSRS2 (da Costa), and Autofib (Ellis *et al.* 1996) all support this approximate factor of two increase in the local normalization. The support for a similar normalization as that of the Stromlo/APM results comes from the LCRS (Lin *et al.*).

The great difficulty with these and other surveys (in particular the Las Campanas Redshift Survey) is the inability to detect low-surface-brightness galaxies. These deficiencies are discussed in the papers by Marzke, by Lin, and by Small, Sargent, & Hamilton. The recent surveys by Dalcanton (1996) and by Sprayberry *et al.* (1995) have clearly demonstrated the existence of such beasties. The general conclusion of these surveys, and that of Zamorani *et al.*, is that galaxy luminosity function steepens at the faint end and that this is due exclusively to the contribution of low-surface brightness emission-line galaxies (hence blue). Previous surveys have usually pre-selected high-surface brightness (or intrinsically bright) galaxies.

Springel presents a new technique for deriving luminosity functions based upon the non-parametric maximum likelihood method incorporating possible luminosity evolution. The selection function is concomitantly derived and this technique is applied to the 1.2 Jy IRAS survey data and similar results with that of previous analyses are derived.

### 3.4. PAIRWISE-VELOCITY DISPERSION

One critical test of CDM is the distribution of galaxies' velocities relative to each other. With a COBE quadrupole anisotropy normalization of the power spectrum, standard CDM predicts a rather *hot* dispersion of $\sim 1000$ km $\sec^{-1}$. One way around this is to introduce biasing. Another way is to assume that the density of matter is less, *i.e.*, the Universe is open. The derivation of Davis and Peebles (1983) of the pairwise-velocity dispersion based upon CfA1 results suggested that this dispersion was $\sim 350$ km $\sec^{-1}$.

Recent results of more complete galaxy samples have derived dispersions at least twice the result of Davis and Peebles. The advantage of both the Norris Survey and that of the combined SSRS2/CfA2 surveys (Marzke *et al.* 1995) are the substantially larger volumes in which galaxies were surveyed than that the original CfA1. Unfortunately, the value of the dispersion is *critically* dependent upon which galaxies are included in the derivation. Removing the Abell clusters from the Norris Survey results reduces the dispersion by $\sim 300$ km $\sec^{-1}$, to something that agrees with that of Davis and Peebles. The net conclusion that should be derived from these results is that there needs to be a more robust statistic for measuring the hotness of the galaxy *gas*.

### 3.5. POWER SPECTRUM ANALYSIS

Derivations of the power spectra can yield useful constraints on the structure formation mechanism (see, *e.g.*, the papers by Lin and Schuecker). Several groups presented their respective results. The Las Campanas Redshift Survey power spectrum analysis is described in the paper by Lin (spatial) and that by Landy and by Tucker (both surface analyses). For those wave vectors that overlap with those of previous surveys such as that of SSRS2/CfA2, IRAS 1.2 Jy, and QDOT the LCRS results are a confirmation. The results from the Norris Survey also included power spectrum analysis confirming the existence of the $\sim 250$ Mpc periodicity in the redshift distribution found by Broadhurst *et al.* (1990).

3.6. CLOSING

I wish to thank the other members of Program Committee for their help in organizing a meeting enjoyed by all present. The members of the committee included Roger Blandford, Simon White, Peter Schneider, and Margaret Geller. It is Margaret Geller to whom I owe a great deal of gratitude for her advice. Unfortunately, she could not attend the meeting. I am deeply indebited to Dr. Ulrich Hopp for a careful reading of all the manuscripts and preparing the table of contents.

Finally, I wish to thank the Max-Planck-Gesellschaft and the Deutscheforschungsgemeinschaft for their funding of this meeting and to the staff of Schloß Ringberg, especially to the manager Herr Axel Hörmann, for their professional assistance in making this meeting a memorable and useful experience for the participants.

Donald Hamilton
*Fakultät der Physik, Technische Universität München &*
*Institut für Astronomie und Astrophysik, Universität München*
*Bogenhausen, 1 May 1998*

# LARGE-SCALE STRUCTURE AND THE NEARBY UNIVERSE

LUIZ NICOLACI DA COSTA
*European Southern Observatory*
*Karl-Schwarzschild Strasse 2, D-85748 Garching bei München, Germany*

**Abstract.** We review recent results obtained from nearby redshift and redshift-distance surveys to argue the importance of studies of the local universe to our understanding of structure formation and evolution. In particular, data from dense and complete spectroscopic surveys combined with data from peculiar-velocity measurements may play a key role for understanding galaxy biasing. Significant progress on this front may soon become possible using 4-m class telescopes.

## 1. Introduction

HST and Keck have had an enormous impact in many areas of astronomy, in particular shifting much of the attention of cosmologists to the high-redshift universe. Breakthroughs like the identification of a population of normal-looking galaxies at redshifts $z \gtrsim 3$ and the investigation of their spatial distribution, studies of the star-formation history of galaxies and their chemical evolution, studies of the three-dimensional distribution of absorption systems, and hints for the presence of massive clusters at high-redshift have been some of the main topics of recent interest.

However, in spite of these exciting new developments and the impact that they will have in our understanding of galaxy and large-scale structure formation and evolution, they should not overshadow the contribution that studies at small ($z \lesssim 0.05$) and intermediate ($z \lesssim 0.2$) redshifts will continue to make, some of which unique in nature. There are several reasons for that. First, redshift surveys of the nearby universe characterize the large-scale structure and galaxy properties at the present epoch and thus serve as a reference for evolutionary studies. Second, wide-angle, dense sampling surveys are ideal for comparisons with N-body simulations. Third, redshift-



*D. Hamilton (ed.), The Evolving Universe,* 11–21.

distance surveys can only be used locally to recover the underlying mass distribution from measurements of the peculiar velocity field of galaxies. This is currently the most powerful method to probe mass fluctuations on intermediate scales ( $\lesssim 100h^{-1}$ Mpc, where $H_o = 100h$ km s$^{-1}$ Mpc$^{-1}$), complementing studies of weak-lensing on cluster scales and CMBR anisotropies on very large scales. Fourth, by combining the data from redshift and redshift-distance surveys one can study the relative distribution of galaxies and mass and the dependence of the internal properties of galaxies on the local density. Such studies provide the means for a direct study of galaxy biasing mechanisms. Finally, the data can also be used to constrain cosmological parameters over a wide range of scales by using, for instance, redshift distortions, the galaxy power-spectrum and large scale motions to derive estimates of $\beta = \Omega^{0.6}/b$, where $\Omega$ is the cosmological density parameter and $b$ the linear biasing factor.

In §2 we discuss the nature of the three-dimensional galaxy distribution as revealed by complete nearby surveys. In §3 we discuss the nature of the mass distribution recovered from maps of the peculiar-velocity field of galaxies as derived from an $I$-band Tully-Fisher survey of late spirals. In §4 we discuss how these results may be combined to improve our understanding of galaxy biasing. A brief summary is presented in §5.

## 2. Nearby Redshift Surveys

### 2.1. DISTRIBUTION OF GALAXIES

Wide-angle and complete redshift surveys like CfA2 [15] and SSRS2 [5] [8] provide a unique and so far unmatched database which combines dense sampling and an almost full three-dimensional view of the present-day galaxy distribution out to a moderate depth ($cz \lesssim 15{,}000$ km s$^{-1}$). Such characteristics make these samples of great value as they allow us to study the galaxy distribution over a wide range of scales. Furthermore, the large number of galaxies ( $\gtrsim 15{,}000$ galaxies in the combined CfA2-SSRS2 sample) makes it possible to subdivide the sample in a variety of ways, an essential feature for more detailed studies of the clustering properties of galaxies of different types.

The projected distribution of galaxies in the SSRS2 is shown in Figure 1. The SSRS2 consists of $\sim$ 5,500 galaxies with $m_B \leq 15.5$ probing different directions on the sky. The SSRS2 south covers about 1.1 steradians of the southern galactic cap ($b \leq -40°$) in the declination range -40° $\leq \delta \leq$ -2.5 ° [5]. The SSRS2 north [8] covers 0.6 steradians in the northern galactic cap ($b \geq 35°$, $\delta \leq 0°$). At the equator the SSRS2 joins the CfA2 providing a contiguous sky coverage over about two steradians in each galactic cap. These surveys provide a unique sampling of the galaxy distribution out to

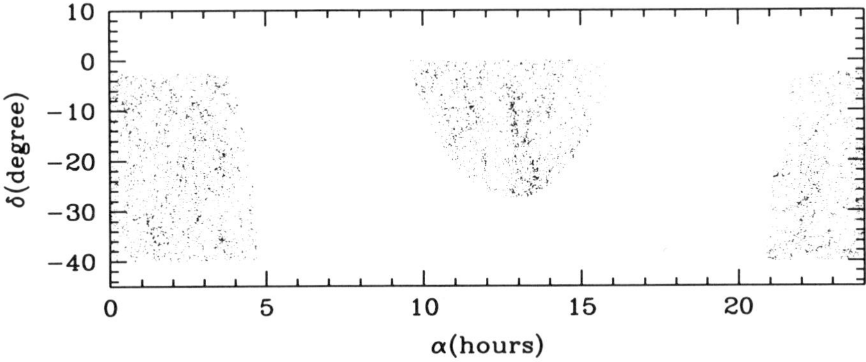


*Figure 1.* Projected distribution of SSRS2 sample.

z $\sim$ 0.05 covering more than one-third of the sky. When combined, they probe galaxy separations of the order of 300 $h^{-1}$ Mpc. Since the surveys probe different directions of the sky, sampling different structures, they can also be used to estimate directly sample-to-sample variations for different statistics.

In Figures 2 and 3 we show redshift maps displaying a cross-section of the local galaxy distribution obtained by combining the CfA2 and SSRS2 data, which probe opposite directions on the sky. From the redshift maps alone one finds that large coherent structures appear to be a common feature of the galaxy distribution. Walls and voids, 5000 km s$^{-1}$ in diameter, are seen in every region large enough to contain them. The qualitative picture that emerges is one in which the galaxy distribution consists of a volume-filling network of voids. Support for this picture comes from more detailed analysis of the geometrical properties of the galaxy distribution using an automatic void-finding algorithm [12]. Application of this method to the SSRS2 south shows that voids occupy at least 50% of the volume and walls only about 25%. The mean diameter of the voids is about 40 $h^{-1}$ Mpc, consistent with the scale inferred by visual inspection of the redshift maps. We should emphasize that the SSRS2 is a quiet region [24] [28], where redshift distortions are small and do not significantly affect these results.

We have also computed the power-spectrum (PS) in redshift space of the combined CfA2-SSRS2 sample [6]. Our results indicate that the PS is still rising on scales $\lesssim$200 $h^{-1}$ Mpc with no strong evidence for a turn over. These results have been confirmed by similar analysis of other optical and infrared-selected samples. A good fit for the observed PS in redshift space can be obtained with a CDM $\Omega h$ =0.2 model with minimal bias. It is worth noting that the sample-to-sample variations are within our estimated errors and are probably due to differences in the number of clusters in each

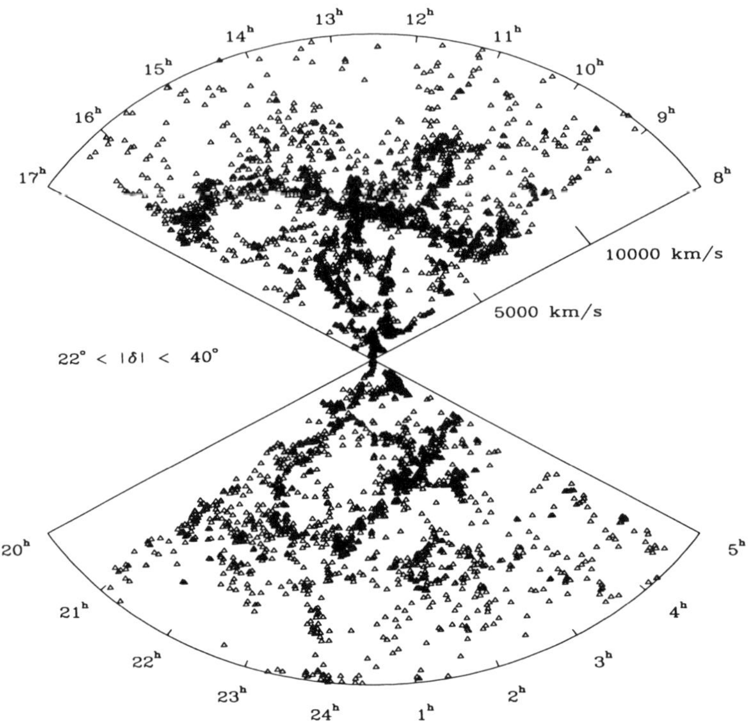


*Figure 2.* Right ascension versus radial velocity for the SSRS2 south (lower portion) and CfA2 north (upper portion) galaxies with cz$\leq$ 15,000 km s$^{-1}$.

sample which varies significantly from sample to sample.

## 2.2. GALAXY PROPERTIES

### 2.2.1. *Luminosity Function*

A puzzling question which remains unresolved is the discrepancy between the normalization of the local luminosity function (LF) based on deep surveys [29] and that derived from nearby surveys. It has been argued that errors in the magnitude system used to derive the local samples could be partially responsible for the discrepancy. The large variations in the Schechter parameters for the CfA2 north and south samples have been used as an argument in favor of this hypothesis. However, using the recently completed SSRS2 north [8] one finds that the LF for SSRS2 north is remarkably similar to that previously determined for the SSRS2 south. This implies that the problem does not lie in the magnitude system used for the nearby samples. Other possibilities are the existence of a local hole

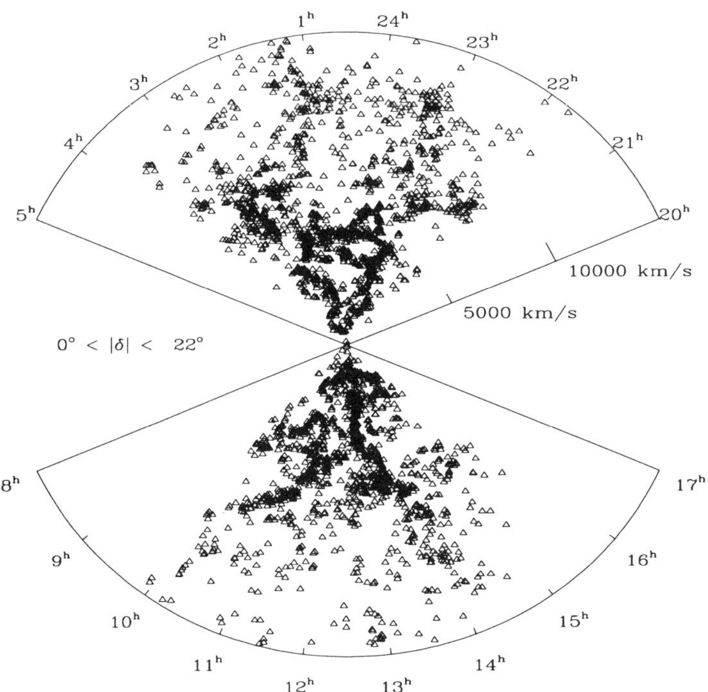


*Figure 3.* Right ascension versus radial velocity for the SSRS2 north (lower portion) and CfA2 south (upper portion) galaxies with cz$\leq$ 15,000 km s$^{-1}$. The absence of galaxies near the edges of the diagram is caused by the imposed galactic latitude limit of the SSRS2 north.

in galaxy distribution, recent rapid evolution of the luminosity function, or the existence of a large population of low surface-brightness galaxies not included in the present catalogs. Note that the good agreement between the LFs computed for the SSRS2 north and south samples suggests that if there is a hole it must be large.

The luminosity function of galaxies has also been examined as a function of color and morphology using the SSRS2 galaxies. The analyses show that even locally one observes an excess of blue galaxies at faint magnitudes. It is estimated that the faint-end slope of blue galaxies is $\alpha \lesssim -1.3$ [21]. It has also been found that early- and late-type galaxies have very similar LFs [22], in good agreement with previous measurements from CfA2 [23]. This result suggests that the Stromlo-APM [19] underestimates the abundance of faint, early-type galaxies. More importantly, we confirm the steep faint-end slope of the LF of Irr/Pec galaxies ($\alpha \sim -1.85$ ) obtained from CfA2 [23].

While the steep slope in principle minimizes the required evolution of the LF to account for the results of deep number counts, the low normalization of the SSRS2 still implies large discrepancies between the observed counts and no-evolution predictions.

### 2.2.2. *Clustering Properties*

The large number of galaxies available in the nearby dense surveys has made it possible to examine in greater detail the clustering properties of galaxies of different types. Recent work based on the SSRS2 survey [2] has shown strong evidence for a luminosity bias that is scale independent but depends strongly on the luminosity. More luminous galaxies show a much stronger correlation than sub-$L_*$ galaxies. The result is in marked contrast to the findings based on the Stromlo-APM survey [20]. The reasons for the discrepancy are not yet clear. While several models of galaxy formation predict some kind of luminosity dependence none can reproduce the observed variation.

An interesting spin-off of this analysis has been to find that very bright galaxies ( $\gtrsim 3\ L_*$) show a large correlation length ($\sim$ 15 $h^{-1}$ Mpc), comparable to that observed for clusters. Interestingly, these galaxies are not found preferentially in either clusters or loose groups, and a large fraction show perturbed morphologies. The speculation is that they may reside in more massive dark halos. Work is currently underway to examine in more detail the properties of these very bright galaxies and their environment.

Counts-in-cells analysis has also been used to compute the Void Probability Function $P(0, V)$ and, more generally, the count probability distribution function $P(N, V)$, from which the normalized skewness $S_3$ and kurtosis $S_4$ have been computed for different volume-limited samples [3]. From this analysis it has been found that the high order moments, in particular $S_3$, do not depend on the luminosity in contradiction with the linear biasing scheme, but in good agreement with the predictions of theoretical models for biasing [26]. This analysis illustrates the usefulness of dense local samples for studying galaxy biasing,

Using the combined SSRS2 south and north samples the two-point correlation function has been computed for both volume-limited subsamples and the complete magnitude limited sample in redshift and real space. The main results are: 1) we confirm the luminosity-dependent bias; 2) we find that the sample-to-sample variations are within the estimated errors; 3) we confirm that both portions of the sky surveyed by the SSRS2 are fairly "quiet" with small redshift distortions; 4) we find a remarkable agreement between the correlation parameters obtained from the analysis of the SSRS2 and those from the Stromlo-APM and LCRS surveys, despite the fact that the latter surveys probe a volume significantly larger than the SSRS2.

We have also examined the correlation properties of galaxies as a function of morphological types and colors [28]. By considering volume-limited samples we find that the luminosity dependence is independent of morphological type of the galaxies, with the clustering amplitude increasing for both early- and late-type galaxies in a similar way as the whole sample. We also find that while in redshift space there is an apparent bias of ellipticals relative to spirals, and of red relative to blue galaxies, this is not the case in real space. Taking at face value these results suggest that the clustering properties of galaxies are independent of their morphology and color, at least in the absence of rich clusters as it is the case for the volume probed by the SSRS2.

## 3. Large-Scale Motions

A unique feature of studies of the nearby universe is that it is only locally ( $\lesssim$8000 km $s^{-1}$) that we can expect to recover the mass distribution from measurements of the peculiar velocity of galaxies. Although cosmic flows offer a powerful to probe mass fluctuations the compilation of suitable samples for analysis are extremely demanding as they need all-sky samples and both optical and radio data. As a consequence large redshift-distance surveys have lagged behind redshift surveys and the available samples have been sparse, shallow and with a non-uniform sky coverage.

Recently, two $I$-band Tully-Fisher distance surveys have been completed: the Mathewson *et al.* survey (Mat92 [25]) of spirals of different types in the southern hemisphere and the SFI and SCI surveys [7] [16] [17] consisting of over 1200 Sbc-Sc field galaxies in the region $\delta > -45°$ and $b > |10°|$ and about 800 spirals in the direction of 24 clusters. These surveys are based on $I$-band photometry and measurements of the rotational velocity from either optical rotation curves or the 21 cm line-width. The Mat92 and SFI surveys are the basis of the most extensive all-sky catalogs currently available to map the peculiar velocity field. We should point out that in the near future these surveys of spiral galaxies will be complemented by a large all-sky survey, near completion, of over 1500 early-type galaxies (ENEAR) with estimated distances [1]. Combined they will provide an unprecedented sample for studies of the mass distribution.

### 3.1. MASS DISTRIBUTION

We have used the SFI sample to reconstruct the mass and three-dimensional velocity fields [7] under the assumption that the peculiar motions are induced by the gravity field associated with mass fluctuations. If this is the case, then the flow is fully described by a scalar potential, at least on large scales where linear theory is valid, which can be derived from the observed radial component of the peculiar velocity [4]. Distances to the individual galaxies were estimated using the $I$-band direct Tully-Fisher relation derived by combining the data for 24 clusters in the SCI sample. In order to correct for biases, which include the homogeneous and inhomogeneous Malmquist bias as well as sample selection bias, a Monte-Carlo approach has been used. The method estimates the bias directly using the real space density field of galaxies, reconstructed from redshift surveys, and large mock catalogs of artificial galaxies "observed" using the assumed TF relation and scatter as the real data [14].

One of the main findings is that the derived mass distribution resembles the observed galaxy redshift distribution much more closely than any previous reconstruction [11]. In particular, voids in the galaxy distribution reflect real voids in the mass distribution [12]. Another remarkable feature is that velocity field along the supergalactic plane shows for the first time a bifurcation towards the GA and PP, in much better agreement with the predicted *IRAS* velocity field, as expected if light traces matter.

A detailed comparison between the measured peculiar-velocity field of the SFI galaxies and that derived from the *IRAS* 1.2 Jy galaxies is currently underway [9]. The analysis is based on the expansion of the SFI and the *IRAS* velocity fields by smooth orthonormal functions [27]. Preliminary results show that the SFI and the *IRAS* velocity field agree remarkably well and represent the same underlying velocity field. This can be seen by inspecting the residual velocity field shown in Figure 4, where no coherent features are present, in marked contrast to previous findings [10]. The better match between the measured peculiar-velocity field and that predicted from the galaxy distribution should yield a better estimate of $\beta$.

### 3.2. GALAXY BIASING

As more data on peculiar motions of galaxies are accumulated it becomes possible to study in more detail the relation between galaxy properties and the underlying mass distribution. This is a crucial ingredient for improving our understanding of galaxy biasing. As an example the reconstructed peculiar-velocity field can be used to derive the real space location of all galaxies in the SSRS2 within a volume 6000 km $s^{-1}$ in radius. Galaxies can then be selected in different mass density regimes and the luminosity function can be computed for the different sub-samples. A preliminary analysis along these lines indicates that galaxies in low-density regions have a significantly steeper faint-end slope as compared to the combined sample, being more consistent with theoretical predictions of hierarchical models. The implications of this result are currently being investigated.

Further insight will be obtained once the spectra of the SSRS2 are

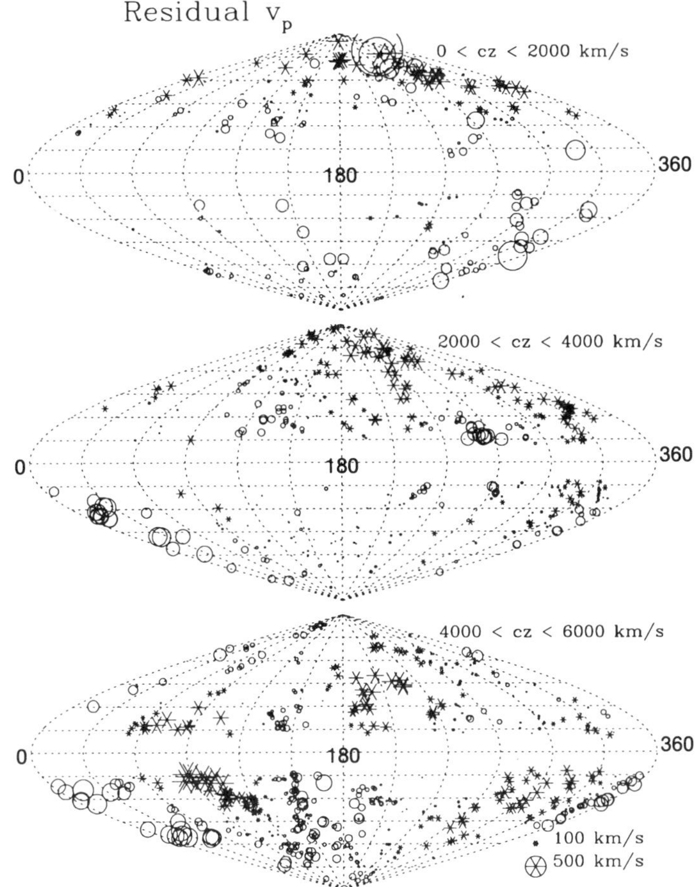


*Figure 4.* Sky projection in galactic coordinates of the residual field $u_{SFI} - u_{IRAS}$ ($\beta = 0.6$) as seen in the LG frame. The open symbols are points that are moving inward and the stars are points flowing outward.

classified using the PC analysis [13]. This will provide a quantitative way to analyze how the internal properties of galaxies vary as a function of the local mass density.

## 4. Summary

While the new 8-m class telescopes will open new windows for cosmological studies, by allowing us to probe the high-redshift universe ($z \gtrsim 1$), one should not overlook the importance of studies of the local universe which can continue to be carried out using smaller telescopes. This is particularly true for measurements of the peculiar velocity of galaxies which have progressed slowly since until now this kind of program has been restricted to 2-m class telescopes.

While most studies of the large-scale distribution of galaxies will rely on data that will become available from surveys like Sloan and 2dF, CfA2 and SSRS2 will continue to be a useful database for more detailed studies concerning the relation between galaxies and the underlying mass distribution. However, further progress on the study of galaxy biasing requires obtaining better quality spectra for galaxies in the existing database and measurements of peculiar velocities for larger samples, probing the local universe more densely. Hopefully, as the 8-m class telescopes come on-line more time will become available in the 4-m for these programs.

## Acknowledgments

I would like to thank all of my collaborators in the SSRS2. I would also like to thank W. Freudling, R. Giovanelli, M. Haynes, J. Salzer and G. Wegner for allowing me to present preliminary results based on the SFI sample prior to publication. I also thank M. Davis, A. Dekel, M. Geller and A. Nusser for many useful discussions.

## References

1. Alonso, M. V., Bernardi, M., da Costa, L., Freudling, W., Pellegrini, P.S., Wegner, G. & Willmer, C.N.A., 1997, *in preparation*
2. Benoist, C. , Maurogordato, S., da Costa, L. N., Cappi, A. & Schaeffer, R., 1996, ApJ, 472, 452
3. Benoist, C. , Cappi, A., Maurogordato, S., da Costa, L. N., Bouchet, F. & Schaeffer, R., 1997, ApJ, *submmited*
4. Bertschinger, E., Dekel, A., Faber, S. M., Dressler. A. & Burstein, D., 1990, APJ, 364, 370
5. da Costa, L. Nicolaci, Geller, M. J.; Pellegrini, P. S., Latham, D. W., Fairall, A. P., Marzke, R. O., Willmer, C. N. A., Huchra, J. P., Calderon, J. H., Ramella, M. & Kurtz, M, 1994, ApJ, 424, L1
6. da Costa, L. N., Vogeley, M. S., Geller, M. J., Huchra, J.P. & Park, C., 1994, ApJ, 437, L1
7. da Costa, L. N., Freudling, W., Wegner, G., Giovanelli, R, Haynes, M. P. & Salzer, J.J. , 1996, APJ, 468, L5
8. da Costa, L. N. *et. al.* 1997a, *in preparation*
9. da Costa, L. N. *et. al.* 1997b, *in preparation*
10. Davis, M., Nusser, A., & Willick, J., 1996, ApJ, 473, 22
11. Dekel, A., 1994, ARAA, 32, 371
12. Elad, H., Piran, T. & da Costa, L. N., 1996, ApJ, 462, L13
13. Folkes, S., da Costa, L. N., Lahav, O., Willmer, C., Freudling, W. & Pellegrini P.S. 1997, *in preparation*
14. Freudling, W., da Costa, L. N., Wegner, G., Giovanelli, R, Haynes, M. P. & Salzer, J.J. , 1995, AJ, 110, 920
15. Geller, M. J. & Huchra, J.P., 1989, Science, 246, 857
16. Giovanelli, R., Haynes, M.P., Salzer, J.J., Wegner, G., da Costa, L.N. & Freudling, W., 1994, AJ, 107, 2036

17. Giovanelli,R., Haynes, M. P.,T. Herter,T., Vogt N., Salzer, J., Wegner, G., da Costa, L. N. and Freudling, W., 1997, AJ, 113, 22
18. Giovanelli,R., Haynes, M. P.,Herter,T., Vogt N., da Costa, L. N, Freudling, W., Salzer, J. J. & Wegner, G., 1997, AJ, 113, 53
19. Loveday, J., Peterson, B., Efstathiou, G, & Maddox, S. J., 1992, ApJ, 390, 338
20. Loveday, J., Maddox, S. J., Efstathiou, G & Peterson, B., 1995, ApJ, 442, 457
21. Marzke, R. O. & da Costa, L. N., 1997, AJ, 113, 185
22. Marzke, R. O., da Costa, L. N., Pellegrini, P.S & Willmer, C.N.A. 1997, *in preparation*
23. Marzke, R. O., Geller, M. J., Huchra, J. P. & Corwin, H. G., 1994, AJ, 108, 437
24. Marzke, R. O., Geller, M. J. da Costa, L. N.& Huchra, J. P., 1995, AJ, 110 477
25. Mathewson, D.S., Ford, V.L. & Buchhorn, M., 1992, APJS, 81, 413
26. Mo, H.J. & White, S.D.M., 1996, MNRAS, 282, 347
27. Nusser, A. & Davis, M., 1995, ApJ, 449, 439
28. Willmer, C.N.A., da Costa, L.N. & Pellegrini, P.S.S., 1997, *in preparation*
29. Zamorani, G. 1997, this volume .

# THE GALAXY LUMINOSITY FUNCTION AT REDSHIFT ZERO: CONSTRAINTS ON GALAXY FORMATION

R.O. MARZKE
*Dominion Astrophysical Observatory*
*Herzberg Institute of Astrophysics*
*National Research Council of Canada*
*5071 W. Saanich Rd., Victoria, BC V8X 4M6*
*Canada*

## 1. Introduction

The luminosity function (LF) of low-redshift galaxies is a primary benchmark for models of galaxy formation. The detailed shape of the LF reflects a host of physical processes ranging from the collapse of dark-matter halos to the complex cycle of gas cooling, star formation, and feedback into the interstellar medium. Additional observables such as morphology, color, and emission-line strength provide similar but independent constraints on the input physics. Along with dynamical measures of galaxy masses, the LF and its dependence on morphology, color and emission-line strength anchor the theory of galaxy formation at redshift zero.

In this review, I compare the luminosity functions derived from several recent redshift surveys and then discuss the implications for semi-analytic models of galaxy formation. I will begin with the shape of the LF averaged over all galaxies, and then the dependence of the LF on galaxy morphology, color, emission-line strength, and environment will be discussed. It will be argued that each of these physical properties provides somewhat different constraints on the process of galaxy formation. I will conclude with a discussion of the very faint end of the field-galaxy luminosity function.

## 2. Two Routes to Galaxy Evolution: Forward and Backward

Theories of the formation and evolution of galaxies separate into two primary categories. The first begins with the properties of present-day galaxies and extrapolates back in time to predict what observers should observe at faint apparent magnitudes and high redshifts. The second begins with a the-



*D. Hamilton (ed.), The Evolving Universe,* 23–39.

ory of structure formation, follows the evolution of both collisionless and dissipative components of the universe and predicts what we should see at all steps along the way. The latter, "forward" approach is more appealing but considerably more difficult; the physical processes driving galaxy formation are complex and are intricately linked over a wide range of scales. Because of this complexity, the favored path has been the more empirical, "backward" approach. Recently, however, heuristic models have been gaining sophistication and are now providing testable predictions. At this early stage, these models are still considered exploratory, but a detailed comparison with observations is certainly warranted and indeed necessary to push the theory forward. In this paper, I will discuss recent progess on the observational side, and I will point out a few of the implications for both the traditional, "backward" approach as well as the semi-analytic, "forward" models of galaxy evolution.

The forward approach was pioneered by Press and Schechter (1974), who computed the evolution of the mass spectrum of collapsed objects in an Einstein-de Sitter universe. Their technique follows the growth of overdense regions until they turn around from the Hubble flow, at which point they collapse and virialize. The time at which the region collapses is determined by the initial overdensity, the size of the region, and the evolution of the background scale factor; the mass spectrum is then determined by the spectrum of initial density fluctuations, the density distribution function, and the background cosmogony. If dark matter particles are cold, structures collapse on small scales and are then incorporated into structures of larger and larger scale as time goes on. For a reasonable range of initial conditions, the mass spectrum predicted at the present epoch is qualitatively similar to the observed LF: at low mass, the mass spectrum is a power law, while at high mass, the spectrum cuts off exponentially.

More detailed comparisons between predicted mass functions and observed LFs require some understanding of the dissipative evolution of the baryonic component. Early efforts by White and Rees (1978) and Rees and Ostriker (1977) laid the groundwork for a semi-analytic approach that combines simple scaling laws for gas cooling and supernova feedback with the Press-Schechter prescription for the evolution of dark halos. Although the details of star formation and feedback are poorly understood, these simple scaling arguments may yield some insight into the gross features of the galaxy LF.

The cold dark matter model has been the focus of a great deal of theoretical effort over the past decade. Although the standard, high-bias CDM model has been ruled out for some time, several variants of the original theory remain viable and provide useful grounds for comparison. A well known problem with models based on hierarchichal structure formation is

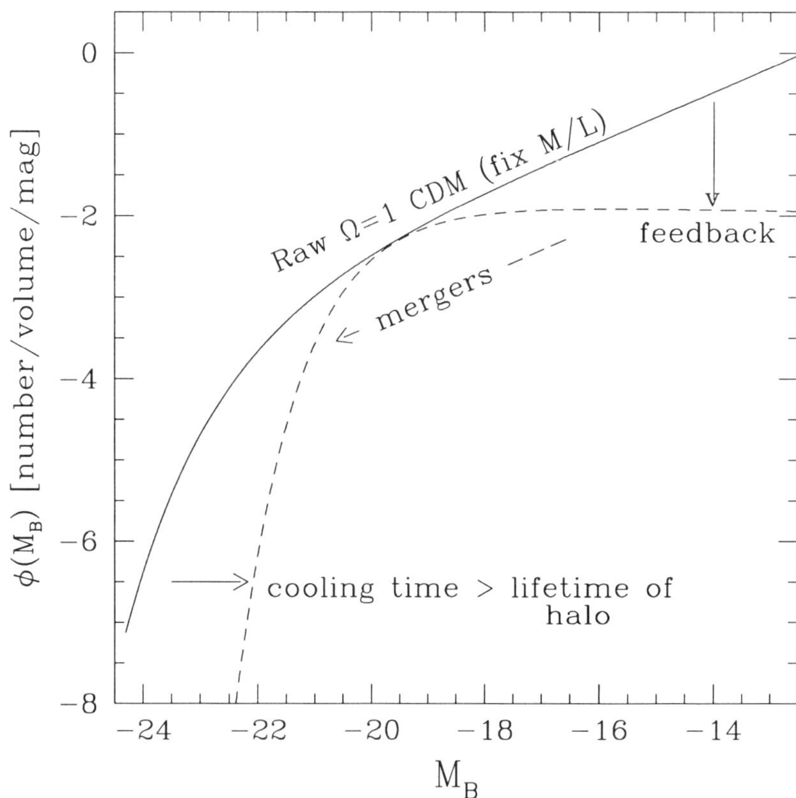


*Figure 1.* Cartoon illustrating how three important processes modify the observed luminosity function from the underlying spectrum of halo masses.

the steep slope at the faint end of the predicted LF. The mass spectrum predicted by the Press-Schechter formalism and reproduced in N-body simulations has a very steep slope at the low-mass end, $d \log N / d \log m \approx -2$. This slope is much steeper than the faint-end slope of the observed LF: $\alpha = d \log N / d \log L \approx -1$. However, as noted by White and Rees (1978), the shape of the LF is dictated as much by the processes of cooling, star formation and feedback as it is by the raw mass spectrum, and it remains to be seen whether this discrepancy is a fundamental limitation of hierarchical models.

The cartoon in Figure 1 depicts some of the physical processes shaping the galaxy LF. Dark-matter halos do not produce visible galaxies unless their gas has had time to cool from the virial temperature of the halo.

**The Local LF: Recent Redshift Surveys**

| Sample | $m_{limit}$ | $\langle z \rangle$ | $N_{gal}$ |
|---|---|---|---|
| * Century Survey<br>Geller *et al.* 1997 | $R \leq 16.1$ | 0.07 | 1,683 |
| Las Campanas Survey<br>Lin *et al.* 1996 | $15.0 \leq r \leq 17.7$ | 0.1 | 23,690 |
| Autofib Bright Survey<br>Ellis *et al.* 1995 | $b_J \leq 20$ | 0.1 | 1,105 |
| * Southern Sky Redshift Survey<br>da Costa *et al.* 1994 | $B_{26} \leq 15.5$ | 0.02 | 3,592 |
| CfA Redshift Survey<br>Marzke *et al.* 1994 | $B(0) \leq 15.5$ | 0.02 | 9,063 |
| Stromlo-APM Survey<br>Loveday *et al.* 1992 | $b_J \leq 17.2$ | 0.06 | 1,782 |

Table 1: A list of recent redshift surveys. Asterisks indicate surveys with new results presented here.

Large, massive halos do not have time to cool before the present epoch, and the bright end of the LF is consequently depressed below the raw mass spectrum. Small galaxies cool very effectively, and thus cooling alone cannot resolve the conflict at the faint end of the LF.

At the faint end of the LF, feedback from generations of stars and from mergers of small galaxies into larger ones drive the evolution. Supernovae and stellar winds can remove the gas from a dwarf galaxy entirely, quenching star formation and leaving the galaxy to fade into obscurity. Feedback thus provides an effective mechanism for decreasing $\alpha$. However, it is important to note that feedback is most effective in galaxies of very low mass, and these are not the galaxies that contribute most to the observational determination of the faint-end slope of the LF.

Mergers between galaxies within a dark halo occur on a dynamical friction timescale, and rough estimates suggest that such mergers are too slow to combat the overcooling problem. However, if merging is much more efficient than the rough scaling arguments predict, then mergers may also play a role (Kauffmann, White & Guiderdoni 1993).

Theory and observation show little sign of converging on the shape of the galaxy LF. The observers' task is to make sure the observed LF fairly represents the entire galaxy population; the theorists' challenge is to decrease the slope of the predicted LF from the fiducial CDM $\alpha \approx -2$ to the observed $\alpha \approx -1$ for galaxies within a few magnitudes of $L_*$. Contrary to many claims in the literature, it is the sharp turnover in the LF around $L_*$ that is the most difficult feature for hierarchical models to reproduce, not the very faint end of the LF.

## 3. Observations: The Shape of the General Luminosity Function

Table 1 lists three redshift surveys in which I have been involved along with three other major surveys which were published at the time of this conference. The six surveys cover a wide range of depths, apparent magnitudes and sampling strategies. In the next few sections, I will compare the results of these surveys and ask which aspects of the observed LF are generally agreed upon and which are still disputed. I will begin by discussing recent results from the Century Survey (Geller *et al.* 1997) and the Southern Sky Redshift Survey (da Costa *et al.* 1994).

Figure 2 shows the LF of the recently completed Century Survey (Geller *et al.* 1997). For comparison, I show the LF of the SSRS2 transposed to the $R$ band using the mean SSRS2 color: $\langle B_{26} - R \rangle = 1.3$. Here, we adjust the normalization of the SSRS2 to the Century Survey in order to compare the shapes. The LFs are nearly identical except for a small difference in $M_*$. The faint-end slope of the fitted Schechter functions are $\alpha = -1.14$ for the Century Survey and $\alpha = -1.16$ for the SSRS2. The agreement in the shape of the LF between samples selected in $B$ and $R$ and over different redshift ranges is reassuring.

For galaxies roughly three magnitudes fainter than $M_*$ and brighter, there remains some disagreement on the shape of the field LF. Figure 3 shows the Schechter function parameters derived from the surveys listed in Table 1. Although there are clear discrepancies, at least one conclusion appears to be robust: the faint-end slope of the LF is unlikely to be steeper than $\alpha \approx -1.25$ (The Schechter parameter $\alpha$, commonly called the "faint-end slope" is determined almost entirely by these relatively bright galaxies, not by very faint dwarfs. I will use the standard terminology and will discuss the truly faint end of the LF at the end of this paper).

This upper limit to the slope of the LF just fainter than $M_*$ is a problem for semi-analytic models of galaxy formation. Feedback reduces the faint-end slope from the raw halo spectrum, but because the effectiveness of feedback decreases as $1/V_c^2$, feedback is not very successful at removing galaxies near $M_*$. This is the primary reason why detailed fits of the semi-

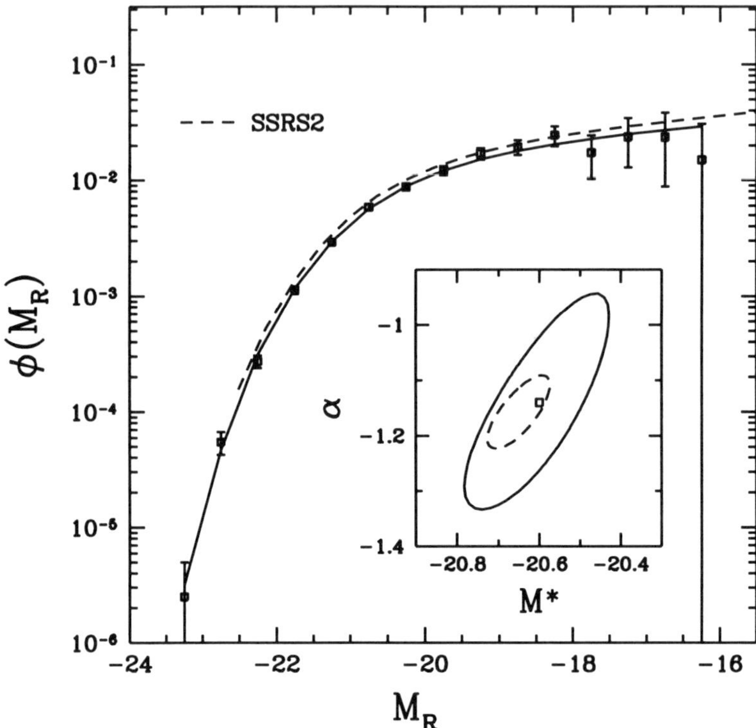


*Figure 2.* The luminosity function of the Century Survey (Geller *et al.* 1997). The dashed line indicates the SSRS2 LF translated to $R$ using $< B - R >= 1.3$. The inset shows confidence intervals for the fitted Schechter parameters (dashed ellipse represents the SSRS2).

analytic LFs to observations fail - it is *not* the very faint dwarfs that cause the discrepancy. Because this part of the LF is exactly the stretch which is best determined in magnitude-limited surveys, this constraint is difficult to escape.

An interesting way out of this conundrum is to hypothesize that standard redshift surveys miss a substantial fraction of the galaxy population, particularly those of low surface brightness (Disney 1976). That low surface-brightness galaxies are missed is well demonstrated (Sprayberry *et al.* 1995, Dalcanton 1997); the challenge is to measure enough redshifts for these galaxies to figure out how they contribute to the overall LF. Such surveys

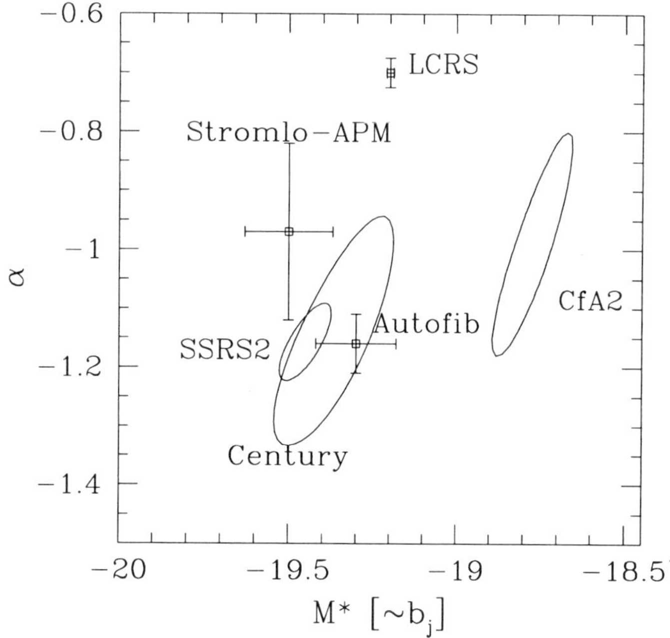


*Figure 3.* Schechter parameters for the surveys listed in Table 1. Values of $M_*$ have been shifted approximately to $b_j$ using mean survey colors.

are underway and will provide a very important step forward in this field.

The upper limit on $\alpha$ seems reasonably secure to surface brightnesses as low as $\mu = 26.5$ mag arcsec$^{-2}$ (Ellis *et al.* 1996). However, Figure 3 also reveals vexing discrepancies among the local surveys. The CfA2 LF (Marzke *et al.* 1994*a*) is offset from the rest by more than half a magnitude in $M_*$. The reason for this offset is not entirely clear; one possibility is a problem with the Zwicky magnitude scale. The LCRS (Lin *et al.* 1996) has an anomalously shallow faint-end slope of $\alpha = -0.7$, strongly discrepant with the flat or slowly increasing slopes measured in all other surveys. The reasons for these discrepancies at the faint end are also unclear; a detailed investigation of the biases in each survey may yield a more consistent picture.

Interestingly, the mean luminosity densities measured in these surveys are all consistent within their error bars: $\langle \mathcal{L} \rangle \approx 2 \times 10^8 h L_\odot \, \mathrm{Mpc}^{-3}$ (Figure 4). This luminosity density is a well-known problem for $\Omega = 1$ CDM

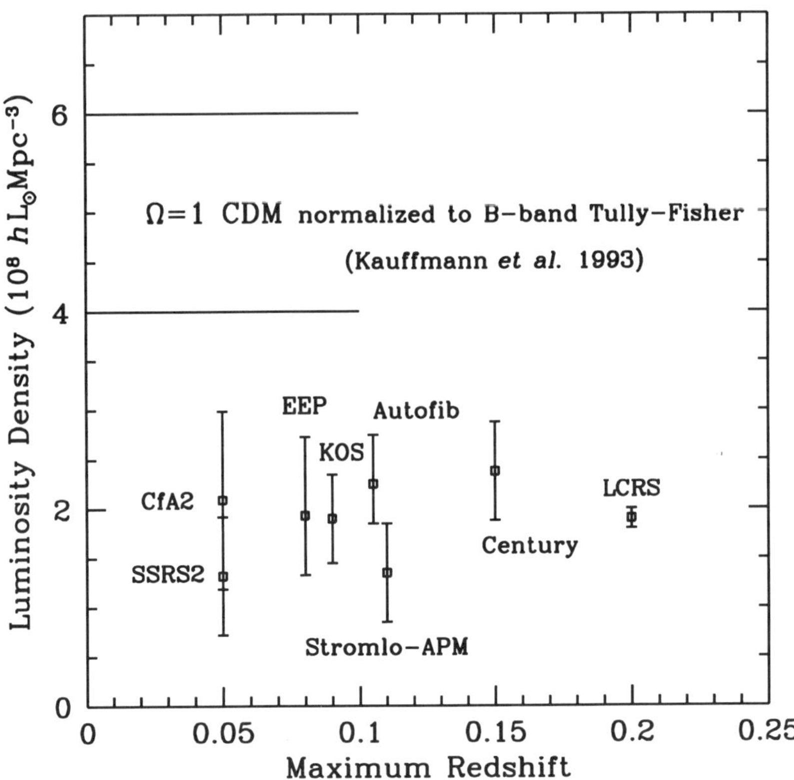


*Figure 4.* A comparison of the luminosity densities measured in several recent surveys versus the survey depth. Also included are two points from earlier surveys: EEP (Efstathiou, Ellis & Peterson 1988) and KOS (Kirshner, Oemler & Schechter 1983). The Century Survey and the LCRS are selected at $R$; the rest are selected at $B$.

when combined with the observed Tully-Fisher relation. If the models are normalized such that that the circular-velocity/luminosity relation follows the $B$-band Tully-Fisher relation, then the predicted $B$-band luminosity density is too high by factors of 2-4 for standard CDM (Lacey *et al.* 1993, Kauffmann *et al.* 1993). The discrepancy is simply a reflection of the abundant low-mass halos in $\Omega = 1$ CDM and their ability to cool their gas on short timescales (Kauffmann & White 1993).

Clearly, $\Omega = 1$ CDM is only one of many possible models, and its inability to reproduce many other features of the galaxy distribution have already constrained it severely. An obvious solution to the problem of the luminosity density is to reduce the total quantity of mass in the universe by reducing $\Omega$. Low-$\Omega$ models yield large-scale structure much more consistent with the observed galaxy distribution and are favored by a wide range of dynamical measures. However, the shape of the $z = 0$ LF in low-$\Omega$ models is nearly a power law over the range of magnitudes we have discussed so far and is thus a very poor fit to the LF around $M_*$ (Kauffmann *et al.* 1993).

## 4. Dependence on Galaxy Type

The dependence of the LF on galaxy type offers important clues to the origin of the Hubble sequence. The two surveys in Table 1 that include accurate morphological data lead to quite different conclusions (see Figure 5). In the Stromlo-APM survey, Loveday *et al.* (1992) measure a flat faint-end slope for spiral galaxies, consistent with the survey as a whole, but they find a steeply decreasing faint-end slope for ellipticals and S0's combined ($\alpha = +0.2$). On the other hand, Marzke *et al.* (1994*b*) used the more detailed morphologies available in the CfA survey and measured flat faint-end slopes for both ellipticals and S0's. For E's and S0's combined, the CfA survey yields $\alpha = -0.97$. In the CfA survey, all types show very similar LFs with the notable exception of the Magellanic spirals and irregulars, which have a very steep faint-end slope, $\alpha = -1.88$ (I will return to the Sm/Im's later in the paper).

In Marzke *et al.* (1994*b*), we argued that the incomplete morphological classification in the Stromlo-APM survey could bias the E/S0 LF. Only 78% of the galaxies in the Stromlo-APM were classifiable; if we make the reasonable assertion that classifiability is determined largely by the size, then any correlations between apparent size and intrinsic luminosity bias the LF computed from this sample. Using the known correlation between size and luminosity for Virgo ellipticals, we showed that faint ellipticals were more likely to be too small for morphological classification given the depth of the survey and the mean seeing conditions. The effect is quite large: even if the true E/S0 LF has a flat faint end, the measured $\alpha$ could easily be as shallow as the one measured in the Stromlo-APM.

da Costa *et al.* (1997) have since accumulated morphological types for all galaxies in the SSRS2, and Marzke *et al.* (1997) have derived the LFs for early and late-type galaxies. The resulting LFs are given in Figure 6. As in the CfA survey, the shapes of the E/S0 and spiral LFs are indistinguishable, and the faint-end slopes are flat in both cases. The flat E/S0 LF has important ramifications for the galaxy counts in deep HST surveys like the Medium Deep Survey (Driver *et al.* 1995, Glazebrook *et al.* 1995) and the Hubble Deep Field (Abraham *et al.* 1996) *et al.* ). A common conclusion drawn from these surveys is that the early-type galaxies follow the

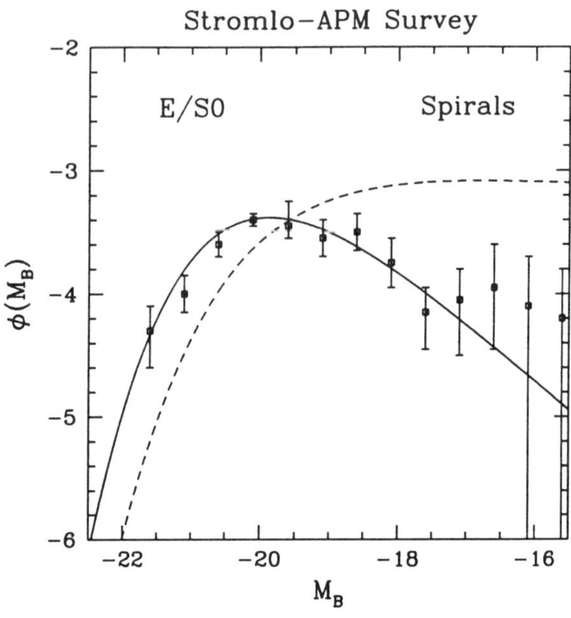


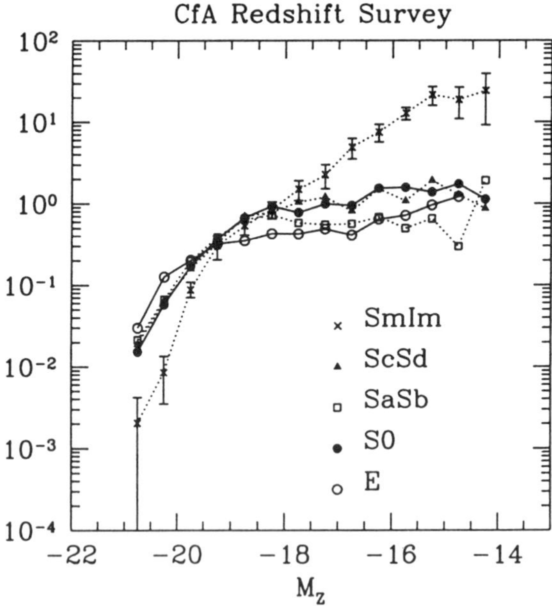


*Figure 5.* A comparison between type-specific LFs in the field. Note the steep decline in the Stromlo-APM early-type LF.

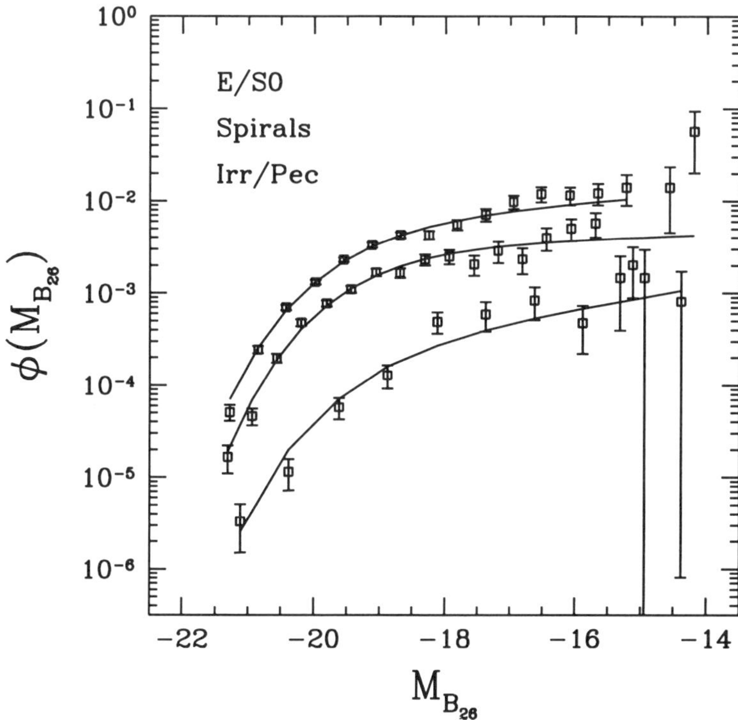


*Figure 6.* The luminosity function for different morphological types in the SSRS2 (Marzke *et al.* 1997).

no-evolution predictions out to quite faint apparent magnitudes. It is sometimes forgotten that this conclusion is based on two ad-hoc adjustments to the local LF: 1) the faint-end of the local E/S0 LF is often assumed to be flat and 2) the local normalization is boosted by a nearly a factor of two. We are now confident that the first assumption has been justified. The second remains problematic. Are these surveys really missing half of the galaxies even at $L_*$? This question is left to the next generation of redshift surveys, most notably the ones targeted at low-surface-brightness galaxies.

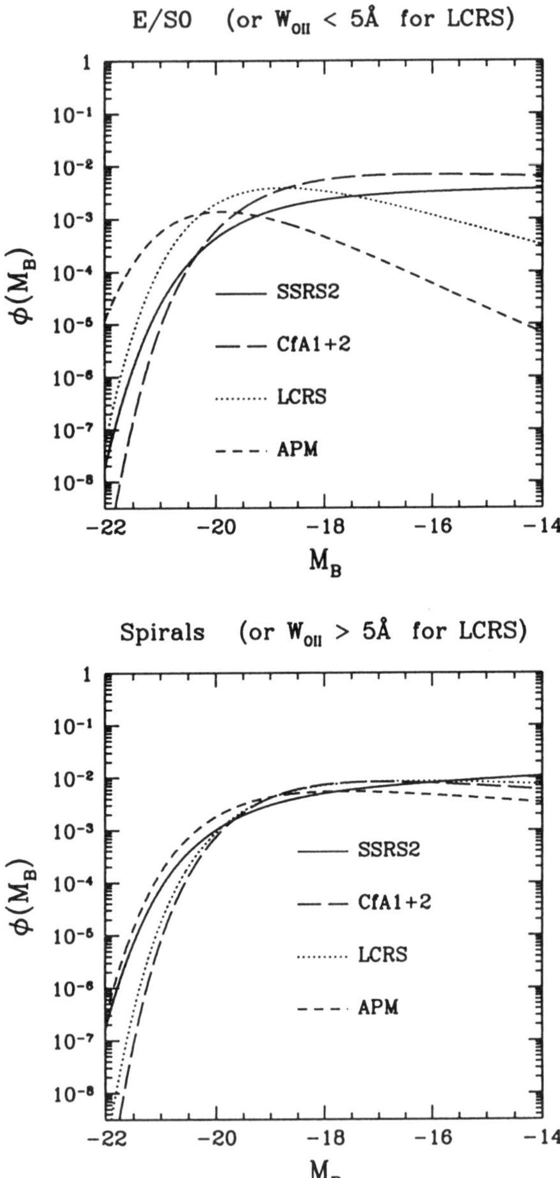


*Figure 7.* A comparison between type-specific LFs in the field. Note the steep decline in the Stromlo-APM early-type LF.

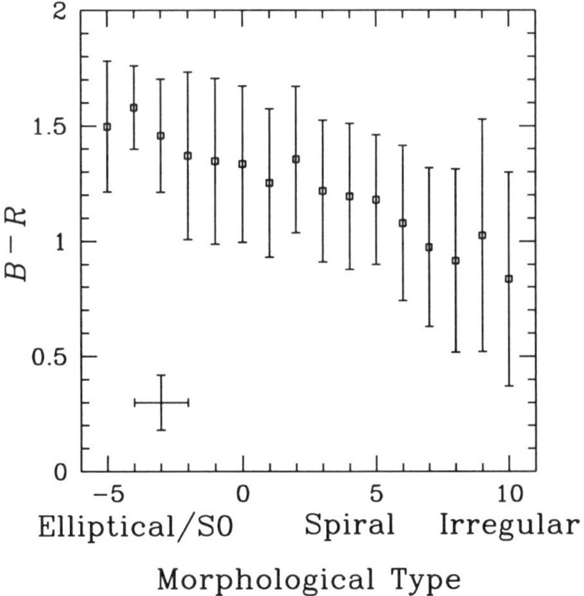


*Figure 8.* The relation between color and morphology derived from 15,000 galaxies in the ESO-LV (Lauberts & Valentijn 1989). The error bar in the lower left corner shows the uncertainty for an individual galaxy.

## 5. Dependence on Other Galaxy Properties

Figure 7 shows the LFs divided by type for the CfA, SSRS2 and Stromlo-APM surveys along with the LF divided by emission-line strength in the LCRS. This figure reiterates the agreement at the faint-end of the CfA and SSRS2 surveys and highlights once again the discrepancy with the Stromlo-APM in the early-type LF.

The addition of the LCRS to this figure raises an interesting question: is it fair to translate the LFs for different morphologies into LFs separated by other physical properties such as emission-line strength or color? The LCRS LF suggests that it might not be. Although emission-line strength is correlated with morphology in the mean, there is considerable scatter about the relation, and the residuals may be correlated with luminosity. For example, it is reasonable to expect E/S0's to have weak emission lines in general, and thus one expects the E/S0 LF to be similar to the LF of galaxies with weak emission. However, many Es and S0s in the CfA and

SSRS2 surveys have strong enough emission lines to put them over the threshold used by Lin *et al.* (1996), and many of them are intrinsically faint. It remains to be seen whether these galaxies are enough to cause the discrepancy seen in Figure 7; reliable conclusions can be drawn only from samples with complete line strengths and morphologies, and such analyses are just getting underway.

The dependence of the LF on galaxy color is another example of the hazards encountered when transforming LFs among morphology, emission-line strength, color, or any other observed property of the galaxy. In Figure 8, we show the relation between color and morphology in the ESO-LV (Lauberts and Valentijn 1989). Although the error in color and morphology is quite large, it is clear that the scatter in the relation is larger than would be expected from the random errors alone. Figure 9 shows the LF computed for galaxies bluer and redder than $(B - R) = 1.3$. In this case, the LFs are quite different from those produced by a direct translation of E/S0 to red and Spiral+Irr to blue.

The dependence of the LF on color yields more direct constraints on models of galaxy formation than the dependence on morphology. A well-known problem with hierarchical models is their inability to reproduce the slope of the color-magnitude relation of elliptical galaxies. Faint ellipticals are observed to be bluer than bright ellipticals, whereas model color-magnitude relations tend to be flat or to tilt in the other direction (White & Frenk 1991, Lacey & Silk 1991, Kauffmann *et al.* 1993). This trend results from the fact that low-mass halos collapse, cool and form stars at very early epochs and are thus very red by the present epoch. However, this problem is not necessarily fundamental; proper accounting of the chemical evolution in these models may alleviate the conflict (Kauffmann 1996, private communication).

With large enough samples now becoming available, the true multivariate distribution of luminosity, color, morphology, and emission-line strength will yield a wealth of clues about the origin and evolution of galaxies.

## 6. The Very Faint End of the Luminosity Function

The distribution of absolute magnitudes is narrowly peaked around $M_*$ in magnitude-limited surveys. The cutoff at the bright end reflects the exponential cutoff in the LF; at the faint end, galaxies are visible only locally, and the observed distribution cuts off as the sampled volume goes to zero. The abundance of truly faint galaxies is thus an elusive target; useful samples of intrinsically faint galaxies emerge only after observers wade through many thousands of $M_*$ galaxy redshifts. Even in samples that are large enough to populate the very faint end of the LF, the volume sampled is

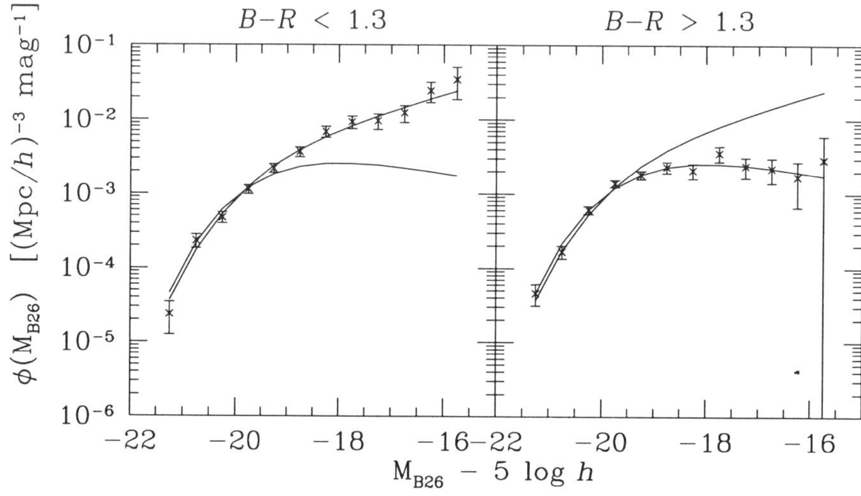


*Figure 9.* The luminosity function for blue and red galaxies in the SSRS2 (Marzke & da Costa 1997).

still much smaller than the volume probed for $M_*$ galaxies and is thus less representative of the universe as a whole. Perhaps not surprisingly, it is these elusive, very faint dwarfs that provide some of the most interesting constraints on models of galaxy formation.

At this time, the faint end of the LF remains very poorly known. The problem is exacerbated by the fact that a substantial fraction of the faintest galaxies have quite low surface brightness, and even if they make it into the photometric catalogs, they are the least likely to have spectroscopic measurements in incomplete samples. Using the largest available fully-sampled redshift survey (CfA2), we found evidence for an upturn in the LF fainter than $M_B \approx -16$ (Marzke *et al.* 1994*a*). Below this limit, the CfA2 LF steepens to nearly $\alpha = -1.8$. We also showed in Marzke *et al.* (1994*b*) that the excess is dominated at the faintest magnitudes by Magellanic spiral and irregular galaxies, highlighting the potential for severe surface-brightness biases at the faint end of the LF. However, the CfA luminosity scale is based on photographic magnitudes obtained using the Schraffierkassette technique (Zwicky *et al.* 1961-68), and uncertainties in the magnitude scale render the slope at faint magnitudes uncertain.

As discussed earlier, the LF is nearly flat between $M_*$ and $M_* - 3$; models require quite efficient feedback to reconcile $\Omega = 1$ CDM with these observations. If feedback is strong enough to match the LF around $L_*$, a steep upturn in the LF at very faint magnitudes becomes quite difficult to explain. Because feedback becomes more effective in halos of lower mass, models with efficient feedback may actually produce too *few* galaxies at low luminosity. A more accurate determination of the very faint end of the LF is clearly desirable.

Although very large redshift surveys are underway, these surveys are not very well suited to the task of measuring the faint end of the LF. These surveys are optimized for analyses of large-scale structure and are therefore tuned to produce as many redshifts as possible in a fixed observing time. Such surveys are best carried out using a sparse sampling strategy and a multi-object spectrograph. Galaxies which are not concentrated enough to provide the required signal-to-noise in a fiber or slit spectrograph are passed over. Because of correlations between surface brightness and absolute magnitude, these selection effects produce strong biases at the very faint end of the LF.

An alternative to the standard, magnitude-limited field redshift survey is a targeted survey of systems of galaxies. Nearby overdensities enhance the efficiency of detecting dwarf galaxies, and if the overdensities are large enough, it is possible to compute LFs from photometry alone by comparing counts in a target field with counts in background fields. This approach has been quite successful in rich clusters (Bernstein *et al.* 1995, Gaidos 1997, dePropris *et al.* 1995). However, because the S/N in the LF is directly proportional to the richness of the system, poor systems of galaxies have not been well studied. Because nearly half of all galaxies live in groups, a measurement of the very faint end of the LF in groups provides a first guess at the field LF and is in itself an interesting constraint on models of galaxy formation. With these ideas in mind, Michael Hudson and I have begun to obtain deep imaging of poor systems of galaxies to measure the LF all the way down to the luminosity of the brightest globular clusters. These observations will soon be underway at CTIO, and we expect first results by the end of 1997.

## References

Abraham, R.G., Tanvir, N.R., Santiago, B.X., Ellis, R.S., Glazebrook, K., & van den Bergh, S., 1996 MNRAS, 279, L47

Bernstein, G.M., Nichol, R.C., Tyson, J.A., Ulmer, M.P. & Wittman, D. 1995 AJ 110, 1507

da Costa *et al.* 1997, in preparation

da Costa, L. N., Geller, M. J., Pellegrini, P. S., Latham, D. W., Fairall, A. P., Marzke, R. O., Willmer, C. N. A., Huchra, J. P., Calderon, J. H., Ramella, M. & Kurtz, M.

J., 1994, ApJ Letters, 424, L1
Dalcanton *et al.* 1997, private communication.
de Propris, R., Pritchet, C.J., Harris, W.E. & McClure, R.D. 1995, ApJ 450, 534
Disney, M.J. 1976 Nature, 263, 573
Driver,S.P., Windhorst, R.A., & Griffiths, R.E. 1995 ApJ, 456, L21
Efstathiou, G., Ellis, R. S., & Peterson, B. A. 1988, MNRAS 232, 431
Ellis, R. S., Colless, M., Broadhurst, T., Heyl, J., & Glazebrook, K. 1996, MNRAS 280, 235
Gaidos, E.J. 1997 AJ, 113, 117
Geller *et al.* 1997, in preparation
Glazebrook, K., Ellis, R. S., Colless, M., Broadhurst, T., Allington-Smith, J., & Tanvir, N. 1995, MNRAS 273, 157
Kauffmann,G., White, S.D.M., & Guiderdoni, B. 1993 MNRAS, 264, 201
Kauffmann, G. & White, S.D.M. 1993 MNRAS 261, 921
Kirshner, R.P., Oemler, A.,Jr., & Schechter, P.L. 1979 AJ, 84, 951
Lacey, C., Guiderdoni, B., Rocca-Volmerange, B., & Silk, J. 1993 ApJ 402, 15
Lacey, C. & Silk, J. 1991 ApJ, 381, 14
Lauberts, A., & Valentijn, E. A. 1989, The Surface Photometry Catalogue of the ESO/Uppsala Survey,(Garching: ESO)
Lin, H., Kirshner, R. P.,Shectman, S., Landy, S. D., Oemler, A., Tucker, D. L., & Schechter, P. 1996, ApJ, 464, 60
Loveday, J., Peterson, B. A., Efstathiou, G., & Maddox, S. J., 1992, ApJ, 390, 338
Marzke *et al.* 1997, ApJL, submitted
Marzke, R. O., & da Costa, L. N. 1997, AJ, 113, 185
Marzke, R. O., Huchra, J. P., & Geller, M. J. 1994*a*, ApJ 428, 43
Marzke, R. O., Geller, M. J., Huchra, J. P., & Corwin, H. G. 1994*b*, AJ 108, 437
Press,W.H., & Schechter,P. 1974,ApJ,187,425
Rees, M.J. & Ostriker, J.P. 1977 MNRAS, 179, 541
Sprayberry, D., Impey, C.D., Bothun, G.D. & Irwin, M.J. 1995, AJ, 109, 558
White, S.D.M. & Rees, M.J. 1978 MNRAS, 183, 341
White, S.D.M. & Frenk, C.S. 1991, ApJ 379, 52
Zwicky,F.,Herzog,E.,Wild,P,Karpowicz,M.& Kowal,C. 1961-1968, Catalogue of Galaxies and of Clusters of Galaxies (Pasadena: Cal. Inst. of Tech.)

# CFA SURVEYS AND THEIR EXTENSIONS

DANIEL G. FABRICANT
*Harvard–Smithsonian Center for Astrophysics*
*Cambridge, Massachusetts*

## 1. Introduction

Our progress in studies of large-scale structure and other areas of observational cosmology is punctuated by the construction of new telescopes and the development of more powerful instruments. I describe here current and future CfA surveys against the backdrop of the instruments that make them possible. The CfA's instrumental capabilities will shortly undergo a dramatic change. Current CfA surveys are largely carried out on 1.5m class telescopes where the spectra are collected individually. In about one year, the converted MMT will be brought into operation with a 6.5 m primary and a spectrograph (fed by optical fibers) that will collect 300 spectra simultaneously in a 1° diameter field.

I begin by introducing four just completed surveys and the instruments that made them possible: Decaspec and FAST. Three of these are redshift surveys: the Century Survey, the 15R Survey, and the final portion of the original CfA Survey, and one is a spectral and photometric survey of nearby field galaxies that will serve as a benchmark for studies of galaxy evolution. I then describe the capabilities of the converted MMT and the suite of new wide-field instruments that we are constructing at the CfA, including Hectospec (a 300 optical fiber robotic positioner and spectrograph), Megacam (a 36 CCD optical imaging array) and Binospec (a dual-beam multiple slitlet spectrograph). I conclude by discussing the initial redshift survey to be conducted with Hectospec.

## 2. Current Surveys and Instruments

### 2.1. INSTRUMENTAL BACKGROUND

Through 1993, the vast majority of redshifts for the CfA surveys were obtained with the Z-machine at the 1.5m Tillinghast Telescope at Mt. Hopkins



*D. Hamilton (ed.), The Evolving Universe,* 41–51.

in Arizona. By the mid 1980's the image-intensified Reticon detectors and the inefficient camera used in the Z-machine had become noncompetitive with modern CCD spectrographs. We began to design a completely new CCD spectrograph in 1990; the result was FAST (**FA**st **S**pectrograph for the **T**illinghast), which was commissioned in January 1994. The FAST/Tillinghast combination has a throughput peaking at ~26% (Fabricant *et al.* 1997), including all losses from the point that light strikes the primary mirror. The peak throughput of the Z-machine/Tillinghast combination was ~2%. FAST has extended the grasp of the Tillinghast by nearly 3 magnitudes, and has allowed the CfA to remain competitive in redshift surveys of bright galaxies before the completion of the MMT conversion.

In addition to FAST, CfA scientists have access to the existing MMT, which has an efficient spectrograph. The current MMT is not well suited to redshift surveys because its 3′ diameter field precludes the multiobject spectral techniques that allow efficient observation of faint galaxies with a 4 m class telescope. Nonetheless, the MMT has been used for the Century Survey, which is described below.

The Century Survey collaboration includes Gary Wegner and John Thorstensen at Dartmouth College. Dartmouth operates the 2.4m Hiltner Telescope on Kitt Peak with Michigan and MIT as part of the MDM consortium. Initially, the Mark III spectrograph at the Hiltner Telescope was to be used to obtain the spectra singly, as at the MMT. However, the field of view allowed by the finder/guider package, 20′, was large enough to consider multiobject spectroscopy of the R~16 Century Survey galaxies. Accordingly, Decaspec, a fiber-optics adapter with 10 individually motorized probes was designed and built (Fabricant and Hertz 1990). The fibers exiting Decaspec feed the existing Mark III spectrograph. Because Decaspec was used to observe the densest portions of the Century Survey (leaving the rest to the MMT), at least six of the Decaspec probes could typically be deployed on a galaxy. Use of the Decaspec reduced the required 2.4m observing time by a factor ~3.

## 2.2. SURVEYS

### 2.2.1. *Century Survey*

As its name implies, the Century Survey covers 100 square degrees of sky in a strip 1° wide in declination (DEC) and 100° long in right ascension (RA). The strip is centered at RA = 12.5 hrs, DEC = 29.5°. The survey includes ~1800 galaxies to a magnitude limit R = 16.1. The median survey redshift is 0.06. Approximately 1/2 of the new survey redshifts were acquired with Decaspec and 1/2 with the MMT spectrograph. The photometry for the survey was obtained from CCD calibrated, digitized POSS plates. The

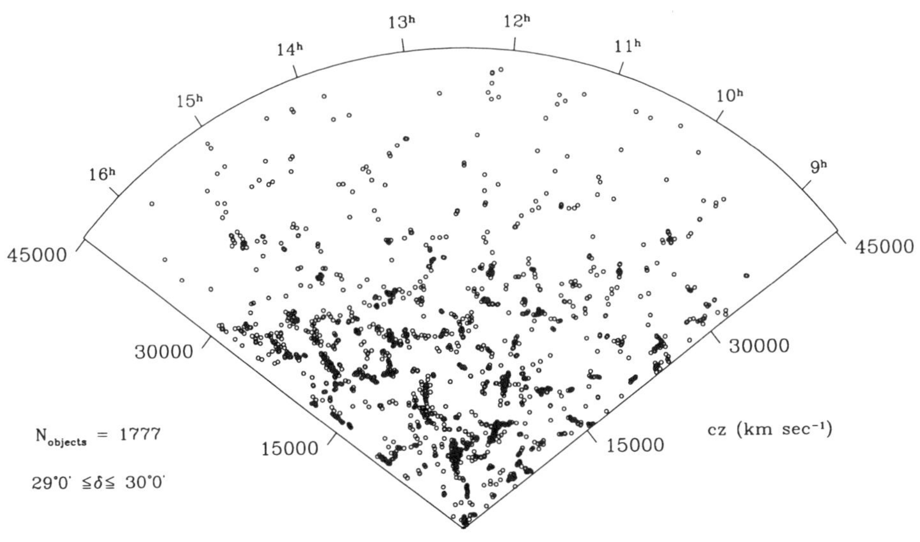


*Figure 1.* The Century Survey.

survey is now complete, and a cone diagram is shown in Figure 1.

### 2.2.2. *15R Survey*

The 15R Survey covers 1000 square degrees to a magnitude limit of R = 15.4. The northern portion of the survey extends over the same RA range as the Century Survey (8.5 hrs to 16.5 hrs) and between DEC 26.5° and 32.5°. The southern portion of the survey extends between RA's of 21 to 3 hrs, and DEC's of 10° and 13°. About 2/3 of the survey redshifts were obtained with FAST, and the remaining 1/3 were obtained with the Z-machine, Decaspec and the MMT. Here, too, the photometry was based on CCD calibrated, digitized POSS plates. The 15R Survey is essentially complete, awaiting only a few dozen spectra.

The FAST spectra obtained for the 15R Survey are of sufficient quality to study the stellar population of the galaxies. Regular observation of flux standard stars allows the measurement of accurate emission line ratios and

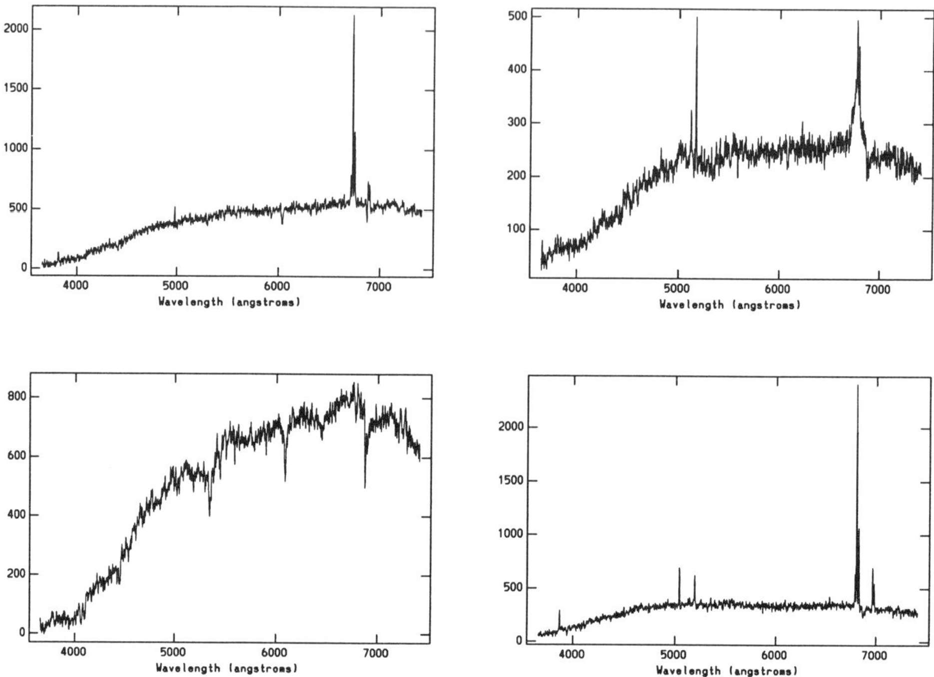


*Figure 2.* Sample 15R spectra. Exposure times are 3 to 6 minutes for these galaxies with $13.7<R<15.0$.

continuum shapes. Barbara Carter, a Smithsonian predoctoral fellow, will base her doctoral thesis on these data. Sample 15R spectra are shown in Figure 2.

### 2.2.3. *Completion of the CfA Survey*

Earlier CfA redshift surveys were based on the Zwicky catalog, which extends to a magnitude limit $B{\sim}15.5$. The last portion of the original survey ($8.5<RA<16.5$ and $10^\circ<DEC<12^\circ$) has been completed with FAST.

### 2.2.4. *Nearby Field Galaxy Survey*

The Nearby Field Galaxy Survey is a spectral, photometric and kinematical survey of 200 bright galaxies ($B<14.5$) selected from the CfA1 redshift survey. This survey is intended to be a benchmark for studies of galaxy evolution. The survey includes a random subsample of ~20% of the CFA1 galaxies that lie outside the Virgo Cluster, weighted approximately by the galaxy luminosity function. Our object is to sample galaxies with a wide range of absolute magnitudes ($-14 < M_B < -23$). To ensure that the sample galaxies fit onto FAST's $3'$ long slit, a variable redshift cutoff at low redshift as a function of $M_B$ is applied.

Images in $U$, $B$ and $R$ as well as integrated and nuclear spectra have been obtained for each galaxy. The spectrograph slit was aligned with the galaxy major axis in each case; the integrated spectra were obtained by scanning the slit $\pm 0.25$ $D_{25}$ along the galaxy minor axis. These integrated spectra sample approximately half the total galaxy light. Rolf Jansen's doctoral thesis at the University of Groningen will be based on these data.

We are currently expanding the survey to collect kinematic data for the sample galaxies. Wherever possible we measure a central (or bulge) velocity dispersion and a rotation curve. These data will form part of Sheila Kannappan's doctoral thesis. Now that kinematic data at intermediate redshifts are becoming available (*e.g.* Franx 1993 and Vogt 1996), it is becoming clear that our limited knowledge about the properties of present day galaxies weakens our conclusions about galaxy evolution.

## 2.3. SURVEY SUMMARY

The CfA surveys introduced above are summarized in Table 1.

TABLE 1. Summary of Four Current CfA Surveys

| Survey Name | Type | Galaxy Sample | Mag. Limit | Extent (Sq. Deg.) | Participants |
|---|---|---|---|---|---|
| Century | redshift | 1800 | R=16.1 | 100 | M. Geller, M. Kurtz, J. Huchra, D. Fabricant & R. Schild (CfA), J. Thorstensen & G. Wegner, (Dartmouth), R. Marzke (DAO) |
| 15R | redshift, spectral | 10000 | R=15.4 | 1000 | M. Geller, M. Kurtz, D. Fabricant (CfA), J. Thorstensen & G. Wegner (Dartmouth), M. Postman & B. McLean (STScI) |
| CfA (final portion) | redshift | 3000 | B=15.5 | 1200 | J. Huchra, M. Geller & D. Fabricant |
| Nearby Field Galaxy | spectral | 200 | B=14.5 | | D. Fabricant & N. Caldwell, (CfA), M. Franx & R. Jansen, (Groningen) |

## 3. MMT Conversion and Wide-Field Instruments

### 3.1. INTRODUCTION TO THE CONVERSION

Future CfA programs in observational cosmology are strongly tied to the conversion of the Multiple Mirror Telescope to use a single 6.5 m primary mirror (West *et al.* 1996). The conversion offers a welcome increase (2.5X) in collecting area, but it will offer an even more significant increase in field of view to a diameter of 1° from the present 3′. This field of view, the largest now planned for the new generation of 8m class telescopes, makes it possible to use an array of such powerful tools as wide-field imaging, spectroscopy with optical fibers, and multiple-slitlet, areal spectrographs. The optical design of the telescope (carried out by Harland Epps) has been tailored to allow wide-field instruments to perform with maximum efficiency.

The wide-field focus of the converted MMT will have a final focal ratio of f/5, and will require a large refractive corrector that is described below. In addition, the telescope will have an f/9 classical Cassegrain focus to service existing instruments, as well as an f/15 focus that accommodates a powerful adaptive optics system. We expect first light with the converted MMT in late 1997 or early 1998.

### 3.2. WIDE-FIELD CORRECTOR

The wide field corrector for the converted MMT can be configured for spectroscopy over a 1° diameter focal surface, or for imaging over a 35′ diameter flat field. The spectroscopic focal surface is a hyperboloid with a sag of ~8 mm, and is highly telecentric for efficient feeding of optical fibers. A nontelecentric corrector will degrade the focal ratio emerging from the fibers, unless each fiber is individually aimed at the exit pupil.

The spectroscopy configuration contains a pair of counter-rotating, zero-deviation prisms to correct atmospheric dispersion. To change to the imaging configuration, these prisms and the adjacent lens are removed and replaced with another lens. The polychromatic (0.33–1.1$\mu$m) RMS image diameters produced by the wide-field system are $<0.6''$ in the spectroscopic configuration and $<0.15''$ in the imaging configuration.

### 3.3. INTRODUCTION TO THE WIDE-FIELD INSTRUMENTS

The Smithsonian Institution plans to build five instruments for the wide-field focus of the converted MMT. Below, I describe those instruments that will be most useful for observational cosmology: Hectospec, Megacam and Binospec. Table 2 gives a capsule summary of each of these three instruments, as well as the wide-field corrector.

TABLE 2. Wide-Field MMT Instruments for Observational Cosmology

| Instrument | Function | Completion Date | Participants |
|---|---|---|---|
| Hectospec | 300 object fiber-fed spectrograph | 1997 | D. Fabricant (PI), E. Hertz (Proj. Eng.) J. Barberis, P. Cheimets, R. Fata T. Gauron, J. Roll & A. Szentgyorgyi |
| Megacam | 36 CCD imager | 1999 | B. McLeod (PI), T. Gauron (Proj. Eng.) D. Fabricant, J. Geary, M. Ordway & M. Pieri |
| Binospec | 150 slitlet spectrogaph | 2000 | D. Fabricant (PI), R. Fata (Proj. Eng.) E. Hertz & J. Barberis |
| Wide-Field Corrector | refractive corrector | 1997 | D. Fabricant (PI), R. Fata (Proj. Eng.) J. Barberis |

## 3.4. HECTOSPEC

### 3.4.1. *Introduction*

Hectospec (Fabricant, Hertz, and Szentgyorgyi 1994) will be the first of the wide-field instruments to be completed. We plan to commission Hectospec just after first light with the wide-field corrector, in mid 1998. Hectospec consists of a robotic fiber positioner connected to a large bench-mounted spectrograph by a 25 m bundle of 300 optical fibers. The layout of the robotic positioner follows the basic design introduced by Parry and Gray (1986). The fibers enter the focal surface radially and each terminates in a 3 by 6 mm button. The button contains a tiny prism to allow the fiber to receive light from the telescope (incident normal to the focal surface), as well as a small, but powerful, magnet to hold the button to the steel focal surface. A pair of five-axis robots, working in tandem at opposite sides of the focal surface, move the fibers between observing configurations. The robots reposition the 300 fibers in approximately 300 seconds.

The spectrograph has a 250 mm collimated beam diameter, and uses an Epps camera of the style developed for the HIRES spectrograph (Vogt 1994). The peak system efficiency (telescope, spectrograph, fibers, and detectors) is expected to exceed 20%. Hectospec will obtain spectra of sufficient quality to measure redshifts at its limit of R$\sim$21 in $\sim$60 minutes.

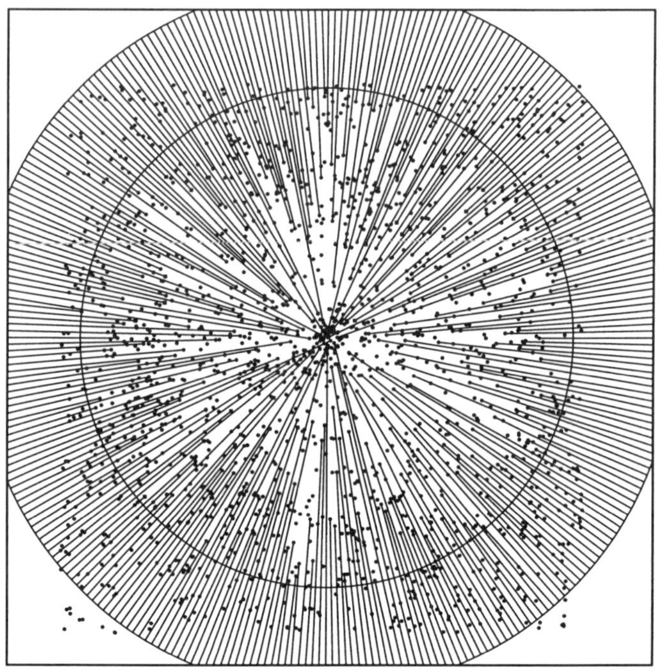

*Figure 3.* A typical Hectospec setup. Here 264 fibers are deployed on galaxies in the field surrounding A2141, the remaining 36 measure the sky background.

### 3.4.2. *Observing with Hectospec*

The Hectospec team has put a good deal of effort into refining algorithms for efficient fiber to target matching. The earliest work was carried out by David Becker, then a Harvard undergraduate. His work was refined and extended by John Roll. Here, I show simulations using a field of galaxies centered on A2141, plotted from data kindly supplied from the POSS-2 survey by S. Djorgovski and R. deCarvalho. There are 1143 galaxies to g$\sim$20.5 within the 1° diameter accessible to Hectospec. A typical Hectospec setup is shown in Figure 3. After four such setups, all but 66 of the original 1143 galaxies are observed. On average, 269 of the Hectospec fibers are deployed on a target galaxy, leaving about 10% of the fibers to measure the sky level. The fibers run nearly radially, and we find that efficient targeting does not require fiber crossings.

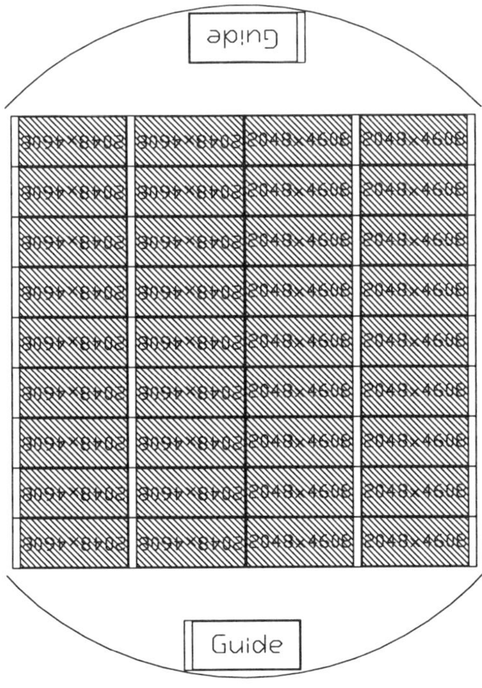


*Figure 4.* Megacam with 36 2048×4608 pixel CCDs.

### 3.5. MEGACAM

Megacam is a 36 CCD array that fills essentially the entire useful imaging field at the converted MMT, ~22′ by 22′. The array is shown in Figure 4. In the early 1990's, when Harland Epps and I laid out the imaging corrector configuration, I incorrectly imagined that it might be 15 years before we could consider tiling the focal surface with CCD's. In the *R* band, Megacam will reach R=25 at a signal to noise ratio of 3 in 300 seconds.

### 3.6. BINOSPEC

Binospec is a multiobject spectrograph that uses multiple slitlets rather than optical fibers. Binospec complements Hectospec by allowing spectroscopy of galaxies fainter than Hectospec's magnitude limit of R~21. Hectospec's limit is determined by our ability to measure and subtract sky background light. Precise sky subtraction with fibers is complicated by the fact that the fiber throughput into the design focal ratio of the spectrograph

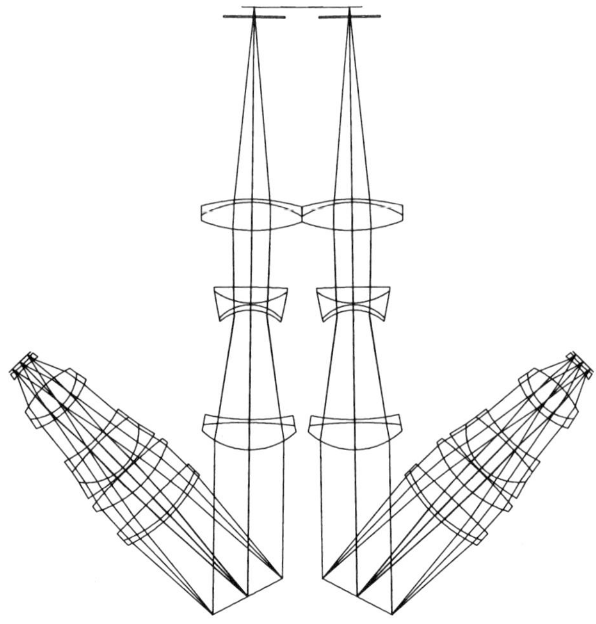

*Figure 5.* Binospec optical layout. The converted MMT's focal surface is at the top, followed by 3 collimator lens groups, a diffraction grating and 5 camera lens groups. A fold mirror will be inserted following the second collimator lens group to allow a more compact instrument.

varies slightly as the fiber link moves to follow the tracking telescope. In order to measure redshifts from very faint galaxies using absorption lines, the superior sky subtraction offered by a direct areal spectrograph is crucial. Binospec is a dual beam spectrograph, with the two beams deployed on areas of the sky separated by $\sim 10'$. Each beam of the spectrograph allows $15'$ of total slit length, and for each beam the slits can be deployed in a region that is $6'$ wide. Figure 5 shows the Binospec optics. The capabilities of Binospec and Hectospec are compared in Table 3.

## 4. Conclusion

The instruments under construction for the converted MMT will allow CfA observers to play a leading role in the study of the evolution of large-scale structure at redshifts to $z \sim 1$, as well as galaxy evolution to $z \sim 1$ and beyond. Our first step will be to conduct a relatively modest redshift survey with

TABLE 3. Comparison of Hectospec and Binospec

| Characteristic | Hectospec | Binospec |
|---|---|---|
| Field of View | $1^\circ$ dia. | $2\times6'\times15'$ |
| Number of Objects | 300 | 150 |
| Spectral Coverage | 4150 Å | 5000 Å |
| Blue Cutoff | 3600 Å | 3900 Å |
| Nominal Fiber/Slit Width | 1.5″ | 1.0″ |
| Spectral Resolution (FWHM) | <6.5 Å | <6 Å |
| Limiting Magnitude | R$\sim$21 | R$\sim$23 |
| Exposure Time to Reach Limit | $\sim$1 hr. | $\sim$6 hr. |

the Hectospec, probing to a median z$\sim$0.3. It appears that we will be able to survey 16 square degrees to R$\sim$20.5 in a single dark run, obtaining the redshifts of $\sim$25,000 galaxies.

## References


Fabricant, D. and Hertz, E. (1990) The Decaspec: A Fiber Optics Adapter for Multiobject Spectroscopy, *SPIE Proc.*, **1235**, pp 747–753.

Fabricant, D., Hertz, E. and Szentgyorgyi, A. (1994) Hectospec: A 300 Optical Fiber Spectrograph for the Converted MMT, *SPIE Proc.*, **2198**, pp 251–263.

Fabricant, D., Cheimets, P., Caldwell, N. and Geary, J. (1997) The FAST Spectrograph for the Tillinghast Telescope, *PASP*, in press.

Franx, M. (1993) Constraining Galaxy Evolution and Cosmology from Galaxy Kinematics - First Observations at z=0.18, *PASP*, **105**, p 1058.

Parry, I. and Gray, P. (1986) An Automated Multiobject Fibre Optic Coupler for the Anglo-Australian Telescope, *SPIE Proc.*, **627**, pp 118–124.

Vogt, S.S. *et al.* (1994) HIRES: the High-Resolution Echelle Spectrometer on the Keck 10-m Telescope, *SPIE Proc.*, **2198**, pp 362–375.

Vogt, N., Forbes, D., Philips, A., Gronwall, C., Faber, S., Illingworth, G. and Koo, D. (1996) Optical Rotation Curves of Distant Field Galaxies: Keck Results at Redshifts to Z$\sim$1, *ApJ Letters*, bf 465, p 15.

West, S. *et al.* (1997) Towards First Light for the 6.5-m MMT Telescope, *SPIE Proc.*, **2871**, in press.

# THE SPATIAL DISTRIBUTION OF EMISSION-LINE GALAXIES

C.C. POPESCU[1,3], U. HOPP[2], H. ELSÄSSER[1]
[1] *Max-Planck Institut für Astronomie, Königstuhl 17, D-69117 Heidelberg, Germany*
[2] *Universitätssternwarte München, Scheiner Str.1, D-81679 München, Germany*
[3] *The Astronomical Institute of the Romanian Academy, Str. Cutitul de Argint 5, 75212 Bucharest, Romania*

## 1. Introduction

Galaxy redshift surveys carried out over the last decade have revealed that bright galaxies are organized in sheets and filaments enclosing large empty regions, the voids, which occupy more than 80 per cent of the whole space. Still under debate is the question whether all galaxies follow this non-uniform distribution or if less luminous galaxies are more equally distributed. From an observer's point of view, the emptiness of the voids may be a result of observational bias. From the theoretical point of view, there has been the prediction (*e.g.*, Kaiser 1986, Bardeen 1986) that luminous galaxies were formed preferentially in high-density regions of the Universe, thus giving us a biased view of the large-scale distribution of matter.

Several studies of the spatial distribution of galaxies were carried out to overcome some of these biases (Bothun *et al.* 1986, Eder *et al.* 1989, Binggeli *et al.* 1990, Thuan *et al.* 1991). The results of these surveys did in general not favor the biasing galaxy formation and no void population was found. More recently, a new search for faint galaxies in voids was accomplished, with the intention to overcome the limitations of previous surveys in magnitude, diameter and surface brightness (Hopp *et al.* 1995, Kuhn *et al.* 1997). Other studies based on emission-line objects (Moody *et al.* 1987, Weistrop & Downs 1988, Weistrop 1989, Salzer 1989, Rosenberg *et al.* 1994) have suggested that a few emission-line galaxies have been found in the voids.

Having in mind that no definitive conclusion was drawn from previous studies and no faint void population was found in literature, we have un-



D. Hamilton (ed.), The Evolving Universe, 53–57.

dertaken a program to search for dwarf galaxies in voids (Popescu 1996, Popescu *et al.* 1996). We especially selected nearby voids which were very well defined in the distribution of normal galaxies. Since there was a hint that emission-line galaxies were found in voids, we chose to search for emission-line objects, but with the aim of finding mainly dwarf HII galaxies or BCDs.

## 2. Selection of candidates and follow-up observations

We used published cone diagrams to identify nearby ($v_R \leq 7500$ km $s^{-1}$) voids. The candidates were selected on the objective prism IIIa-J plates taken in the frame of the Hamburg Quasar Survey (HQS) (Hagen *et al.* 1995). Automated search software was applied to the the low resolution digitized data to select spectra based on two parameters, the slope of the continuum and the "luminosity"of the integrated spectra (in counts). The selected candidates were afterwards rescanned with high resolution and the final selected spectra were visually inspected for the presence of emission-lines. The objects are mainly selected based on the strong [OIII] $\lambda$5007 line, which appears as a distinctive peak near the green head of the objective prism spectra. Often the [OII] $\lambda$3727 can be identified too.

The follow-up spectroscopy of our candidates was mainly done at the 2.2 m telescope at the German-Spanish Observatory in Calar Alto, Almeria (Spain). We used both the standard Boller & Chivens Cassegrain Spectrograph as well as the CAFOS 22 (Calar Alto Faint Object Spectrograph). All our candidates were observed which allowed us to obtain a complete sample of ELGs. The final catalog consists on 239 objects, of which 227 were ELGs. A few galaxies with absorption and some quasars enter our catalog only as failures of our selection procedure.

Our sample contains objects as faint as $B_{lim} = 20.5$, and also detects intrinsically faint objects, down to $M_B = -12.0$. This indicates that the present survey goes deeper than other surveys and is, therefore, adequately for a search for faint objects in voids. The redshift values range between $0 < z < 0.1$, with a peak at z=0.015, which show that we found a sample of relatively nearby galaxies.

Most of our objects seem to be typical HII galaxies, as indicated in a diagnostic diagram log([OIII] $\lambda 5007/H_\beta$) vs log([SII] $\lambda\lambda 6716+6731/H_\alpha$) (see Fig. 1). The solid line is the dividing line between active galactic nuclei (upper right) and HII galaxies (lower left), as taken from Osterbrock (1989).

## 3. Cone-diagrams

The spatial distribution of the ELGs is illustrated in a cone diagram (Fig. 2) of one of our surveyed region. Only the southern strip of this region (7°

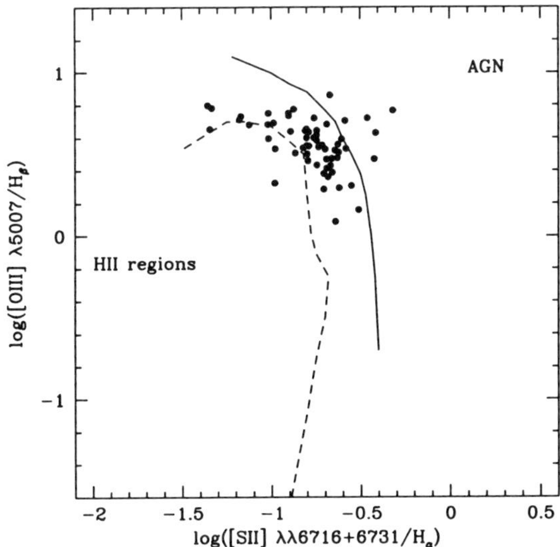


*Figure 1.* The diagnostic diagram log([OIII] $\lambda$5007/H$_\beta$) *vs* log([SII] $\lambda\lambda$6716 + 6731/H$_\alpha$ for our sample of ELGs. The solid line is the dividing line between active galactic nuclei (upper right) and HII galaxies (lower left), as taken from Osterbrock (1989). Most of our objects lie in the region of HII galaxies.

in declination) is projected into the diagram. We have plotted our ELGs together with the comparison catalogue (the CfA galaxies brighter than 15.5). We have also plotted the CfA galaxies fainter than 15.5, even though they do not form a complete sample, but with the intention to have a first impression of how they correlate with the ELGs. The most remarkable feature of the cone is the "Great Wall", which crosses our diagram from 6500 km s$^{-1}$ at $\alpha = 12^{\rm h}00^{\rm m}$ to 9000 km s$^{-1}$ at $\alpha = 15^{\rm h}30^{\rm m}$. In front of the Great Wall there are two very well defined nearby voids: one is centered at $\alpha \sim 13^{\rm h}15^{\rm m}$, v $\sim$ 3000 km s$^{-1}$ and we call it Void 1 and the other one is just in front of the Great Wall and we call it Void 2.

Our galaxies seem to follow the structures described above as well. At a closer inspection one can discover that there are some galaxies that lie in these foreground voids. In Void 1 there are two ELGs that have the nearest bright CfA neighbor at a distance of 3.85 $h^{-1}$ Mpc and 6.63 $h^{-1}$ Mpc, respectively. They are among the best candidates we found in voids. Two further faint CfA galaxies are also present in the void. In Void 2 we found an "Arch" of 7 ELGs, that seem to divide the void into three smaller voids.

The Arch is also populated by three faint CfAs while a fourth one closes the Arch at lower redshifts. We have also found a large number of ELGs at higher redshifts, where the CfA catalogue becomes practically non-existent. This indicates that the ELGs have a better power to trace the luminous matter at further distances than the normal galaxies.

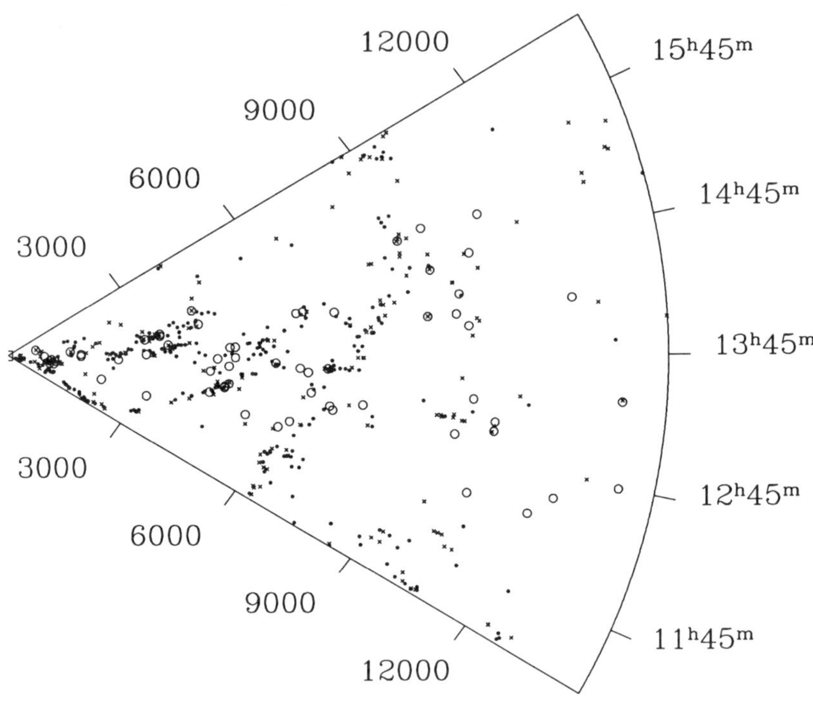


*Figure 2.* Wedge-plot of redshift ($cz$ in km s$^{-1}$ ) - right ascension out to a redshift of 15000 km s$^{-1}$. The CfA $m < 15.5$: small dots, N=276 , the CfA $m > 15.5$: crosses, N=165, the ELGs: open circles, N=63. The wedge is a 7° wide strip in declination centered on $\delta = 36°30'$. The cone contain a larger angle in right ascension than our surveyed region, which is only between ($12^h$, $15^h30^m$).

The absolute magnitudes of our void galaxies range between $-12.7 < M_B < -18.5$ which means that all of them are dwarfs. These galaxies are therefore intrinsically faint objects, quite different from the galaxies that were found in the Bootes void (Weistrop *et al.* 1995), which were mainly M* galaxies or brighter. Nevertheless, the distribution of absolute magnitudes of our parent sample starts to drop around $M_B = -17$ and only a few galaxies are found in the range $-12 < M_B < -15$. They may constitute the tip of an iceberg of an even fainter population. We suggest that the void galaxies have also a tendency for clustering. This result is far from being secure, but some of our void galaxies seem to associate with faint late-type CfAs. For example our Arch of 7 ELGs in Void 2 is followed by four faint CfAs. The Arch seem to divide that bigger void in 3 smaller voids. And not at the end, 75% from our isolated galaxies have their nearest neighbors among themself. These results are close to the observational results of an HI survey in the Bootes void (Szomoru 1996). Lindner *et al.* (1996) also suggest a hierarchical distribution of galaxies and voids, in the sense that superclusters and clusters delimit bigger voids, bright galaxies some smaller voids and faint galaxies delimit very small voids, inside the bigger ones.

From our estimates of the expected number of void galaxies we conclude that we did not find an underlying homogenous void population.

## References

Bardeen, J.M. 1986, in *Inner Space/Outer Space*, ed. E.W. Kolb, M.S. Turner, D. Lindley, K. Olive, and D. Seckel (University of Chicago Press, Chicago), p.212
Binggeli, B., Tarenghi, M., Sandage, A. 1990, A&A 228, 42
Bothun , G.D., Beers, T.C., Mould, J.R., Huchra, J.P. 1986, ApJ 308, 510
Eder J.A., Schombert J.M., Dekel A., Oemler A. 1989, ApJ 340, 29
Hagen, H.-J., Groote, D., Engels, D., Reimers, D. 1995, A&AS 111, 195
Hopp, U., Kuhn, B., Thiele, U., Birkle, K., Elsässer H. 1995, A&AS 109, 537
Kaiser, N. 1986, in *Inner Space/Outer Space*, see ref. Bardeen (1986), p.258
Kuhn, B., Hopp, U., Elsässer, H. 1997, A&A 318, 405
Lindner, U., Einasto, M., Einasto, J., Freudling, W., Fricke, K., Lipovetsky, V., Pustilnik, S., Izotov, Y., Richter, G. 1996, A&A 314, 1
Moody, J.W., Kirshner, R.P., MacAlpine, G.M., Gregory, S.A. 1987, ApJ 314, L33
Osterbrock, D.E. 1989, Astrophysics of Gaseous Nebulae and Active Galactic Nuclei, University Science Books, California, p349.
Popescu, C.C. 1996, Ph.D thesis, Heidelberg
Popescu, C.C., Hopp, U., Hagen, H., Elsässer H. 1996, (Paper 1) A&AS 116, 43
Rosenberg, J.L., Salzer, J., Moody, J.W., 1994, AJ 108, 1557
Salzer J.J. 1989, ApJ 347, 152
Szomoru, A., van Gorkom, J.H., Gregg, M.D., Strauss, M.A. 1996, AJ 111, 2150
Thuan, T.X., Alimi, J-M, Gott, J.R., Schneider, S.E. 1991, ApJ 370, 25
Weistrop D. 1989, AJ 97, 357
Weistrop D., Downs R.A. 1988, ApJ 331, 172
Weistrop, D., Hintzen, P., Liu, C., Lowenthal, J., Cheng, K,-P., Oliversen, R., Brown, L., Woodgate, B. 1995, AJ 109, 981

# PROPERTIES OF GALAXIES IN AND AROUND VOIDS

ULRICH HOPP
*Universitätssternwarte München*
*Scheiner Str. 1, D 81679 München*
*email: hopp@usm.uni-muenchen.de*

**Abstract.** Two surveys for intrinsically faint galaxies towards nearby voids have been conducted at the MPI für Astronomie, Heidelberg. One selected targets from a new diameter limited ($\Phi \geq 5''$) catalog with morphological criteria while the other used digitized objective prism Schmidt plates to select mainly H II dwarf galaxies. For some 450 galaxies, redshifts and other optical data were obtained. We studied the spatial distribution of the sample objects, their luminosity function, and their intrinsic properties.

Most of the galaxies belong to already well known sheets and filaments. But we found about a dozen highly isolated galaxies in each sample (nearest neighbor distance $\geq 3h_{75}^{-1}$ *Mpc*). These tend to populate additional structures and are not distributed homogeneously throughout the voids. As our results on 'void galaxies' still suffer from small sample statistics, I also tried to combine similar existing surveys of nearby voids to get further hints on the larger structure and on the luminosity function of the isolated galaxies. No differences in the luminosity function of sheet and void galaxies could be found.

The optical and infrared properties of both samples are in the normal range for samples dominated by late-type dwarfs. Follow-up H I studies show that the isolated dwarfs in both samples have unusual high amount of neutral gas for a given luminosity.

## 1. Introduction

One of the surprizing results of the first redshift surveys was the detection of empty regions in space, the voids. Some theories predicted that a homogeneously distributed population of dwarf galaxies may exist inside these voids (Dekel & Silk 1986), others predicts at least filaments of matter



*D. Hamilton (ed.), The Evolving Universe*, 59–70.

crossing the empty regions of the nearby universe (see Colberg *et al.*, these proceedings). In 1990, we started a survey for intrinsically small and faint galaxies towards nearby voids with the idea that some selection effects introduced by the catalogs and observers might have prevented the detection of a void population (Hopp & Kuhn 1995). Obviously, dwarf galaxies of all types with their faint absolute magnitudes, often rather low surface brightness (SB), and small or even vanishing apparent diameters on Schmidt survey plates are easily missed in all kind of surveys. Indeed, it is known that dwarfs dominate the universe in number ($\geq 65\%$[1]), but form only a minor contribution ($\leq 10\%$) to the catalogs ('Zwicky', UGC) on which the early redshift surveys like the CfA (Huchra *et al.* 1990) and its southern counterpart (da Costa *et al.* 1994) were based.

Early attempts to clarify whether or not dwarfs are distributed like the giants in the standard catalogs failed as they were limited by the standard catalogs (*e.g.*, Thuan *et al.* 1987). They were not able to detect real dwarf galaxies even in the distance of the nearest voids (Binggeli *et al.* 1990, Hopp 1994). But these surveys already showed that one has to deal with really faint objects[2], and, therefore, has to study the nearby voids ($v_R \leq 10^4$ $km\ s^{-1}$). In this sense, the famous Boötes void (Kirchner *et al.* 1981) which attracted otherwise very important survey work (*e.g.*, Weistrop *et al.* 1995, Szomoru *et al.* 1996), is too distant. Indeed, most of the so-far identified $\sim 60$ galaxies in the Boötes void are not dwarfs.

Here, I will report about the two Heidelberg-void surveys. I will especially describe the properties of the very isolated galaxies which we identified in and around voids. Meanwhile, other independent surveys were done, mainly by searching emission line galaxies (hereinafter ELG; see Popescu *et al.* 1996, for references), or galaxies of low surface brightness (LSBG, *e.g.*, Bothun *et al.* 1986, Mo *et al.* 1994, Roennback & Bergvall 1996, Bothun *et al.* 1997, Impey *et al.* 1996). Two surveys were dedicated to morphological selection (Eder *et al.* 1989, Binggeli *et al.* 1990). I will compare to these other surveys and show that a combination of the catalogs to a common data base has great potential in studying the matter distribution in and around voids. All surveys together may serve as an ideal data base to reconstruct the mass distribution from the peculiar motion of the galaxies (following *e.g.*, da Costa *et al.* 1996), especially, when the gaps between them will be filled. Throughout this paper, I use $H_0 = 75$ $km\ s^{-1}\ Mpc^{-1}$.

[1] At least in the 10 Mpc sample of Kraan-Korteweg & Tammann (Schmidt & Boller 1992, Karachentsev 1994).

[2] *e.g.*, $M_B \geq -16^m, m_B \geq 18^m, SB_0 \geq 22^m/\Box'', r_s \leq 1kpc \sim 3''$.

## 2. Two Heidelberg surveys

The shape of the luminosity function of galaxies yields magnitude limited galaxy catalogs which are dominated in number by $L^*$ galaxies (Dickey 1988). Increasing the magnitude limit of a catalog thus largely increases the number of $L^*$ galaxies at large distances while it adds only a few to the nearer dwarfs. The strategy of a general redshift survey like the LCRS or the SLOAN[3] was, therefore, neither efficient nor feasible for our purpose of finding dwarfs in the redshift distance of the nearest voids. Similar arguments hold for diameter-limited catalogs (Binggeli *et al.* 1990). Selection effects in SB can strongly influence the results of any survey (Disney 1976, Bothun *et al.* 1997). Thus, we started dedicated surveys towards a few selected nearby voids. We tried to optimize the selection for dwarf galaxies and LSBG in and around these voids (Hopp 1994, Hopp & Kuhn 1995). The data were reported in Hopp *et al.* (1995, paper 1 hereafter), and Popescu *et al.* 1996 (paper 2) while the spatial distribution of the samples in discussed in Kuhn *et al.* 1997 (paper 3), and Popescu *et al.* 1997 (paper 4).

### 2.1. VOID SELECTION

When we started, few void catalogs were available. We used, therefore, the available cone diagrams (mostly from CfA1), and selected those voids which have a diameter of $\geq 20$ $Mpc$, and are completely free of Zwicky and UGC galaxies. We selected four fields where $\mid b \mid \geq 40^o$. Our fields cover a total of 10 voids up to $10^4$ $km\ s^{-1}$ (see paper 1, 2). Later, we learned that our naively defined voids agree quit well with more objective definitions (Kauffmann & Fairall 1991, Saunders *et al.* 1991, Slezak *et al.* 1993, Lindner *et al.* 1995).

### 2.2. SELECTION EFFECTS

The redshift surveys available at that time (CfA, SSRS) were based on traditional catalogs (Zwicky *et al.* 1961-65, Nilson 1973, Lauberts 1982) which are limited either in apparent magnitude of B $\leq 15.5^m$ or in diameters $\Phi \geq 1.0'$. They are most severely limited in SB due to the POSS I or ESO Quick-B capabilities. A discussion of the structural properties of very nearby dwarfs forces one to include much smaller diameters and also objects of lower SB (Hopp 1994; footnote 2).

We obtained deep Calar Alto 3.5m prime focus images towards the center of three of our fields and establish a diameter-limited catalog down to

[3]For a description of these two major redshift surveys see the contributions by the two teams in these proceedings.

$\Phi \geq 5''$. In the surroundings of these deep images, I surveyed the POSS down to $\Phi \geq 21''$. Figure 1 in paper 1 shows that the resulting sample fits nicely to the diameter distribution of the UGC, going to smaller values. From this calatalog, we selected dwarf candidates by their morphology or low SB. The follow-up redshift survey revealed a high percentage ($\sim 70\%$) of ELG in this sample (paper 1) and a higher-than-average rate of ELGs among the isolated galaxies. We decided, therefore, to start a second survey for ELG's in the Hamburg Quasar Survey data base (Hagen *et al.* 1995) which is decribed in detail by Popescu (these proceedings, paper 2). The ELG survey is only seeing limited in $\Phi$, but galaxies with large $\Phi$ in their emission regions are hard to detect on objective prism plates. Both Heidelberg void surveys cover the range $15 \leq B \leq 20.5^m$ ($20.0 \leq \mu_B[m/\square''] \leq 24.5$). The magnitude completeness limit of the morphological survey is $B \sim 19.0^m$. The leading selection effect for the ELG sample is the equivalent width of the emission line used for selection, in our case [OIII]5007. The limit is 0.8 nm (see Popescu, this volume, and paper 4).

## 2.3. THE MORPHOLOGICAL SURVEY

This survey covered three fields towards nearby voids. A total of $\sim 175$ redshifts were obtained together with B and R CCD images of most of the galaxies (paper 1). The SB(r) of the galaxies in a subsample was obtained by Vennik *et al.* 1996 (paper 5). Half of the galaxies have redshifts above the limits of the CfA which is the bright comparison sample in our study and are, therefore, useless for our goal. The average surface density of all objects is about $1/\square^o$. Cone diagrams of all fields and a detailed discussion of the nearest neighbor distance $D_{NN}$ distribution are the essentials of paper 3.

## 2.4. THE EMISSION LINE GALAXY SURVEY

The survey was obtained in four fields, some identical to the fields of the morphological survey. A total of $\sim 250$ redshifts were obtained. For most of the galaxies, we have R or B CCD images, for some both colors, which are now under study. Most of the spectra are good enough to discuss the abundances of the objects (in prep.). The data were presented in paper 2 while the spatial distribution and the details of the selection effects are discussed in paper 4 and by Popescu in these proceedings. This survey was pretty successful in finding dwarf galaxies below $v_r \leq 10^4\ km\ s^{-1}$ (see Fig. 2 of Hopp & Kuhn 1995).

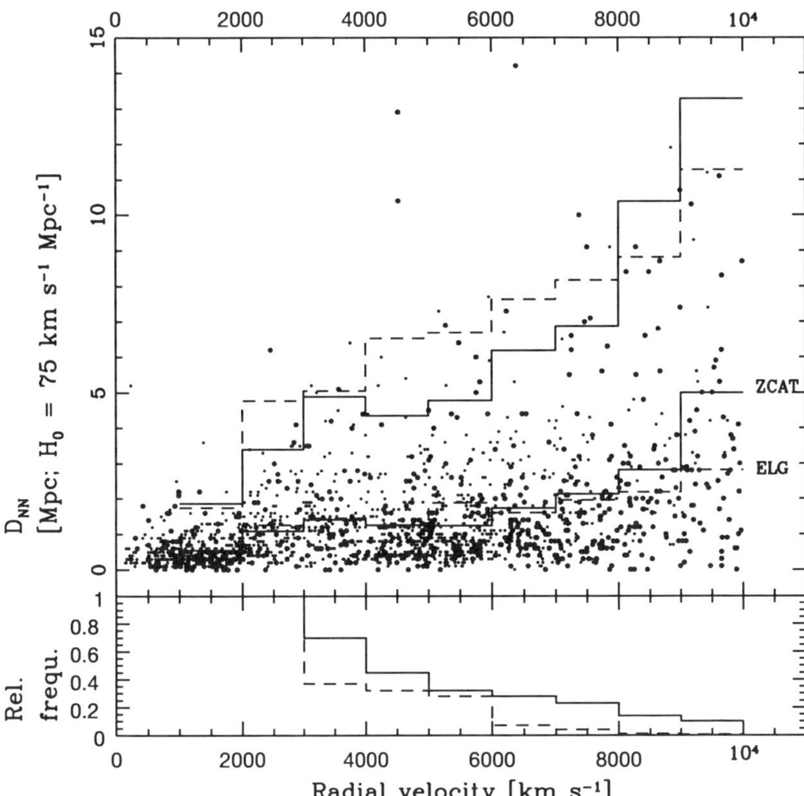


*Figure 1.* *Top:* Nearest-neighbor distance $D_{NN}$ distribution as a function of radial velocity for the MDS galaxies (big symbols) and the CfA galaxies from the same volume (small). The lower lines indicate the mean values in redshift bins of $10^3$ $km\ s^{-1}$ for the CfA (solid) and the MDS (dashed), the upper the mean values plus $3\sigma$. *Bottom:* Relative frequency of galaxy numbers in MDS and CfA normalized to the bin at 2500 $km\ s^{-1}$. Obviously, the sampling decreases in both samples with increasing redshift in a similar manner. Thus, the absolute values of $D_{NN}$ at low and high redshifts cannot be directly compared. We restricted, therefore, the MDS to the interval 3000 – 6000 $km\ s^{-1}$ for comparison of high and low density galaxies with a cut between these two regimes at $3.0h_{75}^{-1}$ $Mpc$.

## 3. Spatial distribution

For a detailed discussion, including cone diagrams, we refer to papers 3, 4 as well as to Popescu in these proceedings. We were able to detect about 25 highly isolated galaxies. We call a galaxy highly isolated when $D_{NN} \geq$

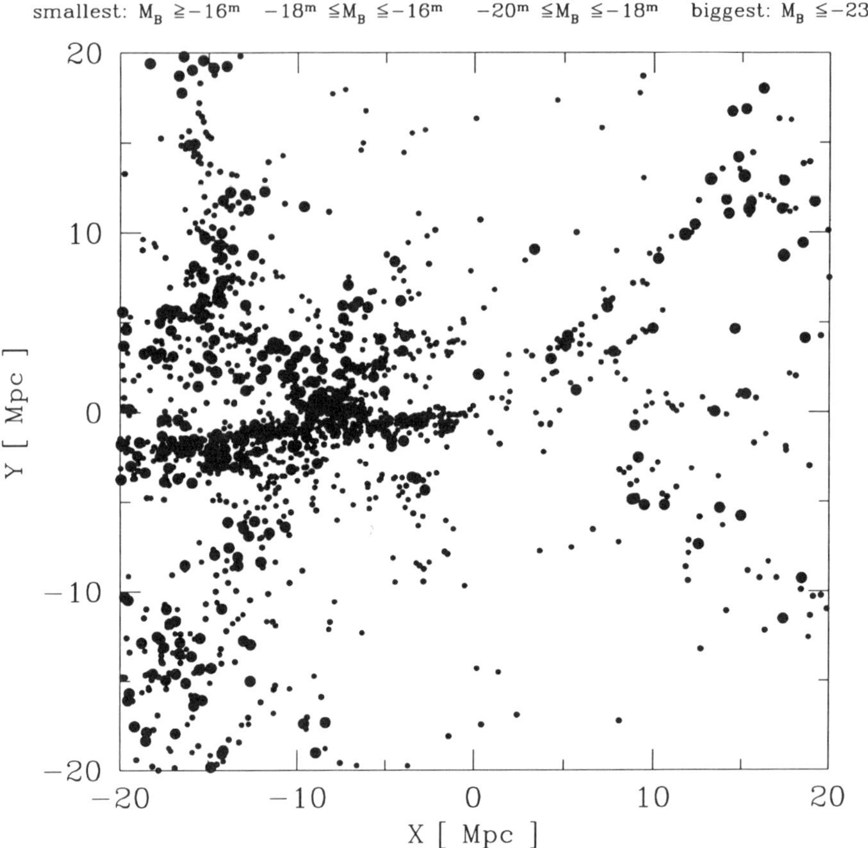


*Figure 2.* Spatial distribution of the merged dwarf sample and the CfA galaxies in a cube of $20h_{75}^{-1}$ *Mpc* side length. Third coordinate is projected. The symbol sizes indicates the ***absolute*** magnitude of the galaxies. Besides the well-known rich structures which are populated by galaxies of all luminosities, there exist regions which are only populated by low-luminosity galaxies.

$3.0h_{75}^{-1}$ *Mpc* and $cz \leq 10^4$ $km$ $s^{-1}$ (Fig. 1).[4] All other galaxies ($\sim$ 95%) follow well known sheets and filaments or are situated in the background where the sampling is too sparse for any density estimates.

Many of the highly isolated galaxies tend to populate the rims of the voids. This may simply reflect the fact that late type dwarfs have the lowest clustering power and drop off with a shallower gradient from the density peaks of the sheets.

[4]One should not forget that $D_{NN}$ can be measured *only* to the next *cataloged* galaxy!

A few are located more to the central parts of the voids. These galaxies show some tendency to align themself in very sparsely populated chains or filaments which devide bigger voids into smaller entities (paper 4). Filaments which are populated only by low luminosity galaxies are an observational argument in favour of hierarchical structure formation (Vogeley *et al.* 1996). Unfortunately, we have only very few cases at hand in our surveys, but similar conclusion were based on other ELG surveys (Lindner *et al.* 1996). We estimated the number density contrast between the sheet galaxies and those few in the voids to be $\Delta\rho/\rho \leq 0.1$ (paper 3, 4).

### 3.1. SIMILAR SURVEYS

Having so few isolated galaxies in our samples, I tried to combine our surveys with other surveys which are similar in selection functions, covered field, volume, velocity and magnitude range. These are the SBS survey (Pustil'nik *et al.* 1995), the UM and Case Surveys by Salzer (1989) and Rosenberg *et al.* (1994), respectively. For a comparison of these different surveys see paper 2. I call the merged dwarf sample MDS. Naturally, this merging of catalogs yields an inhomogeneous sample (Hopp 1997). But this exercize was mainly done to convince myself that a survey which will cover the whole nothern galactic cap and link all the above mentioned surveys, including the two Heidelberg void surveys, will be the right way to proceed. A project (under the name Hamburg-SAO survey) was already started (Ugryumov *et al.* 1997). We obtained some further 75 redshift of ELGs which are already included in the MDS.

From the same fields as the surveys, I extracted the CfA galaxies as a comparison sample. In total, the MDS covers $4750\square^o$ and contains 1012 (and 877 CfA) galaxies within $v_R \leq 10^4\ km\ s^{-1}$. All individual surveys show a few isolated galaxies as discussed by their authors.

Contrary to the authors of the ELG surveys, those who analysed the spatial distribution of LSBG (*e.g.*, Eder *et al.* 1989, Mo *et al.* 1994, Schombert *et al.* 1997) found that while these LSBG are avoiding high density regions, they else still follow the structures occupied by the giants. As there is still some discrepancy in the interpretation[5] , and as the selection function surely differs from those of the ELG samples, I did not include these surveys into the MDS so far.

Most of the MDS galaxies follow the distribution outlined by the CfA galaxies, namely the well-known structures which are defined by the massive

[5]This comes as a little surprize as the LSBG cone diagrams shows also very isolated objects and statistical results are similar. I was always left with the impression that the discrepancy is more of semantic nature depending on the topic the authors like to address. ELG and LSBG surveys both rule out the high-biasing scenario that (irregular) dwarfs fill the voids.

galaxies. About 140 of the MDS galaxies are very isolated with $D_{NN} \geq 3.0h_{75}^{-1}$ $Mpc$ (Fig. 1). These isolated galaxies are not randomly distributed, but they occupy the rims of the voids or form additional filaments which seems to be populated only by low luminosity galaxies (Fig. 2; Hopp 1997, paper 4). Similar results have been found by Lindner *et al.* (1996) and Vogeley *et al.* (1994).

## 4. Properties of isolated galaxies

### 4.1. SURFACE PHOTOMETRY

As far as we finished the analysis, most of our objects are disk galaxies with rather blue colors, quite common for a dwarf sample. The mean scale length for the dwarfs is $2.1 \pm 0.8h_{75}^{-1}kpc$ (paper 5). The isolated galaxies do not differ in their structural properties. In the theoretical SB *vs.* scale-length diagram shown by Dalcanton, our sheet and isolated galaxies distribute between the diameter limited UGC sample and the Virgo cluster sample in the regime of relatively small scale length and over nearly the whole range in SB (Dalcanton *et al.* 1997). Again, no difference is obvious between isolated and sheet galaxies. Most objects have central SB well below the 'Freeman disk'.

### 4.2. LUMINOSITY FUNCTION

I calculated the luminosity function for our sample as for the other ELG samples (Fig 3) after applying $V/V_m$ test completeness corrections for a volume-limited sample. They all are pretty similar and show a steep faint end slope. Then, I calculated the luminosity function for the MDS within $3000 \leq cz \leq 6000\ km\ s^{-1}$ as the completeness in this redshift range drops only slowly for dwarfs (Fig. 1). The calculation was done separately for low- and high-density regions (dividing line: $D_{NN} = 3.0h_{75}^{-1}\ Mpc$). Within the rather poor statistics for the low density regime ($\sim 140$ galaxies), the low- and high-density luminosity functions are identical.

### 4.3. HIGH HI ABUNDANCES

We obtained HI observations for subsamples of both Heidelberg void surveys, mostly with the Effelsberg 100m antenna. The detection rate was high ($\sim 67\%$ and $\geq 50\%$, respectively) and many galaxies show a high amount of HI gas. The line profiles indicate in most cases a dwarf galaxy. Especially, we found a trend that the HI-mass to blue luminosity ratio of dwarfs for a given luminosity increases with decreasing galaxy density, from the Virgo cluster to sheets to voids (Huchtmeier *et al.* 1997). A first analysis

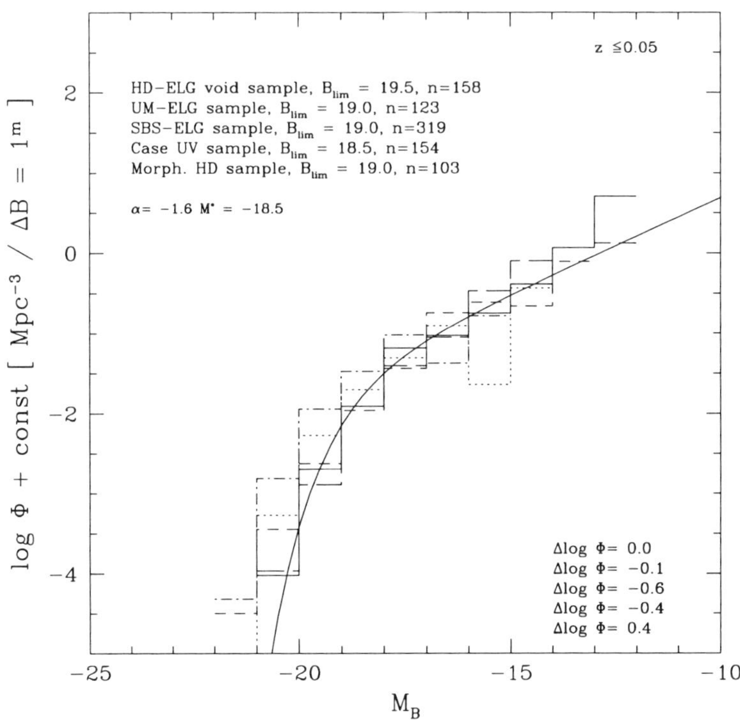


*Figure 3.* Luminosity function derived from our surveys and several similar sample from the literature (histogram). The line shows a Schechter function with the indicated parameters for comparison, it is not a fit. All these field dwarf samples tend to show a steep faint end slope. The differences between the samples is negligible within the given statistical and systematic errors.

of the ELG data also supports this finding. No significant deviation from the Tully-Fisher relation was found so far, including the void galaxies.

### 4.4. OTHER FREQUENCIES

With the aid of NED[6] and the recently publicly available ROSAT point source catalog (Voges *et al.*, 1996), we checked whether the galaxies found in our two surveys were detected at other frequencies, too. No galaxies from the morphological approach were detected by IRAS, they do not shine up

[6]The NASA/IPAC extragalactic database (NED) is operated by the Jet Propulsion Laboratory, California Institute of Technology, under contract with the National Aeronautics and Space Administration.

in the ROSAT point source catalog, nor does NED list any radio continuum detection.

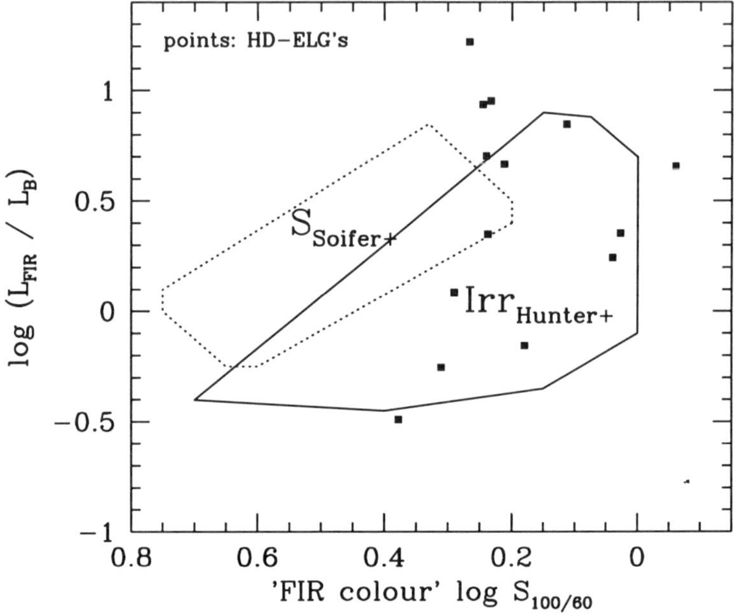


*Figure 4.* Plot of FIR color index in 60 $\mu$ through 100 $\mu$ *vs.* ratio of FIR luminosity to B luminosity. Data from the IRAS point source catalog for those of the Heidelberg-ELG sample which were detected by IRAS in both bands. The lines indicate the distribution regions for spiral galaxies according to Soifer *et al.* (1987) and for irregular galaxies according to Hunter *et al.* (1989). Most of our galaxies occupy the region of irregular dwarfs.

On the other hand, about 15% of the ELG galaxies were detected by IRAS in the 60$\mu$ band, but none of the detected sources is highly isolated. Some of them were also detected at 100$\mu$ or at 25$\mu$. In Fig. 4, we show a FIR-B color-color plot where the detected objects are indicated together with the general distribution for spirals and irregular dwarfs. These detections again show that our ELGs occupy mostly the dwarf parameter space. Only one single source may have been detected by ROSAT. Seven 20 cm FIRST detections are available from NED.

## 5. Conclusions

I presented our two Heidelberg void surveys which each identified about a dozen of highly isolated galaxies inside the voids. Some of them are distributed along the rims of the voids while others populate filaments across the voids. Some voids remain completely empty. Most of the galaxies are dwarfs with quite normal optical properties, but high HI mass. Our results agree well with the results of other surveys which detected dwarfs in and around voids. From a combination of all ELG surveys we found no evidence for a difference in the luminosity function of isolated and sheet galaxies. The number density contrast between sheets and the voids remains high ($\Delta\rho/\rho \leq 0.1$) despite the newly identified void galaxies.

I like to thank for many useful discussion Drs. R. Bender, B. Binggeli, J. Einasto, H. Elsässer, H.-J. Hagen, W.K. Huchtmeier, B. Kuhn, C.C. Popescu, S. Pulstil'nik, J. Roennback, J. Salzer, and J. Vennik. Especially, I remember with a deep and warm feeling the discussions with the late Valentin Lipovetzky who died the Sunday just before this meeting. I like to dedicate this paper to the memory of Valentin. It's a pleasure to thank Ulrich Thiele and the Calar Alto crew for the support during the many observations. The author acknowledges the support of the SFB 375 of the Deutsche Forschungsgemeinschaft.

## References

Binggeli,B., Tarenghi,M., Sandage,A. 1990, *A&A*, **228**, 42.
Bothun,C.D., Beers,T.C., Mould,J.R., Huchra,J.P. 1986, *ApJ*, **308**, 510
Bothun,C.D., Impey,C., McGaugh,S., 1997, *PASP* in press
da Costa, L.N., Geller,M.J., Pellegrini,P.S., Latham,D.W., Fairall,A.P., Marzke,R.O., Willmer,C.N.A., Huchra,J.P. 1994, *ApJ*, **424**, L1
da Costa, L.N., Freudling,W., Wegner,G., Giovanelli,R., Haynes,M.P., Salzer,J.J., 1996, *ApJ*, **468**, L5
Dalcanton,J., Spergel,D.N., Summers,F.J. 1997, *ApJ* preprint
Dekel,A., Silk,J. 1986, *ApJ*, **303**, 39
Dickey,J.M. 1988, *Astron.Soc.Pac.Conf.Ser.*, **5**, 9
Disney,M.J. 1976, *Nature*, **263**, 573
Eder,J.A., Schombert,J.M., Deckel,A., Oemler,A. 1989, *ApJ*, **340**, 29
Hagen,H.-J., Groote,D., Engels,D., Reimers,D. 1995, *A&AS*, **111**, 195
Hopp, U. 1994, in: G.Meylan and P.Prugniel (eds.), *Dwarf Galaxies*, ESO Conf. Proceedings, **49**, 37
Hopp,U. 1997, *Proc. Symp. IAU*, **179**, in press
Hopp,U., Kuhn,B., Thiele,U., Birkle,K., Elsässer,H., Kovachev,B. 1995, *A&AS*, **109**, 537 **(paper 1)**
Hopp,U., Kuhn,B. 1995, *Reviews in Modern Astronomy*, **7**, 277
Huchra,J.P., Geller,M.J., de Lapparent,V., Corwin,H.G. 1990, *ApJS*, **72**, 433
Huchtmeier,W.K., Hopp,U., Kuhn,B. 1997, *A&A* accepted
Hunter,D., Gallagher,J.S., Rice,W.L., Gillett,F.C. 1989, *ApJ*, **336**, 152
Impey,C.D., Sparberry,D., Irwin,M.J., Bothun,G.D. 1996, *ApJS*, **105**, 209
Karachentsev, I.D. 1994, *A&Ap Transactions*, **6**, 1
Kauffmann,G., Fairall,A.P. 1991, *MNRAS*, **248**, 313
Kirchner,R.P., Oemler,A., Schechter,P.L., Schectman,S.A. 1981, *ApJ*, **248**, L57
Kuhn,B., Hopp,U., Elsässer,H. 1997, *A&A*, **318**, 405 **(paper 3)**
Lauberts,A. 1982, *The ESO/Uppsala Survey of the ESO (B) Atlas*, ESO, München.
Lindner,U., Einasto,M., Einasto,J., Freudling,W., Fricke,K., Lipovetsky,V., Pustilnik,S., Izotov,Y., Richter,G. 1996, *A&A*, **314**, 1

Mo,H.J., McGaugh,S., Bothun,G.D. 1994, *MNRAS*, 267, 129
Nilson,P. 1973, *Uppsala General Catalog of Galaxies*, Nova Acta Reginae Soc. Sci. Uppsalinisis Ser. V: A Vol. **1** (UGC).
Popescu,C.C., Hopp,U., Hagen,H.-J., Elsässer,H. 1996, *A&AS*, **116**, 1 (**paper 2**)
Popescu,C.C., Hopp,U., Elsässer,H. 1997, *A&A* accepted, (**paper 4**)
Pustil'nik,S.A., Ugryumov,A.V., Lipovetsky, A.A., Thuan,T.X., Guseva,N.G. 1995, *ApJ*, **443**, 499
Roennback,J., Bergvall,N. 1996, *A&A*, **302**, 353
Rosenberg,J.L. *et al.* 1994, *AJ*, **108**, 1557
Salzer,J.J. 1989, *ApJ*, **347**, 152
Saunders,W., Frenk,C., Rowan-Robinsom,M., Efstathiou,G., Lawrence,A., Ellis,R., Crawford,J., Xia,X.Y., Parry,I. 1991, *Nature* **349**, 32
Schmidt, K.H., Boller, T. 1992, *AN*, **313**, 329
Schombert,J.M., Pildes,R.A., Eder,J.A. 1997, *ApJS* accepted
Slezak,E., de Lapparent,V., Bijaoui,A. 1993, *ApJ*, **409**, 517
Soifer,B.T., Houck,J.R., Neugebauer,G. 1987, *ARA&A*, **25**, 187
Szomoru,A., van Gorkom,J.H., Gregg,M.D., Strauss,M.A. 1996, *AJ*, **111**, 2150
Thuan,T.X., Gott,J.R., Schneider,S.E. 1987, *ApJ*, **315**, L93
Vennik,J., Hopp,U., Kovachev,B., Kuhn,B., Elsässer,H. 1996, *A&AS*, **117**, 261 (**paper 5**)
Vogeley,M.S., Geller,M.J., Park,C., Huchra,J.P. 1994, *AJ*, **108**, 745
Voges,W., Aschenbach,B., Boller,Th., Bruninger,H., Briel,U., Burkert,W., Dennerl,W., Englhauser,J., Gruber,R., Haberl,F., Hartner,G., Hasinger,G., Kürster,M., Pfeffermann,E., Pietsch,W., Predehl,P., Rosso,C., Schmitt,J.H.M.M., Trümper,J., Zimmermann,H.-U. 1996, *A&A*, in press
Weistrop,D., Hintzen,P., Liu,C., Lowenthal,J., Cheng,K.-P., Oliversen,R., Brown,L., Woodgate, B. 1995, *AJ*, **109**, 981
Ugrymov,A., Engels,D., Lipovetsky,V., Hopp,U., Richter,G., Izotov,Y.I., Kniazev,A.Y., Popescu,C.C. 1997, *Proc. Symp. IAU*, **179**, in press
Zwicky, F., Herzog,E., Wild,P. 1961-1968, *Catalog of Galaxies and Clusters of Galaxies*, Speich, Zürich (CGCG).

# LARGE-SCALE FLOWS FROM THE MARK III TULLY-FISHER DATA

ADI NUSSER
*Max-Planck-Institut für Astrophysik*
*Karl-Schwarzschild-Str. 1*
*85740 Garching, Germany*

**Abstract.** We use the Mark III Tully-Fisher data of 2900 spiral galaxies to derive an estimate of the large-scale peculiar velocity field. A comparison of this field with the 1.2-Jy IRAS gravity field yields the parameter $\beta = \Omega^{0.6}/b$, where $b$ is the linear bias factor relating the distribution of the IRAS galaxies to the dark matter density fluctuations. By working in redshift space and the inverse Tully-Fisher relation, this comparison is essentially free of Malmquist biases. The velocity and gravity fields are expanded in the same set of smooth functions and therefore our estimate of $\beta$ and the assessment of the agreement between the fields are free of ambiguities resulting from possible differences in the smoothing of the fields. We find a general good agreement, especially for data within 3000 $\mathrm{km\,s^{-1}}$, between the velocity and gravity for $\beta \sim 0.6$. The statistical $1\sigma$ error in our estimate of $\beta$ is 0.1. However, the fields do not agree in detail, precluding a firm determination of $\beta$ from these data sets at present.

## 1. Introduction

Peculiar velocities of galaxies are a direct probe of the nature of the underlying dark matter in the Universe. Under the assumption that the large-scale structure has formed via gravitational amplification of small initial density fluctuations, the observed velocity field yields a prediction for the present matter-density fluctuations given an assumed value for $\Omega$. Furthermore, quasi-linear gravitational instability theory can be used to reconstruct the primordial density fluctuations from the observed flows. This enables us to address fundamental questions about the nature of the initial fluctua-



*D. Hamilton (ed.), The Evolving Universe,* 71–81.

tions and their relation to $\Omega$. For example, Nusser & Dekel (1993) have demonstrated that unless $\Omega$ is around unity the initial density fluctuations were non-gaussian. Here we concentrate on comparing the peculiar velocity field from Tully-Fisher data with the gravity field computed from the distribution of galaxies in redshift space. Unlike quasi-linear methods for estimating $\Omega$ based on the observed velocity field alone combined with some assumption about the initial fluctuations such as following Gaussian statistics (Nusser & Dekel 1993, Berbardeau *et al.* 1995), linear methods involving peculiar velocities and redshift surveys do not yield estimates of $\Omega$ independent of the bias factor $b$. The basic idea behind velocity-gravity comparisons is the following: assuming a linear biasing relation between the large-scale fluctuations in the number density of galaxies and the mass fluctuations, we obtain from linear theory

$$\mathbf{v_L}(\mathbf{r}) = \frac{H_0\beta}{4\pi\bar{n}} \sum_i \frac{1}{\phi(r_i)} \frac{\mathbf{r}_i - \mathbf{r}}{|\mathbf{r}_i - \mathbf{r}|^3} + \frac{H_0\beta}{3}\mathbf{r} \,, \tag{1}$$

where $\bar{n}$ is the true mean galaxy density in the sample, $\beta \equiv \Omega^{0.6}/b$, and where $\phi(r)$ is the radial selection function (Yahil *et al.* 1991). Note the sum in (1) is to be computed in real space, whereas the galaxy catalog exists in redshift space. Note also that the result is insensitive to the value of $H_0$, as the right hand side has units of velocity. We shall henceforth quote all distances in units of $\mathrm{km\,s^{-1}}$. The density $\bar{n}$ and selection function $\phi(r)$ can be estimated from the redshift catalog used in the analysis, so that comparison of the measured velocity field to the predicted velocity field $\mathbf{v_L}(\mathbf{r})$ gives us a measure of $\beta$. The value of $\beta$ is not the only ultimate outcome of such comparisons. With the advent of large peculiar velocity data sets, such comparisons can be indispensable in constraining the biasing relation (galaxy formation) and assessing the universality of intrinsic properties of galaxies, *e.g.,* the Tully-Fisher relation.

Several comparisons of velocity and gravity fields have been made with older datasets (*e.g.*, Hudson 1994; Kaiser *et al.* 1991; Yahil 1988; Strauss and Davis 1988) but these analyses were all meant to be preliminary and none of their conclusions were compelling. These studies did not included a proper treatment of the correlated noise in the analysis. Nusser & Davis (1994) computed the dipole component of peculiar velocity field on distant shells as predicted from the *IRAS* 1.2 Jy and compared it with the observed velocity dipole obtained from the POTENT compilation (Dekel *et al.* 1990) of the Mark III data (Willick *et al.* 1996). This comparison is especially useful since the *IRAS* dipole relative to the Local Group motion is entirely determined by the structure internal to that shell. This comparison showed a good alignment between the two dipole fields and yielded $\beta = 0.55 \pm 0.05$. This analysis however is not suitable for inspecting details of the flow patterns. Shaya *et al.* (1995) have compared the peculiar velocities of a set of 298 spiral galaxies within a redshift of 3000 $\mathrm{km\,s^{-1}}$ to the gravity field derived by least action analysis of the 1138 mass tracers derived from the *Nearby Galaxies Catalog* (Tully 1988). Their preliminary analysis gives a value $\Omega_0 = 0.2 \pm 0.2$, but they have not included covariance in the errors of the predicted mass model, nor have they realistically included possible influence from the mass distribution at redshifts $z > 3000$ $\mathrm{km\,s^{-1}}$, nor is it clear how well their analysis compares with the linear theory techniques used by others. More recently, Willick *et al.* (1996) made a non-linear maximum likelihood analysis of the Mark III data and the *IRAS* 1.2-Jy gravity. They restricted their consideration to data within $cz = 3000$ $\mathrm{km\,s^{-1}}$ and concluded $\beta = 0.49 \pm 0.07$. consistent with the Nusser & Davis (1994) result from the dipole analysis. Their approach does not naturally provide visual point by point comparison of the the fields.

A somewhat more direct way to estimate $\beta$ can be done at the density level. Because of the covariance of the *IRAS* gravity field a comparison of the mass fluctuations as inferred from the flows versus the *IRAS* density field has considerable merits. Assuming the flow is irrotational, the POTENT algorithm reconstructs a three dimensional flow from the observed radial flow field, and then computes derivatives of this flow to infer the underlying mass field (Nusser *et al.* 1991). This then can be compared directly to the *IRAS* density field (or to any other survey). Such a comparison was done by Dekel *et al.* (1993), using an earlier version of the *IRAS* catalog, and the Mark II catalog of peculiar velocities for 493 objects (Faber and Burstein 1988). On the basis of this density-density comparison, Dekel *et al.* (1990) derived $\beta = 1.3 \pm 0.3$. More recently, Hudson *et al.* (1995) have compared POTENT mass densities reconstructed from the Mark III data (Willick *et al.* 1996) with the optical density field of Hudson (1994), finding $\beta = 0.74 \pm 0.13$. One difference between the velocity-gravity and density-density comparisons is that, in the presence of observational errors, the velocity field probes scales larger than those which the density field is sensitive to. Another difference is that the mass-fluctuations which are derived by POTENT increase with decreasing $\beta$. According to the quasi-linear approximation (Nusser *et al.* 1991) employed by POTENT, the mass-fluctuation field inferred from the velocities diverges for low $\beta$, thus invalidating the density-density comparison. Therefore the density-density comparison can not formally rule out very low values of $\beta$. However, the fluctuations in the velocity field predicted from the observed distribution of galaxies are smaller for lower $\beta$ and the applicability of the linear approximation is guaranteed. Therefore, the velocity-velocity comparison provides a better tool for assessing the agreement between the fields for low $\beta$. The velocity-gravity comparison is therefore somewhat "orthogonal" to the density-density comparison.

In this paper we shall compare the observed radial velocity field derived from the Mark III Tully-Fisher data with the radial gravity field inferred from the 1.2 Jy IRAS redshift survey. We shall use the method of orthogonal mode expansion as described by Nusser and Davis (1995), which we shall henceforth label as the Inverse Tully-Fisher (ITF) method. We follow the the methods and analysis described in Nusser and Davis (1995) and Davis, Nusser and Willick (1996) (henceafter ND and DNW). In Section 2 we briefly review the ITF method for deriving peculiar velocities and discuss how to fit the velocity and gravity fields with the same set of functions to ensure identical smoothing for both fields. Section 3 gives details of the comparison of the Mark III velocity field with the *IRAS* inferred gravity field. Section 4 presents a summary and conclusions.

## 2. Reconstruction of Peculiar Velocities using the inverse Tully-Fisher Relation

Any of the available methods (*e.g.*, Yahil *et al.* 1991, Fisher *et al.* 1995b) for generating *IRAS* predicted peculiar velocities from the redshift space distribution of galaxies can be used. However, the method of Nusser & Davis (1994) is particularly convenient, as it is easy to implement, fast, and requires no iterations. Moreover, this method closely parallels the derivation of peculiar velocities from Tully-Fisher data (ND, DNW) briefly outlined bellow.

Given a sample of galaxies with measured circular velocity parameters, $\eta_i$, apparent magnitudes, $m_i$, and redshifts, $z_i$, the goal is to derive an estimate for the smooth underlying peculiar velocity field. We assume that the circular velocity parameter, $\eta$, of a galaxy is, up to a random scatter, related to its absolute magnitude, $M$, by means of a linear inverse Tully-Fisher (ITF) relation, *i.e.*,

$$\eta = \gamma M + \eta_0. \tag{2}$$

We use the inverse relation because samples selected by magnitude, as most are, will not be plagued by selection Malmquist bias effects when analyzed in the inverse direction (Schechter 1980, Aaronson *et al.* 1982, Tully 1988, Lynden-bell 1991). We also work with a velocity model that is parameterized in terms of the observable redshifts and therefore our estimation of the underlying velocity field is free of the spatial Malmquist biases which arise when working in measured distance space. ND consider a very general model and write the absolute magnitude of a galaxy, $M_i = M_{0i} + P_i$, where $M_{0i} = m_i + 5\log(z_i) - 15$ and $P_i = 5\log(1 - u_i/z_i)$, where $m_i$ is the apparent magnitude of the galaxy, $z_i$ is its redshift in units of $\mathrm{km\,s^{-1}}$, and $u_i$ its radial peculiar velocity in the LG frame. In general, one can write the function $P_i$ in terms of an expansion over orthogonal functions,

$$P_i = \sum_{j=0}^{j_{max}} \alpha^j \tilde{F}_i^j, \tag{3}$$

with orthonormality conditions, $\sum_{i=1}^{N_g} \tilde{F}_i^j \tilde{F}_i^{j'} = \delta_K^{j,j'}$, with the zeroth mode defined by $\tilde{F}_i^0 = 1/\sqrt{N_g}$, where $N_g$ is the number of galaxies in the sample. The zeroth mode describes a Hubble-like flow in the space of the data set which is clearly degenerate with the zero point of the ITF relation. Here we arbitrarily set $\tilde{\alpha}^0 = 0$. The best fit parameters, $\alpha^j$, the slope, $\gamma$, and the zero point $\eta_0$, are found by minimizing the $\chi^2$ statistic

$$\chi^2 = \sum_i \frac{(\gamma M_{0i} + \gamma P_i + \eta_0 - \eta_i)^2}{\sigma_\eta^2}. \tag{4}$$

where $\sigma_\eta$ is the *rms* scatter in $\eta$ about the ITF relation.

The choice of the basis functions, $\tilde{F}_j$, the expansion of the modes can be made with considerable latitude. The functions should obviously be linearly independent, smooth, and and close to a complete set of functions up to a given resolution limit. ND chose spherical harmonics $Y_l^m$ for the angular wavefunctions (Fisher *et al.* 1995b) and derivatives of spherical Bessel functions, $j_l[y(z)]$ for the radial basis functions, where the transformation from $z$ to $y$ in the argument of these Bessel functions is designed to make them oscillate nonuniformly with depth in order to match the spatial distribution of the TF data. The use of the coordinate $y$ instead of $z$ significantly reduces the number of the fit parameters necessary to describe the underlying velocity field in terms of our velocity model (DNW). As a demonstration of the use of the variable $y$, we tested the Mark III catalog which lists 2237 galaxies within $cz = 6000$ km s$^{-1}$, details of which will be described below. We took the linewidth scatter to be $\sigma_\eta = 0.05$ (Willick *et al.* 1996). For no velocity model, (*i.e.*, setting all coefficients $\alpha^j = 0$), the scatter diagram of the Tully-Fisher regression of observed versus predicted linewidths $\eta$ yields $\chi^2 = 3516$. Fitting a flow model of 69 degrees of freedom ($l_{max} = 3$, $n_{max} = 4$) in terms of of the coordinate $z$ leads to a much reduced $\chi^2 = 2667$, while doing the same fit of 69 modes in terms of $y = [ln(1 + z/z_*)]^{1/2}$ with $z_* = 1000$ km s$^{-1}$, gives $\chi^2 = 2644$, a modest but significant improvement. Further tests on simulations with smaller $\sigma_\eta$ demonstrate that the effect of the coordinate transformation is considerably more pronounced. There is clearly room for refinement here, but this choice of radial variable is sufficient for our purposes, especially since we shall use identical basis functions for the expansion of the velocity and gravity fields.

Using the machinery for computing a gravity field described in Nusser & Davis (1994), one can generate a linear theory predicted peculiar velocity $v_L$ for any point in space as a function of its redshift for any value of $\beta$. We must ensure that the smoothing scales of the ITF and *IRAS* predicted peculiar velocities are matched to the same resolution. Therefore we expand the *IRAS* predicted peculiar velocity in terms of the modes used in the ITF velocity model (3). This will filter out higher frequency modes in the *IRAS* field, $v_L$, that are not described by the resolution of our basis functions. However, it is important to recall that the modes are chosen to be orthogonal to the mode $P_i$ = constant, which would describe a smooth Hubble flow. In the fitting for the ITF modes, pure Hubble flow is absorbed into a shift of the zero point $\eta_0$ and the orthogonality is ensured. Within a given set of test points occupying a volume smaller than that used to define the gravity field, it is possible for $v_L$ to have a non-zero component of a Hubble flow like, which must be removed before we tabulate the mode coefficients. This component of the Hubble flow is not trivial in amplitude, and can be a 10% correction on the effective Hubble constant within simulated Mock catalogs.

## 3. Comparison of Mark III versus IRAS

In the comparison of the Mark III velocity and *IRAS* gravity fields discussed below, we shall use the Mark III sample only within a redshift limit of 6000 $\mathrm{km\,s^{-1}}$. The basis functions, $\tilde{F}_j$, for our velocity model where chosen such that the radial and the transverse resolutions are approximately the same (cf. DNW for details). The functions were constructed to have varying resolution with depth. In order to match the sparseness of the Mark III data the resolution was 680 $\mathrm{km\,s^{-1}}$ at $z = 1000$ $\mathrm{km\,s^{-1}}$, and 3400 $\mathrm{km\,s^{-1}}$ at $z = 4000$ $\mathrm{km\,s^{-1}}$ and varied as $dz/dy$. The total number of the resulting fit parameters, $\alpha^j$, was 56. Figure 1 shows the correlation function of the $\eta$ residuals, before and after the modal expansion. This plot demonstrates the residuals to have zero coherence on scales larger than the resolution scale of the mode expansion. It appears as though the mode expansion has done its job. The randomness of the $\delta\eta$ scatter suggests that the 56 mode expansion is sufficient for this dataset.

The resulting ITF velocity field of the Mark III galaxies is shown in slices of redshift space in Figure 2. This velocity maps is in the LG frame. In the nearby zone, the flow is dominated by infall to Virgo ($l = 284°$, $b = 74°$) and Ursa Major (145°, 65°) in the North, and modest outflow from Fornax (237°, −54°) in the south. In the middle redshift zone, the dipole pattern persists, and is more pronounced. This is composed partially by reflex dipole of the motion of the LG, plus backside infall from behind

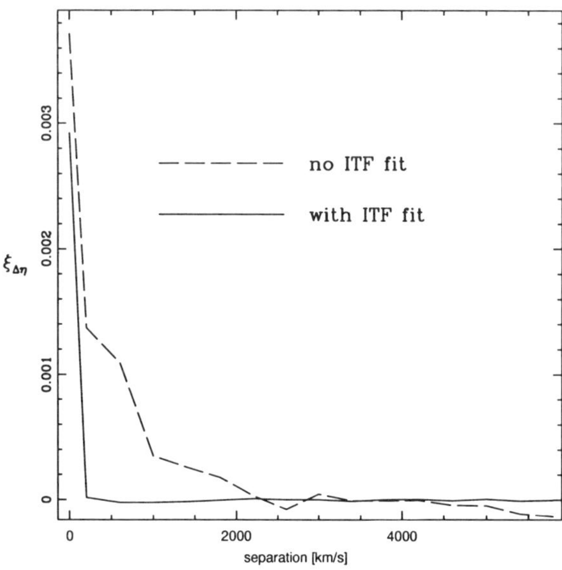


*Figure 1.* The auto-correlation function of the $\eta$ residuals of the Mark III galaxies versus redshift space separation. The dashed curve results when the 56 mode coefficients are set to 0, while the solid curve is after the 56 modes are fit to the Mark III data.

the Virgo supercluster. The Hydra cluster seen at (270°, 26°) is observed to be falling toward us, while the foreground of the Centaurus region (310°, 20°) is moving away from us. In the most distant slice of redshift space, the dipole pattern is further enhanced, with the entire Southern galactic sky, including the Perseus-Pisces region (150°, −15°) and the Pavo-Indus-Telescopium region (320°, −15°) flowing away from us, the Centaurus cluster flowing away rather substantially, but the background of Hydra and the Northern galactic cap flowing toward us. All of this is a combination of the reflex of the motion of the LG, plus the motion of the individual regions. Note the very strong shear in the observed ITF field along the boundary $l = 300$, $b > 0$, which divides the Hydra and Centaurus regions, and amounts to a change of radial peculiar velocity of approximately 900 km s$^{-1}$ in a transverse interval of 4100 km s$^{-1}$, or 20% of a Hubble flow differential motion over a substantial volume! This is quite a major gradient, one that appears to have been confirmed by distance estimates of elliptical galaxies in Centaurus (Lynden-Bell *et al.* 1988, Dressler 1994). In fact the true gradient is limited by the angular resolution of our modal expansion, since the reported outflow of the Centaurus clusters is 500–2000 km s$^{-1}$ in the LG frame, whereas the ITF measured flows for the entire Centaurus cluster are in the range 350–450 km s$^{-1}$ in the LG frame. For comparison, Figure

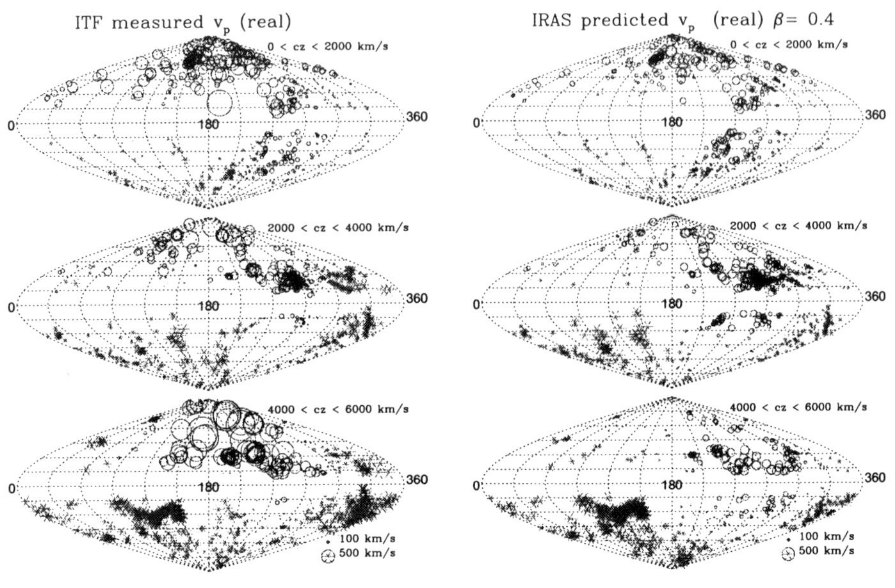


*Figure 2.* The $u_{itf}$ (left) and $u_{iras}$ (right) sky projection as seen in the LG frame for the Mark III galaxies, in galactic coordinates. The *IRAS* predicted flow $u_{iras}$ is for $\beta = 0.4$. The open circles are points that are flowing inward and the stars are points flowing outward. The size of the symbols is proportional to the flow velocity, with a key showing 500 km s$^{-1}$ flow. Note the three separate panels of redshift interval. Note also the dominance of the dipole mode, especially for the outer shell. This is the signature of the reflex motion of the LG.

3 shows the *IRAS* predicted field for $\beta = 0.4$. Figure 4 shows the difference of the fields in Figure 2 and 3. It is important to remember that the ITF velocity and *IRAS* gravity fields have been processed completely independently of each other. To a first approximation, the agreement between these Figures is a remarkable confirmation of the consistency between the gravity and velocity fields of our local Universe. The *IRAS* galaxies must have some close relationship to the underlying mass distribution. The qualitative alignment of the fields is consistent with the large-scale structure having been formed by the process of gravitational instability and the universality of the Tully-Fisher relation.

On the other hand, the residual field is not nearly as clean as seen in the mock catalogs in Figure 5, and is dominated by strong dipole patterns that increase linearly with redshift. Inspection of the *IRAS* maps for various

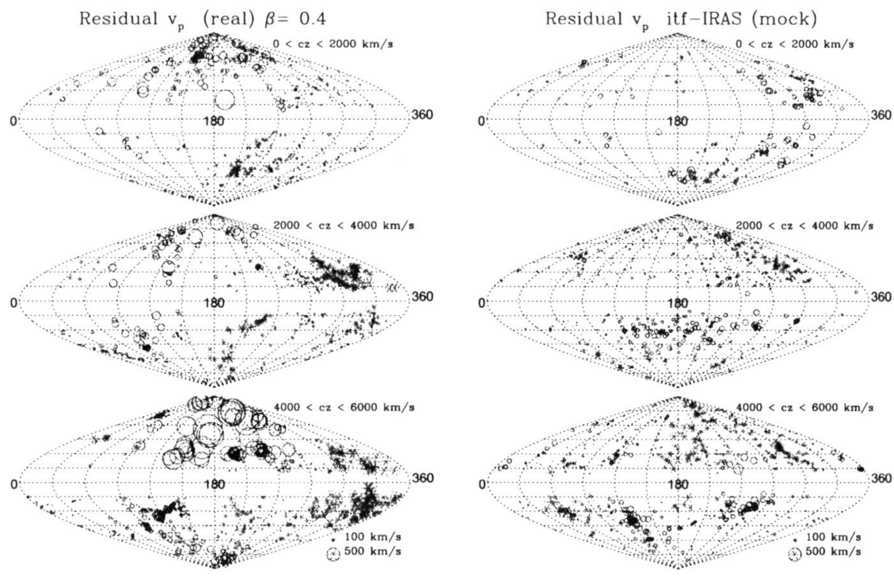


*Figure 3.* The sky projection of the residuals $u_{itf} - u_{iras}$: *Left* for real Mark III and *IRAS* for $\beta = 0.4$. This is simply the difference between the fields in Figure 2. *Right* for a mock catalog. Note the small amplitude of the mock catalog residual on the right and the absence of significant dipole contribution to it.

values of $\beta$ show that they never reproduce the strong shear seen in the ITF maps between the Hydra and Centaurus regions. They never reproduce the strong infall measured by ITF for the $l = 180°$, $b = 45°$, $cz = 5000\ \mathrm{km\,s^{-1}}$ region. For $\beta \leq 0.3$, the local supercluster flow is much too small, and the flow toward Perseus-Pisces is too small. For $\beta = 0.4$, the residual field in the local zone is still completely coherent, but the flow toward Perseus-Pisces and the average of Hydra and Centaurus is approximately correct. For $\beta = 0.7$ the relative motion toward Virgo matches the ITF measurement ($-340\ \mathrm{km\,s^{-1}}$ in the LG frame) but the flow away from Perseus-Pisces and toward the Great Attractor complex is now too large.

## 4. Summary and Conclusions

We have presented a method for comparing the velocity field derived from Tully-Fisher type data with the gravity field derived from full-sky redshift surveys of galaxies, such as the *IRAS* surveys. The chief strength of the method is that the two fields are filtered through the same set of low resolution modes, so that one is assured they indeed have the same resolution. In spite of working in the 'inverse' direction, we are able to produce pictures of the measured fields.

Although the method works extremely well in mock catalogs, with modal coefficients fully consistent with the expected noise, we find a much poorer agreement for the real data. This discrepancy is not consistent with random errors, and may be indicative of a systematic error in one or both of the datasets. We derive a "best" value of $\beta = 0.4 - 0.6$, in reasonable agreement with the value derived from the analysis of Willick *et al.* (1996), though considerably smaller than the values ($\beta \simeq 1.0$) typically obtained from density-density comparisons using the POTENT algorithm. Because the $\chi^2$ for the fit of the ITF velocity field to the *IRAS* gravity field is 100 for 55 degrees of freedom, the fields cannot be said to agree in detail. We therefore urge extreme caution in concluding that $\beta$ has been measured in large-scale flows. It is worth emphasizing that there is qualitative agreement between the velocity and gravity fields, particularly at distances $\lesssim 3000\ \mathrm{km\,s^{-1}}$. However, until the discrepancies on larger scales can be resolved, and all methods give consistent answers, we must prudently consider the value of $\beta$ to be an open question.

Future work should greatly improve this situation, as more data, and more precise data, become available. The Mark III catalog is not optimal for the analysis presented here, since the sky coverage is relatively nonuniform. More encouraging results are already achieved in work in progress (da Costa *et al.* 1997) where the analysis of ND and DNW is applied to a newly compiled homogeneous sample of Tully-Fisher data. Details of this project will be reported elsewhere.

I thank my collaborators Marc Davis and Jeffrey Willick.

## References

Aaronson, M. Huchra, J., Mould, J., Schechter, P., & Tully, R. B. 1982, *Astrophys. J.* 258, 64

Bernardeau, F., Juszkiewicz, R., Dekel, A. & Bouchet, F.R. 1995, *M.N.R.A.S.* 274, 20

da Costa, L., Nusser, A., Freudling, W., Giovanelli R., Haynes, M.P., Salzer, J.J., & Wegner, G. 1997, in preparation

Davis, M., Nusser, A. and Willick, J.A. 1996, *Astrophys. J.* 473 22

Dekel, A., Bertschinger, E., & Faber, S. M. 1990, *Astrophys. J.* 364, 349

Dekel, A., Bertschinger, E., Yahil, A., Strauss, M., Davis, M., & Huchra, J. 1993, *Astrophys. J.* 412, 1

Fisher, K.B., Huchra, J., Strauss, M.A., Davis, M. , Yahil, A., & Schlegel, D. 1995a, *Astrophys. J. (suppl.)* 100, 69

Fisher, K.B., Lahav, O., Hoffman, Y., Lynden-Bell, D., & Zaroubi, S. 1995b, *M.N.R.A.S.* 272, 885

Hudson, M. 1994, *M.N.R.A.S.* 266, 475
Hudson, M.J., Dekel, A., Courteau, S., Faber, S.M., & Willick, J.A. 1995, *M.N.R.A.S.* 274, 305
Kaiser, N., Efstathiou, G. Saunders, W., Ellis, R., Frenk, C., Lawrence, A. & Rowan-Robinson, M. 1991, *M.N.R.A.S.* 252, 1
Lynden-Bell, D. 1991, in *Statistical Challenges in Modern Cosmology*, eds. Babu, G.B & Feigelson, E.D.
Nusser, A., & Dekel, A. 1993, *Astrophys. J. (Lett.)* 405, 437
Nusser, A., & Davis, M. 1994, *Astrophys. J. (Lett.)* 421, L1
Nusser, A., & Davis, M. 1995, *M.N.R.A.S.* 276, 1391
Nusser, A., Dekel, A., Bertschinger, E., & Blumenthal, G. R. 1991, *Astrophys. J.* 379, 6
Schechter, P. 1980 *Astrophys. J.* 85, 801
Tully, R. B. 1988, *Nature* 334, 209
Willick, J.A., Courteau, S, Faber, S., Burstein, D., Dekel, A., & Kolatt, T. 1996a, *Astrophys. J.* 457, 460
Willick, J.A., Strauss, M.A., Dekel, A. & Kolatt, T. 1996, astro-ph/9610202

# THE LUMINOSITY FUNCTION AND MEAN GALAXY DENSITY FROM THE ESO SLICE PROJECT (ESP)

G.ZAMORANI[1,2], E.ZUCCA[1,2], G.VETTOLANI[2], A.CAPPI[1], R.MERIGHI[1], M.MIGNOLI[1], G.M.STIRPE[1], H.MACGILLIVRAY[3], C.COLLINS[4], C.BALKOWSKI[5], V.CAYATTE[5], S.MAUROGORDATO[5], D.PROUST[5], G.CHINCARINI[6], L.GUZZO[6], D.MACCAGNI[7], R.SCARAMELLA[8], A.BLANCHARD[9], M.RAMELLA[10]

[1] *Osservatorio Astronomico di Bologna, Bologna*
[2] *Istituto di Radioastronomia del CNR, Bologna*
[3] *Royal Observatory Edinburgh, Edinburgh*
[4] *School of EEEP, John–Moores University, Liverpool*
[5] *Observatoire de Paris, DAEC, Meudon*
[6] *Osservatorio Astronomico di Brera, Merate*
[7] *Istituto di Fisica Cosmica e Tecnologie Relative, Milano*
[8] *Osservatorio Astronomico di Roma, Monteporzio Catone*
[9] *Université L. Pasteur, Observatoire Astronomique, Strasbourg*
[10] *Osservatorio Astronomico di Trieste, Trieste*

**Abstract.**

The ESO Slice Project (ESP) is a galaxy redshift survey nearly complete to the limiting magnitude $b_J = 19.4$ and consists of 3342 galaxies with reliable redshift determination. The ESP survey, spanning a volume of $\sim 5 \times 10^4 \ h^{-3}$ Mpc$^3$ at the sensitivity peak ($z \sim 0.1$), provides an accurate determination of the "local" luminosity function and the mean galaxy density.

We find that, although a Schechter function (with $\alpha = -1.22$, $M^*_{b_J} = -19.61 + 5 \log h$ and $\phi^* = 0.020 \ h^3$ Mpc$^{-3}$ ) is an acceptable representation of the luminosity function over the entire range of magnitudes ($M_{b_J} \leq -12.4 + 5 \log h$ ), our data suggest the presence of a steepening of the luminosity function for $M_{b_J} \geq -17 + 5 \log h$ . Such a steepening at the faint-end of the luminosity function, well fitted by a power law with slope $\beta \sim -1.6$, is almost completely due to galaxies with emission lines: in



*D. Hamilton (ed.), The Evolving Universe,* 83–97.

fact, dividing our galaxies into two samples, *i.e.*, galaxies with and without emission lines, we find significant differences in their luminosity functions. In particular, galaxies with emission lines show a significantly steeper slope and a fainter $M^*$.

The amplitude and the $\alpha$ and $M^*$ parameters of our luminosity function are in good agreement with those of the AUTOFIB redshift survey (Ellis *et al.* 1996). However, our amplitude is significantly higher, by a factor $\sim 1.6$ at $M \sim M^*$, than that found for both the Stromlo-APM (Loveday *et al.* 1992) and the Las Campanas (Lin *et al.* 1996) redshift surveys. Also the faint-end slope of our luminosity function is significantly steeper than that found in these two surveys.

The galaxy number density for $M_{b_J} \leq -16 + 5 \log h$ is well determined ($\bar{n} = 0.08 \pm 0.015$ $h^3$ Mpc$^{-3}$ ). Its estimate for $M_{b_J} \leq -12.4 + 5 \log h$ is more uncertain, ranging from $\bar{n} = 0.28$ $h^3$ Mpc$^{-3}$ , in the case of a fit with a single Schechter function, to $\bar{n} = 0.54$ $h^3$ Mpc$^{-3}$ , in the case of a fit with a Schechter function and a power law. The corresponding blue luminosity densities in these three cases are $\rho_{LUM} = (2.0, 2.2, 2.3) \times 10^8\ h$ L$_\odot$ Mpc$^{-3}$, respectively.

Large over– and under– densities are clearly seen in our data. In particular, we find evidence for a "local" under–density ($n \sim 0.5\bar{n}$ for $D_{comoving} \leq 140$ $h^{-1}$ Mpc ) and a significant overdensity ($n \sim 2\bar{n}$) at $z \sim 0.1$. When these radial density variations are taken into account, our derived luminosity function reproduces very well the observed counts for $b_J \leq 19.4$, including the steeper than Euclidean slope for $b_J \leq 17.0$.

## 1. The ESO Slice Project

The ESO Slice Project (ESP, $b_J \leq 19.4$, Vettolani *et al.* 1997) is aimed to fill the gap between shallow, wide angle surveys and very deep, one–dimensional pencil beams, allowing a robust estimate of the faint-end slope and normalization of the luminosity function derived from a large, uniform and complete sample of galaxies. The redshift distribution of the galaxies of this sample, which peaks at $z \sim 0.1$, is deep enough to allow the sampling of a statistically representative distribution of the structures in a region of the Universe where the evolutionary effects are not expected to be important.

The ESP survey extends over a strip of $\alpha \times \delta = 22^o \times 1^o$, plus a nearby area of $5^o \times 1^o$, five degrees west of the main strip, in the South Galactic Pole region. The position was chosen in order to minimize the galactic absorption effects ($-60^o \lesssim b^{II} \lesssim -75^o$). The target objects, with a limiting magnitude $b_J \leq 19.4$, were selected from the Edinburgh–Durham Southern

Galaxy Catalog (EDSGC, Heydon–Dumbleton *et al.* 1988).

The right ascension limits are $22^h30^m$ and $01^h20^m$, at a mean declination of $-40^o15'$ (1950). In this area we observed a total of 4044 objects, corresponding to $\sim 90\%$ of the parent photometric sample, which contains 4487 objects: the objects we observed were selected to be a random subset of the total catalog with respect to both magnitude and surface brightness. The total number of confirmed galaxies with reliable redshift measurement is 3342, while 493 objects turned out to be stars and one object is a quasar at redshift $z \sim 1.174$. No redshift measurement could be obtained for the remaining 208 spectra.

The volume of the survey is $\sim 5 \times 10^4 \ h^{-3}$ Mpc$^3$ at $z \sim 0.1$, corresponding to the sensitivity peak of the survey, and $\sim 1.9 \times 10^5 \ h^{-3}$ Mpc$^3$ at $z \sim 0.16$, corresponding to the effective depth of the sample.

The absolute magnitudes are computed as

$$M = m - 25 - 5 \log D_L(z) - K(z) \tag{1}$$

where the luminosity distance $D_L$ is given by the Mattig (1958) expression (throughout the paper we adopt $H_o = 100$ km/s Mpc$^{-1}$ and $q_o = 0.5$) and $K(z)$ is the K–correction.

Details on the adopted $K$–correction are given in Zucca *et al.* (1997), where a full discussion of our results on the luminosity function and mean galaxy density is presented.

## 2. The luminosity function

For most samples of field galaxies the luminosity function is well represented by a Schechter (1976) form

$$\phi(L, x, y, z) dL \, dV = \phi^* \left(\frac{L}{L^*}\right)^{\alpha} \mathrm{e}^{-L/L^*} d\left(\frac{L}{L^*}\right) \, dV \tag{2}$$

where $\alpha$ and $L^*$ are parameters describing the shape of the function and $\phi^*$ contains the information about the normalization; these parameters have to be determined from the data. We have computed the luminosity function using both a parametric method (STY, Sandage *et al.* 1979) with the assumption of a Schechter function and a non–parametric method based on a modified version of the C–method of Lynden–Bell (1971).

The normalized Schechter luminosity function derived with the STY method is shown in Fig.1a (dashed line) together with the data points obtained with the non–parametric C–method. The error bars on these points correspond to the statistical (*i.e.*, Poissonian) errors. While the Schechter function is an excellent representation of the C–method data points for $M_{b_J} \lesssim -16 + 5 \log h$ , at fainter magnitudes it lies below all the points

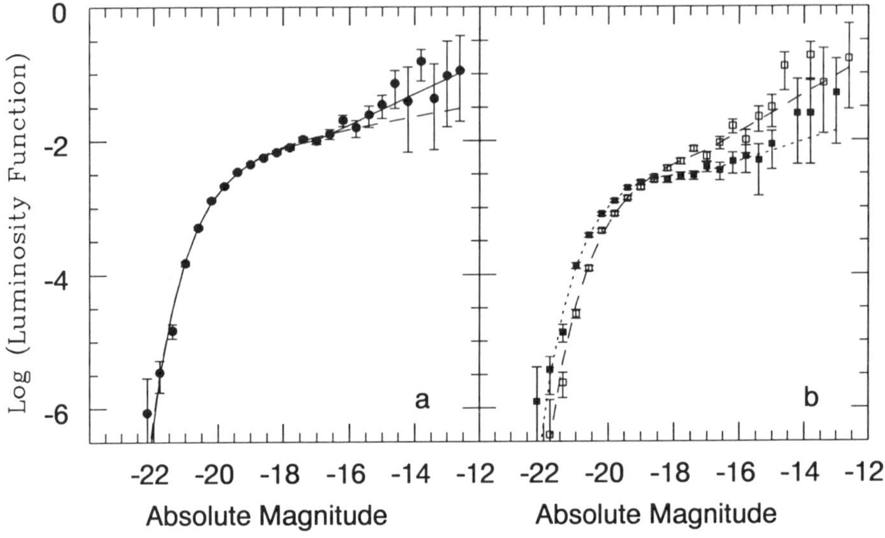


*Figure 1.* a) Normalized luminosity function (galaxies $h^3$ Mpc$^{-3}$ in bins of 0.4 magnitudes) for 3342 ESP galaxies brighter than $M_{b_J} = -12.4 + 5\log h$ . The solid circles are computed with a modified version of the C–method (error bars represent $1\sigma$ Poissonian uncertainties), while the fits are obtained with the STY method. Dashed line: single Schechter function; solid line: Schechter function with power law. b) The same as panel a), but for galaxies with (open squares and dashed line) and without (filled squares and dotted line) emission lines. For clarity only the fit with Schechter function with power law is shown.

down to $M_{b_J} = -12.4 + 5\log h$ . We have therefore modified the model function, adopting a Schechter function for $L > L_c$ and a power law for $L \leq L_c$. In this case the low luminosity part of the luminosity function is described by:

$$\phi(L, x, y, z) dL\, dV = A^* \left(\frac{L}{L^*}\right)^{\beta} d\left(\frac{L}{L^*}\right)\, dV \qquad (3)$$

with $\beta$ and $L_c$ being two additional free parameters in the fitting procedure, while the normalization $A^*$ is fixed by requiring continuity of the two functions at $L = L_c$.

The fit with this two–law function (solid line in Fig.1a) is almost indistinguishable from the single Schechter function at bright magnitudes, and reproduces very well also the faint part of the luminosity function. Statistically, as judged from the decrease of the $\mathcal{S}$ function, the improvement of the two–law fit with respect to a single Schechter function is significant at about

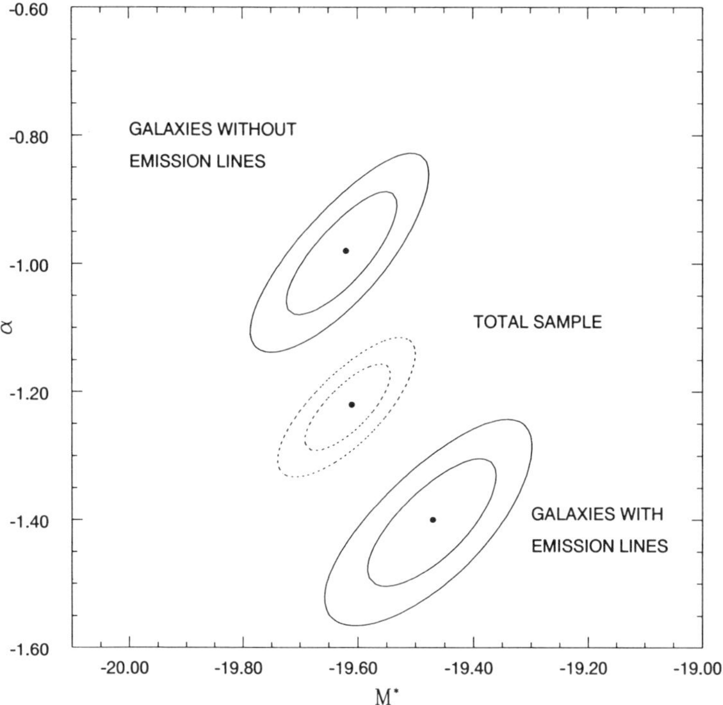


*Figure 2.* Confidence ellipses at $1\sigma$ and $2\sigma$ levels for the parameters $\alpha$ and $M^*$, in the case of a fit with a single Schechter function, for the total sample and for galaxies with and without emission lines.

$2\sigma$ level; therefore, we can conclude that our data suggest the presence of a steepening of the luminosity function for $M_{b_J} \gtrsim -17 + 5\log h$ .

The derived best fit parameters are listed in the first two lines of Table 1. The errors given for the case of a single Schechter function are the projections onto the $\alpha$ and $M^*_{b_J}$ axes of the $1\sigma$ confidence ellipse (see Fig.2) for two interesting parameters ($\Delta\mathcal{S} = 2.30$).

The errors for the two–law function are not reported in the table because, given the intrinsic correlations among the four fitted parameters, it is not very meaningful to give the projection on the four axes of the global four–dimensional error volume. However, we analyzed all the possible combinations of pairs of parameters, considering in turn each pair as "interesting" parameters (Avni 1976). From this analysis we find that the errors in $\alpha$ and $M^*$ are very similar to, although slightly larger than, those computed for the case of a single Schechter function, while $M_c$ is not well constrained by the data, with a $1\sigma$ error larger than one magnitude. No solution with $\beta \geq \alpha$ (*i.e.*, no steepening at the faint-end of the luminosity

TABLE 1. Parameters of the luminosity function

| Sample | $\alpha$ | $M^*_{b_J}$ | $\phi^*$ ( $h^3$ Mpc$^{-3}$ ) | $\beta$ | $M_c$ |
|---|---|---|---|---|---|
| Galaxies in the total sample | $-1.22^{+0.06}_{-0.07}$ | $-19.61^{+0.06}_{-0.08}$ | $0.020 \pm 0.004$ | | |
| (3342 galaxies) | $-1.16$ | $-19.57$ | 0.021 | $-1.57$ | $-16.99$ |
| Galaxies with emission lines | $-1.40^{+0.09}_{-0.10}$ | $-19.47^{+0.10}_{-0.11}$ | $0.010 \pm 0.002$ | | |
| (1575 galaxies) | $-1.34$ | $-19.42$ | 0.010 | $-1.70$ | $-16.94$ |
| Galaxies without emission lines | $-0.98^{+0.09}_{-0.09}$ | $-19.62^{+0.08}_{-0.10}$ | $0.011 \pm 0.002$ | | |
| (1767 galaxies) | $-0.90$ | $-19.57$ | 0.012 | $-1.38$ | $-17.22$ |

function) is consistent with the data at the $2\sigma$ confidence level.

The faint-end steepening is almost completely due to galaxies with detectable emission lines. In fact, dividing the galaxies into two samples, *i.e.*, galaxies with and without emission lines (1575 and 1767 galaxies respectively), we find very significant differences in their luminosity functions. Note that, given the typical signal to noise ratio of our spectra, detection of a line implies an equivalent width larger than about 5 Å. Fig.1b shows the normalized luminosity functions for galaxies with (open squares and dashed line) and without emission lines (solid squares and dotted line): for clarity only the fit for the two–law function (Schechter function with power law at low luminosity) is plotted. It is clearly seen from the figure that galaxies with emission lines show a significantly steeper faint-end slope and a slightly fainter $M^*$ (see best fit parameters in Table 1). The volume density of galaxies with emission lines is lower than the volume density of galaxies without emission lines at bright magnitudes, but becomes higher at faint magnitudes. Fig.2, which shows the $1\sigma$ and $2\sigma$ confidence ellipses of the parameters $\alpha$ and $M^*$ (single Schechter function case) for the total sample and for galaxies with and without emission lines, clearly demonstrates that the difference between the two subsamples is highly significant.

A similar difference in the best fit parameters of galaxies with and without emission lines has been found in the LCRS (see Fig.10 in Lin *et al.* 1996), although for each subsample their best fit slope is significantly flatter than the corresponding slope in our survey.

Finally, in order to check the possible existence of evolutionary effects

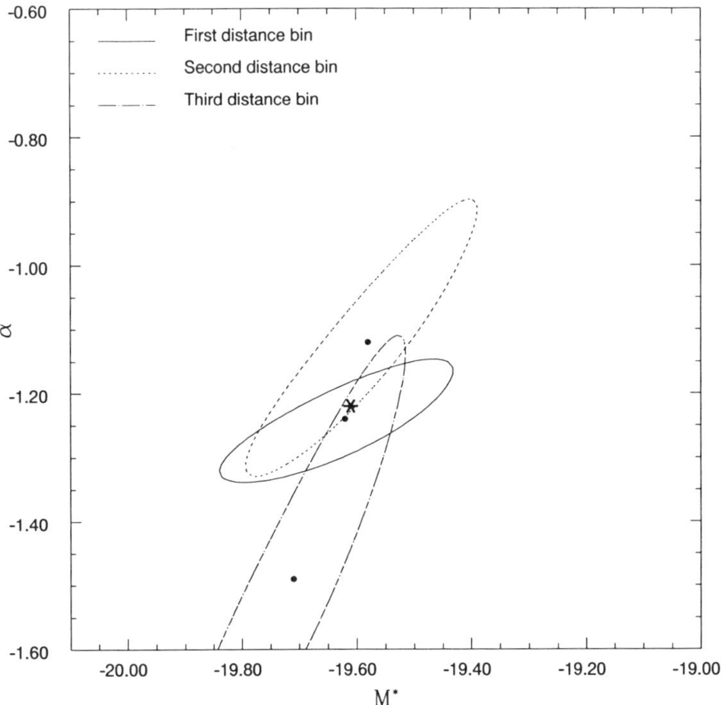


*Figure 3.* Confidence ellipses at $1\sigma$ level for the parameters $\alpha$ and $M^*$ of the luminosity functions in three different distance bins. The asterisk refers to the best fit parameters for the total sample.

also at our moderate redshifts, we have divided the ESP galaxies in three comoving distance bins, chosen in order to have three samples with a comparable number of objects ($D_{com} \leq 250\ h^{-1}$ Mpc ; $250\ h^{-1}$ Mpc $< D_{com} \leq 360\ h^{-1}$ Mpc ; $D_{com} > 360\ h^{-1}$ Mpc ). The $1\sigma$ confidence ellipses of the ($\alpha$, $M^*$) parameters for the three derived luminosity functions are reported in Fig.3. From the overlap of the ellipses we can say that no evidence for evolution is seen in our data, in agreement with the conclusion of Ellis *et al.* (1996) that the bulk of the evolution sets in beyond $z \sim 0.3$.

## 3. The mean galaxy density

Given a magnitude limited sample, an unbiased estimator for the mean number density of galaxies $\bar{n}$ is

$$\bar{n} = \frac{\sum_i n_i W(z_i)}{\int F(z) W(z) \frac{dV}{dz} dz} \tag{4}$$

where $W(z)$ is a weighting function and $F(z)$ is the selection function of the sample. Davis & Huchra (1982; DH82) have discussed a number of different estimators for $\bar{n}$, corresponding to different choices for the weighting function $W(z)$.

The density radial profile, corrected for the small redshift incompleteness of our survey, is shown in Fig.4a as a function of comoving distance. In each distance bin the density has been computed using the $n_3$ estimator of DH82, which corresponds to a constant weighting function in eq.(**??**). The error bars represent $1\sigma$ Poissonian uncertainties. The width of the various distance shells has been chosen in order to have $\sim 150$ expected galaxies in each bin. The solid line represents the value of the global $\bar{n}$ derived from the total sample and the dashed lines indicate the $\pm 1\sigma$ uncertainty on this value. The median of the $\bar{n}$ values in the shells is $0.29 \pm 0.04$ $h^3$ Mpc$^{-3}$, in excellent agreement with the global value $\bar{n} = 0.28 \pm 0.05$ $h^3$ Mpc$^{-3}$. Although the densities derived from most of the distance shells are within $\pm 2\sigma$ from the global mean density of the sample, at least three regions have densities which differ significantly ( $\gtrsim$ a factor of two) from the mean density. These regions (two underdense regions at $D_{com} \lesssim 140$ $h^{-1}$ Mpc and $D_{com} \sim 230$ $h^{-1}$ Mpc , and an overdense region at $D_{com} \sim 290$ $h^{-1}$ Mpc ) are clearly visible also in Fig.4b, which shows the observed distance histogram and the distribution expected for a uniform sample. The relatively large density fluctuations seen in our data are clearly due to the fact that our survey is a slice with a narrow width in one direction. Note, however, that over–densities and under–densities of about a factor two are seen also in wider angle surveys, such as the SSRS2 (Marzke & da Costa 1997), the CfA2 (Marzke *et al.* 1994) and the LCRS (Lin *et al.* 1996) surveys.

The results shown in Fig.4a refer to the case of a fit with a single Schechter function for galaxies brighter than $M_{b_J} = -12.4 + 5\log h$ . Qualitatively, the results for different limiting absolute magnitudes or for the case of the fit with a Schechter function and a power law are similar to those shown in Fig.4a, except for different normalizations. The galaxy number density for $M_{b_J} \leq -16 + 5\log h$ is well determined ($\bar{n} = 0.08 \pm 0.015$ $h^3$ Mpc$^{-3}$ ) and is essentially independent from the adopted fitting law for the luminosity function. Its estimate for $M_{b_J} \leq -12.4 + 5\log h$ is more uncertain, ranging from $\bar{n} = 0.28 \pm 0.05$ $h^3$ Mpc$^{-3}$ , in the case of a fit with a single Schechter function, to $\bar{n} = 0.54 \pm 0.10$ $h^3$ Mpc$^{-3}$ , in the case of a fit with a Schechter function and a power law. The corresponding luminosity densities in these three cases are $\rho_{LUM} = (2.0, 2.2, 2.3) \times 10^8$ $h$ $L_\odot$ Mpc$^{-3}$, respectively: the similarity of these values indicates that the galaxies in the faint-end of the luminosity function contribute strongly to the *number* density, but change only slightly the global *luminosity* density.

From these number densities we have then derived $\phi^*$; the obtained

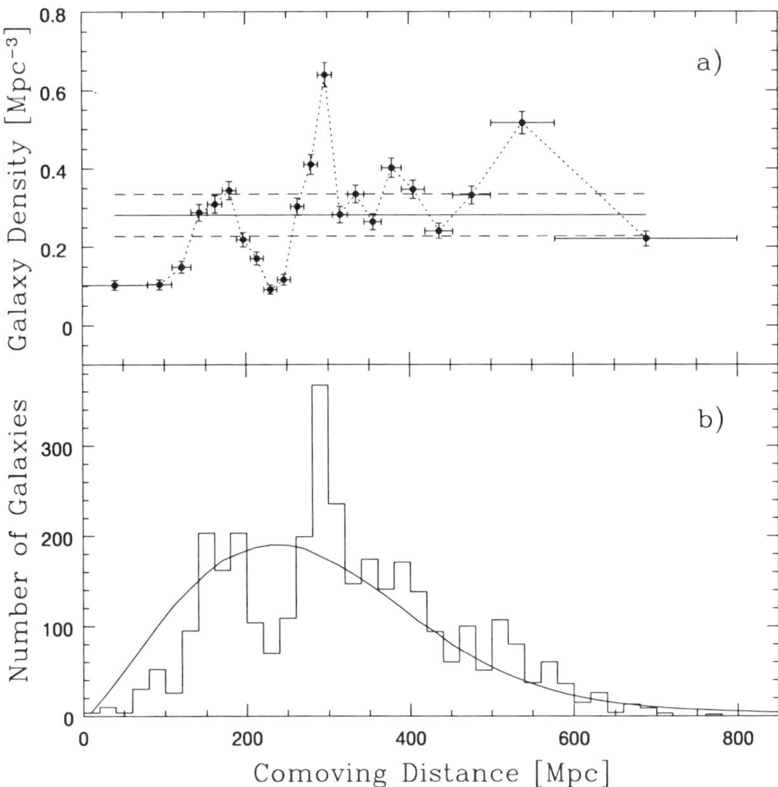


*Figure 4.* a) Galaxy density as a function of the comoving distance: distance shells have been chosen in order to have a number of expected object $\geq 150$; error bars represent $1\sigma$ Poissonian uncertainties. The solid straight line corresponds to the mean value $\bar{n}$, dashed lines represent $1\sigma$ uncertainties. The values refer to galaxies with $M_{b_J} \leq -12.4 + 5 \log h$ , in the case of a fit with a Schechter function. b) Comoving distance histogram of ESP galaxies: the solid line is the distribution expected for a uniform sample.

values are reported in Table 1.

## 4. Discussion and comparison with previous results

In Table 2 we list the parameters from the most recent luminosity function determinations available in the literature. From this table it is clear that there are significant differences between the various surveys. We first note the discrepant value of $M^*$ found in the CfA2 survey (Marzke *et al.* 1994) with respect to all the other samples, which cannot be entirely explained by the different passband used. The $\alpha$ and $M^*$ parameters derived from the SSRS2 survey (Marzke & da Costa 1997) are in good agreement with our

TABLE 2. Parameters of the luminosity function from various samples

| Sample | $N_{gal}$ | $m_{lim}$ | $M_{min}$ | $\alpha$ | $M^*$ |
|---|---|---|---|---|---|
| ESP | 3342 | $b_J = 19.4$ | $-12.4$ | $-1.22^{+0.06}_{-0.07}$ | $-19.61^{+0.06}_{-0.08}$ |
| CfA2 | 9063 | $m_Z = 15.5$ | $-16.5$ | $-1.0 \pm 0.2$ | $-18.8 \pm 0.3$ |
| SSRS2 | 3288 | $m_{B(0)} = 15.5$ | $-14$ | $-1.16^{+0.08}_{-0.06}$ | $-19.45 \pm 0.08$ |
| Stromlo-APM | 1658 | $b_J = 17.15$ | $-15$ | $-0.97 \pm 0.15$ | $-19.50 \pm 0.13$ |
| LCRS | 18678 | $r \sim 17.5$ | $-17.5$ | $-0.70 \pm 0.05$ | $-20.29 \pm 0.02$ |
| AUTOFIBa | 588 | $b_J = 24$ | $-14$ | $-1.16^{+0.05}_{-0.05}$ | $-19.30^{+0.15}_{-0.12}$ |
| AUTOFIBb | 665 | $b_J = 24$ | $-16$ | $-1.41^{+0.12}_{-0.07}$ | $-19.65^{+0.12}_{-0.10}$ |

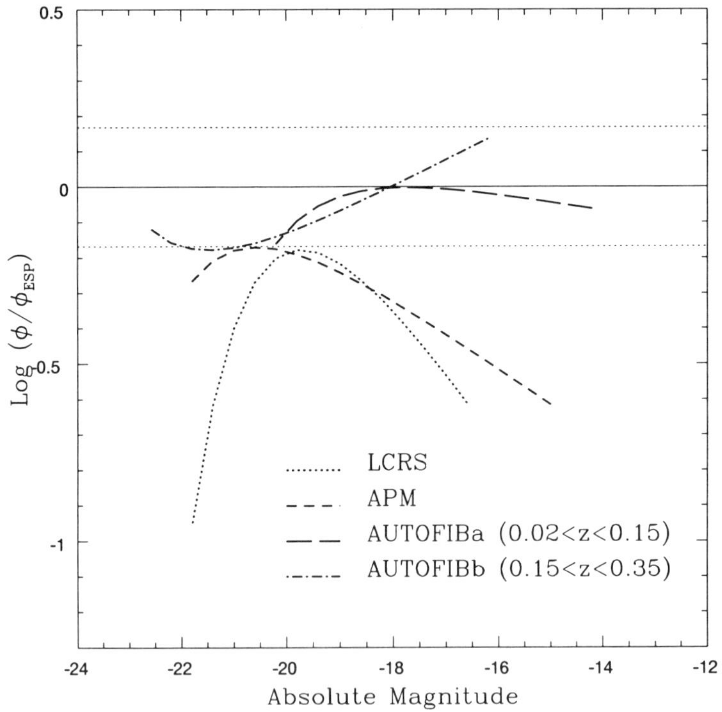


*Figure 5.* Ratio between various luminosity functions and the ESP luminosity function. In order to be consistent with the other luminosity functions, we have used for the ESP galaxies the fit obtained with a single Schechter function. The dotted straight lines represent an approximate $\pm 2\sigma$ uncertainty on the ratio derived from the uncertainties on the normalizations of the various surveys. For sake of clarity, the two shallow surveys (CfA2 and SSRS2) are not plotted.

results. However, the amplitude of the SSRS2 luminosity function is such that the ratio between the SSRS2 and the ESP luminosity functions is in the range $0.4 - 0.5$ for $-20 < M < -14$.

In Fig.5 we compare the luminosity functions derived from different surveys; in order to better visualize the differences, we show in the figure the ratio between the various luminosity functions and the ESP luminosity function.

From this figure we find that:

i) The AUTOFIB luminosity functions (Ellis *et al.* 1996) in their first two redshift bins, which cover approximately the same distance range as our survey, are in good agreement with the ESP luminosity function. Both AUTOFIB luminosity functions are within $2\sigma$ from the ESP luminosity function at all magnitudes.

ii) The Stromlo-APM (Loveday *et al.* 1992) and the LCRS (Lin *et al.* 1996) luminosity functions are significantly different from the ESP luminosity function in both shape (both are flatter than the ESP at low luminosity) and amplitude (both have a lower amplitude by a factor $\sim 1.6$ at $M \sim M^*$).

While a possible explanation for the difference in amplitude with respect to the Stromlo-APM luminosity function is presented later in this section, we are intrigued by the significant difference in amplitude with the LCRS luminosity function, which is derived from galaxies with a redshift distribution very similar to ours. The fact that the LCRS galaxies are selected in the red band can in principle explain, at least in part, their flatter luminosity function on the basis of the fact that at low luminosity most of the galaxies have emission lines and are presumably bluer than the average. The dependence of the local luminosity function on color has been recently discussed by Marzke & da Costa (1997), who indeed found a significantly steeper slope for blue galaxies with respect to the red ones in the SSRS2. This, however, should not affect significantly the amplitude at $M \sim M^*$. It is true that in the LCRS low surface brightness galaxies have been explicitly eliminated from the redshift survey, and, therefore, from the computation of the luminosity function, but the fraction of galaxies eliminated due to this selection effect is estimated to be less than 10% (Lin *et al.* 1996). Another difference between the LCRS and the ESP surveys is in the fraction of stars in the spectroscopic sample of candidate galaxies ($\sim 3\%$ in LCRS and $\sim 12\%$ in ESP). This can in principle be due to a better star–galaxy separation in the Las Campanas photometric data. Alternatively, it is possible that a non negligible number of galaxies, probably the most compact ones, have been classified as stars in the Las Campanas data and therefore have not been observed spectroscopically. Obviously, we have no way to check if something like this has really happened in the Las Campanas data.

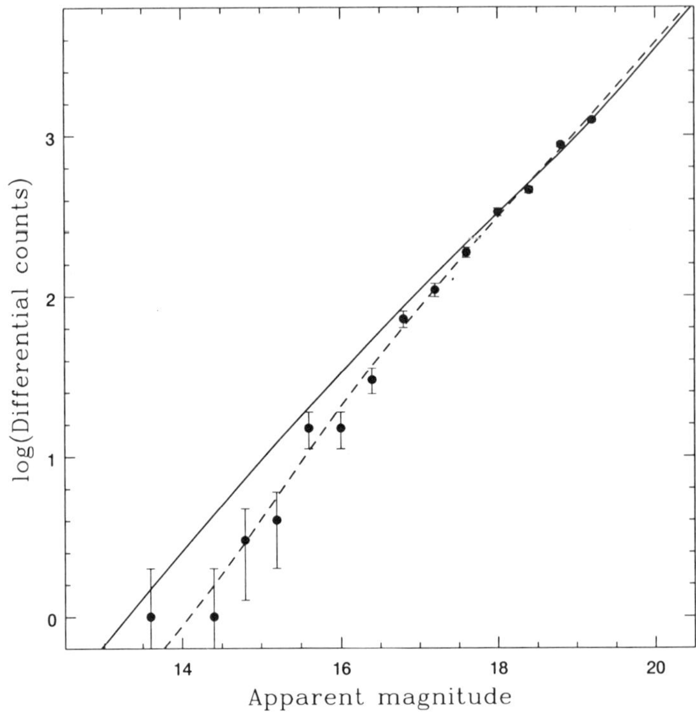


*Figure 6.* Observed galaxy counts compared with those expected from the luminosity function (solid line); the same but taking into account the observed radial density variations (dashed line)

We have used our derived luminosity function to compute the expected galaxy counts in the case of a constant density and compared this prediction (solid line in Fig.6) with the data in our photometric catalog (solid circles with error bars in the same figure). Note that in this magnitude range ($b_J \leq 20$) the counts predicted using the two different fits (single Schechter function or Schechter function with power law) are almost identical, while they differ significantly at magnitudes fainter than $b_J \sim 25$.

It is clear from the figure that, while the counts predicted from the luminosity function are reasonably consistent with the data for $b_J \gtrsim 17$, they are significantly higher at brighter magnitudes. As our counts, with large statistical errors due to the small area, are consistent with the global EDSGC (Heydon–Dumbleton *et al.* 1989) counts and the APM counts are consistent with the EDSGC ones (Maddox *et al.* 1990), this effect is essentially the same which has led Maddox *et al.* (1990) to suggest rapid and dramatic evolution in the galaxy properties for $z \lesssim 0.2$. Such strong "local" evolution is not seen either in our analysis of the luminosity function in three distance intervals (see §2) or in the deeper AUTOFIB redshift survey from which very little, if any, evolution is seen up to $z = 0.35$.

It has been suggested that the low APM counts at bright magnitude may be due to a magnitude scale error in the Stromlo-APM galaxy survey (Metcalfe *et al.* 1995). The possible existence of such an error for the APM and similar catalogs is reinforced by the recent analysis of well calibrated Schmidt plates digitized with the MAMA machine by Bertin and Dennefeld (1997). We cannot exclude a similar problem in our data, although the admittedly limited CCD photometry that we have obtained on bright galaxies in this area (Garilli *et al.* in preparation) does not show any strong magnitude error for galaxies in the magnitude range $16 - 17$.

At least for our data, however, an alternative or additional interpretation is suggested by Fig.4a, which shows significant fluctuations about the mean density. The dashed line in Fig.6 shows the expected counts computed by relaxing the assumption of constant density and allowing the normalization of the luminosity function to change with distance as indicated by the data points in Fig.4a. In this way the agreement between the predicted and observed counts is more than acceptable over the entire magnitude range. In particular, the deficiency of the observed counts at bright magnitudes would be due to the "local" ($D_{com} \leq 140$ $h^{-1}$ Mpc ) under–density. In this respect, it is interesting to note that, as mentioned above, the amplitude of the ESP luminosity function is about a factor of two higher than the SSRS2 one, which has been derived from galaxies with $D_{com} \lesssim 140$ $h^{-1}$ Mpc over a much larger area, which includes the region of our survey. Moreover, an under–density over a similar distance range is seen also in the South Galactic Cap part of the LCRS (see Fig.8a and 8c in Lin *et al.* 1996). Since the LCRS is based on galaxies extracted from a much larger area ($80 \times 6.5$ deg$^2$.), over which our survey region is fully contained, we are led to conclude that the local under–density seen in our data may have a size of more than 100 $h^{-1}$ Mpc in at least the right ascension direction. If such an under–density extends with a similar size also in the declination direction, it could contribute significantly, possibly in addition to magnitude scale errors, to the low APM counts of bright galaxies. In this framework, the difference in amplitude between our luminosity function and the Loveday *et al.* (1992) luminosity function based on APM galaxies with $b_J \leq 17.15$ could be at least partly explained by the presence of a real, giant under–density extending over a significant fraction of their survey.

It is also interesting to note that, since strong "local" evolution is not easily accommodated in the standard evolutionary galaxy models, most models for the faint galaxy counts have often assumed a high normalization of the local luminosity function (see for example Guiderdoni & Rocca–

Volmerange 1990, Pozzetti *et al.* 1996). We find that the normalization of our luminosity function is in good agreement with that usually assumed in these models and, therefore, our result gives *a posteriori* confirmation of the somewhat *ad hoc* assumption adopted in these models.

## 5. Conclusions

In this paper we have presented the results on the luminosity function and the mean density from the ESP galaxy redshift survey; our main results are the following:

1) Although a Schechter function is an acceptable representation of the luminosity function over the entire range of magnitudes ($M_{b_J} \leq -12.4 + 5\log h$ ), our data suggest the presence of a steepening of the luminosity function for $M_{b_J} \geq -17 + 5\log h$ . Such a steepening, well fitted by a power law with slope $\beta \sim -1.6$, is in agreement with what has been recently found by similar analyses for both field galaxies (Marzke *et al.* 1994) and galaxies in clusters (see for instance Driver & Phillipps 1996).

2) The steepening at the faint-end of the luminosity function is almost completely due to galaxies with emission lines: in fact, dividing our galaxies into two samples, *i.e.*, galaxies with and without emission lines, we find significant differences in their luminosity functions. In particular, galaxies with emission lines show a significantly steeper slope and a fainter $M^*$. The volume density of galaxies with emission lines is lower than the volume density of galaxies without emission lines at bright magnitudes, but becomes higher at faint magnitudes.

3) The amplitude and the $\alpha$ and $M^*$ parameters of our luminosity function are in good agreement with those of the AUTOFIB redshift survey. On the other hand, our amplitude is a factor $\sim 1.6$ higher, at $M \sim M^*$, than that found for both the Stromlo-APM and the Las Campanas redshift surveys. Also the faint-end slope of the luminosity function is significantly steeper for the ESP galaxies than that found in these two surveys.

4) We find evidence for a local under–density, extending up to a comoving distance $\sim 140 \ h^{-1}$ Mpc . The volume probed by the ESP within such a distance is smaller than the volume of a typical void with 50 $h^{-1}$ Mpc diameter, and in principle this observed nearby under–density could be due to the specific direction of the survey piercing through a local void. Our data do not allow to characterize this low density region in terms of size and shape. When the radial density variations observed in our data are taken into account, our derived luminosity function reproduces very well the observed counts for $b_J \leq 19.4$, including the steeper than Euclidean slope for $b_J \leq 17.0$. If this under–density extends over a much larger solid angle than that covered by our survey, it could, at least partly, explain the

low amplitude of the Stromlo-APM luminosity function.

5) A similar explanation can not justify the significant difference in amplitude between the ESP and the LCRS luminosity functions, because the two samples cover essentially the same redshift range. One possibility, which has yet to be verified, is that a non-negligible number of galaxies are missing from the original CCD photometric catalog of the LCRS.

## References

Avni, Y. 1976, ApJ, 210, 642
Bertin, E., Dennefeld, M. 1997, A&A ,317, 43
Davis, M., Huchra, J.P. 1982, ApJ, 254, 437 [DH82]
Driver, S.P., Phillipps, S. 1996, ApJ, 469, 529
Ellis, R.S., Colless, M., Broadhurst, T., Heyl, J., Glazebrook, K. 1996, MNRAS, 280, 235
Guiderdoni, B., Rocca–Volmerange, B. 1990, A&A, 227, 362
Heydon–Dumbleton, N.H., Collins, C.A., MacGillivray, H.T. 1988, in *Large–Scale Structures in the Universe*, ed. W. Seitter, H.W. Duerbeck and M. Tacke (Springer–Verlag), p. 71
Heydon–Dumbleton, N.H., Collins, C.A., MacGillivray, H.T. 1989, MNRAS, 238, 379
Lin, H., Kirshner, R.P., Shectman, S.A., Landy, S.D., Oemler, A., Tucker, D.L., Schechter, P.L. 1996, ApJ, 464, 60
Loveday, J., Peterson, B.A., Efstathiou, G., Maddox, S.J. 1992, ApJ, 390, 338
Lynden–Bell, D. 1971, MNRAS, 155, 95
Maddox, S.J., Sutherland, W.J., Efstathiou, G.P., Loveday, J., Peterson, B.A. 1990, MNRAS, 247, 1p
Marzke, R.O., & da Costa, L.N. 1997, AJ, 113, 185
Marzke, R.O., Huchra, J.P., Geller, M.J. 1994, ApJ, 428, 43
Mattig, W. 1958, Astron. Nachr., 284, 109
Metcalfe, N., Fong, R., Shanks, T. 1995, MNRAS, 274, 769
Pozzetti, L., Bruzual, G., Zamorani, G. 1996, MNRAS, 281, 953
Sandage, A., Tamman, G., Yahil, A. 1979, ApJ, 232, 252
Schechter, P. 1976, ApJ, 203, 297
Vettolani, G., *et al.* 1997, A&A in press [paper I]
Zucca, E., *et al.* 1997, A&A in press [paper II]

# THE MUENSTER REDSHIFT PROJECT (MRSP)

## *THE POWER SPECTRUM & THE DECELERATION PARAMETER*

PETER SCHUECKER
*Astronomisches Institut Münster*
*Wilhelm-Klemm-Straße 10, D-48149 Münster, Germany*
–
*Astrophysikalisches Institut Potsdam*
*An der Sternwarte 16, D-14482 Potsdam, Germany*

## 1. Introduction

The Muenster Redshift Project is a wide-field survey, where large numbers of low-dispersion spectra are analyzed for cosmological investigations (MRSP; see Duerbeck *et al.* 1989, Seitter 1992). The project differs in several aspects from traditional surveys because the redshifts $z$ are estimated with ($1\sigma$) errors of 10 percent at the maximum redshift $z = 0.3$. Conceptually, the MRSP shares common properties both with two and with three-dimensional surveys: With the MRSP survey strategy, many objects can be studied within large volumes in comparatively short times – this is the major aspect of two-dimensional surveys. Several thousand reference objects are used for accurate wavelength, magnitude, and redshift calibrations so that the large number of observed survey galaxies permits (via deconvolution) very precise determinations of cosmological quantities – this shows that on large scales the MRSP corresponds to a three-dimensional survey.

The present paper illustrates the reliability of this new approach. With redshift samples of $10^5$ galaxies, the redshift space power spectrum of the comoving galaxy number density fluctuations is measured on scales between 40 and $600\,h^{-1}$ Mpc ($H_0 = 100\,h\,\mathrm{km\,s^{-1}\,Mpc^{-1}}$). The same sample is used to determine the cosmic deceleration parameter $q_0$, and to constrain the normalized cosmological constant $\lambda_0$. Comparatively small $z$ values and red photometric magnitudes assure that biased object selection, galaxy evolution, and errors in the cosmic $K$-correction do not affect the measurements in uncontrolled ways.



*D. Hamilton (ed.), The Evolving Universe,* 99–109.

## 2. Data

Schmidt plates are digitized with the microdensitometer PDS 2020 GM$^{\rm plus}$ at the Astronomisches Institut Münster. Film copies of direkt ESO/SERC IIIa-F Schmidt plates are used to identify the galaxies (Naumann, Ungruhe & Seitter 1993) and to measure their $r_{\rm F}$ magnitudes. CCD sequences on at least one plate of each overlapping pair are used for the calibrations (*e.g.*, Cunow 1993, Cunow & Ungruhe 1995; observations in progress). For galaxies with $14.0 \leq r_{\rm F} \leq 18.0$ mag the comparison of isophotal and total magnitudes shows that their differences are consistent with the random errors of the individual magnitudes ($\sigma_m = 0.15$ mag), and that no significant systematic magnitude-dependent differences remain (Schuecker *et al.* 1997). Within the given magnitude range the isophotal magnitudes of the MRSP galaxies are thus regarded as total magnitudes.

To date, the MRSP redshift sample includes $9 \times 10^5$ galaxies, $r_{\rm F} \leq 19.5$ mag, with measured redshifts, $z \leq 0.3$, obtained from digitized film copies of very low dispersion objective prism UK Schmidt plates (245 nm/mm at $H_\gamma$). Random redshift errors are $\sigma_z = 0.031$, as obtained from 1,200 reference galaxies with $z \leq 0.3$ (Schuecker 1993, 1996).

The present analysis is based on 89,125 galaxies with magnitudes $14.0 \leq r_{\rm F} \leq 18.0$ mag and redshifts $0 \leq z \leq 0.2$, because these intervals lie well within the extrem survey limits, and include the reference galaxies used to calibrate the survey (see Fig. 4 in Schuecker 1996). Thus, all error estimates are realistic. The redshift sample covers an area of 750 square degrees located around the South Galactic Pole. We use the cosmic $(K+E)$ correction in the parameterized form $K(z) = K \cdot z\,[\rm mag]$, with $K = 1.0$, typical for populations of passively aging galaxies within the given $r_{\rm F}$ and $z$ limits (Schuecker *et al.* 1997).

Differential number counts, $dN(r_{\rm F})$ and $dN(z)$, and the distribution of comoving galaxy number densities, $n(z)$, computed with $q_0 = 0.5$, convinced us that the chosen sample is not significantly biased by systematic incompleteness (Schuecker, Ott, & Seitter 1996a). The luminosity function of the sample is estimated with a non-parametric density-independent method and with a parametric method. Both results are very similiar and can be combined to a Schechter-type luminosity function with the parameters $M_* = -20.75$ mag and $\alpha = -0.80$ (hereafter called pilot values) for $q_0 = 0.5$. The faint-end slope of the MRSP galaxies indicates incompleteness of galaxies with faint surface brightness. These galaxies are better detected in blue photometric passbands (Schombert *et al.* 1992). Their absence does not affect the conclusions of the present investigation.

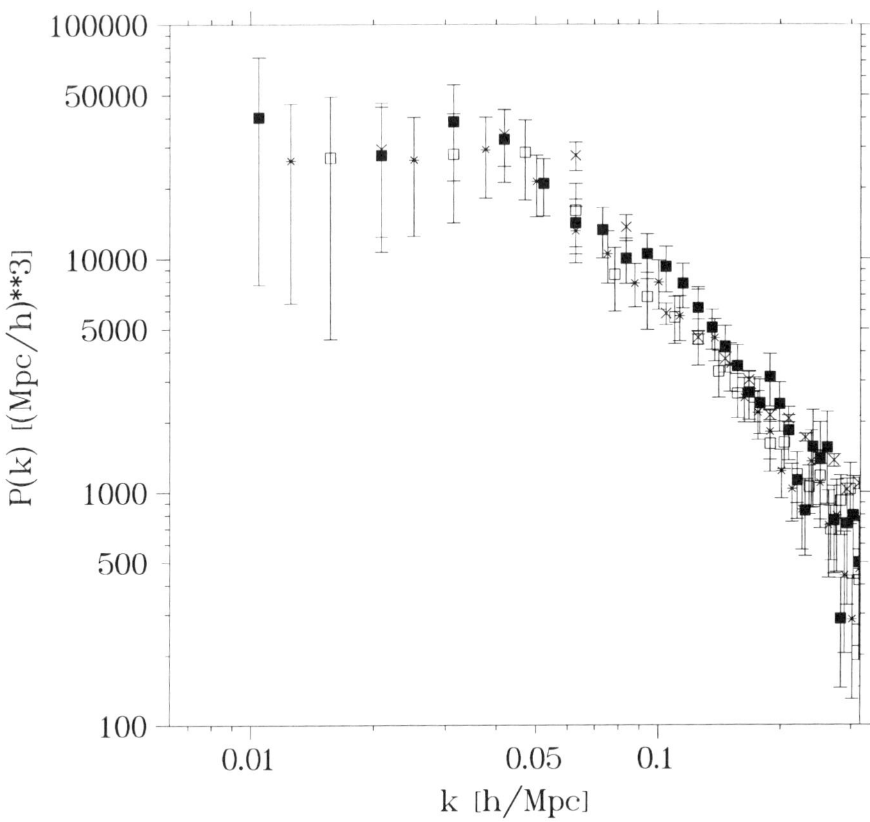


*Figure 1.* Observed deconvolved redshift space power spectra of MRSP-subsamples obtained in different volumes. Filled squares: $(600\,h^{-1}\ \mathrm{Mpc})^3$ with 87,558 galaxies; asterisks: $(500\,h^{-1}\,\mathrm{Mpc})^3$ with 84,446 galaxies; open squares: $(400\,h^{-1}\,\mathrm{Mpc})^3$ with 67,801 galaxies; crosses: $(300\,h^{-1}\,\mathrm{Mpc})^3$ with 44,910 galaxies. For each sample the $(1\sigma)$ error bars are estimated with 100 independent simulations. The simulations show that the MRSP power spectral densities are reliable for wavenumbers $0.01 \leq k \leq 0.15\,h\,\mathrm{Mpc}^{-1}$, corresponding to wavelengths $40 \leq \lambda \leq 600\,h^{-1}\,\mathrm{Mpc}$.

## 3. Power Spectrum

The decomposition of a fluctuation field into plane waves yields very robust although not complete summary statistics, useful for comparisons of observed and simulated galaxy catalogs. The corresponding spectral theory of discrete point processes is described, *e.g.*, by Bartlett (1964) and Ripley (1981). For the continuous Fourier transform (FT) of a con-

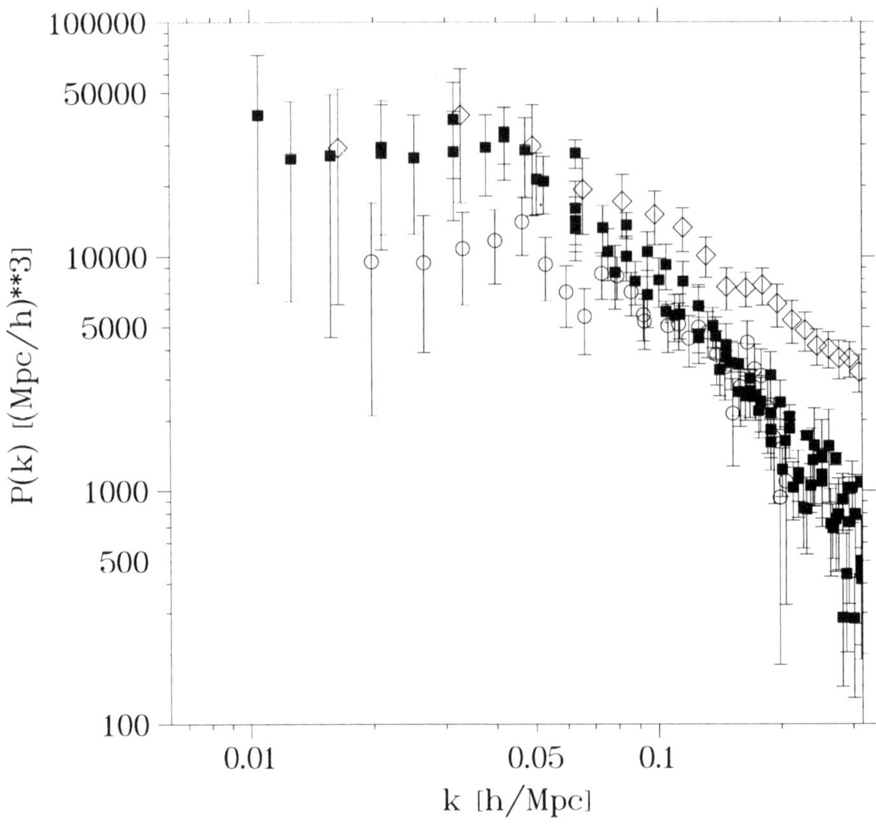


*Figure 2.* Observed redshift space power spectra. Squares: the magnitude-limited MRSP-subsamples as given in Fig. 2; circles: the flux-limited combined QDOT and 1.2-Jy IRAS surveys (Tadros & Efasthiou 1995); diamonds: the volume-limited combined SSRS2 and CfA2 subsample for galaxies with comoving distances $r \leq 130\,h^{-1}$ Mpc (da Costa *et al.* 1994). Error bars correspond to $(1\sigma)$ standard deviations.

tinuous (comoving) fluctuation field, $\delta(\vec{r}) = 1/(2\pi)^3 \int d^3k\ \delta(\vec{k}) e^{-i\vec{k}\cdot\vec{r}}$ and $\delta(\vec{k}) = \int d^3r \delta(\vec{r}) e^{+i\vec{k}\cdot\vec{r}}$, the Fourier-transformed density contrasts $\delta(\vec{k})$ and the power spectral densities $P(k)$ are related by the ensemble average $<\delta(\vec{k})\delta^*(\vec{k'})> = (2\pi)^3 P(k)\ \delta^{\rm D}(\vec{k}-\vec{k'})$, where $\delta^{\rm D}(\vec{k}-\vec{k'})$ is the Dirac delta function. Two discreteness effects modify this relation. (1) The continuous fluctuation field is traced by a discrete galaxy distribution (introducing shot noise), assumed to represent a statistically homogeneous and statistically isotropic (stationary) random point process. (2) The galaxies are sampled

5

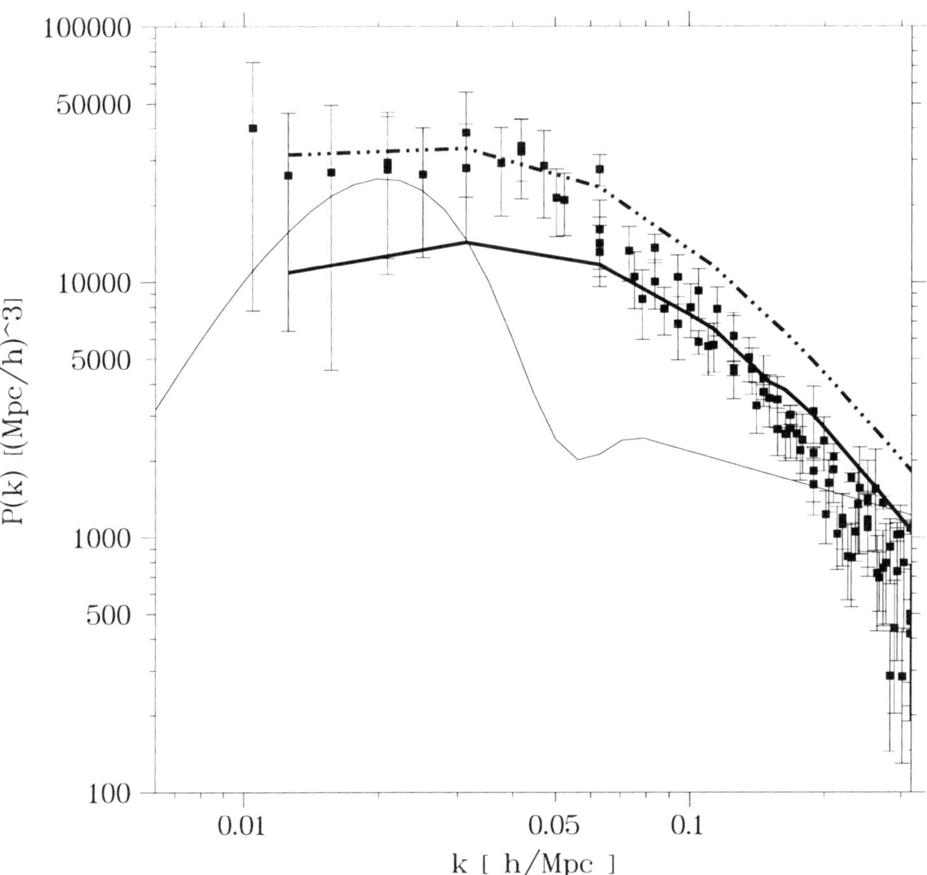


*Figure 3.* Comparison of observed and simulated redshift space power spectra. Squares: combined MRSP power spectrum. Continuous thick line: standard CDM model. Continuous thin line: baryon-dominated isocurvature model. Broke line: Potsdam Broken Scale Invariance model (Amendola *et al.* 1995). Error bars correspond to ($1\sigma$) standard deviations.

in cubic cells of a regular lattice in comoving $\vec{r}$ and in $\vec{k}$ space. This gives the discrete Fourier transform $\delta_{\vec{k}} \approx [\Delta k/(2\pi)]^3 \delta(\vec{k})$ with $(\Delta k)^3 = (2\pi)^3/V$ for the total cubic volume $V$. The corresponding variance of the Fourier amplitudes at discrete wavenumbers $<\delta_{\vec{k}}\, \delta^*_{\vec{k}'}> \; = \left[(2\pi)^{-3}\, P(k)\; (\Delta k)^3 \; + \frac{1}{N}\right] \delta^{\rm K}_{\vec{k},\vec{k}'}$

thus gives the fluctuation power within the $\vec{k}$ space volume element $(\Delta k)^3$ at $\vec{k}$ (one discrete mode), where $\delta^{\rm K}_{\vec{k},\vec{k}'} \approx (\Delta k)^3 \delta^{\rm D}(\vec{k}-\vec{k}')$ is the Kronecker delta. It is seen that $<\delta_{\vec{k}}\,\delta^*_{\vec{k}'}>$ depends on the total number of galaxies $N$ and on the size of the volume $V$. The corresponding relations for non-isotropically filtered (stationary) point processes are given in Schuecker, Ott & Seitter (1996a,b). The filter process is introduced by the redshift error. In the following, the differences between several fundamental statistical quantities derived from error-free and error-affected point distributions are discussed.

(1) Migration effects caused by redshift errors distort the overall redshift number counts $dN(z)$ and thus modify the radial selection function $\phi(z)$ and the luminosity function $\Phi(M)$. For known error distributions $p(z-z')$, the relation between the error-free selection function $\phi(z)$ and the error-affected function $\phi_{\rm e}(z)$ can be approximated by

$$\phi_{\rm e}(z) \approx \frac{1}{z^2}\int_0^\infty dz'\, z'^2\, \phi(z')\, p(z-z')\,, \quad z>0\,. \tag{1}$$

Solutions of the integral can be derived, *e.g.*, for Gaussian error distributions and exponentially decreasing $\phi(z)$.

(2) Redshift errors introduce distortions along the radial direction which formally correspond to the effects of peculiar velocities on non-linear scales. Let $p(k\mu\sigma_r)$ be the FT of the error distribution, given as a function of the wavenumber $k$, of the cosine of the angle between the normal vector of a given perturbation wave and the line-of-side $\mu$, and of a measure of the dispersion of the error distribution in comoving space $\sigma_r$. In this case, the distant observer approximation yields the following relation between the error-free power spectrum $P(k)$ and the error-affected spectrum $P_{\rm e}(k)$:

$$P_{\rm e}(k) = P(k)\int_0^1 d\mu\, |p(k\mu\sigma_r)|^2\,. \tag{2}$$

(3) The error distribution $p(z-z')$ is usually measured in 'radial velocity space'. To apply the formalism described above, $p(z-z')$ must be transformed into comoving space. The transformed profile thus becomes a function of comoving radial distance. Approximate solutions of this problem are discussed in Schuecker, Ott & Seitter (1996b).

(4) For fluctuation measurements a reliable mean particle number density $dN(z)$ is needed as reference level. Effect (1) distorts this reference, introducing systematic errors similar to the Malmquist bias. The statistical estimator of $\delta(\vec{k})$ given in Schuecker, Ott & Seitter (1996a,b) corrects for this effect by comparing the observed structured galaxy distribution with simulated unstructured particle distributions, distorted by the same measuring errors as expected for the observed sample.

The MRSP power spectra shown in Figs. 1,2,3 are obtained with samples of up to 87,558 galaxies (Schuecker, Ott & Seitter 1996a). The maximum spatial extent in comoving $x_c$,$y_c$, and $z_c$ coordinates (scaled to the present hypersphere) is 800, 310, and 520 $h^{-1}$ Mpc, respectively. The resulting spectra are deconvolved with a modified version of the method introduced by Lucy (1974).

The complete reduction process is controlled by several hundred statistically independent random point realizations of different power spectra, computed with the Zel'dovich approximation. The simulation method is described in Schuecker, Ott & Seitter (1996b). The error bars for the MRSP sample shown in Figs. 1,2,3 are obtained from the simulations and represent ($1\sigma$) standard deviations. The simulations show that the power spectral densities are reliable for the comoving wavenumbers $0.01 \leq k \leq 0.15\, h\,\mathrm{Mpc}^{-1}$ corresponding to the wavelengths $40 \leq \lambda \leq 600\, h^{-1}\,\mathrm{Mpc}$.

The MRSP spectrum shows a clear maximum at $150\, h^{-1}$ Mpc (Fig. 1). The spectrum is not strongly affected by the window function of the survey, as indicated by the similarity of the spectra obtained for different survey volumes. The MRSP spectral densities are in general agreement with the results obtained for the combined CfA2 and SSRS2 catalogs and for the combined IRAS ODOT and 1.2-Jy catalogs (Fig. 2). The spectrum rules out baryon-dominated models and is better fitted by models with Cold Dark Matter (Fig. 3).

## 4. Deceleration Parameter

The redshift volume test $V(z)$ is used for the determination of $q_0$ and for constraints of $\lambda_0$. The method has something like a built-in test for the fair-sample hypothesis and is the most direct way of estimating the effects of the overall matter content, luminous and dark, in the universe by counts. The test was first used on truely statistical data by Loh & Spillar (1986). The differential version of the method has a fourfold higher discriminating power between different cosmologies than, *e.g.*, the luminosity distance test (Linder 1988). The present version of the $V(z)$ test uses magnitude-limited samples. The $q_0$ dependency of the radial selection function reduces the sensitivity as compared to the ideal test. The larger numbers of objects in magnitude-limited samples, however, increase the statistical significance of the test and partially offset its smaller sensitivity with respect to $q_0$.

As mentioned above, redshift and magnitude errors distort the radial selection function and the luminosity function. This must be taken into account when observed distributions and models are compared. We use Monte Carlo simulations of unstructured particle distributions (Schuecker, Ott & Seitter 1996b) with all known measuring errors modeled in detail.

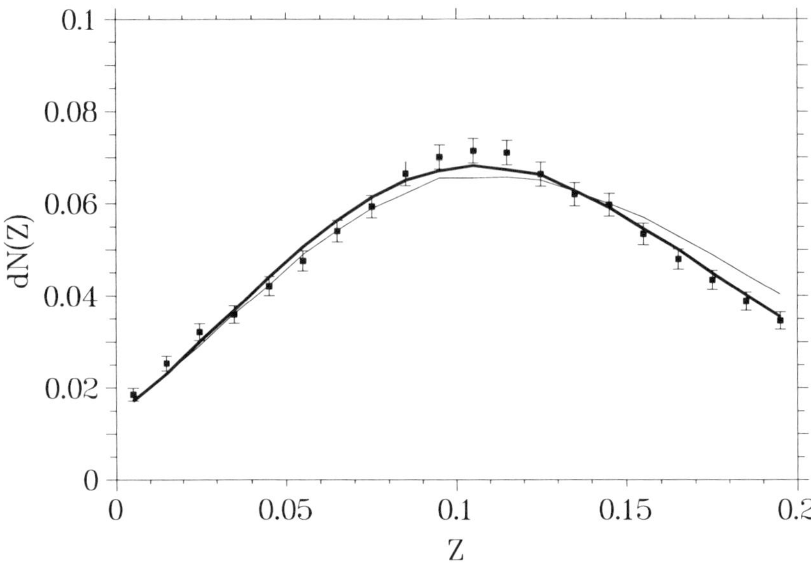


*Figure 4.* $q_0$ test for $\lambda_0 = 0$. Comparison of observed and simulated redshift histograms. Squares: total MRSP sample. Error bars: ($3\sigma$) Poisson errors. Thick line: best fit model with $q_0 = 0.10$. Thin line: model with $q_0 = 3.0$ for comparison.

A first fit of the model to the observed redshift histogram determines the 'local' luminosity function of the survey, using only galaxies with $z \leq 0.1$, assuming $q_0 = 0.5$, and starting with the pilot values of $M_*$ and $\alpha$ given at the end of § 2. The second fit determines the deceleration parameter, using all galaxies within the redshift range $0 \leq z \leq 0.2$, starting with the estimates of $M_*$ and $\alpha$ obtained from the first fit and only slightly modified during the second fit.

Monte Carlo simulations are used to test this method and to measure the confidence intervals of the resulting parameter values (for details see Schuecker *et al.* 1997). No significant systematic errors for $q_0$ values in the range $0.1 \leq q_0 \leq 3.0$ and for different luminosity functions are found. The parameters of the Schechter-type luminosity function obtained with the MRSP sample are very similar to the pilot values. The following discussion concentrates on the values of the cosmological parameters.

For the determination of $q_0$ we use a sample of 89,125 galaxies with $z \leq 0.2$ and $14.0 \leq r_F \leq 18.0$ mag. The normalized redshift histogram (symbols with $3\sigma$ errors) and the best model fit (thick continuous line)

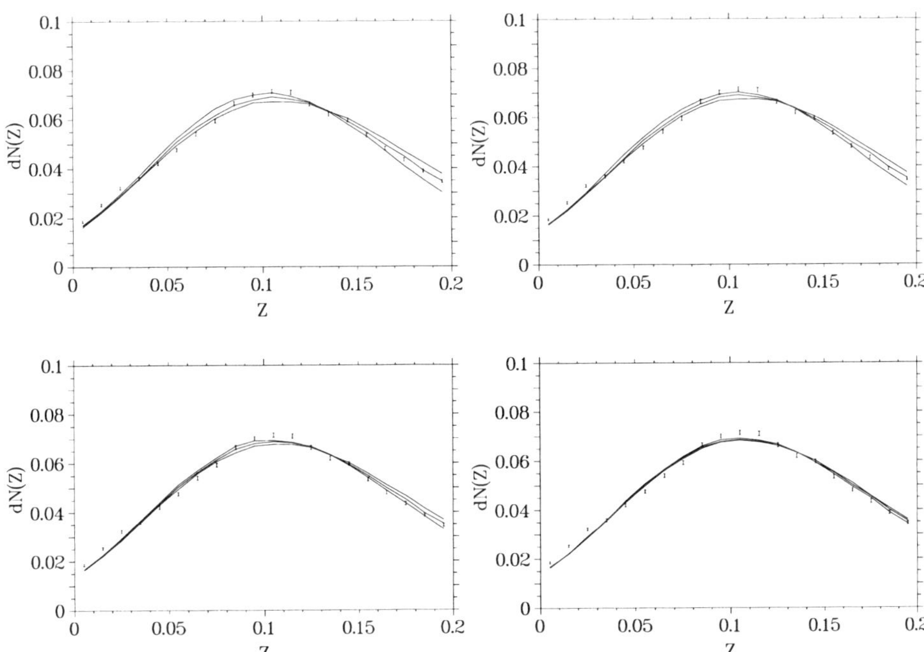


*Figure 5.* $(\Omega_0, \lambda_0)$ tests. Galaxy number counts as a function of redshift. Symbols: observed normalized counts with $(1\sigma)$ standard deviations (expected Poisson errors). Lines: models with the parameters $\Omega_{\rm tot} = \Omega_0 + \lambda_0 = 2.0$ (upper left), $\Omega_{\rm tot} = 1.5$ (upper right), $\Omega_{\rm tot} = 1.0$ (lower left), $\Omega_{\rm tot} = 0.5$ (lower right). The parameter values $(\Omega_0, \lambda_0)$ are related to the model curves in the following way: For $z > 0.13$ the upper curves correspond to models with large $\Omega_0$ and small $\lambda_0$; for $z < 0.13$ this order is reversed.

with $q_0(\text{eff}) = 0.10$ for $\lambda_0 = 0$ is shown in Fig. 4. For comparison, a model curve with $q_0(\text{eff}) = 3.0$ is superposed (thin line). Confidence intervals are estimated with 600 Monte Carlo simulations of the complete measuring process. The final results are:

$$q_0(\text{eff}) = 0.10\,, \quad q_0(\text{eff}) < 0.75 \text{ (95\% confidence limit)}\,. \tag{3}$$

Similar results are obtained for different subvolumes of the total sample. The fits for $\lambda_0 = 0$ are not optimal and possibly biased by extensions of the Sculptor Supercluster at $z = 0.11$ (Schuecker & Ott 1991).

In the following we try to improve the fits by using $\lambda_0$ as an additional free parameter. The effects of $\lambda_0 \neq 0$ are shown in Fig. 5. For $\Omega_{\rm tot} = (\Omega_0 + \lambda_0) \leq 1$, none of the parameter combinations give model curves which cover the majority of data points at $0.13 < z < 0.20$ (30,279 galaxies). The number counts for $z < 0.13$ are probably influenced by fluctuations in the galaxy distribution and are not used for the $(\Omega_0, \lambda_0)$ tests. For the larger values $\Omega_{\rm tot} = 1.5$ and 2.0, the fits suggest the combination $\Omega_0 \approx 0.30$ and

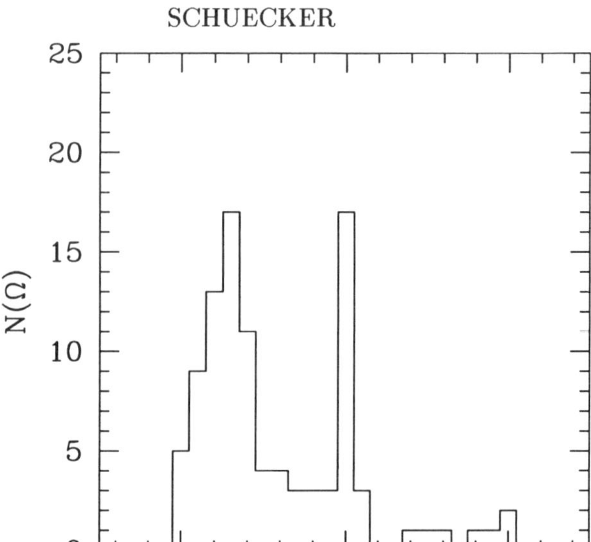


*Figure 6.* Distribution of observed $\Omega_0$ values (assuming $\lambda_0 = 0$) as published between 1990 and 1996 in the major astronomical journals. $\Omega_0$ values are not included when $\Omega_0$ and $\lambda_0$ are determined simultaneously, or when only an upper or an lower limit of the density parameter is given. If both limits are published we took the average of the two values. For methods giving $\Omega_0^{0.6}/b$ we compute $\Omega_0$ by using the biasing parameters $b$ given in Peacock & Dodd (1994) for different galaxy types and for clusters of galaxies. The general shape of the histogram is not affected by this choice.

$\lambda_0 \approx 1.2$ or the combination $\Omega_0 \approx 0.4$ and $\lambda_0 \approx 1.6$. Both cases assign $\lambda_0$ the priority in our universe.

Systematic errors might be introduced by models of galaxy evolution which are not consistent with the standard assumptions, *e.g.*, a different choice of the cosmic $(K + E)$ correction, a $z$-dependent shape of the luminosity function, and an evolution of comoving galaxy number densities. We quantify the evolutionary effects estimated from the redshift dependency of the Schechter parameters of recently published B band galaxy luminosity functions (*e.g.*, Marzke *et al.* 1994, Lilly *et al.* 1995, Ellis *et al.* 1996). The corresponding evolutionary corrections for $q_0$(eff) are

$$q_0(\text{eff}) - q_0 < \left\{ \begin{array}{lcl} +0.191 & : & M_*(z) \\ +0.097 & : & \alpha(z) \\ +0.278 & : & \Phi_*(z) \,. \end{array} \right. \tag{4}$$

They must be regarded as upper limits because evolution is enhanced in the blue compared to the red spectral range. Eqn. (4) indicate that a type of galaxy evolution which is not consistent with our standard assumption of a mild passive galaxy aging would yield overestimates of $q_0$. The MRSP result, $q_0(\text{eff}) = 0.10$, could thus been taken as an upper limit. Independent from the discussion of evolutionary effects, which are supposed to be comparatively small in the given magnitude and redshift ranges, the MRSP data do not support the Einstein-de Sitter world model.

Figure 6 shows the distribution of $\Omega_0 = 2q_0$ values published between 1990 and 1996 in the major astronomical journals. The list is incomplete, although, most likely representative. More than 40 different methods are used to determine $\Omega_0$. Two values dominate the histogram: $\Omega_0 = 0.30$ and $\Omega_0 = 1.0$. The frequency peak at $\Omega_0 = 1.0$ appears to be too narrow compared to the relatively large random errors expected for the corresponding measurements. However, very often the observational material is tested only for consistency with the Einstein-de Sitter case. Several entries at $\Omega_0 = 1$ can thus not be regarded as direct observational evidences.

The MRSP result $\Omega_0 = 2q_0 = 0.20$ is consistent with most of the current estimates implying a low-$\Omega_0$ universe. If the Einstein-de Sitter case is correct, one has to explain why a significant number of the 40 different methods used to estimate $\Omega_0$ give wrong results. The variety of completely different observational approaches yielding $\Omega_0 < 1$ indicates that the case of a low-$\Omega_0$ universe has to be taken seriously.

## References


Amendola, L., Gottlöber, S., Mücket, J.P., & Müller, V., 1995, ApJ, 451, 444

Barlett, M.S., 1964, Biometrica, 51, 299

Cunow, B., 1993, A&A, 268, 491

Cunow, B., & Ungruhe, R., 1995, A&AS, 112,213

da Costa, L.N., *et al.* 1994, ApJ, 437, L1

Duerbeck, H.W., *et al.* 1989, Rev. Mex. Astron. Astrofis., 19, 92

Ellis, R.S., Colless, M., Broadhurst, T., Heyl, J., & Glazebrook, K., 1996, MNRAS, 280, 235

Lilly, S.J., Tresse, L., Hammer, F., Crampton, D., & Le Fevre, O., 1995, ApJ, 455, 108

Linder, E.V., 1988, A&A, 206, 175

Loh, E.D., & Spillar, E.J., 1986, ApJ, 307, L1

Lucy, L.B., 1974, AJ, 79, 745

Marzke, R.O., Geller, M.J., Huchra, J.P., & Corwin, H.G., 1994, AJ, 108, 2

Naumann, M., Ungruhe, R., & Seitter, W.C., 1993, ESO Messenger, 71, 46

Peacock, J.A., & Dodds, S.J., 1994, MNRAS, 267, 1020

Ripley, B.D., 1981, Spatial Statistics, (New York: J. Wiley & Sons)

Schombert, J.M., Bothun, G.D., Schneider, S.E., & McGaugh, S.S., 1992, AJ, 103, 1107

Schuecker, P., 1993, ApJS, 84, 39

Schuecker, P., 1996, MNRAS, 279, 1057

Schuecker, P., & Ott, H.-A., 1991, ApJ, 378, L1

Schuecker, P., Ott, H.-A., & Seitter, W.C., 1996a, ApJ, 472, 485

Schuecker, P., Ott, H.-A., & Seitter, W.C., 1996b, ApJ, 459, 467

Schuecker, P., *et al.*, 1997, ApJ (submitted)

Seitter, W.C., 1992, Proc. Digitized Optical Sky Surveys, eds. H.T. MacGillivary and E.G. Thomson, (Dordrecht: Kluwer), 367

Tadros, H., & Efstathiou, G., 1995, MNRAS, 276, L45

# GEOMETRY AND LARGE-SCALE POWER IN THE LAS CAMPANAS REDSHIFT SURVEY

S.D. LANDY
*Univ. of CA at Berkeley - Department of Astronomy*
*601 Campbell Hall, Berkeley, CA 94720*

The Las Campanas Redshift Survey is currently the largest redshift survey in existence containing over 26,000 galaxies covering approximately 720 square degrees of the sky. In the interest of sampling the galaxy distribution on large scales, *i.e.*, several hundred $h_{100}^{-1}$ Mpc , the survey was designed as a collection of six thin, broad, and deep slices. Each of the slices has dimensions 1.5°in declination by 80°in RA with a mean depth of 300 $h_{100}^{-1}$ Mpc (see Shectman *et al.* 1996 for more detail). These thin, constant declination slices have a nominal size of 350x400x10 $h_{100}^{-1}$ Mpc .

Given this geometry it is evident that on scales above several tens of $h_{100}^{-1}$ Mpc , the survey slices are effectively two-dimensional. For this reason a two-dimensional approach was taken in measuring the large-scale power in Landy *et al.* 1996. This paper reports the measurement of a highly significant amount of excess power on 100 $h_{100}^{-1}$ Mpc scales. In this proceeding, the gross geometry of the survey will be presented together with its associated window function with the goal illustrating the advantages of a two-dimensional approach to analysing large-scale power in the Las Campanas redshift survey. Additionally, an example will be given of the degradation in signal which results from the inclusion of wavevectors whose direction lies significantly out of the 'plane' of a slice. These wavevectors would be included in a full three-dimensional analysis and are responsible for the discrepancy between the 2D result of Landy *et al.* 1996 and the 3D result of Lin *et al.* 1996.

Fig. 1 illustrates the gross geometry of the survey; the two dimensional nature of the survey on large scales is easily seen. In Fig. 2a, as a heuristic we model the survey as a thin plane 350x400x10 $h_{100}^{-1}$ Mpc . The azimuthally averaged window function associated with this slice appears as the thin needle in Fig. 2b. This window function is the convolution kernel which convolves the underlying power spectrum. The window function has some very obvious properties. It is wide in one direction ($k_z$) and thin in the other



*D. Hamilton (ed.), The Evolving Universe,* 111–115.

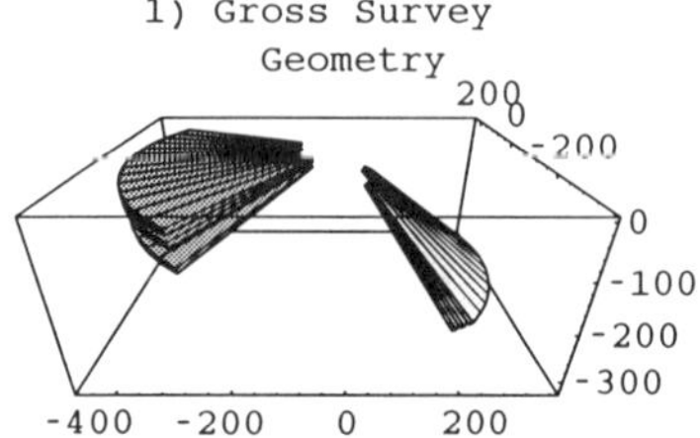




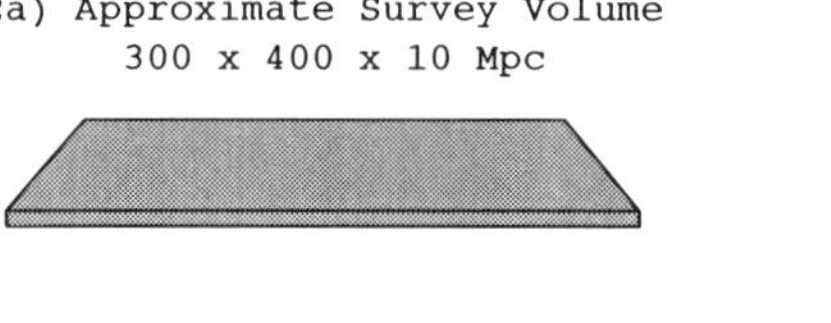


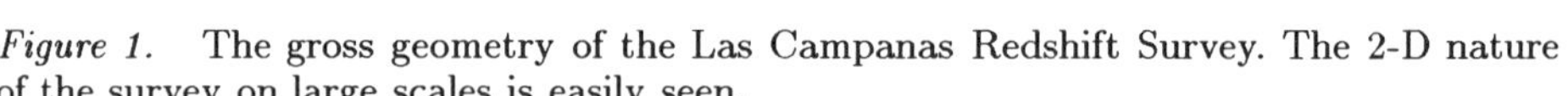

*Figure 1.* The gross geometry of the Las Campanas Redshift Survey. The 2-D nature of the survey on large scales is easily seen.

*Figure 2.* The slices modelled as thin planes is shown in *a* and their associated azimulthally-averaged window functions in *b*.

two $(k_x, k_y)$. This is expected since it is a representation of the inverse of the survey geometry in real space.

In the two dimensional analysis each slice was maximally projected onto the xy plane and the power spectrum measured using only those wavevectors which lay in this plane $(k_x, k_y, k_z = 0)$. A consideration of the survey

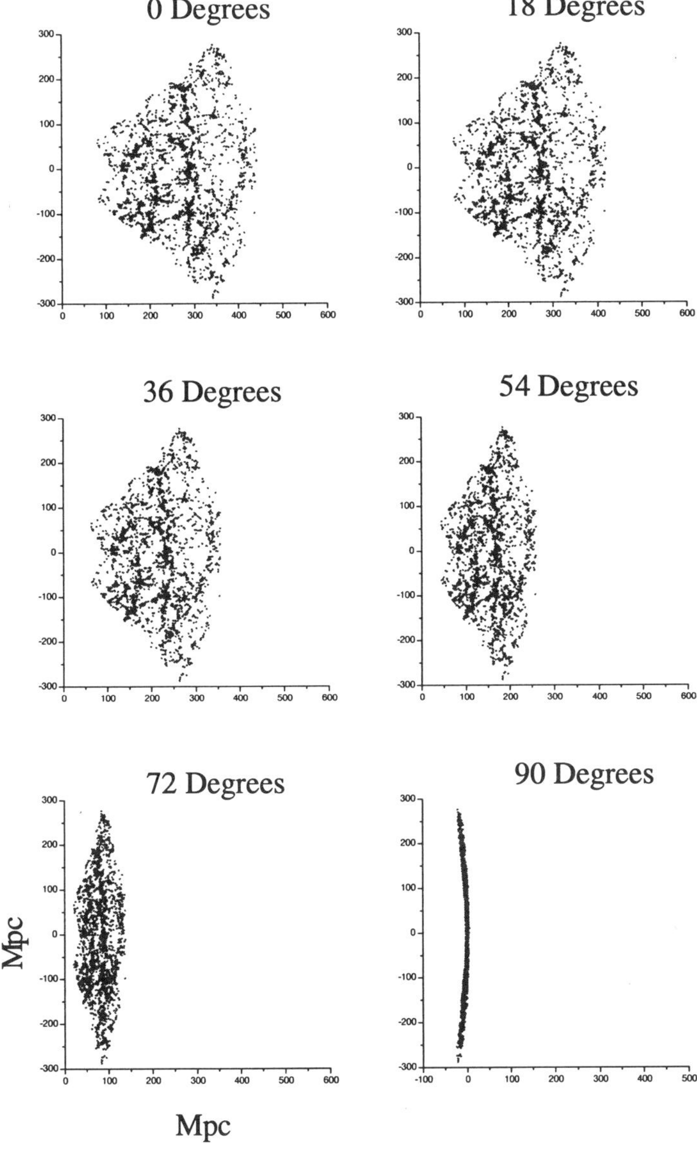


*Figure 3.* Projections of the -12 Dec slice corresponding to wavevectors which lie at that angle to the nominal plane of the slice. This shows the projected galaxy distribution actually being sampled by these wavevectors. At small angles, the distributions offer little new information and at large angles have no information on large scales.

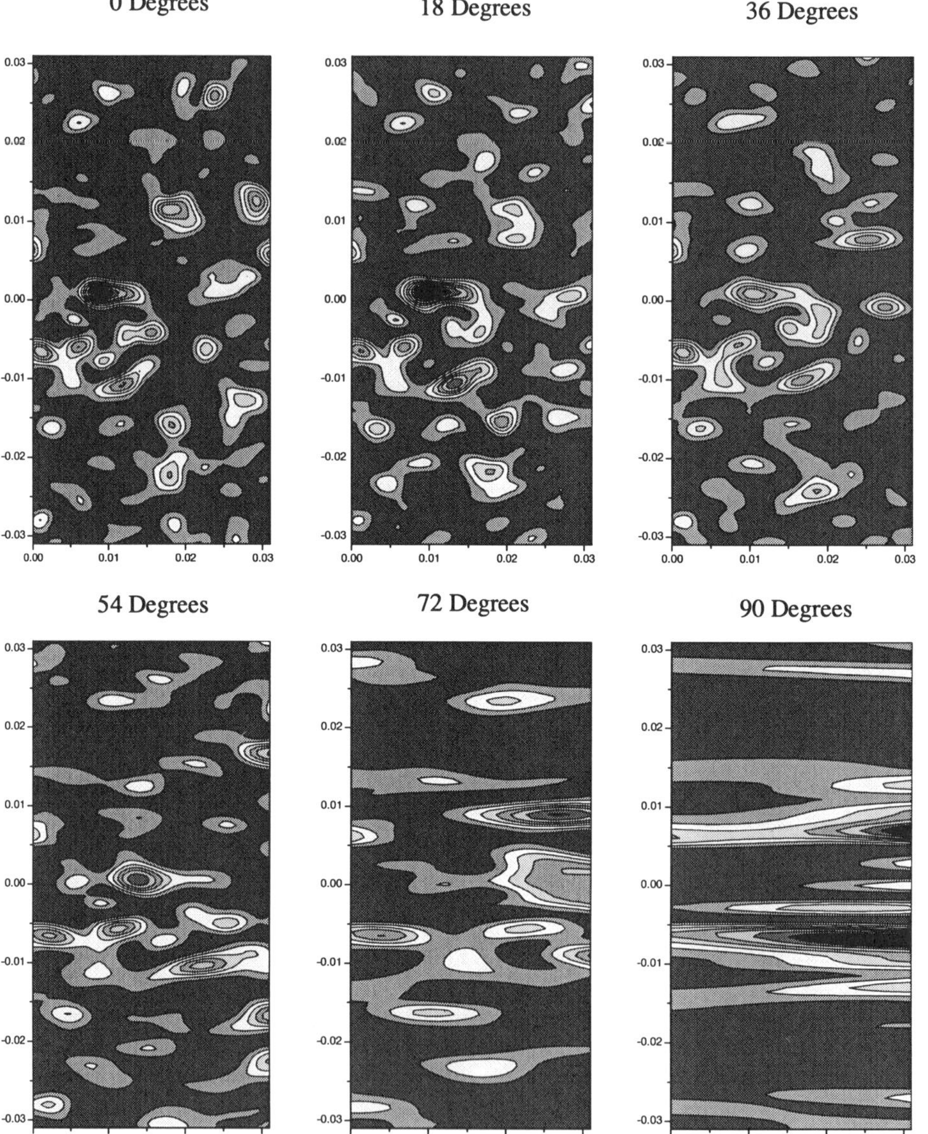


*Figure 4.* The power spectrum of the wavevectors corresponding to the projected distributions in Fig 3. Again the signals are strongly correlated at small angles and are noise at larger angles. This illustrates the advantages of a 2D analysis.

geometry and the shape of the window function allow us to make some very general statements regarding the utility of this approach to measuring the large-scale power spectrum in this survey. First, since the slices are only about 10 $h_{100}^{-1}$ Mpc thick, it seems sensible to exclude any wavevectors from the analysis whose projection perpendicular to this slice is greater than about 10 $h_{100}^{-1}$ Mpc . Secondly, the shape of the window function tells us that any wavevectors which lie only slightly out of the plane will be strongly correlated with those which lie in the plane for any given direction and wavelength. Therefore, limiting the analysis to two dimensions is a reasonable approach and results in little loss of information on large scales.

These general statements can be seen graphically in Figs. 3 and 4. In Fig. 3, a set of projected distributions of the -12°dec slice is shown. These distributions correspond to a two dimensional power spectrum analysis measured at various angles with regard to the plane of the slice. Such sets of projections would be averaged over in a full three dimensional analysis. These graphs show that at small angles the distributions are very similar while at larger angles they contain little or no information on large scales. The power spectrum analogues of these distributions are shown in Fig. 4. Again, at small angles the signals are strongly correlated, with the signal being projected to progressively smaller wavelengths. At large angles, the signal is primarily noise and aliased power. Both of these effects tend to smooth over and wash out the robust signals which lie in the plane of the slice. This shows the advantage of tailoring the power spectrum analysis to the geometry of the survey which is effectively two dimensional on these scales.

## References

Landy, S.D., Shectman, S., Lin, H., Kirshner, R.P., Oemler, A., & Tucker, D. 1996, ApJL, 456, L1

Lin, H., Kirshner, R.P., Shectman, S., Landy, S.D., Oemler, A., Tucker, D. & Schechter, P.L. 1996, ApJ, 471, 617

Shectman, S., Landy, S.D., Oemler, A., Tucker, D.L., Kirshner, R.P., Lin, H., & Schechter, P.L. 1996, ApJ, 470, 172

# THE LAS CAMPANAS REDSHIFT SURVEY RESULTS: THE REDSHIFT-SPACE AUTOCORRELATION FUNCTION

DOUGLAS L. TUCKER
*Fermi National Accelerator Laboratory*
*MS 127*
*P.O. Box 500*
*Batavia, Illinois 60504*
*USA*

**Abstract.** Presented are measurements from the Las Campanas Redshift Survey (LCRS) of the galaxy-galaxy autocorrelation function in redshift-space, $\xi_{gg}(s)$. For separations $2.0h^{-1}$ Mpc $< s < 16.4h^{-1}$ Mpc, $\xi_{gg}(s)$ can be approximated by a power law with slope $\gamma = -1.52 \pm 0.03$ and correlation length $s_0 = 6.28 \pm 0.27h^{-1}$ Mpc. A zero-crossing occurs on scales of $\sim 30 - 40h^{-1}$ Mpc. On larger scales, $\xi_{gg}(s)$ fluctuates closely about zero, indicating a high level of uniformity in the galaxy distribution on these scales. In addition, a selection effect inherent to the LCRS – the 55 arcsec galaxy pair separation limit – is tested and found to have little impact on the measurement of $\xi_{gg}(s)$.

## 1. Introduction

The original goals of the Las Campanas Redshift Survey (LCRS; Shectman *et al.* 1996) were two-fold: firstly, to attempt to sample a 'fair and typical' volume of the nearby Universe in order to constrain the size of the largest structures in the local galaxy distribution, and, secondly, to use this sample to study galaxy clustering on a wide variety of scales. Recently completed, the LCRS contains over 26,000 galaxies. Accurate sky positions and Kron-Cousins $R$-band photometry were obtained with CCD drift scans at the Las Campanas Swope 1-m telescope; spectra were obtained at the Las Campanas Du Pont 2.5-m telescope, originally with a 50-fiber Multi-Object Spectrograph (MOS) and later with a 112-fiber MOS. For observing efficiency, all the fibers are used, but each MOS field is observed only once.



*D. Hamilton (ed.), The Evolving Universe,* 117–121.

Hence, the LCRS is a collection of 50-fiber fields (with nominal apparent magnitude limits of $16.0 \leq R < 17.3$) and 112-fiber fields (with nominal apparent magnitude limits of $15.0 \leq R < 17.7$). Thus, selection criteria vary from field to field, but they are carefully documented and are therefore easily taken into account. Observing each field only once, however, creates an additional selection effect: the individual fibers' protective tubing prevents the observation of galaxy pairs within 55 arcsec of each other. Hence, the cores of rich clusters may be undersampled, potentially causing underestimates in measurements of small-scale galaxy clustering. Here, we will examine the LCRS redshift-space galaxy-galaxy autocorrelation function, $\xi_{\rm gg}(s)$, and the influence on it from the fiber-separation limit. Elsewhere in this volume, Bob Kirshner provides a general description of the survey, Stephen Landy discusses the LCRS 2D power spectrum, and Huan Lin describes both the 3D power spectrum and the luminosity function from the LCRS.

## 2. Method

To account for the survey geometry and for the field-to-field variations in the nominal apparent magnitude limits and in the sampling fraction, catalogues of random galaxies are generated for the same survey volume with the same field-to-field characteristics as the observed LCRS catalogue. (Note that the fiber-separation limit is *not* implemented into the random catalogues.) The redshift distribution of random galaxies is determined via the LCRS luminosity function described in Lin *et al.* (1996), incorporating the subtleties detailed in Section 3.2 of that paper. Then, $\xi_{\rm gg}(s)$ is calculated according to the Hamilton (1993) formalism,

$$1 + \xi_{\rm gg}(s) = \frac{RR(s)}{DR(s)} \times \frac{DD(s)}{DR(s)}, \tag{1}$$

where $DD(s)$, $DR(s)$, and $RR(s)$ are, respectively, the weighted data-data pair count ($\Sigma_{i \neq j} w_i w_j$), the weighted data-random pair count ($\Sigma w_i w_j^{\rm r}$), and the weighted random-random pair count ($\Sigma_{i \neq j} w_i^{\rm r} w_j^{\rm r}$) for the (comoving) separation $s$; $RR/DR$ is a measure of the relative mean density of galaxies in the observed and random catalogues. Aside from small differences in the large-scale normalisation, our results change little if a classic Davis & Peebles (1983) approach is employed (see Fig. 1 below). The Hamilton formalism is preferred, however, since it is less affected by uncertainties in the mean density of galaxies (see also Landy & Szalay 1993).

For the pair counts, a minimum variance weighting is used (see Hamilton 1993). In order to avoid the weights 'blowing up' at the survey's extremal distances, the analysis of $\xi_{\rm gg}(s)$ is confined to those galaxies in the LCRS with velocities of $10,000$ km s$^{-1}$ $\leq cz_{\rm CMB} \leq 45,000$ km s$^{-1}$ and with $R$-band absolute magnitudes of $-22.50 \leq M_R - 5\log h \leq -18.50$ ($H_0 \equiv 100h$ km s$^{-1}$ Mpc$^{-1}$); 19,314 galaxies from the official LCRS sample meet these criteria. The random catalogues typically contain a similar number of galaxies.

## 3. Results

### 3.1. THE OBSERVED AUTOCORRELATION FUNCTION

The results for the observed LCRS $\xi_{\rm gg}(s)$ can be found in Figure 1. For separations $2.0h^{-1}$ Mpc $< s <$ $16.4h^{-1}$ Mpc, the observed LCRS $\xi_{\rm gg}(s)$ can be approximated by a power law with slope $\gamma = -1.52 \pm 0.03$ and correlation length $s_0 = 6.28 \pm 0.27h^{-1}$ Mpc (Fig. 1a, Hamilton formalism). A zero-crossing occurs at $s \sim 30-40h^{-1}$ Mpc. On larger scales, $\xi_{\rm gg}(s)$ fluctuates closely about zero, evidence of a high level of uniformity in the galaxy distribution on these scales (Fig. 1b). Although possible small-amplitude ($\delta\xi_{\rm gg} \approx 0.01 \pm 0.01$) secondary maxima do appear at $\sim 100h^{-1}$ Mpc and at $\sim 200h^{-1}$ Mpc, whether these (and other, larger-scale) relative extrema are characteristic of the galaxy distribution itself or merely statistical fluctuations within the sampled volume remains a matter of debate and is still under investigation. It is interesting to note, however, that these large-scale features – and the differences between the LCRS North and South Galactic Cap samples – are reflected in Landy *et al.* (1996)'s determination of the LCRS 2D power spectrum (their Fig. 2), and that Doroshkevich *et al.* (1997)'s core-sampling analysis of the LCRS reveals a scale of $\sim 100h^{-1}$ Mpc for the typical separation of sheet- or wall-like structures. Furthermore, in an independent sample, Einasto *et al.* (1997) find similar features in the autocorrelation function for clusters in the environments of rich superclusters.

### 3.2. THE FIBER SEPARATION TEST

To test the effects of fiber separation, the fiber separation criterion in each MOS field has been artificially increased from the original 55 arcsec – first to 90 arcsec and then to 120 arcsec. This was done by culling the original LCRS catalogue – field-by-field – of one galaxy in any pair separated on the sky by less than the artificial fiber limit. One can then extrapolate backwards to estimate how the original 55 arcsec separation limit affects the measurement of $\xi_{\rm gg}(s)$. For a more direct test, mock slices have been extracted from Standard Cold Dark Matter (SCDM) simulations (see Tucker *et al.* 1997), to which a 55 arcsec fiber separation limit was imposed in a manner akin to that for a real LCRS slice.

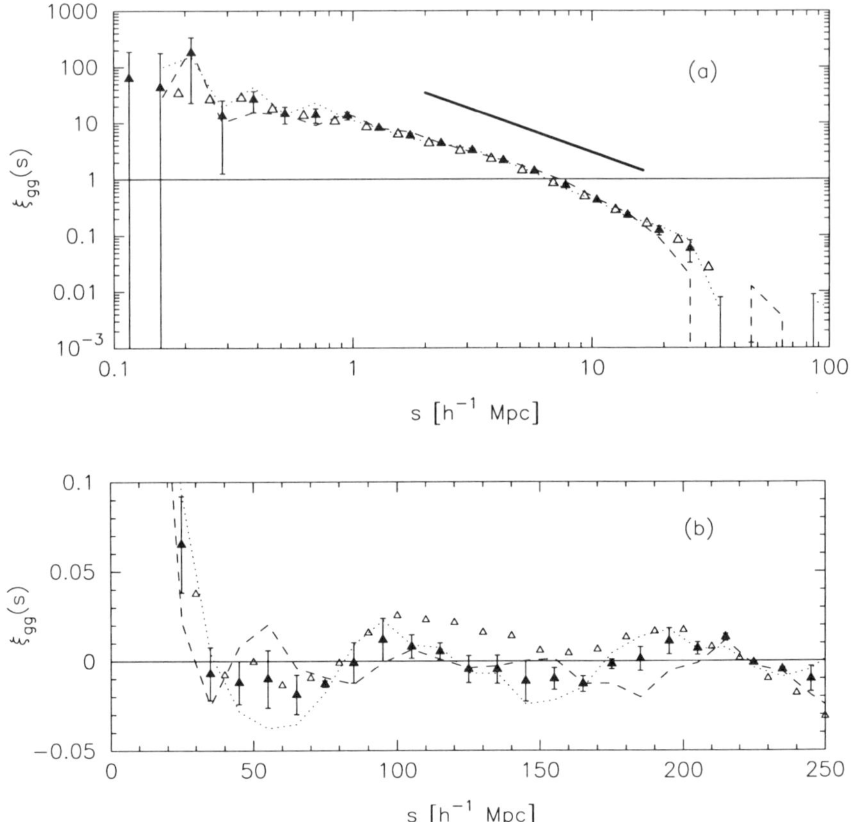


*Figure 1.* The observed LCRS $\xi_{gg}(s)$ for (a) small-to-intermediate scales and for (b) intermediate-to-large scales. The *filled triangles* denote $\xi_{gg}(s)$ for the combined North and South Galactic Cap sample, the *dashed line* for the Northern Cap sample alone, and the *dotted line* for the Southern Cap sample alone; these measurements all use the Hamilton formalism. For comparison, the LCRS $\xi_{gg}(s)$ calculated using the Davis & Peebles formalism is also presented (*unfilled triangles*); here, as with the Hamilton formalism, a minimum variance weighting is used. For clarity, only the *filled triangles* show error bars (error bars were estimated by calculating $\xi_{gg}(s)$ in independent subregions of the LCRS and taking the standard deviation of the mean). Finally, a −1.52 power law, offset, is shown in (a) for the interval $2.0h^{-1}$ Mpc $< s < 16.4h^{-1}$ Mpc (*thick solid line*). [From Tucker *et al.* 1997 (©Blackwell Scientific Publications Ltd).]

From Figure 2, it is clear that the effect is negligible on all but the smallest scales, $s \lesssim 1h^{-1}$ Mpc. At these scales, however, Poisson errors begin to dominate the results, making it difficult to assess the exact magnitude of the effect. Nonetheless, even at separations of $s \sim 0.3h^{-1}$ Mpc, it appears that the LCRS 55 arcsec fiber separation limit results only in a $\sim 10-20\%$ underestimate of the 'true' $\xi_{gg}(s)$. On scales of $1h^{-1}$ Mpc $\lesssim s \lesssim 20h^{-1}$ Mpc, measurements of $\xi_{gg}(s)$ are depressed typically by $\lesssim 5\%$.

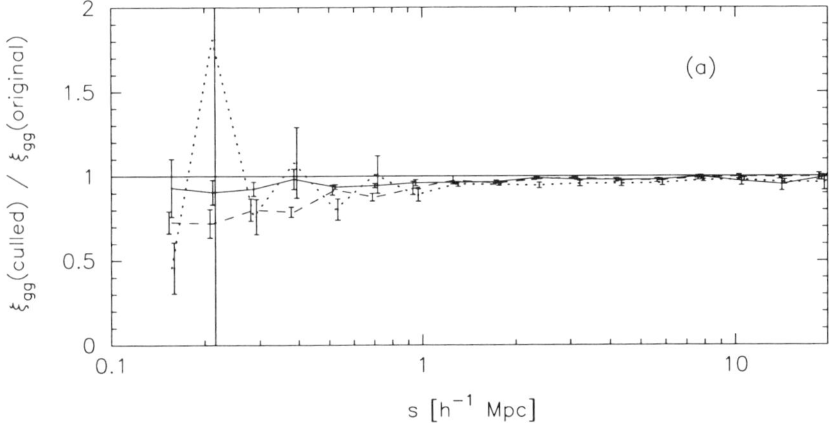


*Figure 2.* The Fiber Separation Test. Plotted is the ratio of $\xi_{gg}(s)$ of a catalogue culled for a given fiber separation limit to that of the original catalogue, $\xi_{gg}$(culled)/$\xi_{gg}$(original). For the LCRS data, $\xi_{gg}$(original) is the LCRS $\xi_{gg}(s)$ from Fig. 1a (Hamilton formalism); this ratio is plotted for the LCRS catalogue culled to 90 arcsec (*dashed line*) and to 120 arcsec (*dotted line*) fiber separation limits. Error bars are the standard deviation of the mean of $\xi_{gg}$(culled)/$\xi_{gg}$(original) for independent subvolumes of the LCRS. For the SCDM mock slices, $\xi_{gg}$(original) is $\xi_{gg}(s)$ with no fiber separation limit; $\xi_{gg}$(culled) is $\xi_{gg}(s)$ for the mock slice culled to a fiber separation limit of 55 arcsec. Plotted is the mean of this ratio based upon 5 individual mock slices (*solid line*), where the error bars are the standard deviation of the mean. [From Tucker *et al.* 1997 (©Blackwell Scientific Publications Ltd).]

## Acknowledgments

I wish to thank Blackwell Scientific Publications Ltd, publishers of MNRAS, for granting me permission to include within this contribution text and plots recently published in Tucker *et al.* 1997, MNRAS, 285, L5.

## References


Davis M., Peebles P. J. E., 1983, ApJ, 267, 465

Doroshkevich A. G., Tucker D. L., Oemler A. Jr., Kirshner R. P., Lin H., Shectman S. A., Landy S. D., Fong R., 1996, MNRAS, 283, 1281

Einasto J. *et al.*, 1997, Nature, 385, 139

Hamilton A. J. S., 1993, ApJ, 417, 19

Landy S. D., Szalay A. S., 1993, ApJ, 412, 64

Landy S. D., Shectman S. A., Lin H., Kirshner R. P., Oemler A., Tucker D. L., 1996, ApJ, 456, L1

Lin H., Kirshner R. P., Shectman S. A., Landy S. D., Oemler A., Tucker D. L., Schechter P. L., 1996, ApJ, 464, 60

Shectman S. A., Landy S. D., Oemler A., Tucker D. L., Lin H., Kirshner R. P., Schechter P. L., 1996, ApJ, 470, 172

Tucker D. L., Oemler A., Kirshner R. P., Lin H., Shectman S. A., Landy S. D., Schechter P. L., Müller V., Gottlöber S., Einasto J., 1997, MNRAS, 285, L5

# LAS CAMPANAS REDSHIFT SURVEY RESULTS: THE 3D POWER SPECTRUM & THE LUMINOSITY FUNCTION

HUAN LIN
*Department of Astronomy*
*University of Toronto*
*60 St. George Street, Toronto, ON M5S 3H8*
*Canada*

**Abstract.** Some recent results on the galaxy 3D power spectrum and the luminosity function derived from the Las Campanas Redshift Survey (LCRS) are discussed. We will present the LCRS power spectrum, compare it against results from other local redshift surveys, and use it to constrain a number of cold dark matter structure formation models. Then we will proceed to describe the LCRS luminosity function, as well as briefly discuss some related topics, specifically galaxy spectral classification and the evolution of the galaxy luminosity density.

## 1. Introduction

The Las Campanas Redshift Survey (LCRS) consists of over 26000 galaxies, with average redshift $z = 0.1$, selected from an $R$-band CCD photometric catalog covering over 700 square degrees on the sky. The survey geometry is that of 6 $1.5° \times 80°$ "slices," 3 each in the north and south galactic caps. The goals of the survey are to obtain fair and representative measures of the clustering, luminosity, and other properties of galaxies in the local universe. A description of the LCRS may be found in the main survey paper, Shectman *et al.* Also, Stephen Landy and Douglas Tucker describe their work on the LCRS 2D power spectrum and correlation function in their respective contributions. Here, I will describe work that I have been most closely involved with, specifically the LCRS 3D power spectrum in § 2, and the LCRS luminosity function and related topics in § 3. Additional information about the LCRS, including papers and the



*D. Hamilton (ed.), The Evolving Universe,* 123–133.

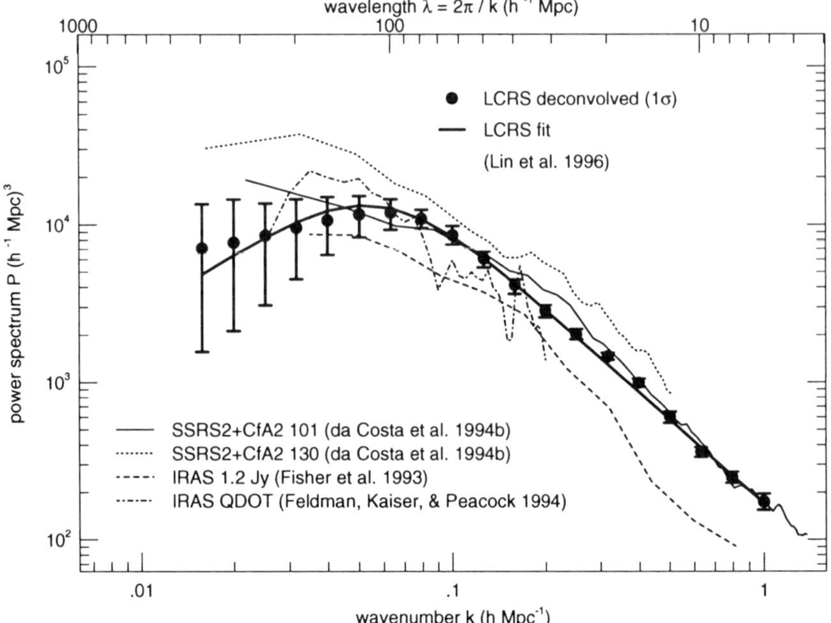


*Figure 1.* The LCRS deconvolved 3D power spectrum and fit are compared to power spectra from the SSRS2+CfA2, IRAS 1.2 Jy, and IRAS QDOT surveys.

publicly available redshift catalog, may be found at the LCRS home page "http://manaslu.astro.utoronto.ca/~lin/lcrs.html."

## 2. The 3D Power Spectrum

The details of the computation of the LCRS 3D power spectrum are given in Lin *et al.* (1996b).; here we will just highlight the main results. The power spectrum $P(k)$ is computed in redshift space for a magnitude-limited sample of 19305 LCRS galaxies, over a wide range of scales $\lambda = 2\pi/k = 5 - 400\ h^{-1}$ Mpc. Care is taken to account for convolution effects resulting from the essentially 2D geometry of the survey. The resulting deconvolved LCRS $P(k)$ is plotted in Fig. 1, and a convenient fit is given by

$$P(k) = \frac{2\pi^2}{k^3}\frac{(k/k_0)^{n_0}}{1+(k_c/k)^{n_c}}\ , \quad \begin{array}{rcl} k_0 & = & (0.17 \pm 0.01)\ h\ \mathrm{Mpc}^{-1} \\ n_0 & = & 1.2 \pm 0.1 \\ k_c & = & (0.06 \pm 0.01)\ h\ \mathrm{Mpc}^{-1} \\ n_c & = & 3 \pm 1 \end{array} \quad . \tag{1}$$

We see that for scales $\lambda \lesssim 100\ h^{-1}$ Mpc, the power spectrum is very well determined, with a power law behavior $P(k) \propto k^{n_0-3} = k^{-1.8\pm0.1}$ on small to intermediate scales $5 - 30\ h^{-1}$ Mpc. The overall amplitude is also well

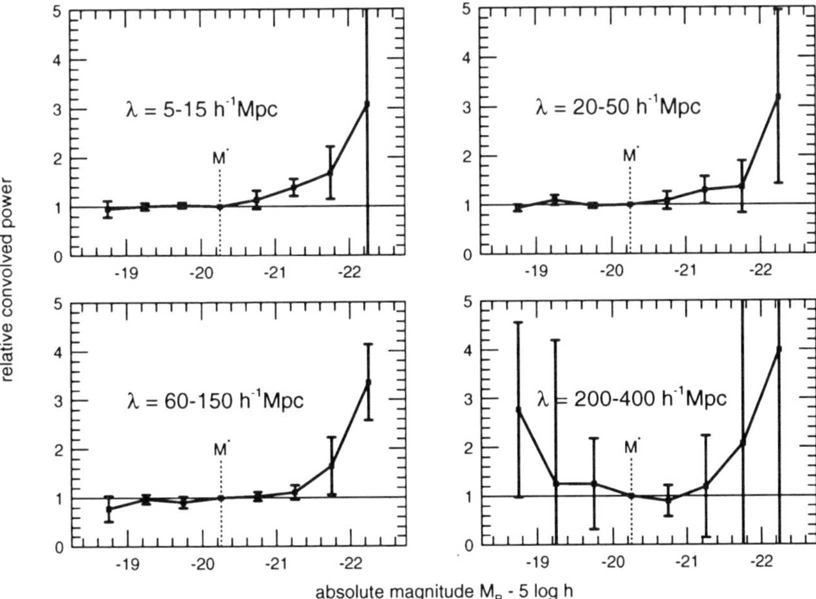


*Figure 2.* The relative clustering power as a function of absolute magnitude $M_R$ for LCRS galaxies. The clustering power is normalized to one at $M_R = M_R^* = -20.3+5\log h$.

measured and corresponds to $\sigma_8 = 1.0 \pm 0.1$ in redshift space. There is an apparent peak at $\lambda = 2\pi/k_c \approx 100\ h^{-1}$ Mpc, and on larger scales $\lambda \approx 200 - 400\ h^{-1}$ Mpc, $P(k) \propto k^{n_0-3+n_c} = k^{1\pm1}$, but here the errors get inevitably large and our results can accommodate a flat to declining $P(k)$ as $k$ decreases.

Fig. 1 also compares the LCRS $P(k)$ to those from three other large redshift surveys: SSRS2+CfA2 (da Costa *et al.* 1994b), IRAS 1.2 Jy (Fisher *et al.* 1993, and IRAS QDOT (Feldman, Kaiser, and Peacock 1994). Note first that the match between LCRS and the canonical SSRS2+CfA2 101 sample is quite good over the whole range of scales shown. Thus it is reassuring that these large, optically-selected, and *independent* surveys are measuring very similar amounts of galaxy clustering in the local universe, though we note that the LCRS probes a significantly larger and deeper volume than SSRS2+CfA2. On the largest scales $\lambda \gtrsim 100\ h^{-1}$ Mpc, the various samples also appear consistent with each other, although the errors are large there. On the other hand, for smaller scales $\lambda \lesssim 100\ h^{-1}$ Mpc, where $P(k)$ is better determined, there is clear evidence that clustering differences exist. Specifically, we find that: (1) the SSRS2+CfA2 130 sample, which is restricted to bright galaxies with $M \lesssim M^* - 1$, shows stronger clustering than SSRS2+CfA2 101 or LCRS, which contain mainly fainter galaxies; and (2) the IRAS samples, which are infrared-selected and have proportionately more spiral galaxies, show weaker clustering. The former

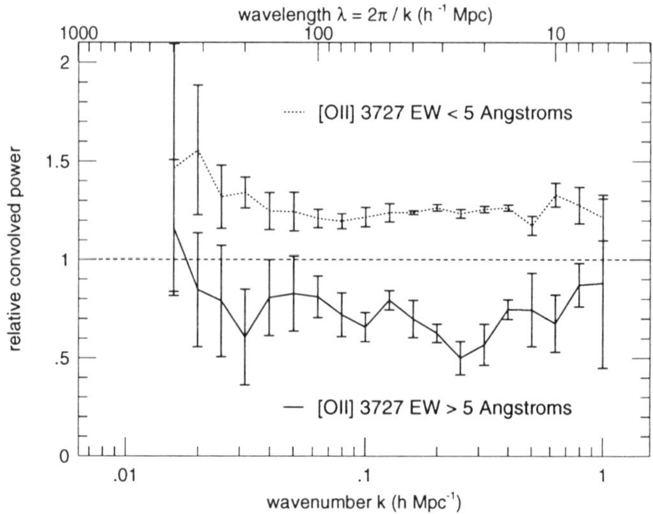


*Figure 3.* The relative clustering power of LCRS emission-line and non-emission-line samples, as defined by [OII] 3727 equivalent width (EW). The clustering power is normalized to one for the full sample.

case argues that clustering varies with galaxy luminosity, whereas the latter case manifests the well-known clustering dependence on morphology (*e.g.*, Davis and Geller 1976). Both these type-dependent clustering effects also appear in the LCRS data, and can be examined better using the larger LCRS sample. In particular, Fig. 2 plots the relative clustering power as a function of absolute magnitude for LCRS galaxies. We find that clustering varies little from $M \approx M^* + 2$ to $M^* - 1$, but brightwards there is about 50% more clustering power, confirming the results from SSRS2+CfA2. The relatively weak dependence of clustering on luminosity argues that $\sigma_8 \approx 1$ for the *mass* distribution (Kauffmann, Nusser, and Steinmetz 1996), and combined with the observed LCRS galaxy $\sigma_8 = 1$, implies that $M^*$ and fainter galaxies in the LCRS should be close to unbiased tracers of the mass.

In Fig. 3 we plot the relative clustering power as a function of scale for LCRS emission-line and non-emission-line galaxies, as defined by [OII] 3727 equivalent width. Our emission-line galaxies will tend to be later-type spirals, and they show 30% weaker clustering *vs.* the full LCRS sample, similar to the behavior of IRAS galaxies. In contrast, our non-emission-line sample will include more early-type, elliptical galaxies, and these show 25% stronger clustering *vs.* the full sample. Clearly, there are type-dependent clustering differences on small to intermediate scales that are robustly seen from one survey to another, and these differences will be relevant for models of galaxy formation and for comparisons against results at higher redshifts.

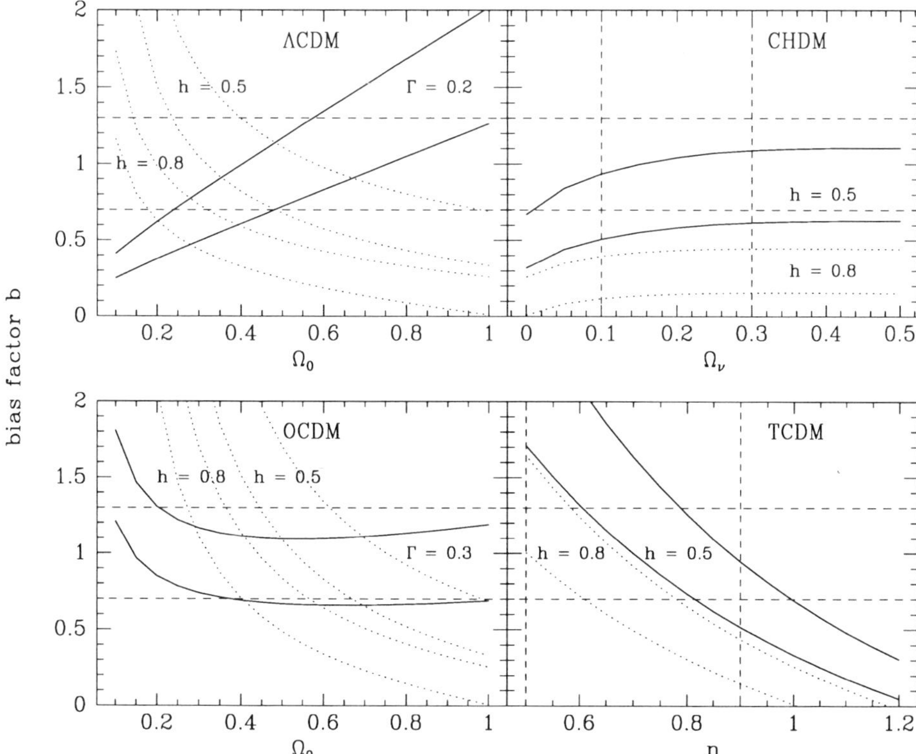


*Figure 4.* The bias factor $b$ of LCRS galaxies relative to various COBE-normalized CDM models. See text for details.

Finally, we use the LCRS results to put constraints on several classes of cold dark matter (CDM) models. For simplicity, we restrict attention to large scales $\lambda > 40\ h^{-1}$ Mpc in, at least approximately, the linear regime. On those scales, we also assume a constant linear bias factor $b$ and the Kaiser redshift-space amplification formula, so that the redshift-space galaxy $P(k)$ is just proportional to the real-space, linear mass power spectrum, as given by various analytical models (see Lin *et al.* 1996b and references therein). We then fit the LCRS results to these CDM models, normalized to COBE, and derive the resulting bias factor $b$ of LCRS galaxies relative to the mass. Four basic CDM variants are considered: (1) ΛCDM, flat CDM with nonzero cosmological constant $\Omega_\Lambda = 1 - \Omega_0$; (2) OCDM, open CDM with $\Omega_0 < 1$; (3) CHDM, flat CDM with massive neutrinos of density $\Omega_\nu$; and (4) TCDM, flat CDM with a tilted primordial spectral index $n$. For each of these CDM classes, Fig. 4 plots the $2\sigma$ error bounds on $b$ for the LCRS, as a function of $\Omega_0$, $\Omega_\nu$, or $n$, and for fixed Hubble constants $h = 0.5$ and $h = 0.8$. For ΛCDM and OCDM we find a best fit shape parameter $\Gamma = 0.3 \pm 0.1$, for CHDM we find $\Omega_\nu = 0.2 \pm 0.1$, and for TCDM we get $n = 0.7 \pm 0.2$; these bounds are indicated by corresponding curves or vertical lines in Fig. 4. Horizontal lines are also drawn at $b = 0.7$ and 1.3 to indicate a "reasonable" range of bias values; *cf.* the earlier type-dependent clustering differences. When combined with cluster abundance constraints (not shown), the LCRS data favor the following parameter choices:

$$\begin{array}{llll} \Lambda\mathrm{CDM}: & \Omega_0 \approx 0.4-0.5 & \Gamma \approx 0.2 & h \approx 0.5 \\ \mathrm{OCDM}: & \Omega_0 \approx 0.4-0.6 & \Gamma \approx 0.3 & h \approx 0.5-0.8 \\ \mathrm{CHDM}: & \Omega_0 = 1 & \Omega_\nu \approx 0.2-0.3 & h \approx 0.5 \\ \mathrm{TCDM}: & \Omega_0 = 1 & n \approx 0.7-0.8 & h \approx 0.5 \end{array} \tag{2}$$

Though we have constrained the available parameter space, clearly there are a number of viable models that survive, and we have not yet exhausted all the theoretical possibilities. Hopefully, the observational constraints should improve as the next generation of very large redshift surveys like 2DF and Sloan (A. Szalay, this volume) come on line, and are combined with improved microwave background measurements, particularly from planned satellite missions.

## 3. The Luminosity Function, Galaxy Populations, and Galaxy Evolution

We turn now to the LCRS luminosity function (LF) and related topics. The LF for a sample of 18678 LCRS galaxies has been measured (the largest sample for which this has been done) in a "hybrid" Kron-Cousins $R$-band, resulting in Schechter parameters $M_R^* = -20.29 \pm 0.02 + 5 \log h$, $\alpha = -0.70 \pm 0.05$, and $\phi^* = 0.019 \pm 0.001\ h^3\ \mathrm{Mpc}^{-3}$ (see Lin *et al.* 199a for details). Fig. 5 compares our $R$-band results against $B$-band results from several other large local redshift surveys (using an empirical 1.1 mag shift to match $R$ to $B$). Earlier this conference, Ron Marzke has already given a detailed review of LF's from different surveys, so I will make only a few comments here. Most importantly, I think it is remarkable that the two largest surveys by *volume* in Fig. 5, LCRS and Stromlo-APM (Loveday *et al.* 1992), have such similar luminosity functions, both in normalization and shape. (Note also that although the LCRS faint-end slope is shallow, $\alpha = -0.7$, a flat slope $\alpha = -1$ is a better description for $M_R \gtrsim -18 + 5 \log h$.) Hopefully this means that we are approaching a fair measurement of the local luminosity function. Thus the higher normalization of the CfA LF (Marzke, Huchra, & Geller 1994) or the steeper $\alpha = -1.2$ for SSRS2 (da Costa *et al.* 1994a), may be the result of the smaller volumes and samples of those surveys (for CfA, the exact $M^*$ shift used is also important). On the other hand, the impact of low surface brightness galaxies on the LCRS (and Stromlo-APM) results is still an open question: specifically, whether the effect is

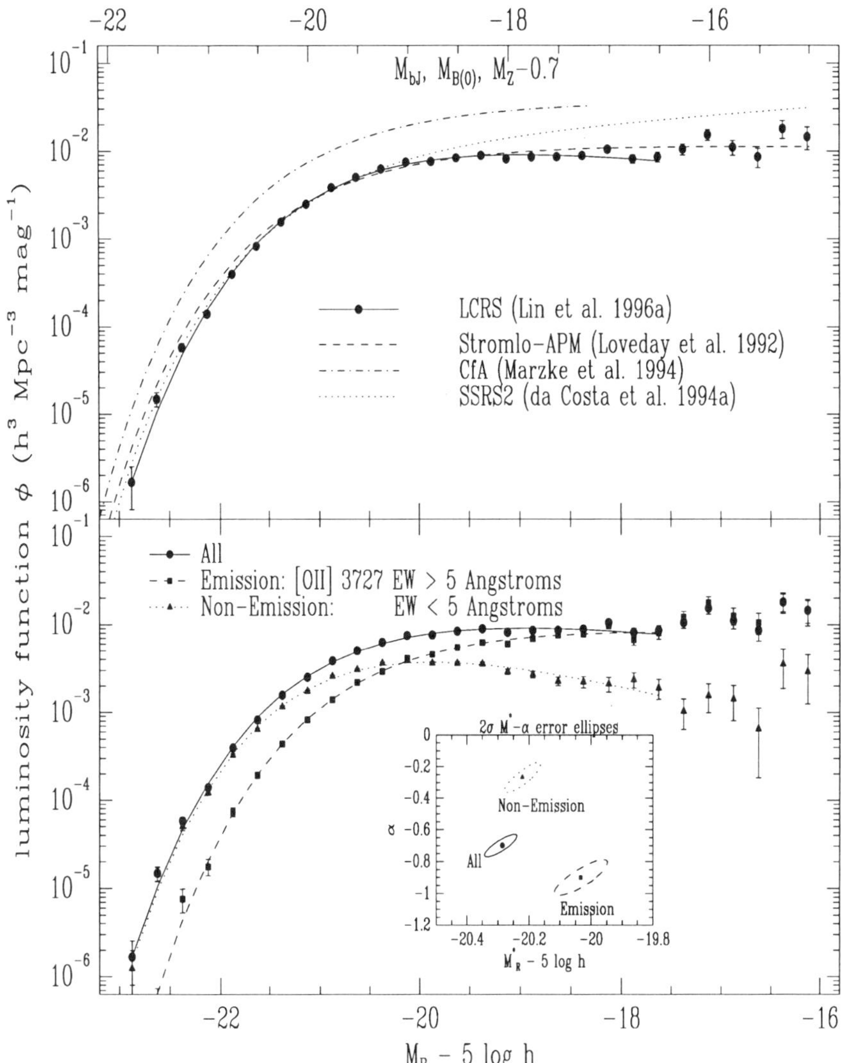


*Figure 5.* *Top:* Comparison of the LCRS $R$-band LF with the $B$-band LF's from the Stromlo-APM, CfA, and SSRS2 samples. *Bottom:* Comparison of the LF's of emission-line and non-emission-line galaxies in the LCRS. The inset shows the $2\sigma$ $M^*$ *vs.* $\alpha$ error ellipses.

relatively small (of order 20% or so and primarily at the faint end), or quite important (of order a factor of two). Work is in progress to better understand the influence that surface brightness selection may have on the LCRS luminosity function.

The large sample size and spectral database of the LCRS also allow us to explore the luminosity function and other properties of different galaxy

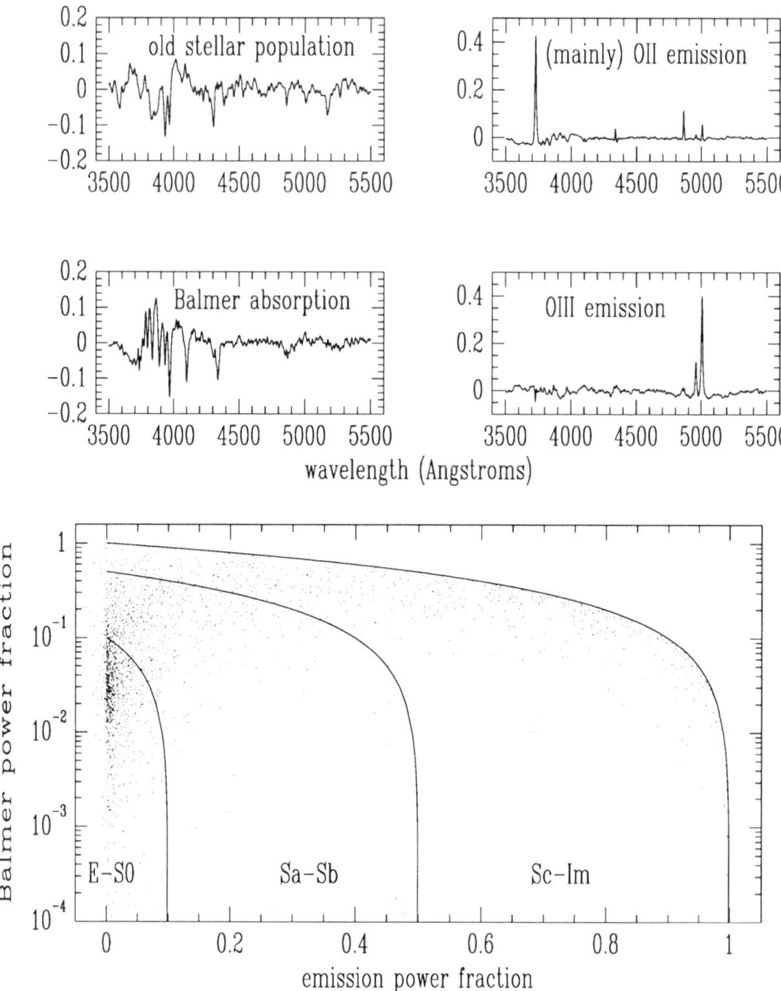


*Figure 6.* *Top:* The 4 spectral "basis vectors" used for galaxy spectral classification. *Bottom:* The spectral classification diagram for a sample of 2400 LCRS galaxies.

populations in some detail. Fig. 5 offers an example, showing the clear difference in the LF's of emission-line and non-emission-line galaxies in the LCRS. The trend that emission-line-strong, later-type, and bluer galaxies dominate the faint end of the LF, while emission-line-weak, ealier-type, and redder galaxies prevail at the bright end, appears to be borne out in a number of other local redshift surveys (*e.g.*, in the ESP sample, G. Zamorani, this volume, and see the review by R. Marzke).

Fig. 6 shows another example, the application of a spectral classification technique (due to W. Press & G. Rybicki) to the LCRS data. In this method one fits each galaxy spectrum to a linear combination of the four

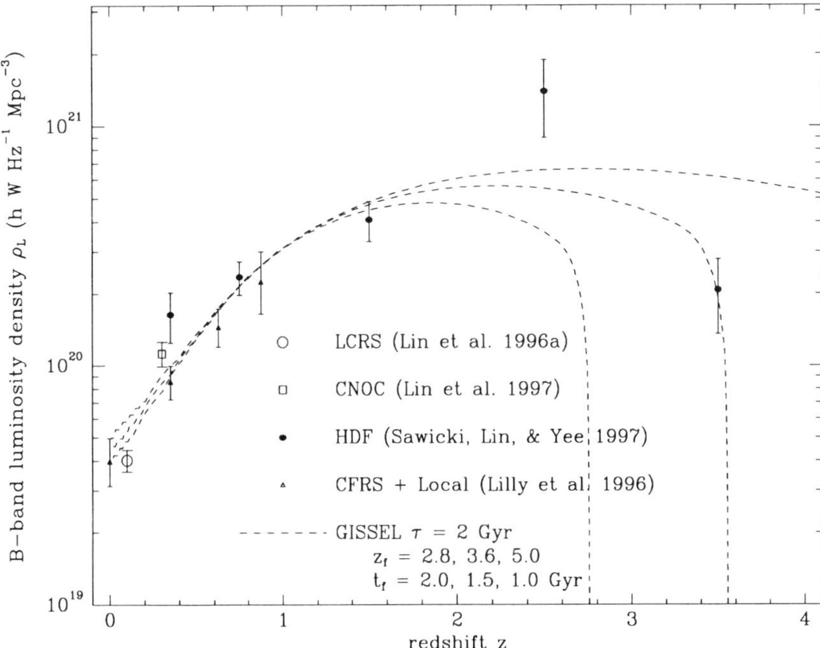


*Figure 7.* $B$-band luminosity density *vs.* redshift derived from the LCRS, CNOC, CFRS, and HDF samples. See text for details on the GISSEL models.

spectral "basis vectors" (derived from the LCRS data itself) plotted in Fig 6, and one then obtains a galaxy spectral classification from the fitted coefficients. Fig. 6 also shows the resulting classifications for a sample of about 2400 LCRS galaxies (just 1/10 of the total possible). Here the classifications "E-S0," "Sa-Sb," and "Sc-Im" have been roughly determined using the Kennicutt atlas galaxies (Kennicutt 1992). The technique offers a potentially powerful and robust means for classifying galaxies from different redshift surveys, and work is in progress to refine it using the LCRS data. Moreover, work has also begun to apply it to both field and cluster galaxy data from the CNOC redshift survey sample (see Ray Carlberg's contribution and Yee, Ellingson, & Carlberg 1996), currently the largest one at *intermediate* redshifts $z \sim 0.2 - 0.6$. The goal will be to understand the evolution of luminosity, spectral, clustering, and other properties of different galaxy populations using the LCRS and CNOC databases.

Finally, we describe an example using the LCRS (and CNOC) data to help constrain the evolution of the galaxy luminosity density as a function of redshift. Fig. 7 shows the luminosity densities derived from the LCRS, CNOC (Lin *et al.* 1997), and CFRS (Lilly *et al.* 1996) surveys, over redshifts $z = 0 - 1$. To go to higher redshifts, we use the luminosity function derived from application of photometric redshift techniques (Sawicki, Lin, & Yee

1997) to the *UBRI* color data for the Hubble Deep Field (HDF) sample (Williams *et al.* 1996). Granted, caution is warranted in using photometric redshifts at high $z$. However, we have tried to be careful, and our results are corroborated by the available HDF spectroscopic redshift data up to $z \approx 3.5$, as well as by the agreement of our LF's with those from the CNOC and CFRS samples at intermediate redshifts $z \lesssim 1$ (see Sawicki, Lin, & Yee 1997 for details). Our HDF luminosity densities are then shown in Fig. 7, plotted up to $z = 4$. Taken at face value, Fig. 7 indicates that the galaxy luminosity density steadily increases with redshift, peaking at $z \approx 2 - 3$, and finally declining at $z \approx 3 - 4$. A simple interpretation of the results is that the initial, major epoch of star formation in the universe occurred for $z \approx 3 - 4$. This is illustrated using several simple galaxy spectral evolution models, derived using the GISSEL library (Bruzual and Charlot 1993), where we take star formation to exponentially decline with decay time $\tau = 2$ Gyr (roughly that of early spirals), and where star formation turns on at $t_f =$ 1.0, 1.5, or 2.0 Gyr after the Big Bang (choosing $\Omega_0 = 1$ and current age $t_0 =$ 14.5 Gyr). The $t_f = 1.5$ Gyr model, with corresponding formation redshift $z_f = 3.6$, appears to match the data best. Clearly, the models are overly simplistic, and caution is also advised in applying photometric redshift results at high redshifts, but nevertheless the results are intriguing. We are planning further work with infrared and morphological data for the HDF to improve our analysis. Hopefully, with the large databases available from surveys like LCRS at low redshifts and CNOC at intermediate redshifts, and from emerging samples like HDF at very high redshifts, we will soon come to a much better characterization and understanding of the evolution of galaxies and their luminosity and clustering properties.

## References

Bruzual A., G., & Charlot, S. 1993, ApJ, 405, 538
da Costa, L. N., *et al.* 1994a, ApJ, 424, L1
da Costa, L. N., Vogeley, M. S., Geller, M. J., Huchra, J. P., & Park, C. 1994b, ApJ, 437, L1
Davis, M., & Geller, M. J. 1976, ApJ, 208, 13
Feldman, H. A., Kaiser, N., & Peacock, J. A. 1994, ApJ, 426, 23
Fisher, K. B., Davis, M., Strauss, M. A., Yahil, A., & Huchra, J. P. 1993, ApJ, 402, 42
Kauffmann, G., Nusser, A., & Steinmetz, M. 1996, MNRAS, in press (astro-ph/9512009)
Kennicutt, R. C. 1992, ApJ, 388, 310
Lilly, S. J., Le Fèvre, O., Hammer, F., & Crampton, D. 1996, ApJ, 460, L1
Lin, H., Kirshner, R. P., Shectman, S. A., Landy, S. D., Oemler, A., Tucker, D. L., & Schechter, P. L. 1996a, ApJ, 464, 60
Lin, H., Kirshner, R. P., Shectman, S. A., Landy, S. D., Oemler, A., Tucker, D. L., & Schechter, P. L. 1996b, ApJ, 471, 617
Lin, H., Yee, H. K. C., Carlberg, R. G., & Ellingson, E. 1997, ApJ, 475, 494
Loveday, J., Peterson, B. A., Efstathiou, G., & Maddox, S. J. 1992, ApJ, 390, 338
Marzke, R. O., Huchra, J. P., & Geller, M. J. 1994, ApJ, 428, 43

Sawicki, M. J., Lin, H., & Yee, H. K. C., 1997, AJ, 113, 1
Shectman, S. A., Landy, S. D., Oemler, A., Tucker, D. L., Lin, H., Kirshner, R. P., & Schechter, P. L. 1996, ApJ, 470, 172
Williams, R. E., *et al.* 1996, AJ, 112, 1335.
Yee, H. K. C., Ellingson, E., & Carlberg, R. G. 1996, ApJS, 102, 269

# THE CNOC CLUSTER SURVEY

## $\Omega$, $\sigma_8$, $\phi(L, z)$ *RESULTS, & PROSPECTS FOR* $\Lambda$ *MEASUREMENT*

R. G. CARLBERG, H. K. C. YEE, H. LIN, C. W. SHEPHERD AND P. GRAVEL
*University of Toronto, Toronto ON, M5S 3H8 Canada*

E. ELLINGSON
*University of Colorado, CO 80309 USA*

S. L. MORRIS, D. SCHADE, J. E. HESSER, J. B. HUTCHINGS,

J. B. OKE
*National Research Council of Canada, Herzberg Institute of Astrophysics, Dominion Astrophysical Observatory, Victoria, BC, V8X 4M6, Canada*

R. ABRAHAM
*Institute of Astronomy, Cambridge CB3 OHA, UK*

M. BALOGH, G. WIRTH, F. D. A. HARTWICK, C. J. PRITCHET
*University of Victoria, Victoria, BC, V8W 3P6, Canada*

AND

T. SMECKER-HANE
*University of California, Irvine, CA 92717, USA*

**Abstract.** Rich galaxy clusters are powerful probes of both cosmological and galaxy evolution parameters. The CNOC cluster survey was primarily designed to distinguish between $\Omega = 1$ and $\Omega \simeq 0.2$ cosmologies. Projected foreground and background galaxies provide a field sample of comparable size. The results strongly support a low-density universe. The luminous cluster galaxies are about 10-30% fainter, depending on color, than the comparable field galaxies, but otherwise they show a slow and nearly parallel evolution. On the average, there is no excess star formation when galaxies fall into clusters. These data provide the basis for a simple $\Lambda$ measurement using the survey's clusters and the field data. The errors in $\Omega_M$, $\Omega_\Lambda$, $\sigma_8$ and galaxy evolution parameters could be reduced to a few percent with a sample of a few hundred clusters spread over the $0 < z < 1$ range.



*D. Hamilton (ed.), The Evolving Universe,* 135–153.

## 1. The $\Omega$ Problem

The total mass-to-light ratio of field galaxies multiplied with the field luminosity density gives, by definition, the mean mass density of the universe, $\rho_0$ (Oort 1958; Gunn 1978). The complications in the measurement are to provide unbiased mass estimates and to measure any luminosity difference between the galaxies for which total system masses are known and field galaxies. Small, dense systems, like individual galaxies and small groups, exhibit *luminosity segregation* where the luminosity distribution is more concentrated than the mass distribution (Bahcall, Lubin & Dorman 1995). The largest virialized structures — rich clusters — indicate $\Omega \simeq 0.2 - 0.4$, but these measurements may be biased by luminosity segregation and differential evolution of the very red cluster galaxies as compared to bluer field galaxies. On even larger scales, where overdense structures are still expanding, but retarded from the Hubble flow, bulk flows of galaxies measure the parameter $\Omega^{0.6}/b$, where $b$ is the assumed linear bias, $b = (\delta n/n)/(\delta\rho/\rho)$, relating the tracer galaxy number densities and the mass field density. The flow measurements tentatively indicated $\Omega \simeq 1$ (Dekel 1994; Strauss & Willick 1995), but recent increases in sample size and new analyses indicate possible compatibility with $\Omega \simeq 0.2 - 0.4$ (Fisher *et al.* 1994; Davis, Nusser & Willick 1996; Willick *et al.* 1996). The value $\Omega = 1$ is of considerable interest as the "natural" value (the "Dicke coincidence") for a universe that has expanded many e-folds, and as the value originally predicted by inflationary cosmology (Guth 1981; Bardeen, Steinhardt & Turner 1983; Turner 1995). If $\Omega = 1$ and all the dark matter falls into clusters along with galaxies (as it must if it is collisionless and cold), then rich clusters, which draw their mass from regions $10h^{-1}$ Mpc in radius, must have a total virialized mass-to-light ratio, $M_v/L$, about 2.5 to 5 times higher than the standard virial analysis gives. The cluster data did not rule out such high $M/L$ values in the field, due to unconstrained systematic errors in both $M_v$ and $L$.

The Canadian Network for Observational Cosmology (CNOC) sample was designed to create a dataset that allows complete internal control of most aspects of the cluster $\Omega$ estimate, especially luminosity segregation and differential evolution of cluster galaxies relative to the field. The cluster sample was chosen from the Einstein Medium Sensitivity Survey Catalog of X-ray clusters (Gioia *et al.* 1990; Henry *et al.* 1992; Gioia & Luppino 1994) to have a high X-ray luminosity, which helps guarantee that the clusters are reasonably virialized, suggests that they will contain many galaxies, and allows other cosmological measurements. The sample was augmented with A2390 to fill an RA gap. The clusters are at $z \sim 0.3$ so they have a significant redshift interval over which the surrounding field galaxies are nearly uniformly sampled in redshift. Observations were made at CFHT in 24 assigned nights in 1993 and 1994, of which 22 were usefully clear. The primary catalogs contain Gunn $r$ magnitudes and $g - r$ colors for 25,000 objects, plus radial velocity measurements at an average accuracy of 100 $\mathrm{km\,s^{-1}}$(Yee, Ellingson & Carlberg 1996) for a subsample of 2600. All results are reported for $H_0 = 100\,\mathrm{km\,s^{-1}\,Mpc^{-1}}$ and $\Omega_0 = 0.2$, $\Omega_\Lambda = 0$.

The overall CNOC program is encapsulated as follows:

- select clusters from an unbiased X-ray catalog of large $z$ range,
- observe clusters out to a radius with overdensity of $200\rho_c$ or lower,
- measure virial mass-to-light ratio,
- test and correct virial mass,
- test for $z$ dependence of $M_v/L$, measure $\phi(L, z)$ for clusters,
- measure $\phi(L, z)$ for the field sample,
- measure $n(> M, z)$, the cosmological density of clusters, and,
- measure the evolution of field clustering and clustering velocities.

The main results from this ongoing program of investigation are summarized below but discussed in detail in a series of papers. The observational methods, which are designed to efficiently measure redshifts of galaxies in the primary sample, are discussed in Yee, Ellingson & Carlberg (1996). The measurement of global quantities, such as the velocity dispersion and virial mass-to-light ratio, is done in Carlberg *et al.* (1996). The average mass and light profiles are compared in Carlberg, Yee & Ellingson (1997) to measure the biases of the virial mass-to-light ratio and correct the luminosities of cluster galaxies to field galaxy values. Carlberg *et al.* (1997a) shows that the same mass profile can be recovered from two independent and very different subsamples of the cluster data. The average mass profile is compared to a theoretical prediction in Carlberg *et al.* (1997b), finding an impressive agreement. The sample was designed such that the amplitude of the primordial density fluctuation spectrum could be measured on cluster scales, with the results given in Carlberg *et al.* (1997b). Some preliminary results of the CNOC field survey, now comprising about 2/3 of the 5000 primary sample redshifts, are given below.

## 2. CNOC Cluster Masses

The virial mass is $M_v = 3G^{-1}\sigma_1^2 r_v$, where $\sigma_1$ is the line-of-sight velocity dispersion and $r_v$ is the virial radius of the observed galaxies (Binney & Tremaine 1987). Deciding which galaxies in redshift space are cluster members is fundamentally ambiguous. That is, a cluster galaxy with a completely reasonable line-of-sight velocity of 1000 $\mathrm{km\,s^{-1}}$ appears in redshift space at $10h^{-1}\,\mathrm{Mpc}$ from the cluster's center of mass, far outside the virialized cluster and intermingled with field galaxies. This complication increases in severity as the cluster density declines with projected distance from the cluster center where our sample is specifically intended to probe. Our straightforward solution to this problem is to subtract the mean density of field galaxies in redshift space from the redshift space of the cluster (Carlberg *et al.* 1996). The resulting velocity dispersions are about 13% lower than precisely the same cluster data give without background subtraction. Gratifyingly, our velocity dispersions are in excellent agreement with those inferred from the mean X-ray temperatures (Mushotzky & Scharf 1997). Our estimate of $r_v$ uses a "ringwise" estimate, rather than the traditional "pointwise" estimate (Binney & Tremaine 1987). This allows us to include a selection function for the roughly rectangular window on the sky which outlines the fields and it helps to reduce noise in the estimated virial radius. The ringwise estimator overestimates the virial radius of the clusters because it does not account for their flattening. We use the mean of the ratio of the ringwise to the pointwise estimator, which is about 1.28, to make this correction. The ringwise estimator includes a small "softening," of one arcsec, to avoid a divergence when two galaxies are at the same radius.

The total galaxy luminosity of the cluster galaxies in ratio to $M_v$ gives $M_v/L$, an estimate of the total mass-to-light ratio. The average $k$-corrected Gunn $r$ $M_v/L$, for $M_r^k \leq -19$ mag but integrated to infinity with the luminosity function ($M_* = -20.3$ mag, $\alpha = 1.1$), is $289{\pm}50h\,\mathrm{M}_\odot/\,\mathrm{L}_\odot$ at $z = 0.31$. If corrected for pure luminosity evolution at the rate of 1 magnitude of brightening per unit redshift the mean Gunn $r$ $M_v/L$ extrapolates to $374 \pm 54h\,\mathrm{M}_\odot/\,\mathrm{L}_\odot$ at redshift zero. The mean standard error per cluster on average is 40%, although this can be reduced to 25% by eliminating the clusters with large errors. A major result is that $M_v/L$ is the same for all clusters once minimal passive evolution of the cluster galaxies is taken into account (Carlberg *et al.* 1996). For the 15 clusters displayed in Figure 1 $\chi^2 = 16.6$ which is about 28% probable for $\nu = 14$ degrees of freedom, that is, consistent with no intrinsic variation. X-ray (David, Jones, & Forman 1995) and weak lensing analyses (Tyson & Fischer 1995; Squires *et al.* 1996; Smail *et al.* 1997; Fischer & Tyson 1997) generally find results in accord with these.

## 3. Mass Profiles and Virial Mass Bias

Our observations extend over what is expected to be the entire virialized volume of the cluster. Analytic models (Gott & Gunn 1972) and simulations (Cole & Lacey 1996) find that the virialized mass is generally contained inside the surface where the mean interior density is approximately $200\rho_c$, which defines the radius $r_{200}$. The small extrapolation

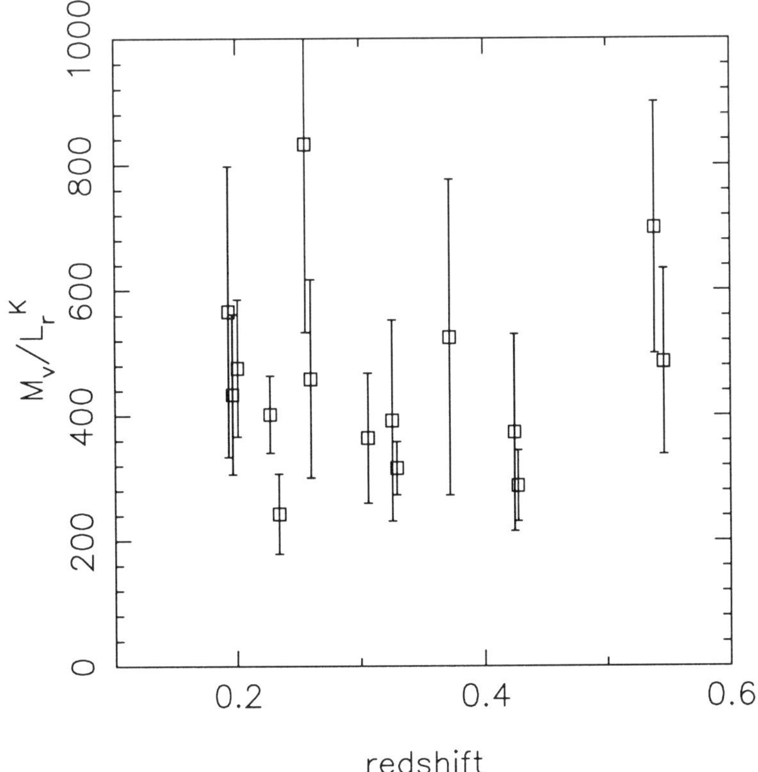


*Figure 1.* The virial mass-to-light ratios as a function of redshift. The luminosities have been $k$-corrected and allow for evolution at the rate of $M_*(z) = M_*(0) - Qz$, with $Q = 1$. The 15 clusters are consistent with having a universal $M_v/L(0) = 374h\,\mathrm{M}_\odot/\,\mathrm{L}_\odot$ with the same dataset giving a closure value $1502h\,\mathrm{M}_\odot/\,\mathrm{L}_\odot$, both with 10% standard errors.

from the observationally derived $r_v$ to $r_{200}$ assumes $M(r) \propto r$, which gives $r_{200} = \sqrt{3}\sigma_1/[10H(z)]$, where $H(z)$ is the Hubble constant at redshift $z$.

The accuracy of the virial mass depends on the tracer galaxies having the same distribution as the mass. We compare the light-traces-mass profile, $M_L(r) = L(r) \times M_v/L$, to the dynamical mass profile, $M_D(r)$, derived from the Jeans Equation,

$$\frac{\sigma_r^2}{r}\left[\frac{d\ln\sigma_r^2}{d\ln r} + \frac{d\ln\nu}{d\ln r} + 2\beta\right] = -\frac{GM_D(r)}{r^2}, \tag{1}$$

where $\beta = 1 - \sigma_\theta^2/\sigma_r^2$ is the velocity anisotropy parameter. This equation does not depend on the density of the tracer galaxies, $\nu(r)$, following the mass density profile, $\rho(r)$.

To create an effectively spherical cluster and to reduce the effects of substructure, all the cluster galaxies are combined into one "ensemble"

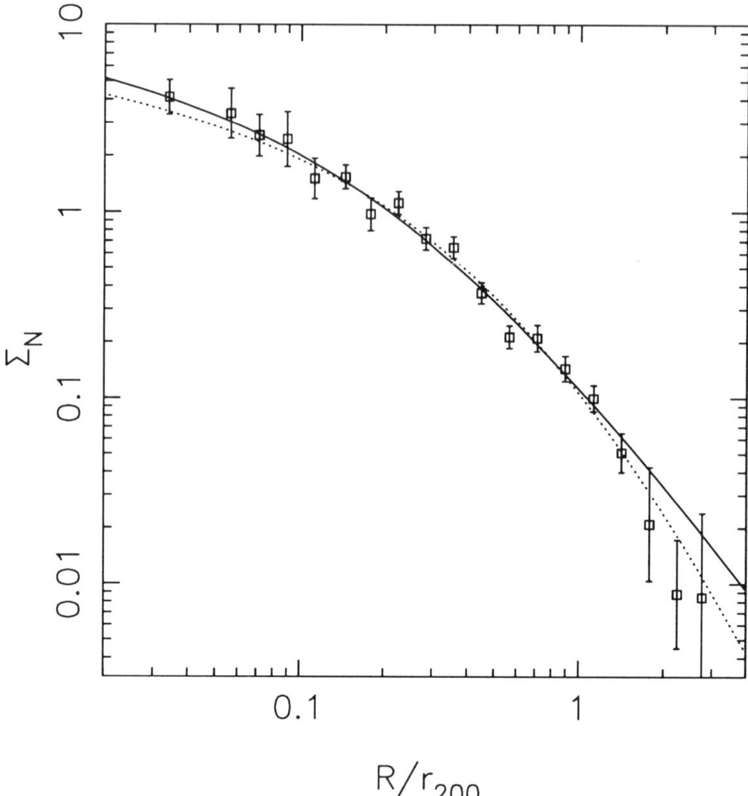


*Figure 2.* The background subtracted surface density profile, fitted with NFW (solid line) and Hernquist (dotted line) functions, both of which are statistically acceptable. The mass is accurately traced with this profile, as shown below. The remarkable result is that the NFW prediction of the scale radius is completely in accord with our measurements.

cluster by normalizing the velocities to $\sigma_1$ and the projected radii to the $r_{200}$ value of each cluster. The two quantities measured from this distribution are the surface number density profile of cluster galaxies, $\Sigma_N(R)$, shown in Figure 2, and the projected velocity dispersion profile, Figure 3. The volume number density profile is modeled as $\nu(r) = A/[r(r+a)^p]$, where $p = 2$ for the Navarro, Frenk & White function (1997, hereafter NFW) and $p = 3$ for Hernquist's (1990) form.

The radial velocity dispersion is modeled as $\sigma_r^2 = B[c_1 r/(1+c_1 r) + c_2]/[1+r/b]$, where $B$ and $b$ are the two parameters adjusted to fit the observed $\sigma_p(R)$. The $c_1$ and $c_2$ parameters are externally fixed to vary the shape of the curve. The velocity anisotropy is taken as an approximate fit to the results of n-body simulations. A range of $\sigma_r(r)$ fits are shown in Figure 3. The resulting ratio, $b_{Mv}(r) = M_D(r)/M_L(r)$ shown in Figure 4, is not very sensitive to the details of the velocity modeling.

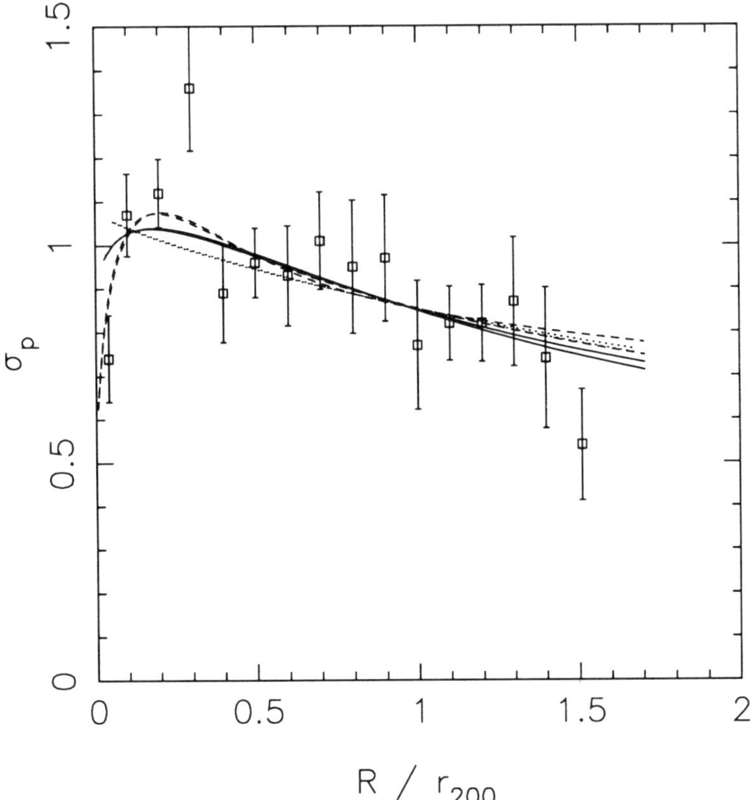


*Figure 3.* The projected velocity dispersion profile and the projection of the fitted profiles, for a range of $\beta$ and $[c_1, c_2]$. See Carlberg *et al.* 1997d for details.

There are two conclusions to be drawn from Figure 4. First, over a range of about two decades of projected density or three decades of volume density, the integrated galaxy distribution traces the mass distribution to an accuracy of about 20% or better. Second, the virial mass is always an overestimate of the true mass contained within that radius, which we estimate to be a factor $0.82 \pm 0.14$. We attribute the mass overestimate of $M_v$ to dropping the surface pressure term at $P_s = 0$ in the virial theorem, *i.e.* $2T + W = 3P_sV$. Simple modeling of an equilibrium cluster shows that truncating the data at $r_{200}$ will cause $M_v$ to be upward biased at the 15-25% level that is inferred from the Jeans Equation.

Splitting the cluster sample into independent blue and red subsamples demonstrates that the clusters are effectively in equilibrium and that the results are robust for drastically different samples. These two subsamples also illustrate the perils of the virial mass estimator. The blue galaxies are about twice as extended and have a 20% higher velocity dispersion than

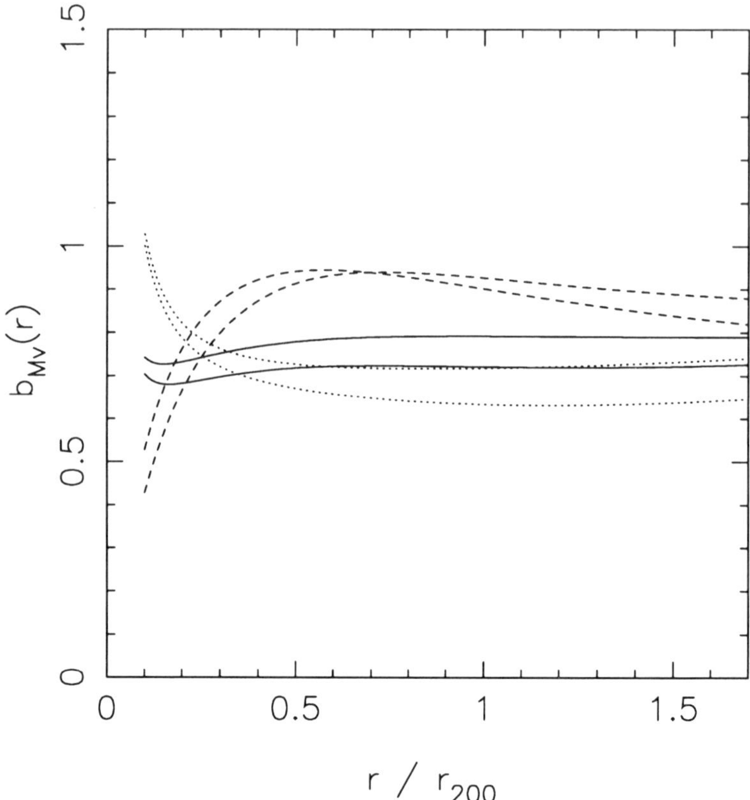


*Figure 4.* The derived ratio of the dynamical mass profile, $M_D(r)$, to $r$ selected galaxy profile, $L(r)$, normalized with the virial mass-to-light ratio evaluated inside $1.5r_{200}$. In each pair of curves the upper line at small radius is for $\beta_m = 0.3$ and the lower for $\beta_m = 0.5$. The dotted line is for $c_1 = 0$, the dashed for $c_2 = 0$ and the solid line is our preferred $c_1 = 8$, $c_2 = 1/2$.

the red galaxies, Figure 5. Consequently the blue galaxies indicate a virial mass about three times larger than the red galaxies give. However each of these subsamples independently gives the same mass profile from the Jeans equation as found from the full sample, which we take as strong support for the assumption that both subsamples are effectively in equilibrium with the cluster potential (Carlberg *et al.* 1997a). It should be noted that galaxies with red colors are somewhat more concentrated than the total mass distribution, so the procedure of comparing the mass profile to the "red-sequence" light profile is likely to give a rising mass-to-light ratio.

The scale radius, $a$, is predicted to be a relatively large 0.20-0.26 of $r_{200}$, for simulations normalized to the observed cluster cosmological density, but otherwise the prediction is relatively insensitive to cosmological parameters (NFW). It is a significant success of the generalized CDM model that the

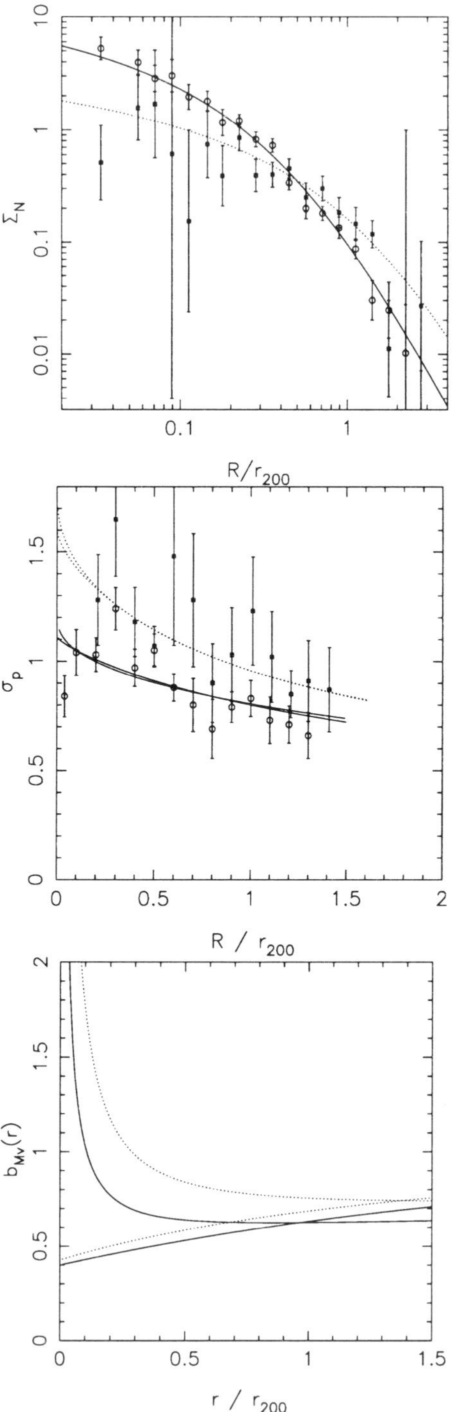


*Figure 5.* The projected number density profiles (upper panel) and velocity dispersion profiles (middle panel) of the blue (solid squares and dotted lines) and red (circles and solid lines) cluster galaxies. They yield *identical* mass profiles (lower panel) near $r_{200}$, where the upper pair of lines are for $\beta = 0$ and the lower pair are for $\beta = 0.5$.

observed scale radius $a = 0.13-0.43$ (95% confidence) is in complete accord with NFW's prediction.

## 4. Normalization of the Density Perturbation Spectrum, $\sigma_8$

The normalization of the density fluctuation spectrum is fundamental to all predictions of structure formation. The parameter $\sigma_8$ is the variance of the linear density perturbation spectrum in $8h^{-1}$ Mpc spheres. The CNOC cluster sample and observational strategy (Yee, Ellingson & Carlberg 1996) were specifically designed to produce data useful for a $\sigma_8$ measurement. The sample's primary advantage is that the cluster masses are accurately known near the virial radius, which is essential for a reliable estimate of the linear mass scale from which the cluster collapsed. We combine our data with similarly selected clusters from the EMSS (Henry *et al.* 1992) and ESO Cluster surveys (Mazure *et al.* 1996) to extend the redshift range of the cosmological density estimates. The cluster cosmological density data are modeled (Press & Schechter 1974) to constrain the $\sigma_8 - \Omega$ pair. It is not observationally straightforward to give accurately the full virialized mass of a cluster, but it is straightforward to extrapolate the mass inside some specified aperture, conventionally the Abell radius, $1.5h^{-1}$ Mpc. The measurement of cluster mass inside a fixed aperture is effectively a mass weighted velocity dispersion or temperature.

The measured redshift change of the cosmological density of clusters, having $\sigma_1 \gtrsim 800\,\mathrm{km}\,s^{-1}$ requires some small corrections for the effects of X-ray selection, which are incorporated in the data of Figure 6. The analysis is greatly simplified because the $L_x - \sigma_1$ relation has no detectable redshift evolution, as is also seen in the X-ray temperatures (Carlberg *et al.* 1997b; Mushotzky & Scharf 1997). The cluster cosmological density data are best described with $\sigma_8 \simeq 0.75 \pm 0.1$ and $\Omega \simeq 0.4 \pm 0.2$ (90% confidence), Figure 7. Taking the $\Omega$ from cluster dynamical analysis as $\Omega = 0.2$, we find $\sigma_8 = 0.95 \pm 0.1$ (90% confidence). The predicted cluster density evolution in an $\Omega = 1$ CDM model exceeds that observed at $z > 0.5$ by more than an order of magnitude, as shown in Figure 6.

## 5. Galaxy Evolution

It is essential for $\Omega$ measurement to understand the differential evolution of cluster galaxies relative to field galaxies. The cluster galaxy population is on the average redder than the field, but the cluster population has an increasing blue fraction with redshift (Butcher & Oemler 1984). The origin of the blue cluster population has three basic possibilities. One possibility is cluster galaxies formed in a different way than field galaxies, with fundamentally different star formation rates at all epochs. How-

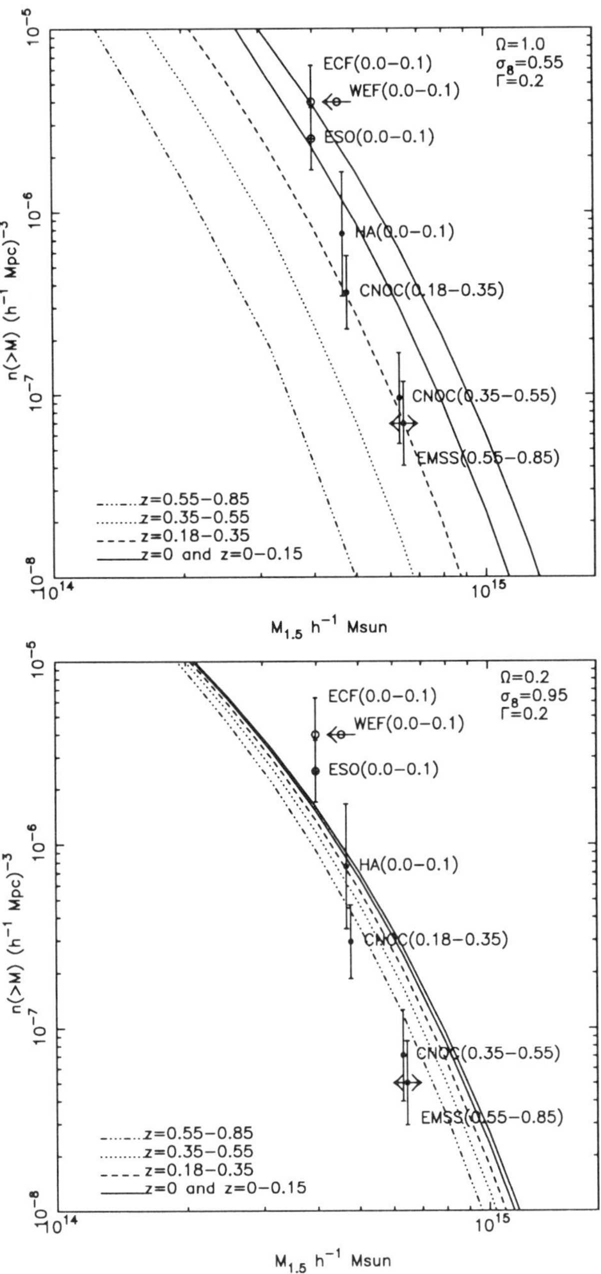


*Figure 6.* The measured cosmological density evolution of galaxy clusters with $\Omega = 1$ (upper) and $\Omega = 0.2$ (lower) predictions.

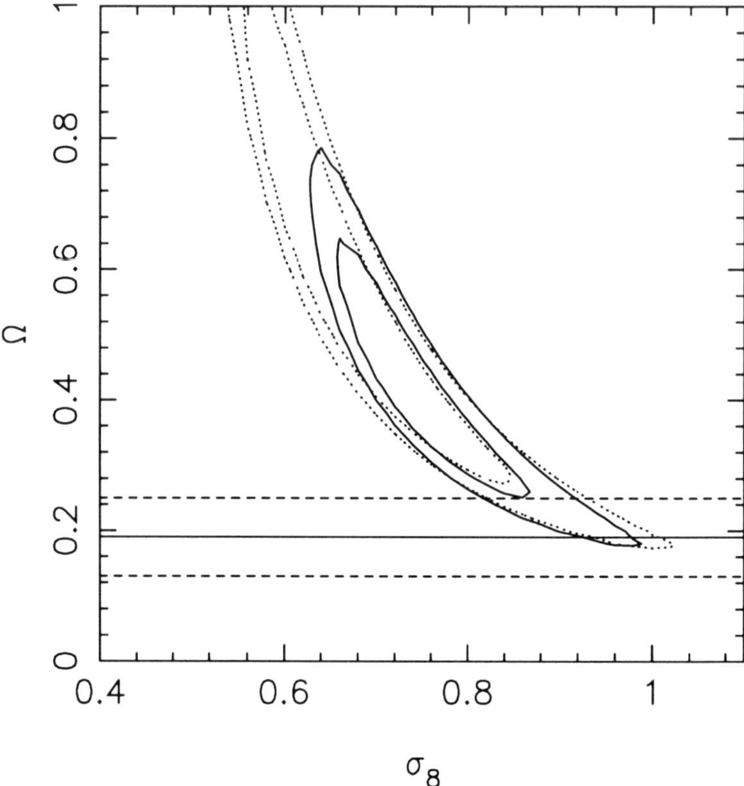


*Figure 7.* A plot of $\chi^2$ for all independent samples (solid lines) and excluding the high redshift EMSS sample (dotted lines). The contours are the 90% and 99% confidence levels. The results of the CNOC analysis, $\Omega = 0.19 \pm 0.06$, with its $1\sigma$ range are indicated.

ever, hierarchical clustering leads one to to expect that cluster galaxies originated in the field, and hence share a similar early star formation history which is altered in some way upon entry into the cluster. Ultimately star formation is greatly suppressed in cluster galaxies: however, there are two routes to this state. Either there was a burst of star formation upon entry into the cluster which largely exhausted the galaxy's gas supply, or, the gas was simply swept out of the galaxy (Baade & Spitzer 1951; Gott & Gunn 1972).

In Figure 8 we show the equivalent width of the [OII]3727 line as a function of radial distance from the cluster center, which is the projected distance in the inner region and the redshift space distance for the field galaxies. The sample is limited at a common $M_r^k \leq -18.5$ mag, with an additional cut to eliminate low signal-to-noise spectra. The measured equivalent width implies a mean star formation rate of approximately $0.05h^{-2}\,\mathrm{M}_\odot\,\mathrm{yr}^{-1}$ in our clusters (Kennicutt 1992). Over $10^{10}$ years this will

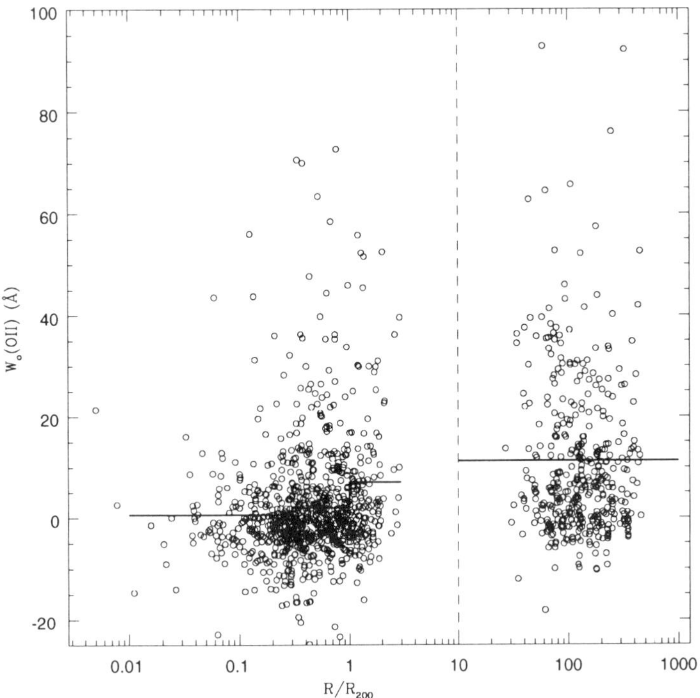


*Figure 8.* The radial dependence of the equivalent width of [OII] for galaxies with $M_r^k \leq -18.5$ (Balogh *et al.* 1997 in preparation). The lines give the mean values in various radial ranges. Nowhere do cluster galaxies show an excess of star formation over the field.

add only a few percent stellar mass to the relatively high luminosity galaxies in our sample. Of particular note is that there is no evidence for a significant increase in star formation rate at any radius. Clusters do contain objects with strong emission lines, mostly near the cluster "edge", but these are rare compared to the field. We conclude that the difference between cluster and field galaxy evolution is that star formation is truncated upon the entry of galaxies into the cluster (Couch & Sharples 1987; Abraham *et al.* 1996).

The evolution of the cluster galaxy population should be dominated by a passively evolving stellar population. The CNOC survey has the advantage that the masses derived for each cluster can be used to normalize the luminosity functions of each cluster so that the redshift evolution of $L/M$ can be measured. The evolution of the parameters of Schechter fits to the luminosity functions of galaxies having $M_B(AB)$ more luminous than approximately $-17$ mag is shown in Figure 9. The $Q$ parameter

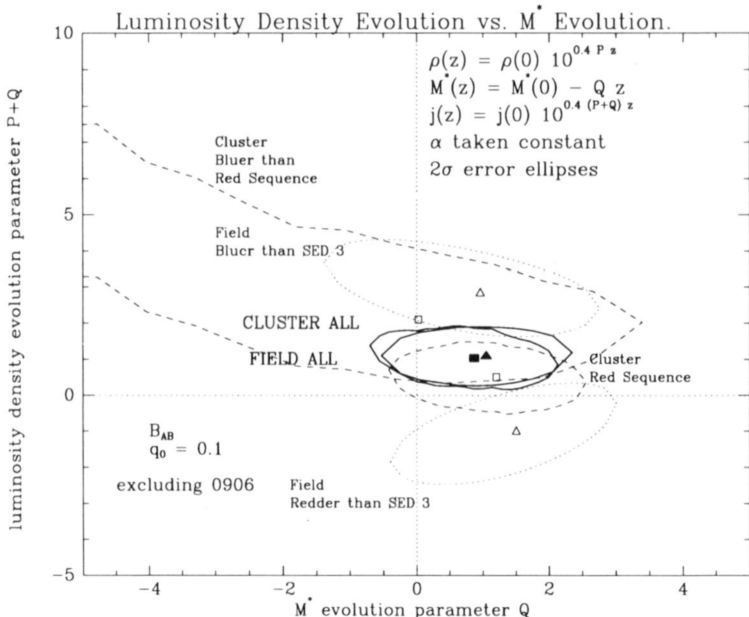


*Figure 9.* The $B_{AB}$-band evolution of $M_*$ and $j$ in magnitudes per unit redshift (Lin *et al.* 1997 in preparation). The solid lines give the 95% confidence contours for the full samples. The dotted and dashed lines are for blue and red subsamples. The cluster data are, unsurprisingly, consistent with passive evolution. The field galaxy evolution is nearly identical to the cluster galaxy evolution.

measures the evolution of $M_*$ with redshift (the faint end slope is held fixed at $\alpha = -0.83$, the best fit value) as $M_* = M_*(0) - Qz$. The evolution of the luminosity density, $j(z)$, is nearly statistically orthogonal to $M_*$. We model $j(z)$ as an evolution in magnitudes, $j(z) = j(0)10^{0.4(P+Q)z}$. These luminosity function and evolution parameters are found using a parametric maximum likelihood technique (Sandage, Tammann & Yahil 1979; Saunders *et al.* 1990) applied to 1838 field galaxies and 1018 cluster galaxies above the sample limits.

We find that to a good approximation the density evolution parameters $P \simeq 0.17$ (consistent with zero) indicating that there is little density evolution of cluster galaxies, as we expect for these high velocity dispersion clusters where merging and star formation are insignificant. We find $Q \simeq 0.87$, which an accord with predictions of purely passive evolution of a predominantly old stellar population of current at ~15 Gyr today. In the field ($\alpha = -0.93$), we find $Q \simeq 1.05$ and $P \simeq 0.04$, nearly identical to the cluster values. Nearly identical values are inferred from this approach to the analysis of the CFRS data (Lilly *et al.* 1995; Lin, *et al.* 1997; Lilly *et al.* 1996). The mean field star formation rate that we infer from

[OII] over the $0.2 \le z \le 0.6$ interval is only about $0.3\,\mathrm{M}_\odot\ \mathrm{yr}^{-1}$ for galaxies more luminous than $0.2L_*$ with a weak redshift dependence. This low mean star formation rate would add about 5-10% more stellar mass to an $M_*$ galaxy (close to the sample mean and median) over this redshift interval, consistent with a mild field luminosity evolution and a modest difference between cluster and field. These results are supported by direct, cosmological model independent, measurements of the surface brightness of the galaxy spheroids and disks (Schade *et al.* 1996a; 1996b).

We conclude that field galaxies evolve roughly in parallel to cluster galaxies, although cluster galaxies have lower luminosity $M_*$, by $0.3 \pm 0.1$ mag in B, or using different procedures, $0.11 \pm 0.05$ mag in $r$. Simple stellar population modeling for truncated star formation shows that this is consistent with the mean color difference of about 0.2 mag. The fading of cluster galaxies relative to the field is an essential, but relatively small, correction in the $\Omega$ estimate. We interpret the Butcher-Oemler effect as likely being the result of blue, star forming, field galaxies falling into clusters at a rate that increases with redshift, although this remains to be quantified and tested.

## 6. A Neo-Classical $\Lambda$ Estimator

Cluster data allow a measurement of the geometry of the universe, using the following procedure. The mass density of the universe, $\rho_0$, is a conserved quantity which is estimated with the product of $M/L(z)$ and $j(z)$. The $M/L(z)$ is estimated from the clusters, with a relatively small correction, 10-30%, to allow for differential evolution with respect to the field. The luminosity density, $j(z)$, involves the volume element, which is strongly $q_0$ dependent. If these quantities are calculated using the values of $\Omega_0$ and $\Omega_\Lambda$ that the real universe has, then $\rho(z) = M/L(z) \times j(z)$ will be constant at $\rho_0$. However, if the "wrong" values are assumed, then the calculated $\rho(z)$ will be dependent on redshift. For instance, if we assume that $\Omega_0 = 1$, but in fact $\Omega_0 = 0.2$, $\Omega_\Lambda = 0.8$, then $\rho(z)$ will rise 76% from z=0.2 to z=0.8, which is easily detectable in samples of the CNOC size but having a larger redshift range. Note that this effect exceeds the entire differential evolution between field and cluster by a factor of two, and that the redshift variation in the differential evolution is even smaller. Although we will correct for galaxy evolution effects, they are small compared to the geometry changes we are trying to measure.

## 7. Discussion and Future Directions

The main goal of the CNOC survey is to use clusters of galaxies to derive a value of $\Omega_0$ with a well determined error budget. The major innovation of our analysis is that it is completely self-contained, with the key assumptions being testable, and that the error estimates are derived from the data themselves. The dominant source of error is random cluster-to-cluster variations, rather than the internal error from individual clusters. The global mass-to-light ratio (in our photometric system, at mean redshift of $z = 0.31$, with k-corrections, but without evolution corrections) of our sample clusters of galaxies is constant within our typical errors of 25% at a value of $289 \pm 50h\,\mathrm{M}_\odot/\,\mathrm{L}_\odot$. Over the same redshift range we measure the closure value, $\rho_c/j$, to be $1136 \pm 138h\,\mathrm{M}_\odot/\,\mathrm{L}_\odot$ (Carlberg *et al.* 1996). After allowing for the $0.11 \pm 0.05$ mag lower luminosities of cluster galaxies and reducing $M_v$ by $0.82 \pm 0.14$, we find that $\Omega_0 = 0.19 \pm 0.06$. In an independent luminosity function analysis, the evolution corrected $B_{AB}$ value of $M_v/L$ leads to $\Omega_0 \simeq 0.23$ with a similar error budget.

There are no variations of the radial mass-to-light profile within the clusters, nor is there any variation with redshift. This requires any additional form of dark matter to be sufficiently hot that it does not fall into clusters, but perhaps participates in large scale flows. This is a very narrow parameter range and produces trouble for the evolution of the density perturbation spectrum. The most likely form of additional mass-energy, if it exists, is in the form of $\Lambda$, which additional cluster plus field observations in a neo-classical volume-redshift test can easily detect.

Two aspects of the cluster mass evolution are particularly satisfying. The derived $\Omega_0$ accurately predicts the observed evolution of cluster cosmological density. With our value of $\Omega_0$ we find that $\sigma_8 = 0.95 \pm 0.1$. Moreover the same $\Omega$ in a cool, dark matter dominated universe, predicts remarkably accurately the "morphology" of the clusters, at least as encapsulated in the NFW results for the mean mass profile. For this low density cosmology the measured rate of both structural evolution, Figure 10 (Carlberg *et al.* 1997c), and galaxy evolution, Figure 9, over our redshift range is consistent with a formation "freeze out" at some $z > 1$, and possibly much higher.

There are considerable opportunities to refine and extend these results in a number of different directions. For the current sample these include improved coverage at 1-3$r_{200}$ to constrain the luminosity profile at large radius, which would lead to a tighter comparison with the core radius predictions and the slope at large radius. The same data would advance the empirical study of how field galaxies are altered upon infall into the cluster. The turnaround radius is at about $5r_{200}$. If the sample extended to projected radii well beyond turnaround, then there would be a clear measurement of the infall into the cluster, which is a "bulk flow", hence measures the $\beta = \Omega^{0.6}/b$ parameter. Such a measurement would directly compare the cluster $\Omega$ estimate with a field $\Omega^{0.6}/b$ measurement to resolve the large scatter present in the current estimates from flows. At these large radii many of

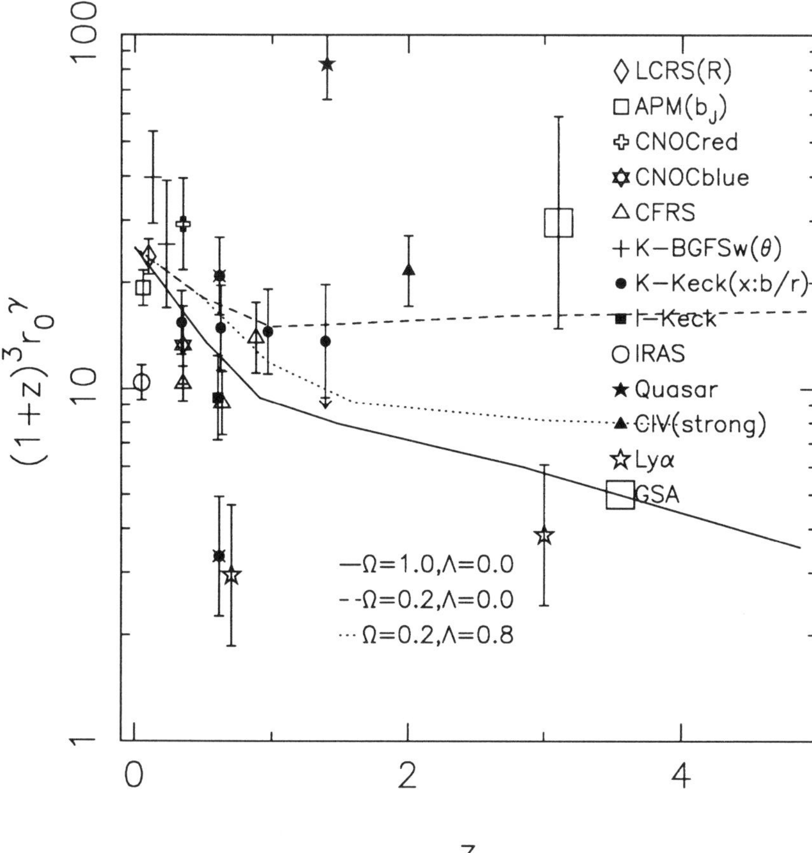


*Figure 10.* The evolution of the physical density of clustering with redshift from various samples. Red galaxies are more correlated than blue ones, notably the K-Keck Carlberg *et al.* 1997 and CNOC2 field (Shepherd *et al.* 1997, in preparation) samples. There is no compelling evidence for any change of physical density of clustered galaxies with redshift.

the galaxies would be distant field galaxies, but their immense value is to provide a good estimate of the unperturbed background density, which is an essential ingredient in the measurement. Including the four high redshift EMSS clusters would allow an $\Omega_\Lambda$ measurement with a precision of about $\Delta\Omega_\Lambda \simeq 0.25$ to be made. With an efficient multi-object spectrograph on a 4 meter telescope, it takes 1-2 nights to observe a cluster at $z \simeq 0.8$.

More ambitiously one can imagine in the near future assembling much larger cluster samples to improve the precision of these cosmological parameter estimates to the few percent level that Cosmic Microwave Background experiments hope to attain. For instance, a sample of 200-300 clusters, with the accompanying field, spread more or less uniformly over the $0 < z < 1$ range would allow $\Omega_M$ to be measured to a precision of about 2-3% and

$\Omega_\Lambda$ to be measured to about 5%. The $\sigma_8$ parameter could be measured to a precision of better than 1%, which would become a very strong constraint on the overall spectrum of fluctuations. The galaxy evolution parameters would be measured at about 5% precision and would challenge stellar population models, and might even begin to rival globular clusters as tools to measure the age of the universe. However, this would also take some advances in modeling precision. The realistic errors are likely to be about twice as large, since new residual systematic errors will be uncovered and removed as the dataset grows. At redshifts beyond about $z = 0.5$ the first problem is to find clusters in some systematic manner. This is mainly a matter of systematic sky surveys of various sorts. It should be noted that, in the $0.2 \leq z \leq 0.55$ range of the EMSS/CNOC survey, we selected only 15 clusters that are sufficiently rich that they are easy to study from 584 square degrees of sky. The implication is that these large, rich, easily-studied clusters are rather scarce on the sky. Nevertheless, once found, clusters of this sort will be exceptionally powerful as direct probes of both cosmological processes and cosmological parameters. Such programs will be feasible with a number of the new generation of telescopes.

## References


Abraham, R. G., Smecker-Hane, T. A., Hutchings, J. B., Carlberg, R. G., Yee, H. K. C., Ellingson, E., Morris, S. L., Oke, J. B., Davidge, T. 1996, ApJ, 471, 694

Baade, W. & Spitzer, L. 1951, ApJ, 113, 413

Bahcall, J. & Tremaine, S. D. 1981, ApJ, 244, 805

Bahcall, N. A., Lubin, L. M., & Dorman, V. 1995, ApJ(Lett), 447, L81

Bardeen, J. M., Steinhardt, P. J. & Turner, M. S. 1983, Phys. Rev. D., 28, 679

Binney, J. & Tremaine, S. 1987, *Galactic Dynamics*, (Princeton Uni. Press: Princeton)

Bruzual, G. & Charlot, S. 1993, ApJ, 405, 538

Butcher, H. & Oemler, A. 1984, ApJ, 285, 423

Carlberg, R. G., Yee, H. K. C., Ellingson, E., Abraham, R., Gravel, P., Morris, S. L., & Pritchet, C. J. 1996, ApJ, 462, 32

Carlberg, R. G., Yee, H. K. C., & Ellingson, E. 1997, ApJ, in press

Carlberg, R. G., Yee, H. K. C., Ellingson, E., Morris, S. L., Abraham, R., Gravel, P., Pritchet, C. J., Smecker-Hane, T., Hartwick, F. D. A., Hesser, J. E., Hutchings, J. B., & Oke, J. B. 1997a, ApJ(Lett), 476, L7

Carlberg, R. G., Morris, S. L., Yee, H. K. C., & Ellingson, E. 1997b, ApJ, in press (astroph/9612169)

Carlberg, R. G., Cowie, L. L., Songaila, A., & Hu, E. M. 1997c, ApJ, in press (astroph/9605024)

Carlberg, R. G., Yee, H. K. C., Ellingson, E., Morris, S. L., Abraham, R., Gravel, P., Pritchet, C. J., Smecker-Hane, T., Hartwick, F. D. A., Hesser, J. E., Hutchings, J. B., & Oke, J. B. 1997d, ApJ(Lett), submitted (astroph/9703107)

Couch, W. J. & Sharples, R. M. 1987, MNRAS, 229, 423

Cole, S. & Lacey, C. G. 1996, MNRAS, in press (astroph/9510147)

David, L. P., Jones, C., & Forman, W. 1995, ApJ, 445, 578

Davis, M., Nusser, A. & Willick, J. A. 1996, ApJ, 473, 22

Dekel, A. 1994, ARAA, 32, 371

Fischer, P. & Tyson, J. A. 1997, AJ, in press (astroph/9703189)

Fisher, K. B., Davis, M., Strauss, M. A., Yahil, A., & Huchra, J. P. 1994, MNRAS, 267, 927
Gioia, I. M. & Luppino, G. A. 1994, ApJS, 94, 583
Gioia, I. M., Maccacaro, T., Schild, R. E., Wolter, Stocke, J. T., Morris, S. L., & Henry, J. P. 1990, ApJS, 72, 567
Gott, J. R. & Gunn, J. 1972, ApJ, 176, 1
Gott, J. R. & Turner, E. L 1976, ApJ, 209, 1
Gunn, J. E. 1978, in *Observational Cosmology*, eds. Maeder, A., Martinet, L, & Tammann, G. 1978 (Geneva Observatory: Sauverny)
Guth, A. 1981, Phys. Rev. D., 23, 34
Henry, J. P., Gioia, I. M., Maccacaro, T., Morris, S. L., Stocke, J. T., & Wolter, A. 1992, ApJ, 386, 408
Hernquist, L. 1990, ApJ, 356, 359
Kennicutt, R. C. Jr. 1992, ApJ, 388, 310
Lilly, S. J., Tresse, L., Hammer, F., Crampton, D., & Le Fèvre, O. 1995, ApJ, 455, 108
Lilly, S. J., Le Fèvre, O., Hammer, F., & Crampton, D. 1996, ApJ(Lett), 460, 1L
Lin, H., Yee., H. K. C., **Carlberg**, R. G., & Ellingson, E. 1997, ApJ, 475, 494
Luppino, G. A. & Gioia, I. M. 1995, ApJ(Lett), 445, 77L
Mushotzky, R. F. & Scharf, C. A. 1997, ApJ(Lett), in press (astroph/9703039)
Mazure, A., Katgert, P., den Hartog, R., Biviano, A., Dubath, P., Escalara, E., Focardi, P., Gerbal, D., Giuricin, G., Jones, B., Le Fèvre, O., Moles, M., Perea, J., & Rhee, G. 1996, AAp, 310, 31
Navarro, J. F., Frenk, C. S. & White, S. D. M. 1997, ApJ, submitted (astro-ph/9611107) NFW
Oort, J. H. 1958, in *La Structure et L'Évolution de L'Univers*, Onzième Conseil de Physique, ed. R. Stoops (Solvay: Bruxelles) p. 163
Press, W. H. & Schechter, P. 1974, ApJ, 187, 425
Sandage, A., Tammann, G. A., & Yahil, A. 1979, ApJ, 232, 352
Saunders, W., Rowan-Robinson, M., Lawrence, A., Efstathiou, G., Kaiser, N., Ellis, R. S., & Frenk, C. S. 1990, MNRAS, 242, 318
Schade, D., Carlberg, R. G., Yee, H. K. C., López-Cruz, O. & Ellingson, E. 1996a, ApJ(Lett), 464, L63
Schade, D., Carlberg, R. G., Yee, H. K. C., López-Cruz, O. & Ellingson, E. 1996b, ApJ(Lett), 465, L103
Smail, I., Ellis, R. S., Dressler, A., Couch, W. J., Oemler, A. Jr., Sharples, R. M., & Butcher, H. 1997, ApJ, 479, 70
Squires, G., Neumann, D. M., Kaiser, N., Fahlman, G., Woods, D., Babul, A., & Böhringer, H. 1996, ApJ, 469, 73
Strauss, M. A. & Willick, J. A. 1995, Physics Reports, 261, 271
Turner, M. 1995, Ann. NY Acad. Sci. 759, 153 (astroph/9703194)
Tyson, J. A. & Fischer, P. 1995, ApJ(Lett), 446, 55L
Willick, J. A., Strauss, M. A., Dekel, A., & Kolatt, T. 1996, ApJ, submitted (astroph/9612240)
Yee, H. K. C., Ellingson, E. & Carlberg, R. G. 1996, ApJS, 102, 269

# RESULTS FROM THE NORRIS SURVEY OF THE CORONA BOREALIS SUPERCLUSTER

TODD A. SMALL
*Institute of Astronomy, Madingley Road, Cambridge, CB3 0HA, UK*

WALLACE L.W. SARGENT
*Palomar Observatory, Caltech 105-24, Pasadena, CA, 91125, USA*

AND

DONALD HAMILTON
*Fakutät für Physik, Technische Universität München*
*Institut für Astronomie und Astrophysik, Universität München*
*Scheinerstraße 1, D-81679 München*

**Abstract.** We present results from our redshift survey of the Corona Borealis Supercluster. Using the Norris Multifiber Spectrograph on the Palomar 200-inch telescope, we have obtained redshifts for 528 galaxies in the supercluster and for 726 field galaxies with redshifts extending to $z \sim 0.7$. In this contribution, we discuss the structure and dynamics of the supercluster and the luminosity functions of the local field and of the supercluster.

## 1. Introduction

Abell, from his survey of galaxy clusters on the Palomar Observatory Sky Survey plates (Abell 1958), was the first to observe and note the existence of clusters of clusters of galaxies. These clusters of clusters of galaxies, since dubbed "superclusters," are among the largest identified objects in the universe. Superclusters are only $\lesssim 10$ times denser than the field. In contrast, the overdensity of an object that has just become virialized is $\sim 200$, and the overdensity of an Abell cluster is $\sim 1000$. Since the dynamical times of superclusters are comparable to the Hubble time, superclusters are not relaxed and should therefore bear the imprints of the physical processes



*D. Hamilton (ed.), The Evolving Universe,* 155–165.

that were dominant during their formation. One hopes that studies of superclusters will ultimately yield information about the nature of density fluctuations in the early universe and may offer clues about the epoch of galaxy formation. In addition, dynamical studies of superclusters provide insights into the distribution of matter on large scales.

The Corona Borealis Supercluster is the most prominent example of superclustering in the northern sky. In a $6° \times 6°$ region centered on right ascension $15^h20^m$, declination $+30°$, there are 12 rich Abell clusters, seven of which have redshifts of $z \approx 0.07$ and define the core of the supercluster. Three of the remaining clusters are at $z \approx 0.11$ and are a part of a background supercluster (see below). Galaxy counts in the field of the supercluster are a factor of three above typical counts in similar high-galactic latitude fields (Weir *et al.* 1995) for $16.5^m \lesssim r \lesssim 19.5^m$. The true extent of the supercluster on the sky is unknown, but Bahcall (1992) has presented circumstantial evidence to argue that while the seven Abell clusters which define the core of the supercluster reside in a region which is $\sim 20h^{-1}$ Mpc ($h$ is the Hubble constant $H_0$ divided by 100 km s$^{-1}$ Mpc$^{-1}$) on a side, the entire supercluster extends for at least $\sim 100h^{-1}$ Mpc in the two directions on the sky. Since it is not practical to survey such a large region on the sky ($\sim 730$ deg$^2$), the aim of our survey is to delineate accurately the structure of the core of the supercluster.

## 2. Observations

The observational details of our survey have been described in detail elsewhere (Small *et al.* 1997a) and will be only briefly reviewed here. Even restricting the survey to the 36 deg$^2$ region which covers the core of the supercluster leaves too large a region to be survey in a finite length of time, and so we planned at the outset to survey 36 fields arranged in a rectangular grid across the 36 deg$^2$ region. Galaxies were selected from photographic plates obtained as part of the Second Palomar Sky Survey. We used the Norris Multifiber Spectrograph (Hamilton *et al.* 1993) mounted at the Cassegrain focus of the Palomar 200-inch telescope for all of our spectroscopic observations. Norris consists of 176 fibers which may be independently positioned over the field of view of 400 arcmin$^2$ and is optimized for spectroscopy of faint galaxies.

As it turned out, we successfully observed 23 of the program fields and 9 additional fields along the ridge of galaxies between Abell 2061 and Abell 2067, yielding redshifts for 1485 galaxies. We have extended our survey with 163 redshifts from the literature, resulting in 1648 galaxy redshifts in the entire survey. 528 of these galaxies lie within the Corona Borealis Supercluster, 352 are in a background supercluster at $z \approx 0.11$ (which we

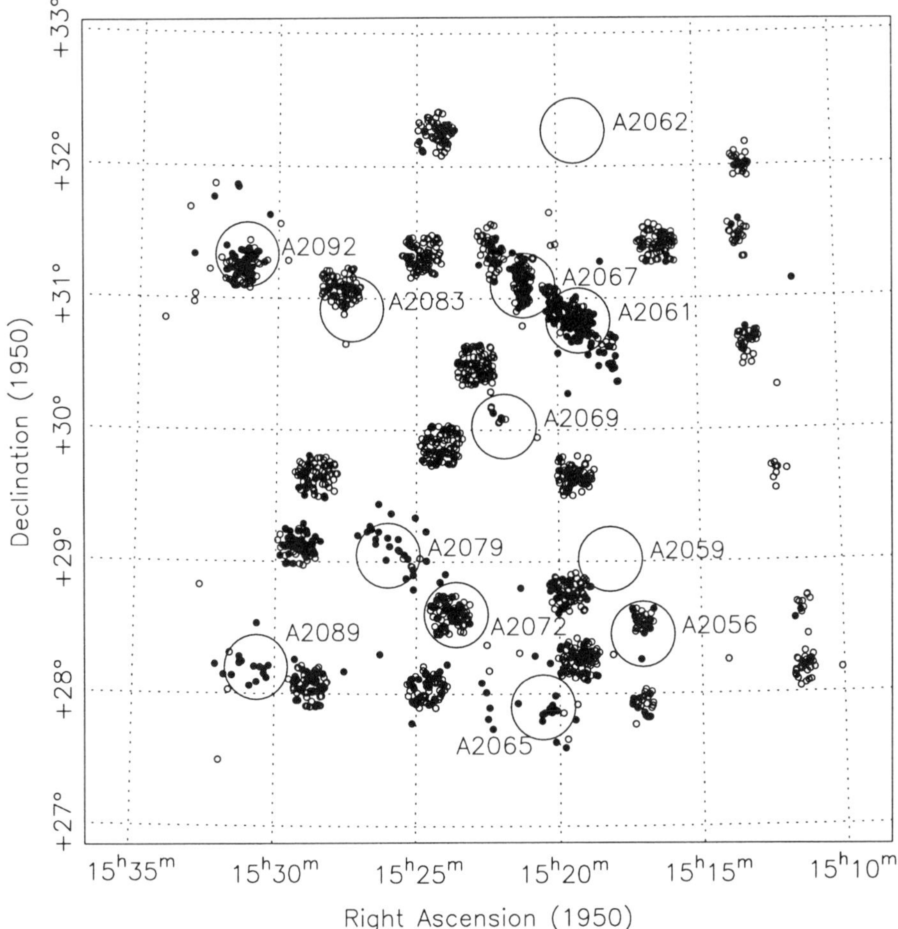


*Figure 1.* The celestial location of all the galaxies with measured redshifts in our survey. Galaxies lying within the Corona Borealis Supercluster are marked with filled circles.

have dubbed the "Abell 2069 Supercluster"), and 726 are in the field. Although we are only substantially complete to $r = 18.5^m$, we have redshifts for galaxies to $r \sim 22^m$. Owing to the overhead associated with deploying and retrieving 176 fibers, we switched fields only once per night. Our exposures on a given field were thus 2-4 hours long, yielding high quality spectra (most at 4Å resolution) for galaxies with $r \lesssim 20^m$. In Figure 1, we plot the the positions on the sky for all galaxies in our survey with measured redshifts. The filled circles are galaxies within the Corona Borealis Supercluster. We have marked the location of the twelve Abell clusters in the field. A2056, A2061, A2065, A2067, A2079, A2089, and A2092 are in the Corona Borealis Supercluster.

## 3. Structure and Dynamics of the Corona Borealis Supercluster

Even projected on the sky, the Corona Borealis Supercluster has an irregular shape. Four of the constituent Abell clusters (A2056, A2065, A2079, and A2089) lie together in the southern part of the supercluster, A2061 and A2067 are right next to each other in the northern part, and A2092 is isolated in the northeastern part. Only in the diamond-shaped region described by A2056, A2065, A2079, and A2089 can an extended region with excess galaxy counts be discerned.

In Figure 2, we plot redshift-right ascension diagrams for all galaxies in our survey with $z < 0.15$. The diagrams are split by declination as indicated in the figure. An inspection of the figure reinforces the impression that the density distribution of the supercluster is far from smooth. The supercluster is sharply delimited along the line-of-sight by foreground and background underdense regions. The well-defined boundaries of the supercluster lead us to restrict our analyses of the supercluster to galaxies with $0.06 \leq z \leq 0.09$.

Since the supercluster is unrelaxed and contains obvious substructure (*e.g.*, the Abell clusters themselves), a traditional dynamical analysis seems of questionable utility. We, therefore, begin our analysis with general considerations of the dynamical state of the supercluster using the evolution of a spherical overdensity as our model (Peebles 1980). However, from dimensional arguments, we expect the mass of the system to be proportion to $v^2 r$, where are $v$ and $r$ are a suitable scale velocity and radius, respectively. We aim to determine the constant of proportionality from simulated superclusters extracted from large $N$-body cosmological simulations (*cf.*, Eisenstein *et al.* 1997).

The overdensity of the supercluster is $\delta_{SC} = (n_{SC}/\bar{n}) - 1 \approx 7$, where $n_{SC}$ and $\bar{n}$ are the mean number densities of galaxies in the supercluster and in the field, respectively. An outer shell of a density perturbation is bound if the mean overdensity within the shell is greater than $\Omega^{-1} - 1$. Ignoring the variation of $\Omega$ from $z = 0$ to $z = 0.07$ and taking $\Omega_0 = 0.3$ as a very rough lower limit on the density parameter, the overdensity of the supercluster must be greater than 2.3 for the supercluster to be bound. Clearly, then, the supercluster is bound. While bound, the supercluster could still be expanding and yet to turn around. The overdensity required for turn-around has been computed for $\Omega \leq 1$ by Silk (1977) and by Regős & Geller (1989). For $\Omega_0 = 0.3$, the turn-around overdensity is $\approx 12$, and so the supercluster is likely to be still expanding, unless $\Omega_0 \gtrsim 0.4$. The eventual collapse of the supercluster as there is only a very limited range of density fluctuations with appreciable contrast ($\delta \gtrsim 1$) which are freely expanding (Peebles 1980).

In order to estimate the mass of the supercluster, we have, in collabora-

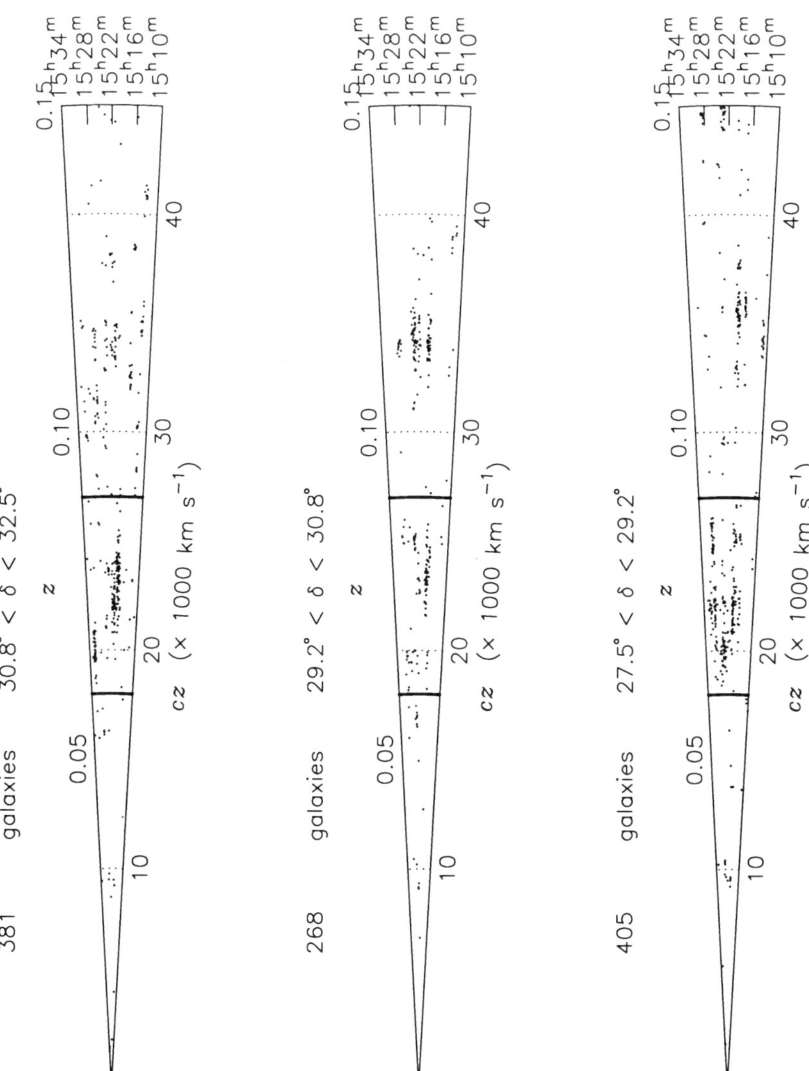


*Figure 2.* Redshift-right ascension cone diagrams for galaxies in our survey with $z < 0.15$, divided into 3 declination slices. The Corona Borealis Supercluster is the prominent overdense region between $z = 0.06$ and $z = 0.09$. The background supercluster lies between $z = 0.10$ and $z = 0.13$.

tion with C.-P. Ma, started a program of $N$-body simulations designed to estimate the constant of proportionality between mass and $v^2r$. We have so far extracted just one simulated supercluster from a simulation of an $\Omega_0 = 1$ CDM universe. The simulated supercluster has a mass of $5.3 \times 10^{15}$ $M_\odot$ (in a volume of $225h^{-3}$ Mpc$^3$) and incorporates 9 clusters with masses greater than $10^{14}$ $M_\odot$. As a first guess at a successful form for the mass estimator, we have simply adopted the virial mass estimator and the projected mass estimator. Perhaps surprisingly, both estimators give measurements of the mass of the supercluster accurate to a factor of two, although the measurements are slightly biased lower than the true mass. We need a more extensive set of simulations before we can state with certainty that the virial mass estimator and the projected mass estimator yield reliable results, but we nevertheless cannot resist the temptation to apply them to the Corona Borealis Supercluster.

When applied to the Corona Borealis Supercluster, both estimators give a mass for the supercluster of $3 \times 10^{16}$ $M_\odot$. By integrating the supercluster luminosity function (see below), we can compute the mean luminosity density of the supercluster and thereby measure the mass-to-light ratio of the supercluster on scales of $\sim 20h^{-1}$ Mpc. The mean luminosity density of the supercluster in the $B_{AB}$ band is $\rho_L(B_{AB}) = (1.9 \pm 1.0) \times 10^9 h L_\odot \mathrm{Mpc}^{-3}$. Taking the solid angle of the survey to be 0.0076 sr (= 25 deg$^2$) and the redshift limits of the supercluster to be $z = 0.06$ and $z = 0.09$, the volume of the survey region is $2.8 \times 10^4 h^{-3} \mathrm{Mpc}^3$. The $B_{AB}$-band mass-to-light ratio of the supercluster is thus $564h(\frac{M}{L})_\odot$. As computed from our own determination of the local luminosity density, the $B_{AB}$-band mass-to-light ratio required to close the universe is $1400h(\frac{M}{L})_\odot$, which yields $\Omega_0 \approx 0.40$ on supercluster scales, or roughly twice the value computed by Carlberg *et al.* (1996) for rich clusters of galaxies. If the reliability of our mass estimators for superclusters is borne out by further simulations, this result implies that the relative amount of dark matter in the universe increases with scale to at least the scale of superclusters.

## 4. The Local Field Galaxy Luminosity Function

As several other contributions to these proceedings have also emphasized, it is very important to have a reliable estimate of the local luminosity function (LF), both for a complete census of the local universe and as a basis for quantitative estimates of galaxy evolution with redshift. It is therefore of concern that there remain considerable uncertainties about the local LF. In particular, there is a factor of two difference in the normalizations of the LF measured in the Stromlo/APM redshift survey (Loveday *et al.* 1992) and in the CfA redshift survey (Marzke *et al.* 1994a). Furthermore, the Autofib redshift survey (Ellis *et al.* 1996), which is selected from the deepest images, also indicates a higher normalization than that determined in the Stromlo/APM redshift survey. The Las Campanas Redshift Survey (LCRS), discussed in these proceedings by Lin, Landy, and Tucker, provides a definitive measurement of the local LF in the $r$ band, but care must be taken in comparing their results to other surveys, which are based on photometry in some variant of the $B$ band. While our survey has far fewer local field galaxies than projects designed explicitly to probe the local universe, we do have photometry in both the $g$ and $r$ bands, which simplifies conversions between the $B$ band and the $r$ band, and we also reach quite faint apparent magnitudes. A complete discussion of the luminosity functions measured in our survey is presented in Small *et al.* (1997b).

Removing the galaxies in the two superclusters in our survey (by deleting all galaxies with $0.06 \leq z \leq 0.13$) leaves a sample of 219 field galaxies with $z \leq 0.2$, total $r$ magnitudes brighter than $20^m$, and core magnitudes (the integrated magnitude in the central 9 sq. arcsec) brighter than $21.7^m$. The median redshift of our local field sample is $z_{med} = 0.15$. As discussed in detail in Small *et al.* (1997a), we believe that by using these limits we avoid systematic errors caused by surface brightness selection effects, although, because of target selection from photographic plates, we are not particularly sensitive to low surface brightness objects. Before we present the local LFs measured in our survey, we must address the question of the normalization of the LFs. Since the region was selected for its extremely high surface density of galaxies, it is fair to be concerned that the mean density of survey, even with the supercluster volumes removed, could be biassed high. We have therefore normalized our local LFs to the $r$-band counts of Weir *et al.* (1995), which are based on POSS-II photographic plates of high galactic latitude fields.

In Figure 3, we have plotted both the $B_{AB}$- and $r$-band local LFs measured in our survey (with the normalization to the Weir *et al.* (1995) counts). The $B_{AB}$-band LF is plotted with respect to the bottom axis, while the $r$-band LF is plotted with respect to the top axis. The top and bottom axes have been offset using the median color of the local galaxies in our survey, $(B_{AB} - r)_{med} = 0.72^m$, which roughly corresponds to an Sb galaxy (*cf.*, Fukugita *et al.* 1995). We have also plotted Schechter (1976) function fits to the $B$-band LFs from the Stromlo/APM survey and the Autofib survey and to the $r$-band LF from the LCRS. (Each of the published Schechter function fits has been convolved with a Gaussian of dispersion $0.25^m$ to account for the magnitude errors in our survey.) We have corrected the LCRS LF to total magnitudes using the 25% isophotal-to-total correction recommended by Shectman *et al.* (1996) and by $0.25^m$ to convert from $R_{LCRS}$ to standard Gunn $r$. The most striking feature revealed by the figure is that,

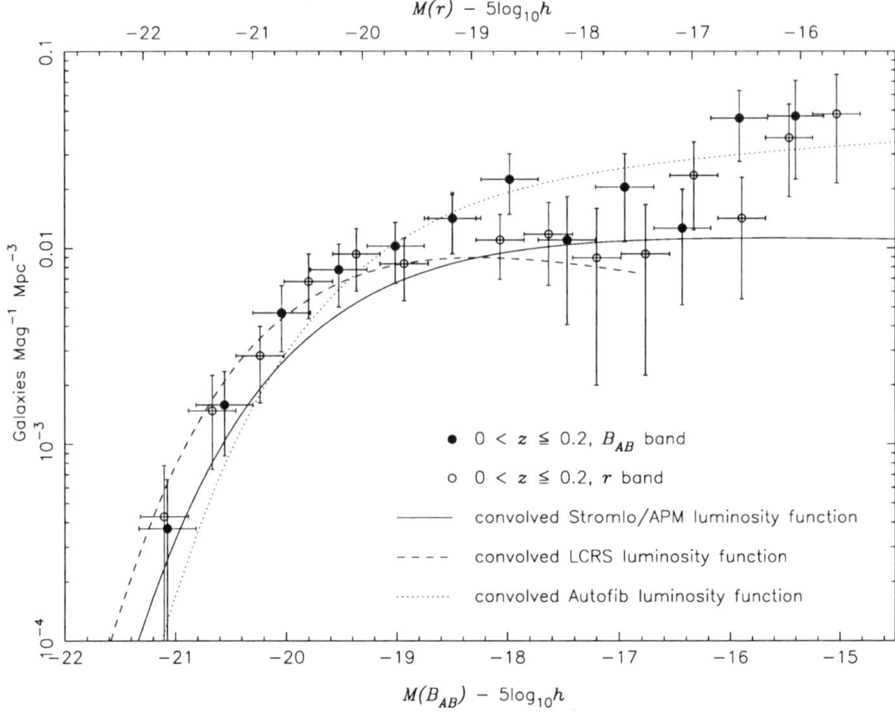


*Figure 3.* $B_{AB}$- and $r$-band luminosity functions for field galaxies with $z \leq 0.2$. We have also plotted Schechter function fits to various published local LFs. The $B_{AB}$-band LFs are plotted with respect to the bottom axis, while the $r$-band LFs are plotted with respect to the top axis. The two axes are offset by $B_{AB} - r = 0.72^m$, which is the median color of field galaxies we measure for local field galaxies.

in contrast to Lin *et al.* (1996), we do *not* conclude that the LCRS LF is consistent with the Stromlo/APM LF. The discrepancy is due the different estimates of the mean color of a local galaxy between the LCRS and our survey. The mean color quoted by Lin *et al.* (1996), determined using the APM catalog as a source of $B_J$ magnitudes for the LCRS galaxies, is that of an elliptical galaxy. Since our $r$-band local LF agrees with the LCRS $r$-band LF, we believe that the most likely source of error is in the photometry of the APM catalog. We also note that our local LFs show a hint, based admittedly on only five galaxies, for an upturn in the LF for galaxies fainter than $M(B_{AB}) > -16^m$. This upturn is consistent with the steep faint ends reported by Marzke *et al.* (1994b) for irregular galaxies and by Sprayberry *et al.* (1997) for low surface brightness galaxies.

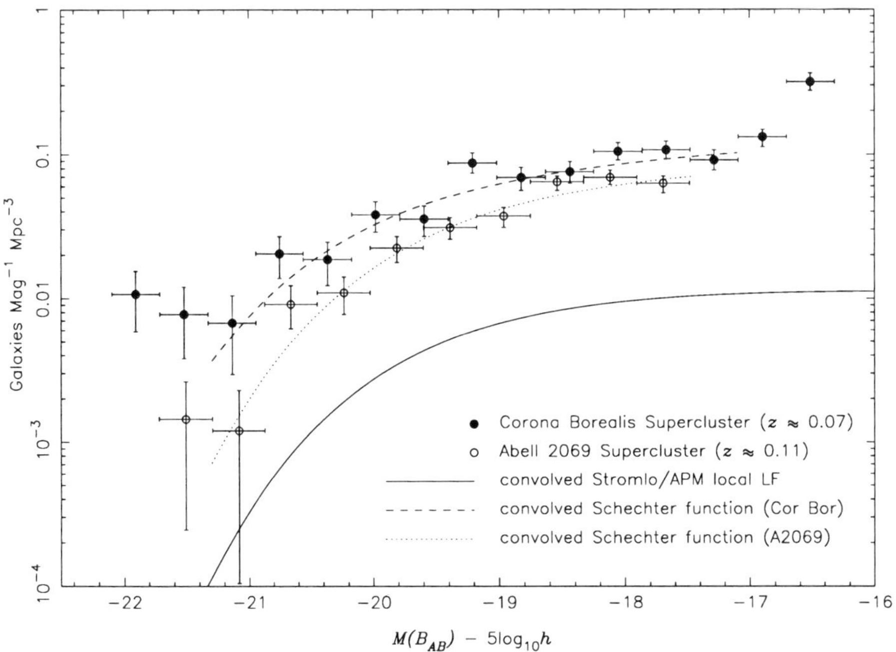


*Figure 4.* Luminosity functions for the Corona Borealis Supercluster (filled circles) and for the Abell 2069 supercluster (unfilled circles). Schechter function fits to the data points with $-21.3^m < M(B_{AB}) < 17.1^m$ are given by the dashed (Corona Borealis) and dotted (Abell 2069) lines. The Schechter function fit to the local LF from the Stromlo/APM is also shown for comparison.

## 5. Supercluster Luminosity Functions

Our data for supercluster galaxies give us the opportunity to measure the LF as a function of environment. Recent measurements of the LF of rich clusters of galaxies have suggested that the slope of the LF faint end may be considerably steeper in clusters than in the field (*e.g.*, De Propris *et al.* 1995). It is not clear, however, how reliable these results are because they mainly rely, since redshifts have not been available, on the subtraction of a background component from the foreground cluster, a procedure prone to systematic errors. Alternatively, the steep slope could be due to low surface brightness galaxies revealed by the typically much fainter surface brightness limits of the cluster imaging surveys relative to the surface brightness limits of the local redshift surveys. By measuring the LFs of the Corona Borealis supercluster and the background Abell 2069 supercluster, we intend to search for differences between the LF of the low density field and that of the higher density ($\sim 5 - 10\times$ more dense than the field) superclusters.

In Figure 4, we plot the LFs of the Corona Borealis Supercluster and Abell 2069 supercluster with filled and unfilled circles, respectively. For comparison, we have also plotted the local field galaxy LF measured in the Stromlo/APM survey. The lines fitted to the data are Schechter function fits for $-21.3^m < M(B_{AB}) < -17.1^m$. The most prominent difference between the field and supercluster LFs is that the supercluster LFs do not continue the exponential decline for galaxies brighter than $M(B_{AB}) \lesssim -21^m$. Both superclusters evidently contain a population of very luminous galaxies, which are presumably just the giant ellipticals that are found in the centers of galaxy clusters. For the Corona Borealis Supercluster, the characteristic absolute magnitude $M^*$ is $\sim 0.5^m$ brighter than in the field, but there is no significant difference between $M^*$ for the Abell 2069 supercluster and the field.

For the Corona Borealis Supercluster LF, the data points for the two faintest magnitude bins hint that the LF may steepen significantly for galaxies fainter than $M(B_{AB}) \sim -17^m$. Since the hint is based on only two data points, which are themselves based on only 29 galaxies, we must be cautious in our interpretation. However, a steepening of the supercluster LF fainter than $M(B_{AB}) \sim -17^m$ would be in accord with observations of the faint end of the LF in galaxy clusters and groups, in which a number of workers have reported steep slopes. Although it would be unwise to draw strong conclusions from our two data points, they do have the virtue of being based on galaxies with measured redshifts.

## References


Abell, G. 1958, ApJS, 3, 211

Bahcall, N. 1992, in Clusters and Superclusters of Galaxies, ed. A. Fabian (Dordrecht: Kluwer), 275

Carlberg, R., Yee, H., Ellingson, E., Abraham, R., Gravel, P., Morris, S., & Pritchet, C. 1996, ApJ, 462, 32

De Propris, R., Pritchet, C., Harris, W., & McClure, R. 1995, ApJ, 450, 534

Eisenstein, D., Loeb, A., & Turner, E. 1997, ApJ, 475, 421

Ellis, R., Colless, M., Broadhurst, T., Heyl, J., & Glazebrook, K. 1996, MN, 280, 235

Hamilton, D. *et al.* 1993, PASP, 105, 1308

Lin, H., Kirshner, R., Shectman, S., Landy, S., Oemler, A., Tucker, D., & Schechter, P. 1996, ApJ, 464, 40

Loveday, J., Peterson, B., Efstathiou, G., & Maddox, S. 1992, ApJ, 390, 338

Marzke, R., Huchra, J., & Geller, M. 1994a, ApJ, 428, 43

Marzke, R., Geller, M., Huchra, J., & Corwin, H. 1994b, AJ, 108, 437

Peebles, P. 1980, The Large-Scale Structure of the Universe (Princeton: Princeton University Press)

Regős, E. & Geller, M. 1989, AJ, 98, 755

Schechter, P. 1976, ApJ, 203, 297

Shectman, S., Landy, S., Oemler, A., Tucker, D., Lin, H., Kirshner, R., & Schechter, P. 1996, ApJ, 470, 172

Silk, J. 1977, A&A, 59, 53
Small, T., Sargent, W., & Hamilton, D. 1997a, ApJS, in press
Small, T., Sargent, W., & Hamilton, D. 1997b, ApJ, in press
Sprayberry, D., Impey, C., Irwin, M., & Bothun, G. 1997, ApJ, in press [astro-ph/9701051]
Weir, N., Djorgovski, S., & Fayyad, U. 1995, AJ, 110, 1

# CLUSTERING AT $Z \geq 0.8$

OLIVIER LE FÈVRE
*DAEC, Paris-Meudon Observatory*
*Pl. J. Janssen, 92195 Meudon Cedex, France*

**Abstract.** We discuss recent observational results describing exploratory steps taken to describe the evolutionary status of the clustering of galaxies in the Universe on scales $\sim 1$ Mpc. The clustering of galaxies is seen to be strongly evolving from the distribution of galaxies in the Canada France Redshift Survey, from epochs when the universe was 40% of its current age. Massive clusters of galaxies are being identified around radio galaxies at $z \sim 1$, indicating that deep potential wells are already in place at these epochs. At yet higher redshifts, tentative evidence for clustering is seen around radio-galaxies. We speculate that we are only now entering a very rich era in which the evolution of the clustering in the universe and its evolution will be examined in details.

## 1. Introduction

Understanding the formation and evolution of large-scale structures is of considerable importance to modern cosmology. While galaxy formation and evolution is receiving much warranted attention, the observational study of the formation and evolution of clusters of galaxies or other large-scale structures is only now developing.

While the observational data on the large-scale structure of the local universe is now substantial, our knowledge of the clustering of galaxies above $z \sim 0.5$ was, until recently, limited to theoretical descriptions illustrated by numerical simulations. With new technology being implemented on the largest telescopes, and in particular the advent of efficient multi-object spectrographs, our observational exploration of the clustering properties of galaxies at high redshifts is now progressing rapidly. I review here some of the recent work I have carried out in the framework of several collaborations.



*D. Hamilton (ed.), The Evolving Universe,* 167–178.

## 2. Canada France Redshift Survey: Evolution of the clustering of galaxies since $z \sim 1$

### 2.1. INTRODUCTION

It is expected that the clustering of galaxies may well change with epoch in an expanding universe in which structures evolve and grow under the action of gravity. As a consequence, the two-point correlation function $\xi(r)$ should change with cosmic epoch, but the exact form of this evolution is at present poorly known.

There have been many studies of the local $\xi(r)$ in surveys such as the CfA, Stromlo-APM, SSRS, IRAS and others (see *e.g.*, Davis & Peebles 1983, Loveday *et al.* 1995, Fisher *et al.* 1994, Benoist *et al.* 1995). They indicate a power-law behaviour with $\xi(r) = (r/r_0)^{-\gamma}$. Values for the correlation length $r_0$ range from 3.8 to 7.5 $h^{-1}$ Mpc with a possible dependence on luminosity and galaxy type (Loveday *et al.* 1995) and $\gamma \sim 1.7$. At higher redshifts, two approaches can be followed to measure $\xi(r)$. The first is to invert the projected angular two-point correlation function $w(\theta)$ through the Limber equation using an observed or predicted redshift distribution N($z$) appropriate to the observed limiting magnitude of the galaxy sample. From the angular correlation function $w(\theta)$, Efstathiou *et al.* (1991) have shown that faint galaxies, which are associated with the "excess" in the blue number counts, are rather weakly clustered. At shallower depths, where spectroscopic surveys are possible, Hudon and Lilly (1995) inverted a measurement of $w(\theta)$ using the redshift distribution based on the CFRS and find a correlation length of $r_0 = 1.9 \pm 0.1 h^{-1}$ Mpc at $z \sim 0.5$.

The alternative approach is to directly compute $\xi(r)$ from the distance information for individual galaxies that is present in deep redshift surveys. The deep $I$-band selected CFRS sample (see Lilly *et al.* 1995a, CFRS-I; Le Fèvre *et al.* 1995, CFRS-II; Crampton *et al.* 1995, CFRS-V) provides for the first time the opportunity to directly evaluate the evolution of the two-point correlation function $\xi(r)$ over the redshift range $0 < z < 1$.

### 2.2. THE CANADA-FRANCE REDSHIFT SURVEY SAMPLE

The statistically complete CFRS sample consists of 943 objects selected in five 10' $\times$ 10' fields to have $17.5 \leq I_{AB} \leq 22.5$, without regard to color or morphology, and with minimal surface brightness selection. The sample is 85% spectroscopically identified, 591 galaxies in the CFRS have secure redshifts. The redshifts extend up to $z \sim 1.3$, with a median redshift $< z >= 0.56$. More than 350 galaxies have $z \geq 0.5$. The field dimension of 10 arcmin corresponds to a comoving dimension of 3.3 $h^{-1}$Mpc at $z \sim 0.5$ and to $5.3 h^{-1}$ Mpc at $z \sim 1$, dimensions comparable to the $z = 0$ correlation length.

The galaxies are distributed in redshift with a "picket–fence" distribution as identified in shallower surveys. We show in Figure 1 that this type of distribution extends over the full dimension of the survey or ~2500 $h^{-1}$ Mpc, with prominent peaks obvious out to $z \sim 1$.

### 2.3. THE EPOCH–DEPENDENT 2–PT CORRELATION FUNCTION $\xi(R,Z)$

The local two-point correlation function is well represented by a power–law (Davis and Peebles 1983)

$$\xi(r) = (\frac{r}{r_0})^\gamma$$

and the epoch–dependent correlation function is usually expressed in terms of an evolutionary parameter $\epsilon$ (Groth and Peebles 1977; Efstathiou *et al.* 1991),

$$\xi(r,z) = \xi(r,0) \times (1+z)^{-(3+\epsilon)}$$

Thus, the correlation length can be written

$$r_0(z) = r_0(0) \times (1+z)^{-(3+\epsilon)/\gamma}$$

To avoid confusion, it should be noted that the correlation length $r_0$(z) is here the correlation length (in physical units) that would be measured by a local observer at the epoch in question. Thus the correlation length will evolve (as $(1+z)^{-1}$) even if the clustering pattern is fixed in comoving space.

In the case of a clustering pattern fixed in comoving coordinates, clustering does not grow with time, and $\epsilon = \gamma - 3 = -1.3$ for $\gamma = 1.7$. When bound gravitational units keep a fixed physical size, the clustering growth is the result of the increasingly diluted galaxy background, and $\epsilon$=0 (Efstathiou *et al.* 1991; Carlberg 1991). For a standard CDM scenario, the mass clustering should grow in the linear regime with $\epsilon \sim 0.7$ on ~1 $h^{-1}$Mpc scales, (Davis *et al.* 1985).

To avoid the effect of peculiar velocities and redshift measurement errors in evaluating the spatial correlation function $\xi(r)$, we have used the projected function $w(r_p)$ (Davis & Peebles 1983), generalized to the observation of galaxies at high redshifts:

$$w(r_p) = \frac{c}{H_0 \times (1+z)^2 \times (1+2q_0 z)^{1/2}} \int_{-\delta z}^{+\delta z} \xi(r_p,\pi)dz \qquad (1)$$

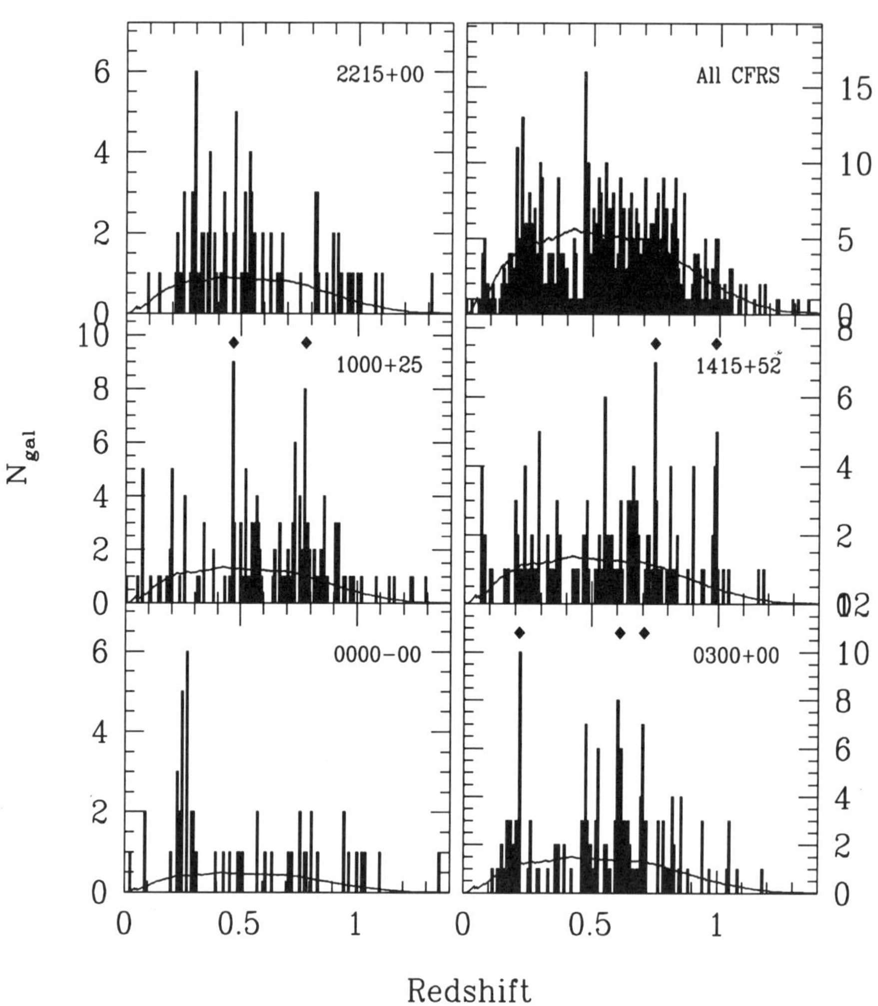


*Figure 1.* The distribution of galaxies in "picket-fence" structure in the Canada-France Redshift Survey.

With $\xi(r) = (r_0/r)^\gamma$ and $r_p << cz/H_0$, $w(r_p)$ becomes (Davis and Peebles 1983)

$$w(r_p) = \frac{\Gamma(1/2)\Gamma[(\gamma - 1)/2]}{\Gamma(\gamma/2)} \times r_0^\gamma \times r_p^{1-\gamma} \qquad (2)$$

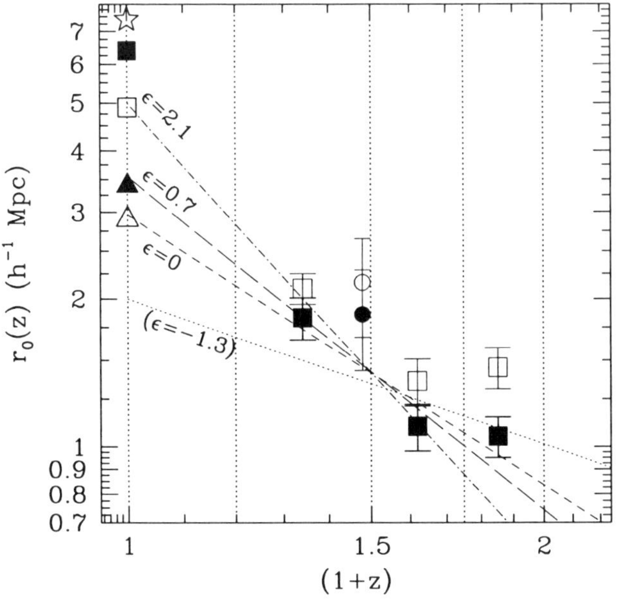


*Figure 2.* The evolution of the correlation length $r_0$ with redshift from the CFRS.

so that computing $w(r_p)$ provides a measurement of $r_0$ and $\gamma$.

In order to examine evolution in the amplitude of the correlation function, the sample was split into three redshift bins: $0.2 \leq z \leq 0.5$, $0.5 \leq z \leq 0.75$, $0.75 \leq z \leq 1$ (the redshift bins match those of our analysis of the evolving luminosity function - see CFRS VI). The resulting $w(r_p)$ are shown in Figure 2. These have been characterized by fitting power-laws with both $r_0$ and $\gamma$ free parameters and, for uniformity, by fitting a power-law with fixed $\gamma$=1.64, following equation (2). In the latter case, $r_0(z{=}0.34)= 1.83\pm0.18$, $r_0(z{=}0.62)= 1.10\pm0.15$ and $r_0(z{=}0.86)= 1.05\pm0.10$ are found.

### 2.4. THE EVOLUTION OF $\xi$(R, $Z$)

Our data indicates that the correlation length is decreasing with redshift, with the amplitude of the correlation function $\xi(r)$ being lower by a factor $\sim 3$ at $z \sim 0.6$ compared to $z \sim 0.3$. Constraints on the apparent value of $\epsilon$ can be derived when our $r_0$ measurements at high redshift are taken together with local measurement (Loveday *et al.* 1995). We discuss here three specific possibilities:

(1) If there was no growth in clustering, *i.e.*, the clustering pattern

is fixed in comoving coordinates, $\epsilon = -1.3$, then our observed clustering at high redshift would imply a local value of $r_0(0) = 2.0 \pm 0.2h^{-1}$Mpc for $q_0 = 0.5$, $r_0(0) = 2.5 \pm 0.2h^{-1}$Mpc for $q_0 = 0$, a clustering strength much lower than for any local population of galaxies yet observed. Given the modest changes in the galaxy luminosity function constructed from this same CFRS sample (CFRS VI and discussion below), the existence of such a currently-unknown galaxy population is extremely unlikely. We, therefore, conclude that some growth of clustering *must* have occurred over this redshift range.

(2) If the physical size of clusters of galaxies had remained fixed with redshift, *i.e.*, the $\epsilon = 0$ case, then $r_0(0) = 3.0 \pm 0.2h^{-1}$ Mpc is implied for $q_0 = 0.5$ ($r_0(0) = 3.9 \pm 0.2h^{-1}$ Mpc for $q_0 = 0.0$). The galaxies we observe at high redshift in the CFRS would have had to evolve with time into the most weakly clustered (low luminosity) galaxies in, for example, the Stromlo-APM survey (Loveday *et al.* 1995) or the SSRS (Benoist *et al.* 1995).

(3) Finally, for the CFRS galaxies to be representative of the local B-band selected samples with roughly L* luminosities would require rapid growth of the clustering, with $\epsilon \geq 1$. For instance, for the CFRS galaxies to evolve into local samples that have $r_0 = 5 \pm 0.15h^{-1}$Mpc (Loveday *et al.* 1995), would require $\epsilon = 2.1 \pm 0.3$ for $q_0 = 0.5$.

The population at $z \sim 0.6$ is expected to be broadly similar to that seen in local B-selected samples of luminous galaxies, but with the likelihood of some luminosity evolution in at least part of the population. This suggests that scenarios similar to those discussed in (2) and (3) above are plausible, probably nearer the latter, and thus we believe that the "true" value of $\epsilon$ is likely to be between 0 and +2. Stronger constraints on the evolution of $\xi(r)$ from a direct measurement will require either much larger samples, $> 10^4$ galaxies, in the same redshift range, and/or the direct measurement of $\xi(r)$ at redshifts significantly in excess of one.

## 3. Discovery of Massive Clusters of Galaxies at $z \sim 1$

### 3.1. INTRODUCTION

To identify clusters at high redshifts is a challenging observational task. At low redshifts, one can rely on the projected 2D galaxy density, as was done successfully from the earliest photographic material (*e.g.*, Abell 1958, Zwicky 1961-1968). However, at high redshifts, $z \geq 0.5$, it becomes increasingly difficult to identify the 2D galaxy overdensity produced by a cluster because the projected foreground and background galaxies are so numerous that the density contrast is severely reduced and chance alignments of groups of galaxies can mimic the appearance of clusters (Frenk *et al.* 1990).

The existence of a distant cluster can be proven only if several lines of evidence are combined, which should include the measurement of a projected overdensity of galaxies in few square arcmin area on the sky, combined with a significant overdensity observed in the redshift distribution of galaxies on velocity scales $\sim$2000 km $s^{-1}$ in order to reduce the contamination by foreground and background interloper galaxies. In addition, evidence for hot gas, through the detection of X-ray emission, and/or evidence for a peaked mass distribution as indicated by weak or strong lensing of background galaxies, provide strong support for the identification of such overdensities as genuine clusters of galaxies. Although a number of candidates have been proposed, only a handful of clusters (or proto-clusters) of galaxies have been unambiguously identified at $z > 0.6$, with the above criteria fulfilled (Dickinson 1996, Luppino & Kaiser 1996). It is of considerable importance for our knowledge of the evolution of large-scale structures to identify more high redshift clusters in order to establish the evolution of their physical properties with redshift.

### 3.2. MASSIVE CLUSTERS AROUND $Z \sim 1$ 3CR RADIO-GALAXIES

Deltorn, Crampton, Dickinson, and Le Fèvre have conducted an exploratory search for clusters of galaxies around known powerful radio galaxies. Since at redshifts around 0.5, 30% of radio galaxies and QSOs are situated in rich clusters (Yee & Green 1984, Hill & Lilly 1991, Dickinson 1996, Le Fèvre *et al.* 1996), a possible search strategy is to look for clusters around high redshift radio-galaxies.

Deep imaging in $V$ and $I$ of fields around 3C265 ($z$=0.811) and 3C184 ($z$=0.996) was followed by multi-slit spectroscopy at the CFHT. The field around 3C184 was later observed with the HST WFPC2.

Significant projected galaxy density excess have been found from the imaging data. Spectroscopic data has allowed to confirm that the excess density is indeed associated to the radio galaxy, and not the result of projection effects, with 15 redshift-confirmed cluster members around 3C265, and 10 redshift-confirmed galaxies at the same redshift as 3C184 (Figure 3).

The velocity dispersion deduced from the 3C184 galaxy cluster redshifts is $634^{+206}_{-102}$km $s^{-1}$, leading to a virial mass $M = 6.16^{+3.94}_{-2.40} \times 10^{14} h_{50}^{-1} M_{\odot}$. In addition, the deep HST images revealed a gravitational arc seen projected at $42 h_{50}^{-1}$kpc away from 3C184. We thus have strong evidence for the presence of a massive cluster at $z \simeq 1$. The mass contained within the arc radius is in the range $[1.20 \times 10^{13} h_{50}^{-1} M_{\odot}, 2.78 \times 10^{13} h_{50}^{-1} M_{\odot}]$ for $z_{arc}$ within the interval 3-1.5 (see Deltorn *et al.*, this volume, and Deltorn *et al.* 1997).

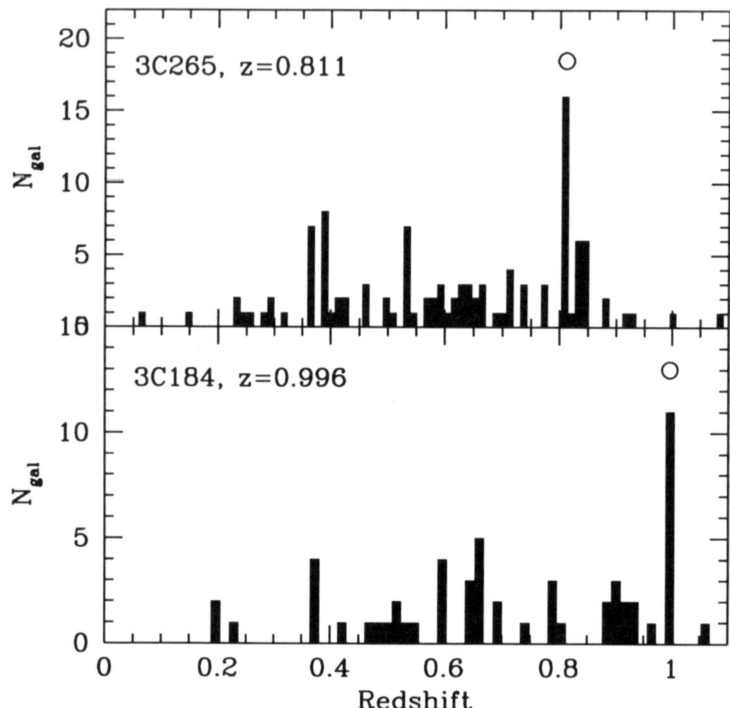


*Figure 3.* Massive clusters of galaxies identified at $z \sim 1$ around radio-galaxies (data from CFHT).

### 3.3. DISCUSSION

The secure identifications of massive bound structures at $z > 0.8$, through spectroscopic observation of cluster members (Dickinson 1996, Deltorn *et al.* 1997), weak gravitational lensing (Luppino & Kaiser 1996, Smail & Dickinson 1995), or X-ray observations (Luppino & Gioia 1995, Castander *et al.* 1994) have shown that those structures might not be as rare as predicted by some cosmological scenarios. As observational capabilities become more accute, the observation of these massive structure at $z \simeq 1$ might come to be in conflict with various cosmological models.

## 4. (Proto-) Cluster at $z = 3.14$

### 4.1. INTRODUCTION

Although a number of galaxies with luminosities $\sim L^*$ are now observed out to redshifts $\sim 3-4$ (*e.g.*, Steidel *et al.* 1996), very few candidate clusters of galaxies have been spectroscopically confirmed at redshifts near or above unity (Le Fèvre *et al.* 1994a, Dickinson 1995, Pascarelle *et al.* 1996, Francis *et al.* 1996). At these very high redshifts, the prevalence of clusters, proto-clusters, or other large-scale structures in the galaxy distribution is as yet unknown, and the evolution of large-scale structures may well be at a critical stage (Peebles *et al.* 1989; Evrard&Charlot 1994; Frenk *et al.* 1996), where observations can directly constrain cosmological models.

### 4.2. *LY-$\alpha$* COMPANION GALAXIES AROUND MRC0316-257, AT $Z = 3.14$

We have observed a $9.2' \times 8.5'$ field around MRC0316-257, a radio-galaxy with $z$=3.14, with the Multi Object imaging Spectrograph (MOS) at CFHT (Le Fèvre *et al.* 1994b). Deep broad band $V$ and $I$ images as well as narrow band images in a filter with central wavelength 5007Å and bandwidth 96Å, containing the $Ly\alpha$ line redshifted to $z \sim 3.14$, were obtained. Multi-slit spectroscopy was subsequently obtained, including two galaxies which exhibit strong excess emission in the narrow band filter, for a total of 51 galaxies in the field.

In addition to the radio-galaxy, we have identified spectroscopically two galaxies with a redshift similar to the radio-galaxy (Figure 4). The first one, "galaxy A", is 44 arcseconds away from the radio-galaxy. Its spectrum shows a strong emission line at 5030.3Å, and faint features with marginal S/N at the same redshift (Fig.4). The identification of the line at 5030Å with $Ly\alpha$ at $z = 3.1378 \pm 0.0028$ seems to be secure, as other alternatives can be ruled out (Le Fèvre *et al.* 1996). The second galaxy, "galaxy B", is located 182 arcseconds from MRC0316-257. Its spectrum shows $Ly\alpha$ and CIV1549Å, and, therefore, gives a secure redshift $z = 3.1351 \pm 0.0028$. Both CIV1549Å, and to a lesser extent $Ly\alpha$ exhibit a broad component, which indicates the presence of an AGN.

Any estimate of the space density of $z \approx 3.14$ galaxies from our observations is necessarily quite uncertain due to the small number of objects which we detect. However, we have been able to tentatively bracket the range of possible space densities by $2.2 \times 10^{-4} < n(h_{50}^3 \mathrm{Mpc}^{-3}) < 3.3 \times 10^{-3}$, with a "best guess" value of $3/1210 = 2.5 \times 10^{-3} h_{50}^3 \mathrm{Mpc}^{-3}$. The population of $3.0 < z < 3.5$ field galaxies identified by Steidel *et al.* (1996) has a comoving space density of $3.6 \times 10^{-4} h_{50}^3 \mathrm{Mpc}^{-3}$ for $q_0$=0.5. At face value, this would imply that our "best guess" space density represents an overdensity of $\sim 7\times$

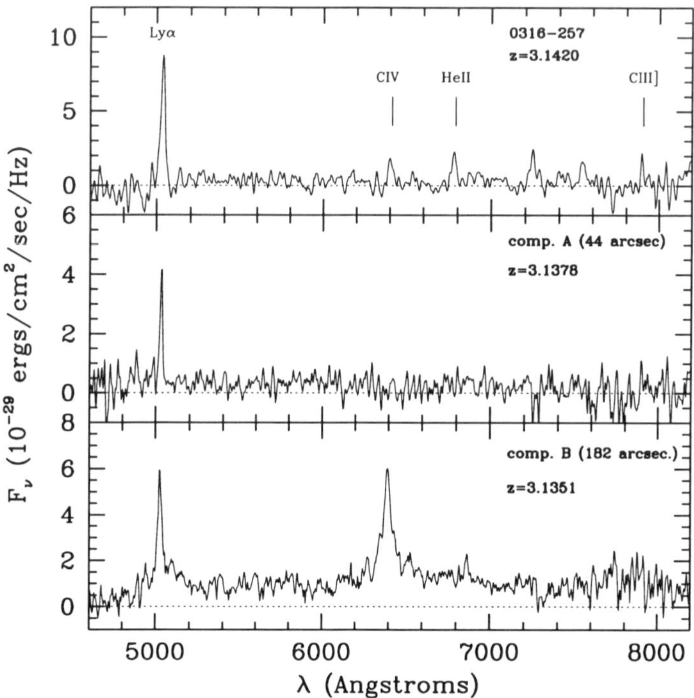


*Figure 4.* Companion galaxies identified around MRC0316-257 at $z$=3.14.

compared to the field population at similar redshift. However, it is difficult to make a direct comparison between our "cluster" space density and the "field" value since our galaxies were selected on the basis of strong Ly$\alpha$ flux, and have brighter continuum magnitudes ($V < 23.8$) than the Steidel *et al.* objects (which have $\mathcal{R} < 25.5$). The estimate given above for the cluster overdensity is, therefore, almost certainly a substantial *underestimate.* Even our derived lower limit of $n > 2.2 \times 10^{-4} h_{50}^3 \mathrm{Mpc}^{-3}$ for the 0316-257 field is substantially higher than the corresponding space density of field galaxies sharing the same continuum magnitudes and Ly$\alpha$ properties, and may indicate the presence of a (proto-) cluster around MRC0316-257. More details can be found in Le Fèvre *et al.* (1996).

New $UBVI$ observations are in progress in this field, in particular to look for Lyman break galaxies at the $z \sim 3.1$ redshift.

## 5. Prospects

The observational understanding of the evolution of the clustering of galaxies since early epochs is only now beginning. This situation is probably equivalent to the observational knowledge of galaxy evolution 10 years ago, which has now developed into the full-blown business that we know.

As new technological developments arise, we can expect to develop the current exploratory vision into a detailled accounting of the properties of large-scale structure at various epochs. Large 8-10m class telescopes equipped with wide field multi-object spectrographs (such as the VLT-VIRMOS project, Le Fèvre *et al.* 1996b), will prove particularly useful to this effect. The 5-10 years to come promise to be very rewarding for the field of clustering evolution.

## 6. Acknowledgements

I thank my collaborators on the various aspects presented here, D. Crampton, J.M. Deltorn, M. Dickinson, F. Hammer, S.J. Lilly, L. Tresse, to have allowed me to present results at this meeting.

## References

Abell, G.O., 1958, ApJS, 3, 211
Benoist, C., Maurogordato, S., da Costa, L., Cappi, A., Shaeffer, R., 1995, Proceedings from the 1995 Moriond Conference, Morogordato, Tran Thanh Van, Eds, Editions Frontières, in press.
Carlberg, R.G., 1991, Ap.J., 367, 385
Castander, F.J., Ellis, R.S., Frenk, C.S., Dressler, A., Gunn, J.E., 1994, Ap.J., 424, L79
Crampton, D., Le Fèvre, O., Lilly, S.J., Hammer, F., 1995, Ap.J., 455, 96 (CFRS-V)
Davis, M., Efstathiou, G., Frenk, C., White, S.D.M., 1985, Ap.J., 292, 371
Davis, M., Peebles, P.J.E., 1983, Ap.J., 267, 465
Deltorn, J.M., Le Fèvre, O., Crampton, D., Dickinson, M., 1997, Ap.J., 483, L1
Deltorn, J.-M., Le Fèvre, O., Crampton, D., Dickinson, M., 1997, in preparation
Dickinson, M., 1995, in "Galaxies in the Young Universe", eds. H. Hippelein, K. Meisenheimer, & H-J. Röser, Springer, p.144.
Dickinson, M., 1996, in "HST and the High Redshift Universe", World Scientific, eds. N. Tanvir, A. Aragon-Salamanca, and J.V. Wall, in press
Efstathiou, G., Bernstein, G., Katz, N., Tyson, J.A., Guhathakurta, P., 1991, Ap.J., 380, L47
Evrard, A.E., & Charlot, S., 1994, Ap.J., 424, L14
Fisher, B.F., Davis, M., Strauss, M.A., Yahil, A., Huchra, J., 1994, MNRAS, 266, 50
Francis, P.J., Woodgate, B.E., Warren, S.J., Moller, P., Mazzolini, M., Bunker, A.J., Lowenthal, J.D., Williams, T.B., Minezaki, T., Kobayashi, Y., Yoshii, Y., 1996, Ap.J., 457, 490
Frenk, C.S., White, S.D.M., Efstathiou, G., Davis, M., 1990, Ap.J., 351, 10
Frenk, C.S., Evrard, A.E., White, S.D.M., Summers, F.J., 1996, Ap.J., preprint
Groth, E., and Peebles, P.J.E 1977, Ap.J., 217, 385
Hill, G.J., Lilly, S.J., 1991, Ap.J., 367, 1
Hudon, D., Lilly, S., 1996, Ap.J., 469, 519

Le Fèvre, O., Crampton, D., Hammer, F., Lilly, S.J., Tresse, L., 1994a, Ap.J., 424, L14
Le Fèvre, O., Crampton, D., Felenbok, P., Monnet, G., 1994b, A&A, 282, 325
Le Fèvre, O., Crampton, D., Hammer, F., Lilly, S.J., Tresse, L., 1995, Ap.J., 455, 60 (CFRS-II)
Le Fèvre, O., Deltorn, J.-M., Crampton, D., Dickinson, M., 1996, Ap.J., 471, L11
Le Fèvre *et al.*, 1996, proc. "Wide Field Spectroscopy", Kontiza *et al.* Eds., in press
Lilly, S.J., Le Fèvre, O., Crampton, D., Hammer, F., Tresse, L., 1995a, Ap.J., 455, 50 (CFRS-I)
Lilly, S.J., Tresse, L., Hammer, F., Crampton, D., Le Fèvre, O., 1995, Ap.J., 455, 108 (CFRS-VI)
Loveday, J., Maddox, S.J., Efstathiou, G., Peterson, B.A., 1995, Ap.J., 442, 457
Luppino, G.A., Gioia, I.M., 1995, Ap.J., 445, L77
Luppino, G.A., Kaiser, N., 1997, Ap.J., 475, 20
Pascarelle, S.M., Windhorst, R.A., Driver, S.P., Ostander, E.J., & Keel, W.C., 1996, Ap.J., 456, L21
Peebles, P.J.E., Daly, R., Juszkiewicz, 1989, Ap.J., 347, 563
Smail, I., Dickinson, M., 1995, Ap.J., 455, L99
Steidel, C.C., Giavalisco, M., Pettini, M., Dickinson, M., Adelberger, K., 1996, Ap.J., 462, L17
Yee, H., Green, 1984, Ap.J., 280, 79
Zwicky, F., Herzoh, E., Wild, P., Karpowicz, M., Kowal, C.T., 1961-1968, Catalog of Galaxies and Clusters of Galaxies, Pasadena, Calif. Inst. Technol.

# A MASSIVE CLUSTER OF GALAXIES AT $Z \simeq 1$

J.-M. DELTORN AND O. LE FÈVRE
*DAEC, Observatoire de Paris-Meudon, Place J. Janssen 5, F-92195 Meudon Cedex, France*

D. CRAMPTON
*Dominion Astrophysical Observatory, National Research Council of Canada, 5071 W. Saanich Road, R.R. 5 Victoria, BC V8X 4M6, Canada*

AND

M. DICKINSON
*Space Telescope Science Institute, 3700 San Martin Drive, Baltimore, MD 21218, USA*

**Abstract.** We report the identification of a cluster of galaxies around the high-redshift radio galaxy 3CR184 at $z = 0.996$. The identification is supported by an excess of galaxies observed in projection in $I$ band images (both in ground-based and HST data), a peak in the redshift distribution comprising 11 galaxies (out of 56 with measured redshifts) in a $\sim$ 2000 km s$^{-1}$ velocity interval, and the observation on HST WFPC2 frames of a gravitational arc seen projected at $42h_{50}^{-1}$kpc away from the central radio galaxy. We thus have strong evidence for the presence of a massive cluster at $z \simeq 1$.

## 1. INTRODUCTION

Clusters at high redshifts (at $z$ >0.8, say) are poorly known systems, essentially because their identification is a challenging observational task. Indeed, the projected 2D galaxy density on which one can rely to define a cluster at low $z$ is not valid any more at higher redshifts due to the confusion induced by the increasing population of field galaxies. Even though several innovative approaches have recently been advanced to search for high-$z$ cluster from projected data only (Postman *et al.* 1996, Dalcanton 1996, Pelló 1997), the unambiguous existence of a distant cluster can only be proven if several lines of evidence are combined, including the observation of a projected



*D. Hamilton (ed.), The Evolving Universe,* 179–184.

excess of galaxies in a surface corresponding to a few typical core radius at the redshift of the presumed cluster, and the measurement of a significant overdensity in the redshift distribution of galaxies in order to reduce the contamination by foreground and background interloper galaxies. In addition, the detection of X-ray emission, and/or evidence for weak or strong lensing of background galaxies, provide strong support for the identification of genuine concentration of mass. Although a number of candidates have been proposed, only a few clusters (or proto-clusters) of galaxies have been securely identified at $z > 0.8$, with the above criteria fulfilled (Dickinson 1996; Luppino & Kaiser 1996). It is, therefore, of considerable importance for our knowledge of large scale structures to identify more high redshift clusters in order to establish the evolution of their physical properties with look-back time. Extending at $z \simeq 1$ the correlation between bright radio glaxies and associated clusters observed at $z \simeq 0.5$ (Hill & Lilly 1991), we present here the identification of a cluster around 3CR184 at $z$=0.996.

## 2. OBSERVATIONS

The field around 3CR184 was observed at CFHT in several runs between 1994 Jan. and 1995 Dec. with the Multi-Object-Spectrograph (MOS). Imaging in $I$ band allowed to select object only on the basis of their magnitude ($I < 22.2$) for spectroscopy. Four multi-slit masks with a total of 122 slits were subsequently designed. The slits were 2" in width and at least 10" in length each, and the R300 grism provided a resolution of 23Å between 5000Å and 1$\mu$m. 56 objects were identified as galaxies, 26 turned out to be galactic stars, one was identified as a QSO and 39 remained unidentified.

HST imaging was also conducted with WFPC2 during HST cycle 5. F814W and F606W images were obtained with a respective total integration time of 11000s and 6600s, leading to a completeness limit of $I = 26$. These high resolution data revealed a gravitational arc 4.9" to the North-East of 3CR184 (Deltorn *et al.* 1997) with $I = 25 \pm 0.4$ and $V - I = 0.3 \pm 0.8$. The presence of this gravitational arc close to the central radio galaxy indicates the proximity of a high concentration of mass.

## 3. EVIDENCES FOR CLUSTERING

Several estimators have been computed in order to quantify the projected excess number of galaxies in the vicinity of 3CR184. By way of example we show in Fig. 1 the density maps computed using the estimator $D_{proj}$ defined in Dressler (1980). Different cuts in color space were applied in order to maximize the contrast toward distant ($z \simeq$1) galaxies leading to a maximum excess of $10\sigma_{bg}$ above the mean galaxy background (where $\sigma_{bg}$

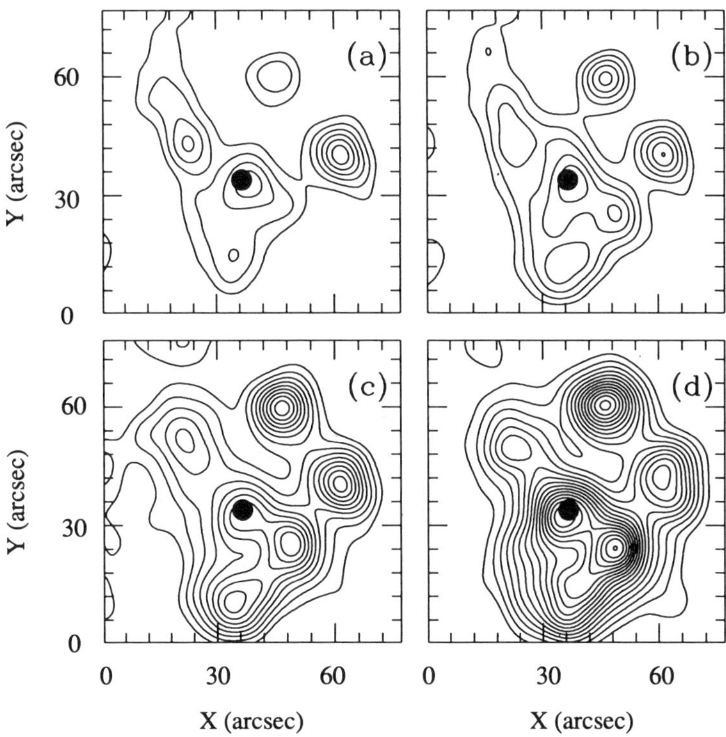


*Figure 1.* Density maps computed using Dressler's density estimator, as obtained from HST images. The first contour corresponds to $D_{proj} = 1.5$ and the following increase by steps of 0.3. The position of the radio galaxy is indicated by a disk. Each image corresponds to a different cut in color: (a) V-I>-0.35, (b) V-I>0.15, (c) V-I>0.65, (d) V-I>1.15. The objects having a high V-I lying on average at higher redshifts, these cuts select an increasingly distant population. Consequently, galaxies with $V - I > 1$ should lie predominantly at redshifts higher than 0.5. The respective excess measured in units of $\sigma_{bg}$ (see text) are (a) $\sigma_{bg} = 5.2$, (b) $\sigma_{bg} = 5.8$, (c) $\sigma_{bg} = 6.5$ and (d) $\sigma_{bg} = 10$. The presence of redder objects around the radio galaxy might indicate a dominant high redshift population participating in the observed projected overdensity. At $z \simeq 1$, 30"$\simeq 260 h_{50}^{-1}\, kpc$, for $q_0 = 0.5$.

corresponds to the square root of the variance of $D_{proj}$ measured outside a radius of 1' centered on 3CR184).

The redshift distribution of galaxies in our spectroscopic sample is shown in Fig. 2. Out of 56 objects, 11 have velocities within $+1198/-750$ kms$^{-1}$ of that of 3CR184, whereas less than one galaxies should be observed in a random sample of field galaxies in the same velocity bin (Crampton *et al.* 1995). The clear overdensity in ($\alpha, \delta$,z) space, together with the observation of a gravitational arc, secures the identification of a cluster of galaxies around 3CR184.

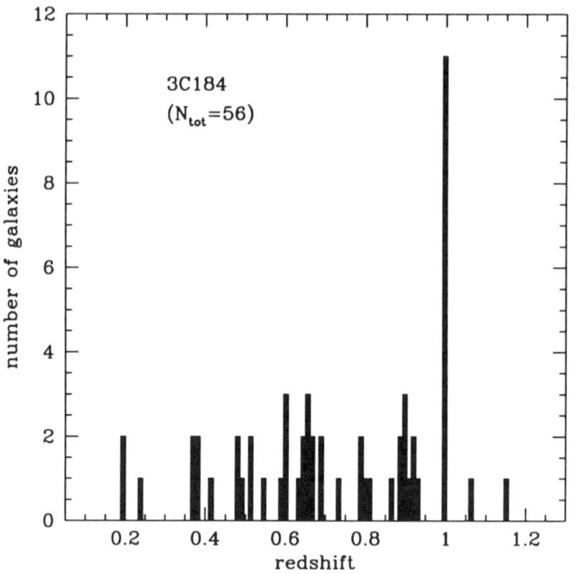


*Figure 2.* Redshift distribution of 56 galaxies in the field of 3CR184 as obtained from MOS spectroscopy at CFHT. Eleven objects are found at $z \simeq 0.996$ in a bin width corresponding to 2000 $kms^{-1}$ at the redshift of the radio galaxy.

## 4. DESCRIPTION OF THE CLUSTER

The observation of a gravitational arc associated with the cluster allows a direct determination of the mass $M_{proj}(\theta_{arc})$ within the perimeter of the arc and, assuming a simple SIS model for the mass distribution, a central velocity dispersion $\sigma_{lens}$. For any value of $z_{arc} \in [1.5, 3]$ we found $M_{proj}(\theta_{arc}) \in [2.78, 1.20]10^{13}h_{50}^{-1}M_{\odot}$, and $\sigma_{lens} \in [990, 650]$ kms$^{-1}$ (assuming the lensed galaxy is coincident with the direction of the cluster center). The measured redshifts gave a velocity dispersion $\sigma_{||} = 634^{+206}_{-102}$km s$^{-1}$, in good agreement with $\sigma_{lens}$. Assuming dynamical equilibrium, isotropy and spherical symmetry of the mass distribution we estimated the deprojected virial mass: $M = 6.16^{+3.94}_{-2.40} \times 10^{14}h_{50}^{-1}M_{\odot}$.

The mass to light ratio can also be computed after correction from the contamination due to field galaxies using the counts of Abraham *et al.* (1996) and knowing that B band at rest roughly corresponds to I at $z \simeq 1$. Assuming a no evolution scenario we find $M/L_B \simeq 56h_{50}$ within the arc radius and $(M/L_B)_{400h_{50}^{-1}} \simeq 200h_{50}$ within $400h_{50}^{-1}kpc$. On the other hand if cluster galaxies follow a luminosity evolution similar to field galaxies (Lilly *et al.* 1995), we then have $M/L_B \simeq 140h_{50}$ within $42h_{50}^{-1}kpc$ and $(M/L_B)_{400h_{50}^{-1}} \simeq 500h_{50}$.

## 5. CONCLUSION

3CR184 might therefore be the host of a massive "cluster", one of the most distant gravitationally bound structure yet identified. A first estimate of the cluster intrinsic physical parameters has been derived, demonstrating the potentialities of an approach combining multi-object spectroscopy of cluster galaxies and gravitational lensing analysis. However, any strong conclusion regarding the dynamical state of the structure will require both a higher number of confirmed cluster galaxies, and to test the assumptions underlying the derivation of the mass or the velocity dispersion (either dynamical equilibrium or spherical symmetry) and the lensing geometry.

The number of high redshift clusters seems to be steadily increasing as observational capabilities become more acute, and the observation of this massive structure at $z \simeq 1$, combined with other observations of very high redshift clusters either through spectroscopic observation of cluster members, weak gravitational lensing, or X-ray observations might come to be in severe conflict with various cosmological scenarios. As the abundance of massive structures provides increasing discrimination with increasing redshift, the identification, without ambiguity, of clusters at $z \simeq 1$ is now beginning to provide useful observational constraints to the cosmological models.

## References

Abraham, R.G., Tanvir, N.R., Santiago, B.X., Ellis, R.S., Glazebrook, K.,van den Bergh, S., MNRAS, in press
Crampton, D., Le Fèvre, O., Lilly, S.J., Hammer, F., 1995, ApJ, 455, 96
Dalcanton, J.J., 1996, ApJ, 466, 92
Deltorn, J.-M., Le Fèvre, O., Crampton, D., Dickinson, M., 1997, preprint
Dickinson, M., 1996, in "HST and the High Redshift Universe", World Scientific, eds. N. Tanvir, A. Aragon-Salamanca, and J.V. Wall, in press
Dressler, 1980, ApJ, 236, 351
Hill, G.J., Lilly, S.J., 1991, ApJ, 367, 1
Lilly, S.J., Tresse, L., Hammer, F., Crampton, D., Le Fèvre, O., 1995, ApJ, 455, 108

Luppino, G.A., Kaiser, N., 1997, ApJ, 475, 20
Pelló, R., Miralles, J.M., Le Borgne, J.-F., Picat, J.-P., Soucail, G., Bruzual, G., 1997, preprint
Postman, M., Lubin, L.M., Gunn, J.E., Oke, J.B., Hoessel, J.G., Schneider, D.P., Christensen, J.A., 1996, AJ, 111, 615

# LINEAR REDSHIFT DISTORTIONS: A REVIEW

A. J. S. HAMILTON
*JILA & Dept. of Astrophysical and Planetary Sciences*
*Box 440, U. Colorado, Boulder, CO 80309, USA;*
*Andrew.Hamilton@colorado.edu;*
*http://casa.colorado.edu/~ajsh*

**Abstract.** Redshift maps of galaxies in the Universe are distorted by the peculiar velocities of galaxies along the line of sight. The amplitude of the distortions on large, linear scales yields a measurement of the linear redshift distortion parameter, which is $\beta \approx \Omega_0^{0.6}/b$ in standard cosmology with cosmological density $\Omega_0$ and light-to-mass bias $b$. All measurements of $\beta$ from linear redshift distortions published up to mid 1997 are reviewed. The average and standard deviation of the reported values is $\beta_{\rm optical} = 0.52 \pm 0.26$ for optically selected galaxies, and $\beta_{IRAS} = 0.77 \pm 0.22$ for *IRAS* selected galaxies. The implied relative bias is $b_{\rm optical}/b_{IRAS} \approx 1.5$. If optical galaxies are unbiased, then $\Omega_0 = 0.33^{+0.32}_{-0.22}$, while if *IRAS* galaxies are unbiased, then $\Omega_0 = 0.63^{+0.35}_{-0.27}$.

## 1. Introduction

The organizers of this thoroughly enjoyable workshop asked me to write a review of redshift distortions aimed primarily at graduate students and others who are not familiar with the field. The review aims at fairly thorough coverage (up to mid 1997) within a rather limited scope: the subject of redshift distortions in the large scale, linear regime. The review does not attempt to cover the large body of work involving the direct measurement of peculiar velocities. The latter has been the subject of recent comprehensive reviews by Strauss & Willick (1995), and by Dekel (1994). Both of those reviews included sections on redshift distortions. Nor does the present review cover nonlinear redshift distortions, except insofar as they affect linear redshift distortions. For an entry to the literature on nonlinear redshift distortions, try Davis, Miller & White (1997).



*D. Hamilton (ed.), The Evolving Universe,* 185–275.

Hubble's (1929) law states that the recession velocity $cz$ of a galaxy is proportional to its distance $d$

$$cz = H_0 d \ , \tag{1.1}$$

with constant of proportionality the Hubble constant $H_0$ (the subscript 0 signifies its present day value). The recession velocity $cz$ of a galaxy can be measured from the redshift $z$ of its spectrum ($c$ is the speed of light), a great deal more easily and accurately than its true distance $d$. This has been a primary motivation for redshift surveys (see e.g. Strauss 1997 for a recent review), which map the Universe in 3 dimensions using the recession velocity $cz$ of each galaxy as a measure of its distance.

Hubble's law is not perfect, however. Galaxies have peculiar velocities $\boldsymbol{v}$ relative to the general Hubble expansion. Thus it is necessary in general to distinguish between a galaxy's **redshift distance** $s$ (conveniently expressed in velocity units)

$$s \equiv cz \tag{1.2}$$

and its true distance $r$ (also conveniently expressed in velocity units)

$$r \equiv H_0 d \ . \tag{1.3}$$

The redshift distance $s$ of a galaxy differs from the true distance $r$ by its peculiar velocity $v \equiv \hat{\boldsymbol{r}}.\boldsymbol{v}$ along the line of sight:

$$s = r + v \ . \tag{1.4}$$

The peculiar velocities of galaxies thus cause them to appear displaced along the line of sight in redshift space. These displacements lead to **redshift distortions** in the pattern of clustering of galaxies in redshift space. Although such distortions complicate the interpretation of redshift maps as positional maps, they have the tremendous advantage of bearing information about the dynamics of galaxies. In particular, the amplitude of distortions on large scales yields a measure of the **linear redshift distortion parameter** $\beta$, which is related to the cosmological density $\Omega_0$, the present day ratio of the matter density of the Universe to the critical density required to close it, by (see §4.1)

$$\beta = \frac{f(\Omega_0)}{b} \approx \frac{\Omega_0^{0.6}}{b} \tag{1.5}$$

in standard pressureless Friedmann cosmology with light-to-mass bias $b$. The goal of much of the current work on redshift distortions is to measure the linear distortion parameter $\beta$, and perhaps, pious hope, if bias can be quantified through nonlinear effects or otherwise, to determine the cosmological density $\Omega_0$ itself.

### 1.1. PLAN

The plan of this review is as follows.

Section 2 attempts to convey visually what redshift distortions look like and why, both §2.1 schematically, and §2.2 observationally.

Section 3 is about power spectra. Sections 3.1 and 3.2 contains definitions, needed for subsequent reference, of correlation functions and power spectra in real and redshift space. Section 3.3 advertises the delights of Hilbert space. Section 3.4 explains why power spectra are better.

Section 4 presents the theory of redshift distortions in the linear regime. The first part, §4.1, explores what $\beta$, the linear redshift distortion parameter, really means. The second part, §4.2, derives the linear redshift distortion operator, which transforms real space into redshift space, for fluctuations in the linear regime. The third and fourth parts, §§4.3 and 4.4, cover two small but important details, the peculiar motion of us, the observers, and the difference between the selection functions in real and redshift space.

Section 5 describes the three different types of method that have been used to date to measure $\beta$ from linear redshift distortions: §5.1, the ratio of real to redshift angle-averaged power; §5.2, the ratio of quadrupole-to-monopole harmonics of the redshift power spectrum; and §5.3, the maximum likelihood approach.

Section 6 interjects an example of measuring $\beta$ from linear redshift distortions, partly as an illustration, partly to bring out the difference between optical and *IRAS*-selected galaxies, and partly to demonstrate the importance of nonlinearities.

Section 7 describes the two methods that have been used to date to deal with nonlinearity: §7.2, a model in which linear redshift distortions are modulated by a random velocity dispersion; and §7.3, the Zel'dovich approximation.

Section 8 compiles measurements of $\beta$ from linear redshift distortions, complete up to the first half of 1997. Section 8.1 summarizes the measurements of $\beta$ in a Table, and gives the average and standard deviation of the measurements, which results are the ones quoted in the abstract. Section 8.2 offers commentary on all the individual measurements. I apologize to authors whose work has been inadvertently omitted, or inadequately portrayed.

Finally, §9 discourses briefly on cosmological redshift distortions, which arise from differences in the geometry of the Universe perceptible at high enough redshift.

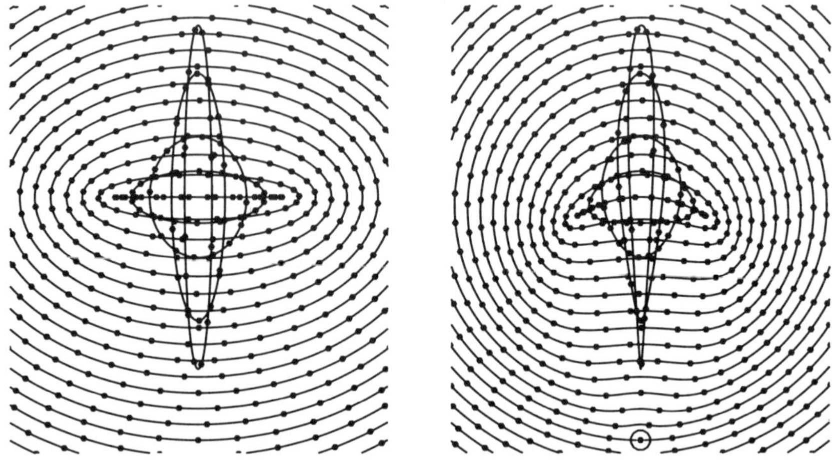

*Figure 1.* A spherical overdensity appears distorted by peculiar velocities when observed in redshift space. On large (linear) scales the overdensity appears squashed along the line of sight, while on small (nonlinear) scales fingers-of-god appear. At left, the overdensity is far from the observer (who is looking upward from somewhere way below the bottom of the diagram), and the distortions are effectively plane-parallel. At right, the overdensity is near the observer (large dot), and the large scale distortions appear kidney-shaped, while the finger-of-god is sharpened on the end pointing at the observer. The observer shares the infall motion towards the overdensity. A similar diagram appears in Kaiser (1987).

The greater part of this review is just that, a review; but there are some new things here and there. The expression for the linear redshift distortion operator $\mathbf{S}^{s\,\mathrm{LG}}$ for the practical case where (a) the redshift overdensity is measured in the Local Group frame, and (b) the selection function is measured in redshift space, also in the Local Group frame, appears here explicitly for the first time, equation (4.81). The analysis of the Stromlo-APM survey reported in §6 has not been published elsewhere.

## 2. What Redshift Distortions Look Like

### 2.1. SCHEMATICALLY

Figure 1 illustrates how a spherical overdensity appears distorted by peculiar velocities along the line of sight, when observed in redshift space. The initial spherical overdensity perturbation here was taken to be a power law with radius, $\delta \propto r^{-1}$, located in an expanding Universe with critical mean density, $\Omega = 1$. The free-fall gravitational collapse of such a spherical pressureless overdensity can be computed analytically (Peebles 1980, §18). The dots (galaxies) started out uniformly distributed in the initial condi-

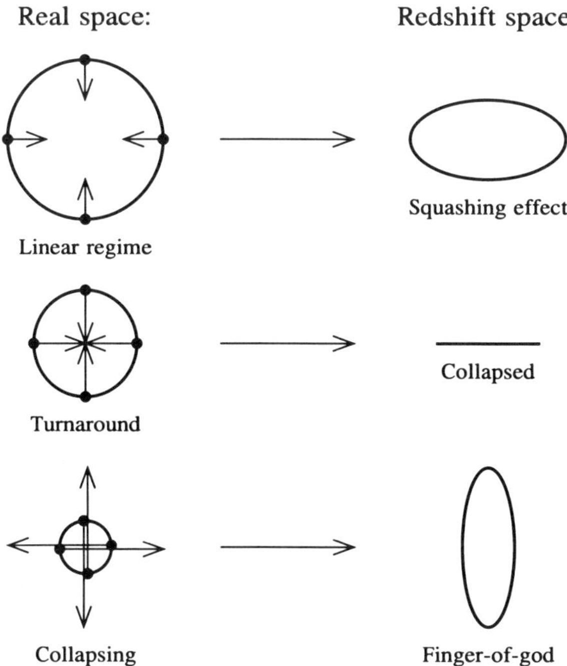


*Figure 2.* Detail of how peculiar velocities lead to the redshift distortions illustrated in Figure 1. The dots are 'galaxies' undergoing infall towards a spherical overdensity, and the arrows represent their peculiar velocities. At large scales, the peculiar velocity of an infalling shell is small compared to its radius, and the shell appears squashed. At smaller scales, not only is the radius of a shell smaller, but also its peculiar infall velocity tends to be larger. The shell that is just at turnaround, its peculiar velocity just cancelling the general Hubble expansion, appears collapsed to a single velocity in redshift space. At yet smaller scales, shells that are collapsing in proper coordinates appear inside out in redshift space. The combination of collapsing shells with previously collapsed, virialized shells, gives rise to fingers-of-god.

tions, being uniformly placed around a series of uniformly spaced concentric shells. Thus the density of dots in Figure 1 indicates the density of galaxies in the collapsing overdensity, as observed in redshift space. Figure 1 omits shells that have collapsed to less than half their radius at turnaround, which shells may be expected to scatter off previously collapsed shells, and to virialize.

Figure 2 shows how peculiar velocities produce the pattern illustrated in Figure 1. On large scales, peculiar infall towards the overdensity causes it to appear squashed along the line of sight. The squashing increases to smaller scales down to the point of turnaround, where the peculiar infall velocity exactly cancels the general Hubble expansion. In the turnaround shell, the near and far parts of the shell appear collapsed to a single radial velocity in redshift space. At smaller scales, shells that have turned around and are collapsing in real space appear turned 'inside out' in redshift space. At even smaller scales (not shown in Figures 1 or 2), collapsed shells are expected to virialize. The combination of collapsing and virialized regions of galaxy clusters gives rise to **fingers-of-god**.

### 2.2. OBSERVATIONALLY

Fingers-of-god are well-known features of redshift surveys. Prominent examples are the fingers-of-god in the Coma cluster (de Lapparent, Geller & Huchra 1986) and the Perseus cluster (Wegner, Haynes & Giovanelli 1993, Figs. 7–10).

The envelope of the finger-of-god in Figure 1 forms a caustic, a surface of infinite density (but finite mass). Such caustics are not obviously seen in real fingers-of-god (Regős & Geller 1989); presumably the caustics are smeared out by subclustering. The structure of the well-studied Coma cluster, for example, is quite complicated (Colless & Dunn 1996).

Visually, the large scale squashing effect is more subtle to discern in real data. Are prominent transverse structures such as the Great Wall (Ramella, Geller & Huchra 1992) enhanced by redshift distortions? Probably yes, at some level (e.g. Praton, Melott & McKee 1997). However, Dell'Antonio, Geller & Bothun (1996) conclude from their analysis of the peculiar velocity field of the Great Wall that any infall is small, $\lesssim 150\,\mathrm{km\,s^{-1}}$.

The large scale squashing effect can however be detected statistically, from the distortion of the redshift space correlation function $\xi^s$, or of its Fourier transform the redshift space power spectrum $P^s$. These quantities are defined formally in the next section, §3. Physically, the redshift correlation function $\xi^s(s_{/\!/}, s_\perp)$ is the mean fractional excess of galaxy neighbors of a galaxy at separations $s_{/\!/}$ and $s_\perp$ parallel and perpendicular to the line of sight, a definition that suffices to understand Figure 3.

Figure 3 shows redshift space correlation functions $\xi^s(s_{/\!/}, s_\perp)$ measured from two sets of redshift surveys, one (QDOT + 1.2 Jy) selected in the infrared, the other (Stromlo-APM) in the optical. The infrared survey is a merger of two *Infrared Astronomical Satellite* (*IRAS*) redshift surveys, the QDOT survey, and the 1.2 Jy survey. The revised QDOT survey (Lawrence *et al.* 1997, in preparation) is a redshift survey of a 1-in-6 subset of galaxies brighter than 0.6 Jy from the *IRAS* Point Source Catalog (PSC). It contains 2376 galaxies over 9.29365 steradians, covering most of the sky above galactic latitude $|b| > 10°$. The *IRAS* 1.2 Jy survey (Fisher *et al.* 1995a; Strauss *et al.* 1992a) is a complete redshift survey of galaxies brighter than 1.2 Jy from the same *IRAS* PSC. It contains 5321 galaxies over 11.02577 steradians, covering most of the sky above galactic latitude $|b| > 5°$. For

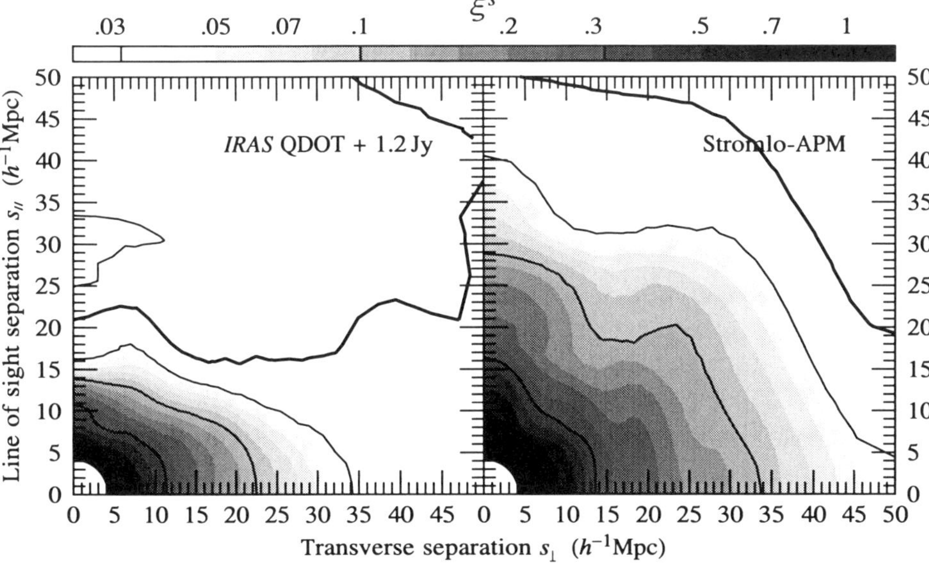


*Figure 3.* Contour plots of the redshift space two-point correlation function $\xi^s$ as a function of separations $s_{//}$ and $s_{\perp}$ parallel and perpendicular to the line of sight in: (left) the *IRAS* QDOT and 1.2 Jy redshift surveys, merged over the angular region of the sky common to both surveys; and (right) the optical Stromlo-APM survey. In each case the region within $25h^{-1}$Mpc of the Milky Way has been excluded, so as to eliminate bias from the local overdensity. A near minimum variance pair weighting has been applied. The thick contour signifies $\xi^s = 0$, and other contours are logarithmically spaced at intervals of 0.5 dex above $10^{-1.5}$ (the left panel also shows one negative contour, at $-10^{-1.5}$). Shading is graduated at intervals of 0.1 dex above $10^{-1.5}$. The correlation function here has been smoothed over pair separation $s = (s_{\perp}^2 + s_{//}^2)^{1/2}$ with a tophat window of width 0.2 dex, and over angles $\theta = \tan^{-1}(s_{\perp}/s_{//})$ to the line of sight with a Gaussian window with a $1\sigma$ width of $10°$.

Figure 3, the surveys were merged over the part of the sky common to both surveys, yielding 5752 galaxies over 9.26740 steradians. Removing the region closer than $25\,h^{-1}$Mpc, to avoid local bias (Hamilton & Culhane 1996, §2.1) left 4826 galaxies. The optical survey is the Stromlo-APM survey (Loveday *et al.* 1996b), which is a redshift survey of a 1-in-20 subset of galaxies brighter $b_J = 17.15$ from the Automatic Plate Measuring (APM) survey (Maddox *et al.* 1990a,b, 1996). The Stromlo-APM redshift survey contains 1790 galaxies over 1.32467 steradians centred roughly around the South Galactic Pole. Eliminating the region closer than $25\,h^{-1}$Mpc left 1725 galaxies.

To ensure the validity of the plane-parallel approximation, the redshift correlation function shown in Figure 3 was computed including only pairs closer than $50°$ on the sky, the line of sight to each pair being defined as the bisector (the angular midpoint on the sky) of the pair. A near minimum variance pair weighting was applied as described by Hamilton (1993b, §5, eqs. [60] & [61]). Further details of the procedure used here to calculate $\xi^s(s_{/\!/}, s_\perp)$ are given in the latter paper.

Figure 3 shows clearly the expected large scale squashing effect, while fingers-of-god show up as an enhancement of the redshift correlation function along the line-of-sight axis. Besides these two expected effects, two other features are apparent. Firstly, the redshift correlation function of the optically selected galaxies is roughly a factor of 2 higher than that of the *IRAS* selected galaxies. Secondly, the fingers-of-god are longer and more prominent in the optical galaxies. Both these features can be attributed to the fact that *IRAS* galaxies, which are dusty, gas-rich spirals, avoid the centres of rich clusters of galaxies, where ellipticals rule.

A contour plot of the redshift space power spectrum from an $N$-body simulation can be found in Cole, Fisher & Weinberg (1994, Fig. 1), while Bromley, Warren & Zurek (1997, Fig. 1) show plots of the redshift space power spectrum as a function of $\theta = \tan^{-1}(k_\perp/k_{/\!/})$ from two large (17 million particle) high resolution simulations.

## 3. Correlation Functions and Power Spectra

Sections 3.1 and 3.2 collect standard definitions of correlation functions and power spectra, in real and redshift space respectively, needed elsewhere in this review. Compared to theoretical correlation functions, defined in §§3.1.1 and 3.2.1, observed correlation functions contain additional noise, the dominant contribution to which is often taken to be Poisson sampling noise, also known as shot noise, discussed in §§3.1.2 and 3.2.2. I adhere throughout this review to what has become the standard convention in the field of galaxy clustering for defining Fourier transforms, notwithstanding the extraneous factors of $2\pi$ that result.

Section 3.3 points out how Hilbert space provides a compact, powerful notation.

Section 3.4 explains what makes power spectra special.

### 3.1. REAL SPACE

A fundamental proposition is that the three dimensional distribution of matter in the Universe constitutes a statistically homogeneous and isotropic random density field. This proposition, a reflection of the proposition that the Universe at large is homogeneous and isotropic — Milne's 1933 Cosmological Principle (see Peebles 1980, §3B) — is powerfully evidenced by the isotropy of the Cosmic Microwave Background (Bennett *et al.* 1996; Górski 1997).

The statistical properties of such a field are completely determined by its irreducible moments, or correlation functions as they are commonly called in the discipline of large scale structure (Peebles 1980, Ch. III). The first two irreducible moments are the mean, a constant, and the covariance, or 2-point correlation function, a function of separation.

In the case of a multivariate Gaussian random field, for which by definition the third and all higher correlation functions vanish, the mean and 2-point correlation function completely specify the statistical properties of the field. If fluctuations generated in the early Universe were the result of a superposition of many independent random processes, as for example quantum fluctuations during inflation, then the Central Limit Theorem guarantees that the fluctuations will form a multivariate Gaussian. Fluctuations on large, linear scales today would then also be Gaussian. Available observational evidence is consistent with fluctuations being Gaussian on linear scales (Stirling & Peacock 1996; Kogut *et al.* 1996).

Once density fluctuations grow large, they cannot remain Gaussian, since density must remain positive. Thus on small, nonlinear scales the third and higher order correlation functions must be non-vanishing. Nonetheless the covariance, the 2-point correlation function, is well-defined in the nonlinear regime, and remains a statistic of fundamental importance.

#### 3.1.1. *The True Density Field of the Universe*

Let $\rho(\boldsymbol{r})$ denote the density of matter at (real, not redshift) position $\boldsymbol{r}$ in the Universe, and let $\bar{\rho}$, a constant in space, denote the mean density. The density $\rho(\boldsymbol{r})$ may be regarded as the density of mass, or of particles, or of galaxies, or of some particular set of objects one is interested in. Whatever the case, the density $\rho(\boldsymbol{r})$ signifies the true density, not the observed density of objects that may be recorded in a real survey (which is considered in §§3.1.2 and 3.2 following), and $\boldsymbol{r}$ is the true position, not the redshift position. The true **overdensity** $\delta(\boldsymbol{r})$ at position $\boldsymbol{r}$ is defined by

$$\delta(\boldsymbol{r}) \equiv \frac{\rho(\boldsymbol{r}) - \bar{\rho}}{\bar{\rho}} \ . \tag{3.1}$$

It is to be noted that although $\delta(\boldsymbol{r})$ is defined to be the 'true' overdensity, it is nevertheless a hypothetical quantity, to be distinguished from a quantity, such as the redshift, angular position, or flux of a galaxy, which is directly observed. This is the usual situation in statistics, where one imagines a hypothetical 'true' population from which observed data are drawn. The properties of statistical homogeneity and isotropy attributed to the 'true' density field are likewise theoretical hypotheses, whose validity can be tested against observed data, though never proven absolutely. The distinction between observed data and theoretical models thereof appears starkly in the maximum likelihood procedure, §5.3.

The most basic statistic that can be constructed from the overdensity (the mean having been subtracted off) is its variance, its second irreducible moment, the **correlation function** $\xi(r_{12})$ (also known as the 2-point correlation function, or 2-point function, or covariance function, or autocovariance function),

$$\xi(r_{12}) \equiv \langle \delta(\boldsymbol{r}_1) \delta(\boldsymbol{r}_2) \rangle \ . \tag{3.2}$$

Equation (3.2) states that the expectation value of the product of overdensities at a pair of randomly positioned points separated by $r_{12}$ in the Universe is $\xi(r_{12})$. More physically, the correlation function $\xi(r_{12})$ is the mean overdensity of neighbors around a random particle (galaxy) (Peebles 1980, §31). The assumption that the density field is statistically homogeneous and isotropic means that the correlation function $\xi(r_{12})$ is a function only of the scalar separation $r_{12} \equiv |\boldsymbol{r}_1 - \boldsymbol{r}_2|$ of the points $\boldsymbol{r}_1$ and $\boldsymbol{r}_2$, not of their overall location or orientation.

The Fourier transform of the overdensity defines the Fourier modes $\hat{\delta}(\boldsymbol{k})$ at wavevector $\boldsymbol{k}$ (hats denote Fourier transforms throughout this review)

$$\hat{\delta}(\boldsymbol{k}) \equiv \int e^{i\boldsymbol{k}.\boldsymbol{r}} \delta(\boldsymbol{r})\, d^3r \ , \quad \delta(\boldsymbol{r}) = \int e^{-i\boldsymbol{k}.\boldsymbol{r}} \hat{\delta}(\boldsymbol{k})\, d^3k/(2\pi)^3 \ . \tag{3.3}$$

The **power spectrum** is by definition the covariance of Fourier modes, which from the definitions (3.3) and (3.2) is equal to the Fourier transform of the correlation function

$$\langle \hat{\delta}(\boldsymbol{k}_1) \hat{\delta}(\boldsymbol{k}_2) \rangle = \int e^{i\boldsymbol{k}_1.\boldsymbol{r}_1 + i\boldsymbol{k}_2.\boldsymbol{r}_2} \xi(r_{12})\, d^3r_1 d^3r_2 \ . \tag{3.4}$$

Since the correlation function $\xi(r_{12})$ is a function only of separation, equation (3.4) reduces to[1]

$$\langle \hat{\delta}(\boldsymbol{k}_1) \hat{\delta}(\boldsymbol{k}_2) \rangle = (2\pi)^3 \delta_D(\boldsymbol{k}_1 + \boldsymbol{k}_2) P(k_1) \ , \tag{3.5}$$

where $P(k)$ is also called the power spectrum,

$$P(k) \equiv \int e^{i\boldsymbol{k}.\boldsymbol{r}} \xi(r)\, d^3r \ , \quad \xi(r) = \int e^{-i\boldsymbol{k}.\boldsymbol{r}} P(k)\, d^3k/(2\pi)^3 \ . \tag{3.6}$$

[1]It is also fine to define the power spectrum as the covariance of modes with one of the modes taken to be the complex conjugate, in which case $\langle \hat{\delta}(\boldsymbol{k}_1) \hat{\delta}^*(\boldsymbol{k}_2) \rangle = (2\pi)^3 \delta_D(\boldsymbol{k}_1 - \boldsymbol{k}_2) P(k_1)$. The equivalence of the two definitions is made clear in §3.3. The advantage of the symmetric choice (3.5) becomes more apparent when dealing with higher order correlation functions, such as the 3-point function $\langle \hat{\delta}(\boldsymbol{k}_1) \hat{\delta}(\boldsymbol{k}_2) \hat{\delta}(\boldsymbol{k}_3) \rangle$.

The 'momentum-conserving' 3-dimensional Dirac delta function $\delta_D(\boldsymbol{k}_1+\boldsymbol{k}_2)$ in equation (3.5) expresses the assumed translation invariance, i.e. statistical homogeneity, of clustering, while the fact that $P(k)$ is a function only of the absolute value $k$ of the wavevector $\boldsymbol{k}$ expresses statistical isotropy.

It is useful also to give here results in spherical transform space, in which the density field is expanded in spherical harmonics about the observer. Let $\delta_{\ell m}(r)$ denote spherical modes in real space

$$\delta_{\ell m}(r) \equiv \int Y_{\ell m}(\hat{\boldsymbol{r}})\delta(\boldsymbol{r})\, do_{\boldsymbol{r}} \ , \quad \delta(\boldsymbol{r}) = \sum_{\ell m} Y^*_{\ell m}(\hat{\boldsymbol{r}})\delta_{\ell m}(r) \tag{3.7}$$

($do_{\boldsymbol{r}}$ denotes an interval of solid angle in real space) and similarly let $\hat{\delta}_{\ell m}(k)$ denote spherical modes in Fourier space

$$\hat{\delta}_{\ell m}(k) \equiv \int Y_{\ell m}(\hat{\boldsymbol{k}})\hat{\delta}(\boldsymbol{k})\, do_{\boldsymbol{k}} \ , \quad \hat{\delta}(\boldsymbol{k}) = \sum_{\ell m} Y^*_{\ell m}(\hat{\boldsymbol{k}})\hat{\delta}_{\ell m}(k) \tag{3.8}$$

($do_{\boldsymbol{k}}$ denotes an interval of solid angle in Fourier space) where $Y_{\ell m}$ are the usual orthonormal spherical harmonics. The spherical transforms in real and Fourier space are related by

$$\begin{aligned} \hat{\delta}_{\ell m}(k) &= i^{\ell} 4\pi \int_0^\infty j_\ell(kr)\delta_{\ell m}(r)\, r^2 dr \ , \\ \delta_{\ell m}(r) &= i^{-\ell} 4\pi \int_0^\infty j_\ell(kr)\hat{\delta}_{\ell m}(k)\, k^2 dk/(2\pi)^3 \end{aligned} \tag{3.9}$$

where $j_\ell(kr)$ are spherical Bessel functions. The reality conditions $\delta^*(\boldsymbol{r}) = \delta(\boldsymbol{r})$, hence $\hat{\delta}^*(\boldsymbol{k}) = \hat{\delta}(-\boldsymbol{k})$, along with the usual properties $Y^*_{\ell m} = (-)^m Y_{\ell,-m}$ and $Y_{\ell m}(-\hat{\boldsymbol{k}}) = (-)^\ell Y_{\ell m}(\hat{\boldsymbol{k}})$ of the spherical harmonics, imply

$$\delta^*_{\ell m}(r) = (-)^m \delta_{\ell,-m}(r) \ , \quad \hat{\delta}^*_{\ell m}(k) = (-)^{\ell+m}\hat{\delta}_{\ell,-m}(k) \ . \tag{3.10}$$

The correlation function of spherical modes in real space is

$$\langle \delta_{\ell_1 m_1}(r_1)\,\delta_{\ell_2 m_2}(r_2)\rangle = (-)^{m_1}\delta^K_{\ell_1\ell_2}\delta^K_{m_1,-m_2}\xi_\ell(r_1,r_2) \tag{3.11}$$

where $\delta^K$ is the Kronecker delta, and the reduced correlation function $\xi_\ell(r_1,r_2)$ is related to the power spectrum $P(k)$ by

$$\xi_\ell(r_1,r_2) = (4\pi)^2 \int_0^\infty j_\ell(kr_1)j_\ell(kr_2)P(k)\, k^2 dk/(2\pi)^3 \ . \tag{3.12}$$

The extraneous minus signs in (3.11), and also in the next equation (3.13), disappear if one takes the complex conjugate on one of the overdensities, as in $\langle \delta \delta^* \rangle$ in place of $\langle \delta \delta \rangle$; see §3.3 for clarification of this point. The correlation function of spherical modes in Fourier space is

$$\langle \hat{\delta}_{\ell_1 m_1}(k_1) \hat{\delta}_{\ell_2 m_2}(k_2) \rangle = (-)^{\ell_1 + m_1} \delta^K_{\ell_1 \ell_2} \delta^K_{m_1, -m_2} (2\pi)^3 \delta_D(k_1 - k_2) k_1^{-2} P(k_1) \tag{3.13}$$

where $\delta_D(k_1 - k_2)$ is the 1-dimensional Dirac delta-function, satisfying $\int \delta_D(k) dk = 1$ integrated over any interval containing the origin $k = 0$. As explained in §3.3, the expression in front of $P(k_1)$ on the right hand side of equation (3.13) is just the unit matrix in $k\ell m$-space, equation (3.42).

### 3.1.2. *The Observed Density Field and Shot Noise*

Real surveys probe only a portion of the density field of the Universe, and what they do survey is liable to be an imperfect representation of the true density field. This subsection discusses what is often assumed to be the dominant source of noise in a survey, which is Poisson sampling noise, often also called shot noise. It is assumed in this subsection that the data lie in real space, not redshift space. Corresponding results in redshift space are given in §3.2.2.

Typically, a galaxy survey does not include all galaxies in a region of space, but only, say, those brighter than some flux limit. Thus a survey generally blends a finer sampling of dim galaxies nearby with a sparser sampling of luminous galaxies farther away. A survey is characterized by its **selection function** $\bar{n}(\boldsymbol{r})$, which is the expected mean number of galaxies at position $\boldsymbol{r}$ given the selection criteria (e.g. the flux limit) of the survey.

The selection function $\bar{n}(\boldsymbol{r})$ of a survey must be measured from it. The problem of measuring the selection function of a survey is reviewed by Binggeli, Sandage & Tammann (1988), and is discussed further in §4.4. Some additional comments on the measurement of the selection function appear in the maximum likelihood section, §5.3, in the paragraph following equation (5.18).

In order to combine heterogeneously sampled regions, the (testable) hypothesis is commonly made that the galaxies observed are drawn randomly from a hypothetical continuous underlying population of galaxies. The observed galaxies then form a **Poisson process** on the underlying population, and the selection function $\bar{n}(\boldsymbol{r})$ is interpreted as specifying the probability (in units of number of galaxies per unit volume) of including a galaxy at position $\boldsymbol{r}$ into the survey.

Let $n(\boldsymbol{r})$ denote the observed number density of galaxies at (real, not redshifted) position $\boldsymbol{r}$ in a survey. The number density $n(\boldsymbol{r})$ is a sum of delta functions, since galaxies come as discrete units. The **observed galaxy**

**overdensity** $\delta_{\rm obs}(\boldsymbol{r})$ is then defined by

$$\delta_{\rm obs}(\boldsymbol{r}) \equiv \frac{n(\boldsymbol{r}) - \bar{n}(\boldsymbol{r})}{\bar{n}(\boldsymbol{r})} \ . \tag{3.14}$$

In the Poisson process model, the observed galaxy overdensity $\delta_{\rm obs}(\boldsymbol{r})$ provides a discretized but unbiased estimate of the true overdensity $\delta(\boldsymbol{r})$, equation (3.1). The subscript obs is retained throughout this section 3 to distinguish the estimate $\delta_{\rm obs}$ from the hypothesized true value $\delta$, but is dropped from most of this review to avoid an overly ponderous notation (correctly, estimated values not only of $\delta$ but also of all other measured quantities should be distinguished from their 'true' values). It should be clear from the context whether what is meant is the theoretical 'true' value of a quantity, or an observational estimate thereof (the theoretical sections 4, 5, and 7 refer largely to theoretical quantities, while the observational sections 6 and 8 report mainly estimates thereof).

In the Poisson process model, the expectation value $C(\boldsymbol{r}_1, \boldsymbol{r}_2)$ of the covariance of observed overdensities is a sum of the true correlation function $\xi(r_{12})$ with a **Poisson sampling noise**, or **shot noise**, term:

$$\langle \delta_{\rm obs}(\boldsymbol{r}_1)\, \delta_{\rm obs}(\boldsymbol{r}_2) \rangle \equiv C(\boldsymbol{r}_1, \boldsymbol{r}_2) = \xi(r_{12}) + \delta_D(\boldsymbol{r}_1 - \boldsymbol{r}_2)[\bar{n}(\boldsymbol{r}_1)]^{-1} \ . \tag{3.15}$$

The form $\delta_D(\boldsymbol{r}_1 - \boldsymbol{r}_2)[\bar{n}(\boldsymbol{r}_1)]^{-1}$ of the Poisson sampling term can be derived from the following argument. First, note that $\int_V n(\boldsymbol{r}_1) n(\boldsymbol{r}_2) d^3 r_1 d^3 r_2 = \int_V n(\boldsymbol{r}_1) d^3 r_1 = 0$ or 1 when the integration is over an infinitesimal volume $V$, which either does not (0) or does (1) contain a galaxy. It follows that the expectation value of the product of densities in the same infinitesimal volume element is $\langle n(\boldsymbol{r}_1) n(\boldsymbol{r}_2) \rangle = \delta_D(\boldsymbol{r}_1 - \boldsymbol{r}_2) \langle n(\boldsymbol{r}_1) \rangle = \delta_D(\boldsymbol{r}_1 - \boldsymbol{r}_2) \bar{n}(\boldsymbol{r}_1)$, whence $\langle \delta_{\rm obs}(\boldsymbol{r}_1) \delta_{\rm obs}(\boldsymbol{r}_2) \rangle = \delta_D(\boldsymbol{r}_1 - \boldsymbol{r}_2)[\bar{n}(\boldsymbol{r}_1)]^{-1}$, which is the shot noise term of equation (3.15) as claimed. The Poisson sampling term reflects the fact that the probability of finding yourself as a neighbor at zero separation is unity. The shot noise becomes infinite in regions outside a survey where the selection function is zero, which makes sense.

Equation (3.15) shows that, in the Poisson process model, the expectation value of the survey covariance is equal to the true correlation function at any finite separation

$$C(\boldsymbol{r}_1, \boldsymbol{r}_2)_{\boldsymbol{r}_1 \neq \boldsymbol{r}_2} = \xi(r_{12}) \ . \tag{3.16}$$

Thus any average $\langle \delta_{\rm obs}(\boldsymbol{r}_1) \delta_{\rm obs}(\boldsymbol{r}_2) \rangle$ of products of pairs of overdensities at any finite separation $r_{12} \equiv |\boldsymbol{r}_1 - \boldsymbol{r}_2| \neq 0$, weighted in any arbitrary a priori fashion, provides an unbiased estimate of the true correlation function $\xi(r_{12})$.

In estimating the correlation function from a survey, the shot noise contribution can be eliminated by excluding from the computation all self-pairs of galaxies (pairs consisting of a galaxy and itself). Alternatively, it may be convenient to include self-pairs, and to subtract off the shot noise as a separate step.

### 3.2. REDSHIFT SPACE

Redshift space quantities (distinguished in this review by a superscript $s$) are defined analogously to real space quantities. Let $n^s(\boldsymbol{s})$ denote the observed number density of galaxies at redshift position $\boldsymbol{s}$ in a redshift survey (the observer is at the origin $\boldsymbol{s} = 0$).

A slightly subtle point in the definition of overdensity $\delta^s(\boldsymbol{s})$ in redshift space arises because, as emphasized by Fisher, Scharf & Lahav (1994), the apparent brightness of a galaxy depends on its true distance $r$, not its redshift distance $s$, so that the selection function of a flux-limited redshift survey is correctly a function $\bar{n}(\boldsymbol{r})$ in real space, not redshift space. This suggests that one might define the observed overdensity in redshift space by $\delta^s_{\rm obs}(\boldsymbol{s}) \equiv [n^s(\boldsymbol{s}) - \bar{n}(\boldsymbol{r})]/\bar{n}(\boldsymbol{r})$. This is possible, but it requires knowing not only the true selection function $\bar{n}(\boldsymbol{r})$ in real space, but also the true distance $r$ to each galaxy, which involves carrying out a full reconstruction of the deredshifted density field (Yahil *et al.* 1991; Fisher *et al.* 1995b; Webster, Lahav & Fisher 1997), whereupon one might as well work with the overdensity $\delta_{\rm obs}(\boldsymbol{r})$ in real space.

There remain two alternative possibilities for defining the redshift overdensity. The first option is to define the observed redshift space galaxy overdensity $\delta^s_{\rm obs}(\boldsymbol{s})$ at redshift position $\boldsymbol{s}$ relative to the real space selection function $\bar{n}(\boldsymbol{s})$ evaluated at the same redshift position $\boldsymbol{s}$

$$\delta^s_{\rm obs}(\boldsymbol{s}) \equiv \frac{n^s(\boldsymbol{s}) - \bar{n}(\boldsymbol{s})}{\bar{n}(\boldsymbol{s})} . \tag{3.17}$$

This is the definition adopted by Kaiser (1987), and also adopted in §4.2.

A drawback of the definition (3.17) is that it can be tricky to estimate the real space selection function $\bar{n}(\boldsymbol{s})$ from a redshift survey, whereas it is relatively straightforward to estimate the redshift space selection function $\bar{n}^s(\boldsymbol{s})$ (see footnote[2]). Thus a second possibility is to define the observed redshift space overdensity $\delta^{ss}_{\rm obs}(\boldsymbol{s})$ [with a double $ss$ superscript to distinguish it from $\delta^s_{\rm obs}(\boldsymbol{s})$] relative to the redshift space selection function $\bar{n}^s(\boldsymbol{s})$

$$\delta^{ss}_{\rm obs}(\boldsymbol{s}) \equiv \frac{n^s(\boldsymbol{s}) - \bar{n}^s(\boldsymbol{s})}{\bar{n}^s(\boldsymbol{s})} . \tag{3.18}$$

[2]Hamilton & Culhane (1996) claim that the real and redshift selection functions agree to linear order, but this is false. See the paragraph just before §4.4.1.

The relation between the real and redshift space selection functions $\bar{n}(\boldsymbol{s})$ and $\bar{n}^s(\boldsymbol{s})$, and the consequent relation between the redshift space overdensities $\delta^s(\boldsymbol{s})$ and $\delta^{ss}(\boldsymbol{s})$, is discussed in §4.4.

### 3.2.1. *The True Density Field in Redshift Space*

As in real space, in redshift space it is necessary to distinguish between the observed redshift overdensity $\delta^s_{\rm obs}(\boldsymbol{s})$, equation (3.17), and the hypothetical 'true' redshift overdensity $\delta^s(\boldsymbol{s})$ whose existence is predicted theoretically. For fluctuations in the linear regime, the theoretical prediction is that the true redshift overdensity $\delta^s(\boldsymbol{s})$ is related to the true unredshifted overdensity $\delta(\boldsymbol{r})$ by equation (4.22), derived in §4.2. Similarly, corresponding to the observed redshift overdensity $\delta^{ss}_{\rm obs}(\boldsymbol{s})$, equation (3.18), is a hypothetical 'true' redshift overdensity $\delta^{ss}(\boldsymbol{s})$ predicted by equation (4.73). Corresponding predictions for the redshift overdensities $\delta^{s\,{\rm LG}}$ and $\delta^{ss\,{\rm LG}}$ measured in the Local Group frame instead of the Cosmic Microwave Background frame are given by equations (4.45) and (4.79). The definitions in this subsection remain valid for all the various flavors of the redshift overdensity: just change the superscripts appropriately.

Unlike the true overdensity $\delta(\boldsymbol{r})$ in real space, which is defined independent of the selection function of the survey, the true overdensity $\delta^s(\boldsymbol{s})$ in redshift space depends on the selection function. This is fine, except that an ambiguity arises for regions outside the survey, where the selection function is zero. Specifically, the true redshift overdensity $\delta^s(\boldsymbol{s})$ depends on the selection function $\bar{n}(\boldsymbol{r})$ through the quantity $\alpha(\boldsymbol{r}) \equiv \partial \ln r^2 \bar{n}(\boldsymbol{r}) / \partial \ln r$ in the redshift distortion operator (4.23). The quantity $\alpha(\boldsymbol{r})$ is zero divided by zero outside the survey, where the selection function is zero, so is ambiguous.

Two comments can be made about this ambiguity in the definition of the true redshift overdensity $\delta^s(\boldsymbol{s})$ outside the survey. Firstly, one can choose to resolve the ambiguity in whatever manner is convenient, and this choice has no effect whatsoever on the comparison between theory and observation. This is because the ambiguity in $\alpha(\boldsymbol{r})$ occurs only at points outside the survey, whereas comparison between theory and observation is made only at points inside the survey. A convenient choice in a flux-limited survey, where the selection function $\bar{n}(r)$, hence $\alpha(r)$, is a function only of depth $r$, not of direction $\hat{\boldsymbol{r}}$ within the angular boundaries of the survey, is to take $\alpha(r)$ to have the same value outside the angular boundaries as within. The advantage of this choice is that it preserves the angular symmetry of the 'true' redshift correlation function about the observer. Beyond the radial extent of the survey, $\alpha(r)$ could be set to zero, or any other convenient choice.

The second comment is that the ambiguity disappears in the plane-parallel, or distant observer, limit, where the $\alpha$ term in the distortion operator goes to zero. In the plane-parallel limit, the 'true' redshift space overdensity $\delta^s(\boldsymbol{s})$ becomes independent of the selection function, and the 'true' plane-parallel redshift space correlation function and power spectrum are correspondingly independent of the selection function. This is the context in which the concept of a redshift space power spectrum is in any case most often considered (§4.2.1).

Suppose hereafter then the true redshift space overdensity $\delta^s(\boldsymbol{s})$ has been defined through all space, the ambiguity outside the boundaries of the survey being resolved if possible as described in the previous two paragraphs. The **redshift space correlation function** $\xi^s(s_{12}, s_1, s_2)$ is defined by

$$\xi^s(s_{12}, s_1, s_2) \equiv \langle \delta^s(\boldsymbol{s}_1) \delta^s(\boldsymbol{s}_2) \rangle \ . \tag{3.19}$$

Redshift distortions partially destroy the symmetry enjoyed by the unredshifted correlation function $\xi(r_{12})$, so that the redshift correlation function $\xi^s(s_{12}, s_1, s_2)$ is a function of the redshift distances $s_1$ and $s_2$ as well as the separation $s_{12}$ of a pair of galaxies. Redshift distortions do however preserve the rotational symmetry of the correlation function about the position of the observer[3].

If the angle between the positions $\boldsymbol{s}_1$ and $\boldsymbol{s}_2$ of the galaxy pair is small enough, then the line-of-sight redshift distortions are effectively plane-parallel, as illustrated in the left panel of Figure 1, and the redshift correlation function $\xi^s$ reduces to a function only of the components $s_{/\!/}$ and $s_\perp$ of the pair separation $\boldsymbol{s}_{12}$ respectively parallel and perpendicular to the line of sight $\boldsymbol{z}$

$$\xi^s(s_{12}, s_1, s_2) \approx \xi^s(s_{/\!/}, s_\perp) \ . \tag{3.20}$$

The Fourier transform of the true redshift space overdensity defines the redshift Fourier modes $\hat{\delta}^s(\boldsymbol{k})$ (here the necessity for $\delta^s(\boldsymbol{s})$ to be defined everywhere, including outside the survey, is apparent)

$$\hat{\delta}^s(\boldsymbol{k}) \equiv \int e^{i\boldsymbol{k}.\boldsymbol{s}} \delta^s(\boldsymbol{s})\, d^3 s \ , \quad \delta^s(\boldsymbol{s}) = \int e^{-i\boldsymbol{k}.\boldsymbol{s}} \hat{\delta}^s(\boldsymbol{k})\, d^3 k/(2\pi)^3 \ . \tag{3.21}$$

The covariance of redshift Fourier modes defines the **redshift space power spectrum**, which is equal to the Fourier transform of the redshift correlation function

$$\langle \hat{\delta}^s(\boldsymbol{k}_1) \hat{\delta}^s(\boldsymbol{k}_2) \rangle = \int e^{i\boldsymbol{k}_1.\boldsymbol{s}_1 + i\boldsymbol{k}_2.\boldsymbol{s}_2} \xi^s(s_{12}, s_1, s_2)\, d^3 s_1 d^3 s_2 \ . \tag{3.22}$$

[3] More correctly, the redshift correlation function has orientation symmetry provided that the selection function is independent of direction over the parts of the sky surveyed (which need not be the whole sky), as in a redshift survey with a uniform flux limit. A selection function that is variable with direction over the sky destroys orientation symmetry about the observer, through the $\alpha(\boldsymbol{r})$ term in the redshift distortion operator (4.23).

The line-of-sight redshift distortions destroy statistical homogeneity, so that the redshift power spectrum is no longer a diagonal matrix. The residual orientation symmetry about the observer[4] implies that the redshift power spectrum is a function only of scalar combinations of its arguments

$$\langle \hat{\delta}^s(\boldsymbol{k}_1)\,\hat{\delta}^s(\boldsymbol{k}_2)\rangle = \hat{\xi}^s(|\boldsymbol{k}_1+\boldsymbol{k}_2|, k_1, k_2) \ . \tag{3.23}$$

In the plane-parallel approximation, however, redshift distortions do preserve statistical homogeneity, and in that case the redshift power spectrum is again a diagonal matrix

$$\langle \hat{\delta}^s(\boldsymbol{k}_1)\,\hat{\delta}^s(\boldsymbol{k}_2)\rangle \approx (2\pi)^3 \delta_D(\boldsymbol{k}_1+\boldsymbol{k}_2) P^s(k_{1/\!/}, k_{1\perp}) \ , \tag{3.24}$$

where $k_{1/\!/}$ and $k_{1\perp}$ are the components of the wavevector $\boldsymbol{k}_1$ respectively parallel and perpendicular to the line of sight $\boldsymbol{z}$, and

$$P^s(k_{/\!/}, k_\perp) \equiv \int e^{i\boldsymbol{k}.\boldsymbol{s}} \xi^s(s_{/\!/}, s_\perp)\, d^3s \ . \tag{3.25}$$

Spherical modes $\delta^s_{\ell m}(s)$ and $\hat{\delta}^s_{\ell m}(k)$ in redshift space are defined analogously to equations (3.7) and (3.8). Redshift distortions preserve rotational symmetry about the observer (but see footnote 3), so the redshift correlation functions remain diagonal with respect to the angular indices $\ell m$. Equations (3.7)–(3.11) take the same form in redshift space, aside from the addition of $s$ superscripts. The relation (3.12) is modified to

$$\xi^s_\ell(s_1, s_2) = (4\pi)^2 \int_0^\infty j_\ell(k_1 s_1) j_\ell(k_2 s_2) \hat{\xi}^s_\ell(k_1, k_2)\, k_1^2 dk_1 k_2^2 dk_2/(2\pi)^6 \tag{3.26}$$

and the correlation function of spherical modes in Fourier space becomes

$$\langle \hat{\delta}^s_{\ell_1 m_1}(k_1)\,\hat{\delta}^s_{\ell_2 m_2}(k_2)\rangle = (-)^{\ell_1+m_1} \delta^K_{\ell_1\ell_2} \delta^K_{m_1,-m_2} \hat{\xi}^s_\ell(k_1, k_2) \ . \tag{3.27}$$

### 3.2.2. *The Observed Density Field and Shot Noise in Redshift Space*

As in real space, in redshift space the observed galaxy density is subject to Poisson sampling noise, or shot noise. In the Poisson process model described in §3.1.2, the expectation value $C^s(\boldsymbol{s}_1, \boldsymbol{s}_2)$ of the covariance of observed redshift space overdensities $\delta^s_{\rm obs}(\boldsymbol{s})$ defined by (3.17) is a sum of the true redshift space correlation function $\xi^s(s_{12}, s_1, s_2)$, equation (4.32), with a Poisson sampling noise term

$$\langle \delta^s_{\rm obs}(\boldsymbol{s}_1)\,\delta^s_{\rm obs}(\boldsymbol{s}_2)\rangle \equiv C^s(\boldsymbol{s}_1, \boldsymbol{s}_2) = \xi^s(s_{12}, s_1, s_2) + \delta_D(\boldsymbol{s}_1 - \boldsymbol{s}_2)[\bar{n}(\boldsymbol{s}_1)]^{-1} \tag{3.28}$$

[4] Again, orientation symmetry is strictly true only if the selection function is uniform within the angular boundaries of the survey, because of the $\alpha(r)$ term in the redshift distortion operator (4.23).

which may be compared to the corresponding equation (3.15) in real space. The coefficient $[\bar{n}(\mathbf{s}_1)]^{-1}$ of the Dirac delta-function in the Poisson sampling term is valid for a flux-limited redshift survey; the coefficient could conceivably be different if there were some criterion other than a flux limit for selecting galaxies into the redshift survey. The form of the Poisson sampling term can be derived by an argument similar to the one in real space which followed equation (3.15). An important part of the argument in redshift space is that the expectation value of the product of redshift densities in the same infinitesimal volume element is $\langle n^s(\mathbf{s}_1) n^s(\mathbf{s}_2)\rangle = \delta_D(\mathbf{s}_1 - \mathbf{s}_2)\langle n^s(\mathbf{s}_1)\rangle = \delta_D(\mathbf{s}_1 - \mathbf{s}_2)\bar{n}(\mathbf{s}_1)$, the last step of which comes from the fact that, in a flux-limited redshift survey, the ensemble-averaged expectation value of the redshift density at a point is equal to the real (not redshift) selection function, $\langle n^s(\mathbf{s})\rangle = \bar{n}(\mathbf{s})$.

Similarly, the expectation value $C^{ss}(\mathbf{s}_1, \mathbf{s}_2)$ of the covariance of observed redshift space overdensities $\delta^{ss}_{\rm obs}(\mathbf{s})$ defined by (3.18) is a sum of the corresponding true redshift space correlation function $\xi^{ss}(s_{12}, s_1, s_2)$, equation (4.83), with a Poisson sampling noise term

$$\langle \delta^{ss}_{\rm obs}(\mathbf{s}_1)\, \delta^{ss}_{\rm obs}(\mathbf{s}_2)\rangle \equiv C^{ss}(\mathbf{s}_1, \mathbf{s}_2) = \xi^{ss}(s_{12}, s_1, s_2) + \delta_D(\mathbf{s}_1 - \mathbf{s}_2)\frac{\bar{n}(\mathbf{s}_1)}{[\bar{n}^s(\mathbf{s}_1)]^2} \ . \tag{3.29}$$

Again, the coefficient $\bar{n}(\mathbf{s}_1)/[\bar{n}^s(\mathbf{s}_1)]^2$ of the Poisson sampling term is valid for a flux-limited redshift survey. The form of the coefficient follows from the same argument as that following equation (3.28).

Equation (3.28) shows that, in the Poisson process model, the covariance of observed redshift space overdensities at any finite separation $\mathbf{s}_1 \neq \mathbf{s}_2$ provides an unbiased estimate of the true redshift correlation function,

$$C^s(\mathbf{s}_1, \mathbf{s}_2)_{\mathbf{s}_1 \neq \mathbf{s}_2} = \xi^s(s_{12}, s_1, s_2) \ , \tag{3.30}$$

similar to the corresponding result (3.16) in real space. Similarly, from equation (3.29), $C^{ss}(\mathbf{s}_1, \mathbf{s}_2)_{\mathbf{s}_1 \neq \mathbf{s}_2} = \xi^{ss}(s_{12}, s_1, s_2)$. As in real space, the shot noise contribution to the survey covariance in redshift space can be removed by excluding all self-pairs, consisting of a galaxy and itself.

### 3.3. HILBERT SPACE

It can be confusing to keep track of the mess of $2\pi$'s and other factors, and the juxtaposition of discrete and continuous quantities, in equations such as (3.13). Hilbert space offers a compact, unifying formalism that takes care of the bookkeeping and provides much insight. It is not the place of this subsection to present a complete account of Hilbert space; rather, the goal is to demonstrate how concepts familiar from quantum mechanics also prove powerful in the present statistical context.

The overdensity can be regarded as a vector $\boldsymbol{\delta}$ in an infinite-dimensional vector space, Hilbert space. This vector has a meaning independent of the basis, i.e. complete set of linearly independent functions, with respect to which it is expanded. Thus for example the overdensity vector $\boldsymbol{\delta}$ has components $\delta_{\boldsymbol{r}}$ [$= \delta(\boldsymbol{r})$] when expanded with respect to a basis of delta functions in real space, and it has components $\delta_{\boldsymbol{k}}$ [$= \hat{\delta}(\boldsymbol{k})$, but now the hat is dropped from $\delta_{\boldsymbol{k}}$ in recognition of the fact $\delta_{\boldsymbol{r}}$ and $\delta_{\boldsymbol{k}}$ are really the same vector] when expanded with respect to a basis of Fourier modes $e^{-i\boldsymbol{k}.\boldsymbol{r}}$; but from a Hilbert space point of view the vector $\boldsymbol{\delta}$ remains unchanged: only the 'coordinate system', the basis of expansion functions, changed. This is in precise analogy to ordinary finite-dimensional vectors, which have a meaning independent of the particular coordinate system used to locate them. Similarly, the covariance matrix

$$\boldsymbol{\xi} = \langle \boldsymbol{\delta}\boldsymbol{\delta} \rangle \tag{3.31}$$

(not an inner product!) can be regarded as a matrix in Hilbert space, with, for example, components $\xi_{\boldsymbol{r}_1\boldsymbol{r}_2}$ in real space, or $\xi_{\boldsymbol{k}_1\boldsymbol{k}_2}$ in Fourier space.

The quintessential property of Hilbert space is the existence of an inner product, or scalar product, of any two vectors $\boldsymbol{a}$ and $\boldsymbol{b}$, which maps the two vectors into a scalar, a (complex-valued, in general) number. This scalar value is independent of any basis with respect to which the vectors may be expressed. Various compact notations exist for writing the scalar product:

$$\langle a | b \rangle = \mathrm{Tr}\, \boldsymbol{a}^\dagger \boldsymbol{b} = a^i b_i \ . \tag{3.32}$$

The first expression in (3.32) is the Dirac bra-ket notation, in which the ket $|a\rangle$ denotes a vector and the bra $\langle a|$ its Hermitian conjugate, the complex conjugate transpose of the vector (the angle brackets $\langle\,\rangle$ here are not to be confused with statistical averages; in the Dirac notation, a vertical line always separates the bra and the ket). The middle expression is matrix notation, in which the vector $\boldsymbol{a}^\dagger$ denotes the Hermitian conjugate of the vector $\boldsymbol{a}$, and Tr signifies the trace. The final expression of (3.32) is the inner product in component form: the quantities $a_i$ are the components of the vector $\boldsymbol{a}$ with respect to some orthonormal basis of functions (see eqs. [3.34], [3.36]), while the quantities $a^i = a_i^*$ with raised index denote their Hermitian conjugates, and there is implicit summation over the indices $i$ (the similarity to tensor notation in general relativity is not coincidental). It is to be understood that the index $i$ may be either discrete or continuous, or a combination of both, and that in the continuous case summation signifies integration with respect to some defined measure of $i$; examples of the latter are seen immediately below. Other variants of the inner product notation also occur, such as the omission of the trace sign

Tr in the third expression, if there is no ambiguity, or the insertion of an explicit summation or integration over $i$ in the last expression. In practice it may be typographically convenient to keep all indices lowered, it being understood that when doubled indices occur, one of the pair is implicitly raised, signifying the Hermitian conjugate.

In a statistically homogeneous 3-dimensional random field, equal volume elements $d^3r$ have equal statistical 'weight', so it is logical to define the inner product of two (complex-valued, in general) functions $a(\boldsymbol{r})$ and $b(\boldsymbol{r})$ as an integral over $d^3r$. With respect to the various bases encountered in §3.1, the inner product of $a(\boldsymbol{r})$ and $b(\boldsymbol{r})$ can be written

$$\begin{aligned}\langle a|b\rangle &= \int a^*(\boldsymbol{r})b(\boldsymbol{r})\,d^3r = \int \hat{a}^*(\boldsymbol{k})\hat{b}(\boldsymbol{k})\,d^3k/(2\pi)^3 \\ &= \int_0^\infty \sum_{\ell m} a^*_{\ell m}(r)b_{\ell m}(r)\,r^2dr = \int_0^\infty \sum_{\ell m} \hat{a}^*_{\ell m}(k)\hat{b}_{\ell m}(k)\,k^2dk/(2\pi)^3 \ .\end{aligned} \tag{3.33}$$

The equality between the real and Fourier integrals, the second and third expressions in equation (3.33), is Parseval's theorem, an apparent triviality in the Hilbert formalism. Note that if the inner product is complex-valued, then the order of the inner product matters: the inner product in the opposite order is its complex conjugate, $\langle b|a\rangle = \langle a|b\rangle^*$.

A basis of functions $\psi_i(\boldsymbol{r})$ is said to be orthonormal if their inner products constitute the unit matrix (see eq. [3.38])

$$\int \psi_i(\boldsymbol{r})\psi^j(\boldsymbol{r})\,d^3r = 1_i^j \ , \quad \sum_i \psi^i(\boldsymbol{r})\psi_i(\boldsymbol{r}') = \delta_D(\boldsymbol{r}-\boldsymbol{r}') \tag{3.34}$$

in which either equality implies the other, given that $\psi_i(\boldsymbol{r})$ constitute a basis, a complete linearly independent set. In index notation, equations (3.34) are

$$\psi_i{}^{\boldsymbol{r}}\psi^j{}_{\boldsymbol{r}} = 1_i^j \ , \quad \psi^i{}_{\boldsymbol{r}}\psi_i{}^{\boldsymbol{r}'} = 1_{\boldsymbol{r}}^{\boldsymbol{r}'} \ . \tag{3.35}$$

The orthonormal functions $\psi_i(\boldsymbol{r}_j)$ can be regarded as the rows of an orthonormal matrix $\Psi_{ij}$, satisfying $\boldsymbol{\Psi\Psi}^\dagger = \boldsymbol{\Psi}^\dagger\boldsymbol{\Psi} = \boldsymbol{1}$. The components $a_i$ of a vector $\boldsymbol{a}$ with respect to the orthonormal basis $\psi_i(\boldsymbol{r})$ are given by

$$a_i = \int \psi_i(\boldsymbol{r})a(\boldsymbol{r})\,d^3r \ , \quad a(\boldsymbol{r}) = \sum_i \psi^i(\boldsymbol{r})a_i \ . \tag{3.36}$$

In index notation, these equations are

$$a_i = \psi_i{}^{\boldsymbol{r}}a_{\boldsymbol{r}} \ , \quad a_{\boldsymbol{r}} = \psi^i{}_{\boldsymbol{r}}a_i \ . \tag{3.37}$$

The rule that raising a lowered index $i$ (or lowering a raised index $i$) means take the complex conjugate works fine for vectors, but the rule becomes less easy to interpret for a matrix, where there is more than one index. In practical calculations it is easier to apply an equivalent set of rules, valid for matrices that are real-valued in real space (as is usually the case for matrices of physical interest, such as the covariance matrix $\langle\boldsymbol{\delta}\,\boldsymbol{\delta}\rangle$, eq. [3.31], or the linear redshift distortion operator $\mathbf{S}$, eq. [4.23]). The rules, which depend on the basis, are that raising (or lowering) an index $i$:

- o in real space does nothing;
- o in Fourier space transforms $\boldsymbol{k}_i \to -\boldsymbol{k}_i$;
- o in $r\ell m$-space transforms $m_i \to -m_i$ and multiplies by $(-)^{m_i}$;
- o in $k\ell m$-space transforms $m_i \to -m_i$ and multiplies by $(-)^{\ell_i+m_i}$.

The unit matrix $1_i^j$, whether for discrete or continuous indices, is defined by the property that

$$1_i^j a_j = a_i \tag{3.38}$$

for any vector $a_i$. From this definition together with the definition (3.33) of the inner product, it is straightforward to determine that the unit matrix in real space is (the following equations [3.39]–[3.42] give both the unit matrix and its expression with both indices lowered)

$$1_{\boldsymbol{r}}^{\boldsymbol{r}'} = \delta_D(\boldsymbol{r}-\boldsymbol{r}')\ , \quad 1_{\boldsymbol{r}\boldsymbol{r}'} = \delta_D(\boldsymbol{r}-\boldsymbol{r}')\ . \tag{3.39}$$

Similarly the unit matrix in Fourier space is

$$1_{\boldsymbol{k}}^{\boldsymbol{k}'} = (2\pi)^3\delta_D(\boldsymbol{k}-\boldsymbol{k}')\ , \quad 1_{\boldsymbol{k}\boldsymbol{k}'} = (2\pi)^3\delta_D(\boldsymbol{k}+\boldsymbol{k}')\ . \tag{3.40}$$

The unredshifted power spectrum (3.5) is thus recognized as being a diagonal matrix, with eigenvalues $P(k)$. In spherical transform space with the radial index in real space, the unit matrix is (abbreviating $\mu = r\ell m$)

$$1_\mu^{\mu'} = \delta_{\ell\ell'}^K \delta_{mm'}^K \frac{\delta_D(r-r')}{r^2}\ , \quad 1_{\mu\mu'} = (-)^m \delta_{\ell\ell'}^K \delta_{m,-m'}^K \frac{\delta_D(r-r')}{r^2}\ . \tag{3.41}$$

In spherical transform space with the radial index in Fourier space, the unit matrix is (abbreviating $\mu = k\ell m$)

$$1_\mu^{\mu'} = \delta_{\ell\ell'}^K \delta_{mm'}^K \frac{(2\pi)^3\delta_D(k-k')}{k^2}\ , \quad 1_{\mu\mu'} = (-)^{\ell+m} \delta_{\ell\ell'}^K \delta_{m,-m'}^K \frac{(2\pi)^3\delta_D(k-k')}{k^2}\ . \tag{3.42}$$

Thus again the unredshifted power spectrum (3.13) in $k\ell m$-space is seen to be a diagonal matrix, with eigenvalues $P(k)$.

There is no need to go further here into the Hilbert space formalism, which can be learned from any quantum mechanical textbook (e.g. Landau & Lifshitz 1958). The advantage of the formalism is that it frees expressions, such as the relation (4.22) between overdensity in real and redshift space, or the expression (5.17) for the Gaussian likelihood function, from being tied to any particular basis, such as real space or Fourier space. If such expressions are written in component form (i.e. the vectors and matrices are written with indices), then any index can be regarded as referring to any basis of functions, possibly a different basis for each index, with the following two provisos: (1) indices and the bases to which they refer should match between the left and right hand sides of an equation; (2) wherever paired indices occur, indicating an inner product, they should be evaluated in the same (arbitrary) basis. In evaluating an inner product, it suffices to bear in mind two rules. First, one of a pair of indices $i$ should always be raised (implicitly, if not typographically) indicating the Hermitian conjugate; in the case of Fourier space for example, this means using $-\boldsymbol{k}_i$ for one index and $+\boldsymbol{k}_i$ for the other (at least for vectors and matrices that are real-valued in real space). Second, if the paired index is continuous, then the volume element of integration is consistently the same, in accordance with the definition of the inner product in the basis used; in Fourier space for example, the volume element is $d^3k_i/(2\pi)^3$, according to equation (3.33).

### 3.4. MERIT OF THE POWER SPECTRUM

Blackman & Tukey (1959, §B.3) in their classic text on power spectra state "The autocovariance function is of little use except as a basis for measuring the power spectrum". The power spectrum of large scale structure in the Universe was apparently first measured by Yu & Peebles (1969) and Peebles (1973).

The fundamental advantage of the (true, unredshifted) power spectrum is that the Fourier modes of a statistically homogeneous random field are uncorrelated: their covariance matrix $\langle\hat{\delta}(\boldsymbol{k}_1)\hat{\delta}(\boldsymbol{k}_2)\rangle$ is a diagonal matrix, equation (3.5). Modes that are uncorrelated are said to be **statistically orthogonal**. The statistical orthogonality of Fourier modes does not imply that they are independent in general, but it does imply that they are independent in the important case of a multivariate Gaussian field, for which by definition the 3-point and all higher correlation functions are zero.

Note that the advantage of the power spectrum over (for example) the spatial correlation function is not that it contains any more information — the power spectrum and the correlation function have precisely the same information content — but rather that the power spectrum parcels the information into uncorrelated chunks.

The statistical orthogonality of the (true, unredshifted) Fourier modes is intimately associated with the assumed translation invariance of the statistical properties of the density field. As is familiar from quantum mechanics, the translation operator, the generator of an infinitesimal translation, is $-i\partial/\partial\boldsymbol{r}$, which is also the momentum operator $\boldsymbol{k}$ (Landau & Lifshitz 1958,

§15). The eigenfunctions of the translation operator $-i\partial/\partial \boldsymbol{r}$ are just the Fourier modes $\hat{\delta}(\boldsymbol{k})$, with eigenvalues $\boldsymbol{k}$. The operator of an overall translation is $-i\partial/\partial \boldsymbol{R} \equiv \sum -i\partial/\partial \boldsymbol{r}$, which is the same as the total momentum operator $\boldsymbol{K} \equiv \sum \boldsymbol{k}$. The sum here is over all relevant coordinates, e.g. the 2 coordinates of the 2-point correlation function. Statistical homogeneity means that the statistical properties, the correlation functions, of the density field are unchanged by an overall translation, $\partial\xi/\partial \boldsymbol{R} = 0$, or equivalently $\boldsymbol{K}\xi = 0$. Thus the correlation functions have zero total momentum, $\boldsymbol{K} = \sum \boldsymbol{k} = 0$, as in equation (3.5).

Similarly, the statistical orthogonality of (true, unredshifted) spherical modes with respect to angular indices $\ell m$, equations (3.11) and (3.13), is intimately associated with statistical isotropy about the observer (or indeed about any other point). The operator of an infinitesimal rotation is the angular momentum operator $\boldsymbol{l}$, whose eigenmodes are the spherical harmonics $Y_{\ell m}$. Statistical isotropy (about some point) implies that the correlation functions have zero total angular momentum (about that point), $\boldsymbol{L} = \sum \boldsymbol{l} = 0$, as in equations (3.11) and (3.13).

Redshift distortions destroy translation invariance, but they preserve isotropy about the observer (but see footnote 3), and hence also preserve the statistical orthogonality of spherical modes with respect to the angular indices $\ell m$. Thus, as emphasized by Fisher, Scharf & Lahav (1994) and Heavens & Taylor (1995), the advantage of the power spectrum can best be preserved in redshift space by working in spherical harmonics.

The statistical orthogonality of Fourier modes windowed through any finite survey (i.e., if the overdensity is multiplied by the, possibly weighted, selection function of the survey) is destroyed by the fact that the survey is not translation invariant, especially at wavelengths approaching the size of the survey. Similarly, the statistical orthogonality of spherical modes windowed through a non-all-sky survey is destroyed by the break down of rotational symmetry. However, the desirability of measuring quantities that are uncorrelated remains. Ways to construct statistically orthogonal modes in real surveys are described by Vogeley & Szalay (1996), Tegmark, Taylor & Heavens (1997), and Tegmark *et al.* (1997), and ways to decorrelate power spectra are described by Hamilton (1997), and Tegmark *et al.* (1997).

## 4. Theory of Linear Redshift Distortions

The theory of linear redshift distortions was greatly clarified in a fundamental paper by Kaiser (1987). Although the existence of large scale redshift distortions and the possibility of using them to measure the cosmological density $\Omega_0$ had been recognized previously (Sargent & Turner 1977; Peebles 1980, §76B), the complete, correct equations describing linear redshift distortions were first derived by Kaiser.

The linear distortion equations are presented in §4.2. First, however, it is useful to discuss what the linear distortion parameter $\beta$, which is the quantity actually measurable from linear redshift distortions, and the focus of much of the fuss, actually means.

### 4.1. THE LINEARIZED CONTINUITY EQUATION

I vividly recall a conversation with Jerry Ostriker (in 1991, I believe) in which I was excitedly explaining how one could measure the cosmological density $\Omega_0$ from redshift distortions in the linear regime. Jerry, a fan of explosive cosmologies (Ostriker & Cowie 1981), was unimpressed. He pointed out that redshift distortions are no more than a consequence of the continuity equation, that the continuity equation is of widespread applicability, and that what redshift distortions measure is not $\Omega_0$ but rather (exact quote) "a dimensionless number of no particular significance".

Jerry's view may be on the blueshifted side of the spectrum of perspectives, but his point is well taken. Physically, to grow an overdense region, galaxies have to move in the general direction of the overdensity. More precisely, *the value of $\beta$ that is measured from redshift distortions in the linear regime is the value that solves the linearized continuity equation*

$$\beta\delta + \boldsymbol{\nabla}.\boldsymbol{v} = 0 \ . \tag{4.1}$$

Actually, an additional proposition enters here, which is that the peculiar velocity field $\boldsymbol{v}$ is irrotational (has zero vorticity), as predicted by gravitational growth theory. These matters are explained below. The reader who wishes to gain insight into the problem is invited to solve the problem illustrated in Figure 4.

When dealing with dynamics in cosmology, it is convenient to work in a **comoving coordinate system**, which expands with the general Hubble expansion of the Universe. To maintain contact with observational reality, it is also useful, as in §1, to express comoving distances in velocity units evaluated at the present day, using the present day values $H_0$ of the Hubble constant and $a_0$ of the cosmic scale factor to relate comoving distance and velocity. Thus let $\boldsymbol{r}$ denote the comoving (unredshifted) position of a galaxy, measured in velocity units at the present time ($\boldsymbol{r}$ is related to Peebles' 1980 §7 comoving distance $\boldsymbol{x}$ by $\boldsymbol{r} = H_0 a_0 \boldsymbol{x}$). For example, the comoving distance to the Coma cluster is $r \approx 7000\,\mathrm{km\,s^{-1}}$, as inferred from its measured redshift; by definition of comoving coordinates, this comoving distance does not change (much) as the Universe expands.

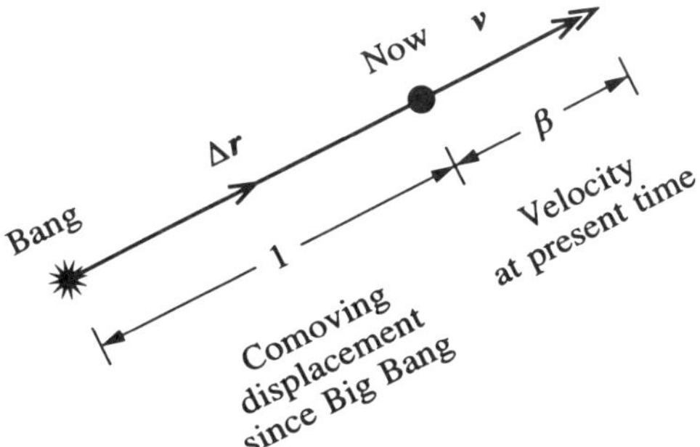


*Figure 4.* *Problem:* From the linearized continuity equation (4.1) show that, in the linear regime, the peculiar velocity $\boldsymbol{v}$ of a particle (galaxy) at the present time is related to its comoving displacement $\Delta\boldsymbol{r}$ (measured in velocity units) since the Big Bang by $\boldsymbol{v} = \beta\Delta\boldsymbol{r}$ . In deriving this result you will need to assume that the density field is initially uniform, and that the velocity is the gradient of a potential. Amongst other things, you should find that in the linear regime particles move in straight lines.

The proper **peculiar velocity** $\boldsymbol{v}$ of a galaxy is its proper velocity in the comoving frame

$$\boldsymbol{v} \equiv \frac{a\, d\boldsymbol{r}}{H_0 a_0 dt} = \frac{d\boldsymbol{r}}{d\tau} \tag{4.2}$$

where $t$ is proper time, $\tau$ is the dimensionless **conformal time** defined by $d\tau \equiv H_0 a_0 dt/a$, and $a(\tau)$ is the cosmic scale factor. One of the advantages of working with conformal time is that objects (e.g. photons) moving at the speed of light $c$ have peculiar velocities $d\boldsymbol{r}/d\tau = c$ at all times.

At this point it is necessary to distinguish carefully between matter, which satisfies a continuity equation, and galaxies, which, thanks to galaxy formation and later merging, may not. The continuity, Euler, and Poisson equations for cold (pressureless) matter (subscripted $M$) in a perturbed Friedmann-Robertson-Walker Universe are, expressed in the comoving coordinate system (Hui & Bertschinger 1996),

$$\frac{\partial \delta_M}{\partial \tau} + \boldsymbol{\nabla}.(1+\delta_M)\boldsymbol{v}_M = 0 \tag{4.3}$$

$$\frac{\partial a \boldsymbol{v}_M}{a\, \partial \tau} + \boldsymbol{v}_M.\boldsymbol{\nabla}\boldsymbol{v}_M = -\boldsymbol{\nabla}\phi \tag{4.4}$$

$$\nabla^2\phi = \frac{4\pi G \bar{\rho}_M a^2 \delta_M}{H_0^2 a_0^2} = \frac{3\Omega_M H^2 a^2 \delta_M}{2 H_0^2 a_0^2} \tag{4.5}$$

where $\boldsymbol{\nabla} \equiv \partial/\partial\boldsymbol{r}$ is the comoving gradient, $\bar{\rho}_M \propto a^{-3}$ is the proper mean matter density, and the cosmological matter density $\Omega_M$ and Hubble constant $H$ both evolve in time. In the linear regime, where $|\delta_M| \ll 1$, the continuity and Euler equations reduce to

$$\frac{\partial \delta_M}{\partial \tau} + \boldsymbol{\nabla}.\boldsymbol{v}_M = 0 \tag{4.6}$$

$$\frac{\partial a \boldsymbol{v}_M}{a\,\partial \tau} = -\boldsymbol{\nabla}\phi \ . \tag{4.7}$$

If the peculiar velocity is decomposed (as can always be done) into its gradient (irrotational, or longitudinal, or scalar) and curl (rotational, or transverse, or vector) parts,

$$\boldsymbol{v}_M = \boldsymbol{v}_M^G + \boldsymbol{v}_M^C \ , \tag{4.8}$$

in which the longitudinal part $\boldsymbol{v}_M^G = \boldsymbol{\nabla}\psi$ is the gradient of some scalar potential $\psi$, while the transverse part $\boldsymbol{v}_M^C = \boldsymbol{\nabla} \times \boldsymbol{A}$ is the curl of some vector potential $\boldsymbol{A}$, then the linearized Euler equation (4.7) becomes

$$\frac{\partial a \boldsymbol{v}_M^G}{a\,\partial \tau} = -\boldsymbol{\nabla}\phi \ , \quad \frac{\partial a \boldsymbol{v}_M^C}{a\,\partial \tau} = 0 \ . \tag{4.9}$$

The second of equations (4.9) shows that the curl part of the velocity decays as $\boldsymbol{v}_M^C \propto a^{-1}$ as the Universe expands, so should be sensibly equal to zero in the absence of a mechanism to generate vorticity. If only gravity operates, then the peculiar velocity in the linear regime should be pure gradient.

When the linearized continuity (4.6), linearized Euler (4.7), and Poisson (4.5) equations are combined, the result is a second order linear differential equation for the overdensity $\delta_M$, which, when further combined with the unperturbed solution for the evolution $a(\tau)$ of the cosmic scale factor, leads to growing and decaying solutions (Peebles 1980, §§10–13; Padmanabhan 1993, §4.5)

$$\delta_M(\boldsymbol{r}, \tau) \propto D(\tau) \tag{4.10}$$

which evolve in time without change of shape. The interesting solution is the unstable, growing solution, and $D(\tau)$ below is taken to be the linear growth factor of the growing mode. The linearized continuity equation (4.6) can then be written

$$\frac{Haf}{H_0 a_0}\delta_M + \boldsymbol{\nabla}.\boldsymbol{v}_M = 0 \tag{4.11}$$

where $f$ is the **dimensionless linear growth rate** of the growing mode

$$f \equiv \frac{H_0 a_0}{Ha}\frac{d \ln D}{d\tau} = \frac{d \ln D}{d \ln a} \ . \tag{4.12}$$

In the standard pressureless matter-dominated cosmology, the dimensionless growth rate $f$ is an analytic function of $\Omega_M$ with a celebrated approximation as a power law[5] (Peebles 1980, eq. [14.8])

$$f(\Omega_M) \approx \Omega_M^{0.6} . \tag{4.13}$$

Lahav *et al.* (1991) give an approximation in the case where there is a cosmological constant[6] so that the total cosmological density is a sum of matter and cosmological constant parts, $\Omega = \Omega_M + \Omega_\Lambda$:

$$f(\Omega_M, \Omega_\Lambda) \approx \Omega_M^{0.6} + \frac{\Omega_\Lambda}{70}\left(1 + \frac{\Omega_M}{2}\right) . \tag{4.14}$$

Evidently the growth rate $f$ depends mainly on the matter density $\Omega_M$, and is only weakly dependent on the cosmological constant.

Equation (4.11) is the continuity equation for the matter; to go from this to the continuity equation (4.1) for galaxies involves one further issue — bias.

For more than a decade it has been apparent that galaxies may not constitute an unbiased tracer of the underlying matter density. If galaxies were an unbiased tracer of the matter, they would, by definition, satisfy $\delta = \delta_M$ (see footnote[7]). That some bias exists, at least in some galaxy populations, follows from the fact that galaxies selected in different ways have correlation functions with different amplitudes (e.g. Peacock & Dodds 1994; Oliver *et al.* 1996; Peacock 1997). For example, the correlation function of optical galaxies is approximately a factor of 2 larger than that of *IRAS* galaxies, a fact evident from Figure 3, which demonstrates that optical galaxies and *IRAS* galaxies cannot both be unbiased tracers (Oliver *et al.* 1996).

The simplest model of bias postulates that the galaxy overdensity $\delta$ is linearly biased by a constant factor, the **linear bias factor** $b$, relative to the underlying mass density $\delta_M$, so that

$$\delta = b\delta_M , \tag{4.15}$$

[5]Occasionally a small fuss is made about the exponent of the power law in this approximation (cf. Lightman & Schechter 1990). A sensible person would use the exact analytic expression for $f(\Omega_M)$ when reporting actual results.

[6]Einstein's cosmological constant $\Lambda$, which could arise from a non-zero vacuum density $\rho_{\rm vac}$, is related to $\Omega_\Lambda$ by $\Lambda = 8\pi G\rho_{\rm vac} = 3H^2\Omega_\Lambda$ .

[7]Since galaxies come as discrete units, one should understand this equation in a probabilistic sense. Galaxies are unbiased if the probability of finding a galaxy in a volume element $dV$ at position $\boldsymbol{r}$ is proportional to the amount of matter $\rho_M(\boldsymbol{r})dV$ in that volume element. The discretized galaxy density $n(\boldsymbol{r})$ is then said to be a Poisson process, with mean proportional to $\rho_M(\boldsymbol{r})$. The statistical properties of such a discretized distribution, i.e. its 2-point and higher correlation functions, coincide with those of the underlying matter density field aside from the addition of delta-function discreteness terms at zero separation (cf. §3.1.2).

whereas galaxy velocities faithfully follow the velocity of matter

$$\boldsymbol{v} = \boldsymbol{v}_M \ . \tag{4.16}$$

The linear biasing model (4.15) was originally motivated by **threshold biasing**, in which galaxies are supposed to form only where the matter density exceeds a certain threshold (Kaiser 1984), and by **peaks biasing** where galaxies form at the peaks of the matter density (Bardeen *et al.* 1986). These models predict that, at least in some regimes, the galaxy correlation function $\xi(r_{12}) \equiv \langle\delta(\boldsymbol{r}_1)\delta(\boldsymbol{r}_2)\rangle$ should be amplified over the matter correlation function $\xi_M(r_{12}) \equiv \langle\delta_M(\boldsymbol{r}_1)\delta_M(\boldsymbol{r}_2)\rangle$ by an approximately constant factor

$$\xi(r_{12}) \approx b^2 \xi_M(r_{12}) \ . \tag{4.17}$$

What is true observationally is that the correlation functions of galaxies selected in different ways differ mainly in their amplitudes, but only weakly in their shapes (Peacock & Dodds 1994; Peacock 1997). Ultimately, it remains unclear how nature chooses to bias galaxies in reality.

The linearized continuity equation (4.11) for the matter, evaluated at the present time, together with the linear bias model (4.15), yield the linearized continuity equation (4.1) for galaxies, where the dimensionless quantity $\beta$ is related to the present day value $f_0$ of the linear growth rate $f$, equation (4.13) or (4.14), and the bias factor $b$, equation (4.15), by

$$\beta = \frac{f_0}{b} \ . \tag{4.18}$$

With the standard formula (4.13) for $f$, this becomes the oft-quoted relation (1.5).

If the matter peculiar velocity $\boldsymbol{v}_M$ is curl-free in the linear regime (see the argument following eq. [4.9]), and the galaxy velocity is unbiased, equation (4.16), then the galaxy peculiar velocity field $\boldsymbol{v}$ is also curl-free in the linear regime. The linearized galaxy continuity equation (4.1) then implies that the galaxy peculiar velocity $\boldsymbol{v}$ is related to the galaxy overdensity $\delta$ at the present day by

$$\boldsymbol{v} = -\beta \boldsymbol{\nabla} \nabla^{-2} \delta \tag{4.19}$$

where $\nabla^{-2}$ is the inverse Laplacian. All measurements of $\beta$ in the linear regime are in effect measurements of the ratio of the peculiar velocity $\boldsymbol{v}$ to the galaxy overdensity $\delta$, the (testable) relation (4.19) being assumed to hold true. This is true not only in the case of large scale redshift distortions, but also for measurements of $\beta$ from direct comparison of overdensities and peculiar velocities (Strauss & Willick 1995), or from comparison of the peculiar motion of the Local Group of galaxies to the dipole anisotropy of galaxies over the sky (Strauss *et al.* 1992b).

Two final points about bias are worth making. Firstly, the linear bias model (4.15) must break down in the nonlinear regime (if $b > 1$, as generally expected), since otherwise equation (4.15) would predict that regions empty of matter, $\delta_M = -1$, would have negative galaxy density, which is impossible. Various models of nonlinear biasing can be contemplated (Fry & Gaztañaga 1993; Mann, Peacock & Heavens 1997).

Secondly, if a linear bias $b_i$ is established so $\delta(\boldsymbol{r}, t_i) = b_i \delta_M(\boldsymbol{r}, t_i)$ at some initial time $t_i$, after which galaxies satisfy continuity (being neither created nor destroyed), so that $[1+\delta(\boldsymbol{r})]/[1+\delta_M(\boldsymbol{r})] \propto n(\boldsymbol{r})/\rho_M(\boldsymbol{r})$ remains constant in Lagrangian elements after the initial time $t_i$, then the bias factor $b = \delta/\delta_M$ evolves. During linear growth, $|\delta_M| \ll 1$, the bias decreases in time while remaining constant in space,

$$b(t) - 1 = \frac{D(t_i)}{D(t)} (b_i - 1) , \tag{4.20}$$

where $D(t)/D(t_i) = \delta_M(\boldsymbol{r}, t)/\delta_M(\boldsymbol{r}, t_i)$ is the linear growth factor. Equation (4.20) shows that the action of continuity is to drive the bias factor closer to (but not necessarily all the way to) unity, as fluctuations grow. If continuity persists into the nonlinear regime, then the bias factor becomes a function also of position,

$$b(\boldsymbol{r}, t) - 1 = \delta_M(\boldsymbol{r}, t_i) \left( \frac{1}{\delta_M(\boldsymbol{r}, t)} + 1 \right) (b_i - 1) \tag{4.21}$$

saturating at $b = 1$ in underdense regions (as $\delta_M(\boldsymbol{r}, t) \to -1$), and at $b = 1 + \delta_M(\boldsymbol{r}, t_i)(b_i - 1)$ in overdense regions (as $\delta_M(\boldsymbol{r}, t) \to \infty$). In the nonlinear regime, streams from several different initial points may overlap at the same final point, so correctly equation (4.21) should be averaged over streams.

The model described in the previous paragraph is of course simplistic. The point is that the bias factor $b$ probably evolves, and that there is a mechanism — continuity — that tends to drive the bias factor closer to unity, that is, to make galaxies less biased as time goes on.

## 4.2. THE LINEAR REDSHIFT DISTORTION OPERATOR

In the linear regime, the overdensity $\delta^s$ in redshift space is related to the overdensity $\delta$ in real space by a **linear redshift distortion operator S**,

$$\delta^s = \mathbf{S}\, \delta \ . \tag{4.22}$$

Conceptually, it is helpful to recognize that *the linear redshift distortion operator* $\mathbf{S}$ *is just a linear operator* (a matrix in Hilbert space), much like the linear operators one encounters in quantum mechanics. In real (as opposed to Fourier, say) space, the distortion operator, derived immediately below, is an integro-differential operator (the inverse Laplacian $\nabla^{-2}$ is the integral part)

$$\mathbf{S} = 1 + \beta \left( \frac{\partial^2}{\partial r^2} + \frac{\alpha(\boldsymbol{r})\partial}{r\partial r} \right) \nabla^{-2} \tag{4.23}$$

where $\alpha(\boldsymbol{r})$ is the logarithmic derivative of $r^2$ times the real space selection function $\bar{n}(\boldsymbol{r})$,

$$\alpha(\boldsymbol{r}) \equiv \frac{\partial \ln r^2 \bar{n}(\boldsymbol{r})}{\partial \ln r} \ . \tag{4.24}$$

The redshift distortion operator (4.23) is valid in the frame of reference of stationary observers, those at rest with respect to the Cosmic Microwave Background (CMB), and provided that the redshift overdensity $\delta^s(\boldsymbol{r})$ is defined relative to the real space selection function $\bar{n}(\boldsymbol{r})$, as in equation (3.17). The modifications to the linear redshift distortion operator $\mathbf{S}$ for the cases where the redshift overdensity is defined in the Local Group frame, and/or relative to the selection function $\bar{n}^s(\boldsymbol{r})$ measured in redshift space, are given by equations (4.46), (4.75), and (4.81).

The linear redshift distortion equations were first derived by Kaiser (1987, §2). The derivation below follows most closely that of Hamilton & Culhane (1996, §2).

The starting point of the derivation of the linear redshift distortion operator (4.23) is a conservation equation for galaxies, which expresses the fact that peculiar velocities displace galaxies along the line of sight, but they do not make galaxies appear or disappear (this conservation equation is not the same as the dynamical continuity equation discussed in §4.1)

$$n^s(\boldsymbol{s})\, d^3 s = n(\boldsymbol{r})\, d^3 r \ . \tag{4.25}$$

In terms of the redshift space overdensity $\delta^s(\boldsymbol{s})$ defined by equation (3.17), the galaxy conservation equation (4.25) is

$$\bar{n}(\boldsymbol{s})[1 + \delta^s(\boldsymbol{s})]\, s^2 ds = \bar{n}(\boldsymbol{r})[1 + \delta(\boldsymbol{r})]\, r^2 dr \ . \tag{4.26}$$

With the relation between redshift position $\boldsymbol{s}$ and real position $\boldsymbol{r}$

$$\boldsymbol{s} = \boldsymbol{r} + v\hat{\boldsymbol{r}} \ , \tag{4.27}$$

equation (4.26) rearranges to

$$1 + \delta^s(\boldsymbol{s}) = \frac{r^2 \bar{n}(\boldsymbol{r})}{(r+v)^2 \bar{n}(\boldsymbol{r} + v\hat{\boldsymbol{r}})} \left( 1 + \frac{\partial v}{\partial r} \right)^{-1} [1 + \delta(\boldsymbol{r})] \ . \tag{4.28}$$

So far equation (4.28) is exact, valid in the linear and nonlinear regimes, at least until orbit-crossing occurs.

The next step is to linearize equation (4.28). In addition to the linear theory assumption $|\delta(\boldsymbol{r})| \ll 1$, which also implies $|\partial v/\partial r| \ll 1$ if the velocity $v$ is given by linear theory, equation (4.31), it is necessary to assume that peculiar velocities $v$ of galaxies are small compared to their distances $r$ from the observer (actually, this assumption can be dropped, if the redshift space overdensity $\delta^s(\boldsymbol{r})$ in the CMB frame is constructed according to equation [4.52]; see §4.3.2),

$$|v| \ll r \ . \tag{4.29}$$

To linear order equation (4.28) then reduces to (note in particular that $\delta^s(\boldsymbol{s}) = \delta^s(\boldsymbol{r})$ to linear order)

$$\delta^s(\boldsymbol{r}) = \delta(\boldsymbol{r}) - \left(\frac{\partial}{\partial r} + \frac{\alpha(\boldsymbol{r})}{r}\right) v \tag{4.30}$$

where $\alpha(\boldsymbol{r})$ is defined by (4.24). In linear theory, the peculiar velocity $\boldsymbol{v}$ is given by equation (4.19), so the line-of-sight component $v$ of the peculiar velocity in the CMB frame is

$$v = -\beta \frac{\partial}{\partial r} \nabla^{-2} \delta \ . \tag{4.31}$$

Inserting this formula into equation (4.30) yields the distortion equation (4.22) with the redshift distortion operator $\mathbf{S}$ given by equation (4.23), as was to be proven.

It follows from equation (4.22) that the redshift correlation function $\xi^s(r_{12}, r_1, r_2)$, equation (3.19), in the linear regime is related to its unredshifted counterpart $\xi(r_{12})$ by

$$\begin{aligned} \xi^s(r_{12}, r_1, r_2) &= \mathbf{S}_1 \mathbf{S}_2 \, \xi(r_{12}) \\ &= \left[1 + \beta\left(\frac{\partial^2}{\partial r_1^2} + \frac{\alpha(\boldsymbol{r}_1)\partial}{r_1 \partial r_1}\right)\nabla_1^{-2}\right]\left[1 + \beta\left(\frac{\partial^2}{\partial r_2^2} + \frac{\alpha(\boldsymbol{r}_2)\partial}{r_2 \partial r_2}\right)\nabla_2^{-2}\right]\xi(r_{12}) \ . \end{aligned} \tag{4.32}$$

This equation is correct for observers who are located randomly in the Universe, who are at rest with respect to the CMB, and who define the redshift overdensity relative to the real space selection function $\bar{n}(\boldsymbol{r})$, as in equation (3.17). As discussed in §3.2.2, in a real redshift survey, where the overdensity field is sampled by discrete galaxies, the correlation function of the observed overdensity field contains an additional shot noise term, equation (3.28).

We, sitting on a galaxy, the Milky Way, are not at a random location, but reside in an overdense region of the Universe, as galaxies are wont to

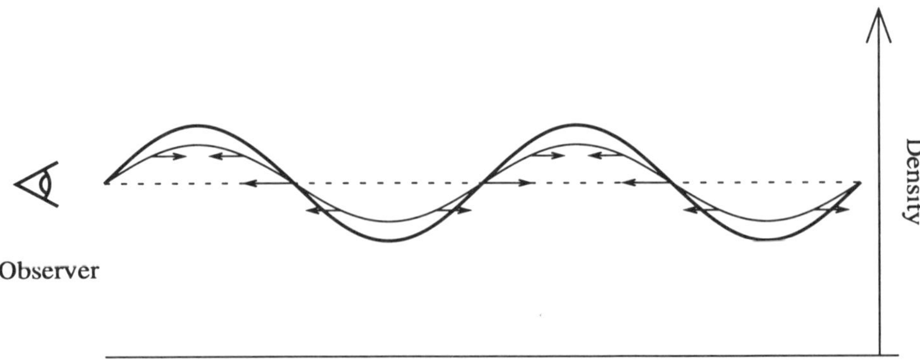


*Figure 5.* A wave of amplitude $\delta(\boldsymbol{k})$ in real space (thin line) appears as a wave with enhanced amplitude $\delta^s(\boldsymbol{k})$ in redshift space (thick line) because of peculiar velocities (arrows). If the wavevector $\boldsymbol{k}$ is along the line of sight, then the amplification factor is $1+\beta$. More generally, a wave that is angled to the line of sight appears amplified in redshift space by a factor $1+\beta\mu_{\boldsymbol{k}}^2$, equation (4.35), where $\mu_{\boldsymbol{k}} \equiv \hat{\boldsymbol{z}}.\hat{\boldsymbol{k}}$ is the cosine of the angle between the wavevector $\boldsymbol{k}$ and the line of sight $\boldsymbol{z}$.

do. To guard against 'local bias', it would a reasonable precaution to excise at least part of the local region when analysing a galaxy survey.

Furthermore, we are not at rest in the CMB. What to do about that is discussed in §4.3.

Finally, it is much more straightforward to measure the selection function $\bar{n}^s(\boldsymbol{r})$ in redshift space than in real space. What to do about that is discussed in §4.4.

#### 4.2.1. *The Linear Plane-Parallel Redshift Distortion Operator*

In the plane-parallel, or distant observer, limit, the linear distortion operator (4.23) reduces to (the superscript $p$ denotes plane-parallel)

$$\mathbf{S}^p = 1 + \beta \frac{\partial^2}{\partial z^2} \nabla^{-2} \tag{4.33}$$

where $z$ is distance along the line of sight. In Fourier space, $(\partial/\partial z)^2 \nabla^{-2} = k_z^2/k^2 = \mu_{\boldsymbol{k}}^2$, where $\mu_{\boldsymbol{k}} \equiv \hat{\boldsymbol{z}}.\hat{\boldsymbol{k}}$ is the cosine of the angle between the wavevector $\boldsymbol{k}$ and the line of sight $\boldsymbol{z}$. Thus in Fourier space the plane-parallel distortion operator reduces to a diagonal operator

$$\mathbf{S}^p = 1 + \beta \mu_{\boldsymbol{k}}^2 \ , \tag{4.34}$$

so that, as illustrated in Figure 5, a Fourier mode $\hat{\delta}^s(\boldsymbol{k})$ in redshift space is simply equal to the unredshifted mode $\hat{\delta}(\boldsymbol{k})$ amplified by a factor $1+\beta\mu_{\boldsymbol{k}}^2$

$$\hat{\delta}^s(\boldsymbol{k}) = (1 + \beta \mu_{\boldsymbol{k}}^2) \hat{\delta}(\boldsymbol{k}) \ . \tag{4.35}$$

It follows from equation (4.35) that, in the plane-parallel approximation, the redshift space power spectrum $P^s(\boldsymbol{k})$ is amplified by $(1+\beta\mu_{\boldsymbol{k}}^2)^2$ over its unredshifted counterpart $P(k)$

$$P^s(\boldsymbol{k}) = (1+\beta\mu_{\boldsymbol{k}}^2)^2 P(k) \ , \tag{4.36}$$

an elegant formula first pointed out by Kaiser (1987, eq. [3.5]).

The simplicity of formula (4.36) arises from the fact that the plane-parallel distortion operator is diagonal in Fourier space, which is intimately associated with the fact that plane-parallel redshift distortions preserve translation invariance. The translation operator $-i\boldsymbol{\nabla}$ commutes with the plane-parallel distortion operator, and therefore the eigenmodes of the translation operator $-i\boldsymbol{\nabla}$, which are precisely the Fourier modes $\hat{\delta}(\boldsymbol{k})$, are also eigenmodes of the plane-parallel distortion operator.

Translation invariance means that redshifted Fourier modes remain statistically orthogonal in the plane-parallel approximation, equation (3.24), preserving the advantage of the unredshifted power spectrum (§3.4).

### 4.2.2. *The Linear Radial Redshift Distortion Operator*

Unfortunately, the full distortion operator (4.23), which I call here the radial redshift distortion operator[8] to distinguish it from the plane-parallel distortion operator (4.33) or (4.34), proves to be quite a bit more complicated to deal with than the plane-parallel operator. The radial operator destroys translation symmetry, so that Fourier modes are no longer eigenmodes of the distortion operator, and the redshift space power spectrum $\langle \delta^s(\boldsymbol{k}_1)\delta^s(\boldsymbol{k}_2)\rangle$ is no longer a diagonal matrix. The linear radial distortion operator in Fourier space is an integral operator, a continuous matrix,

$$\begin{aligned} \hat{\mathrm{S}}(\boldsymbol{k},\boldsymbol{k}') &= \int e^{i\boldsymbol{k}.\boldsymbol{r}}\,\mathrm{S}(\boldsymbol{r})e^{-i\boldsymbol{k}'.\boldsymbol{r}}\,d^3r \\ &= (2\pi)^3\delta_D(\boldsymbol{k}-\boldsymbol{k}')\left(1+\beta\frac{\partial^2}{\partial k^2}k^2\nabla_{\boldsymbol{k}}^{-2}k^{-2}\right) - \beta\hat{\alpha}(\boldsymbol{k}-\boldsymbol{k}')\frac{\partial}{\partial k'}k'\nabla_{\boldsymbol{k}'}^{-2}k'^{-2} \end{aligned} \tag{4.37}$$

where $\boldsymbol{\nabla}_{\boldsymbol{k}} \equiv \partial/\partial\boldsymbol{k}$, and $\hat{\alpha}(\boldsymbol{k}) \equiv \int \alpha(\boldsymbol{r})e^{i\boldsymbol{k}.\boldsymbol{r}}d^3r$ is the Fourier transform of $\alpha(\boldsymbol{r})$. The distortion operator (4.37) yields the redshifted Fourier modes $\hat{\delta}^s(\boldsymbol{k})$ when acting on the unredshifted Fourier modes $\hat{\delta}(\boldsymbol{k})$

$$\hat{\delta}^s(\boldsymbol{k}) = \int \hat{\mathrm{S}}(\boldsymbol{k},\boldsymbol{k}')\hat{\delta}(\boldsymbol{k}')\,d^3k'/(2\pi)^3 \ . \tag{4.38}$$

The form of the linear radial distortion operator in Fourier space is discussed by Zaroubi & Hoffman (1996).

[8]Hamilton & Culhane (1996) call it the spherical distortion operator.

The radial redshift distortion operator does however preserve angular symmetry about the observer[9], so that at least some of the merits of the power spectrum can be retained by working in spherical harmonics. In particular, spherical harmonic modes remain statistically orthogonal with respect to the angular indices in redshift space (§3.4), a point emphasized and taken advantage of first by Fisher, Scharf & Lahav (1994), and then by Heavens & Taylor (1995). Mathematically, the distortion operator $\mathbf{S}$ commutes with the angular momentum operator $\boldsymbol{L}$, so that the eigenmodes of the angular momentum operator, which are the spherical harmonics $Y_{\ell m}$, are also eigenmodes of the distortion operator.

Hamilton & Culhane (1996, §4) point out that the linear radial distortion operator (4.23) is nearly scale-free, to the extent that the quantity $\alpha(\boldsymbol{r})$ is approximately constant. If follows that the distortion operator commutes approximately with the operator $\partial/\partial \ln r$, the logarithmic derivative with respect to depth $r$, or equivalently with the operator $\partial/\partial \ln k = -\partial/\partial \ln r - 3$, the logarithmic derivative with respect to wavenumber $k$. A complete set of commuting Hermitian operators for the spherical distortion operator $\mathbf{S}$ is then, to the extent that $\alpha(\boldsymbol{r})$ is constant,

$$-i\left(\frac{\partial}{\partial \ln r}+\frac{3}{2}\right)=i\left(\frac{\partial}{\partial \ln k}+\frac{3}{2}\right) \ , \quad L^2 \ , \quad L_z \ . \tag{4.39}$$

The operators $L^2$ and $L_z$ are the square and $z$-component (along some arbitrary axis) of the angular momentum operator $\boldsymbol{L} = i\,\boldsymbol{r}\times\partial/\partial\boldsymbol{r} = i\,\boldsymbol{k}\times\partial/\partial\boldsymbol{k}$, which is the same operator in real and Fourier space. The radial eigenfunctions are radial waves in logarithmic depth $r$ or wavenumber $k$, while the angular eigenfunctions are the usual orthonormal spherical harmonics $Y_{\ell m}$. Thus the eigenfunctions of the commuting set (4.39) are spherical waves with radial parts in logarithmic real or Fourier space.

There are two drawbacks to the basis of logarithmic spherical waves proposed by Hamilton & Culhane. First, $\alpha(\boldsymbol{r})$, the logarithmic derivative of $r^2\bar{n}(\boldsymbol{r})$, equation (4.24), is not really constant, although it is slowly varying, and it is typically nearly constant in the nearer part of a redshift survey where the selection function is found empirically to approximate a power

[9]Correctly, the distortion operator preserves angular symmetry to the extent that the selection function is independent of direction, so that the quantity $\alpha(\boldsymbol{r})$ in the distortion operator (4.23) is a function $\alpha(r)$ only of depth $r$, as is true in a survey that is uniformly flux-limited within its angular boundaries. Over unsurveyed regions of the sky (where $\alpha(\boldsymbol{r})$ is zero divided by zero), the distortion operator can be taken to have the same form as over the surveyed part, as discussed in §3.2.1. To avoid confusion, it should be understood that it is the 'true' redshift power spectrum that possesses orientation symmetry. When the power spectrum is windowed through a non-all-sky survey, then the lack of angular symmetry of the window destroys the angular symmetry of the windowed redshift power spectrum.

law. The second problem is that it is not enough that the modes should be eigenfunctions of the distortion operator. It is also desired that the modes be statistically orthogonal, or nearly so. Now the modes would be statistically orthogonal if the power spectrum were, like the distortion operator, scale-free, i.e. a power law $P(k) \propto k^n$, which may not be a bad approximation in reality. However, in any real survey, Poisson sampling noise, §§3.1.2 and 3.2.2, destroys the scale-free symmetry, except in the case that the power spectrum itself is that of Poisson noise, $P(k)$ = constant. It remains to be seen if this idea can be exploited.

The radial redshift distortion operator (4.23) is not Hermitian, unlike the plane-parallel distortion operator (4.33), which is Hermitian. Amongst other things, this means that the eigenfunctions of the radial distortion operator do not all have real eigenvalues. Presumably this means that redshift distortions cause modes close to the observer to undergo a 'phase shift' in passing from real to redshift space. The Hermitian conjugate $\mathbf{S}^\dagger$ of the distortion operator is

$$\mathbf{S}^\dagger = 1 + \beta \nabla^{-2} r^{-2} \frac{\partial}{\partial r} \left( \frac{\partial}{\partial r} - \frac{\alpha(\boldsymbol{r})}{r} \right) r^2 \, . \tag{4.40}$$

Equation (4.40) is valid in the CMB frame. The Hermitian conjugate of the redshift distortion operator in the Local Group frame is given by equation (4.50).

Tegmark & Bromley (1995) derive an expression for the inverse $\mathbf{S}^{-1}$, the Green's function, of the distortion operator for the particular case $\alpha = 2$, corresponding to a volume-limited sample.

## 4.3. PECULIAR VELOCITY OF THE LOCAL GROUP

The redshift distortion operator (4.23) is valid from the point of view of observers at rest with respect to the CMB. The Milky Way however is not stationary, but moving. The linear part of the motion of the Milky Way is the peculiar velocity of the **Local Group** (LG) of galaxies. The essential feature of the Local Group, which contains about 30 galaxies including our own, is that it is the local region of space, about 1 Mpc in radius, that has turned around from the general Hubble expansion of the Universe, and is beginning to fall together for the first time.

In the LG frame, the redshift distance $\boldsymbol{s}^{\mathrm{LG}}$ of a galaxy at true position $\boldsymbol{r}$ relative to the observer is

$$s^{\mathrm{LG}} = r + v - \hat{\boldsymbol{r}}.\boldsymbol{v}^{\mathrm{LG}} \tag{4.41}$$

which differs from the redshift distance $s = r + v$ in the CMB frame by the component of $-\boldsymbol{v}^{\mathrm{LG}}$ along the direction $\hat{\boldsymbol{r}}$ to the galaxy. From conservation of galaxies, $n^{\rm LG}(\boldsymbol{s}^{\rm LG})\, d^3 s^{\rm LG} = n(\boldsymbol{s})\, d^3 s$, one concludes that, at distances much greater than the LG peculiar velocity, $r \gg v^{\rm LG}$, the redshift space overdensity $\delta^{s{\rm LG}}(\boldsymbol{r})$ observed in the LG frame is related to the redshift space overdensity $\delta^s(\boldsymbol{r})$ in the CMB frame by

$$\delta^{s{\rm LG}}(\boldsymbol{r}) = \delta^s(\boldsymbol{r}) + \alpha(\boldsymbol{r}) \frac{\hat{\boldsymbol{r}}.\boldsymbol{v}^{\rm LG}}{r} \ . \tag{4.42}$$

The difference term is a dipole directed along the LG motion $\boldsymbol{v}^{\rm LG}$, modulated by the function $\alpha(\boldsymbol{r})$, equation (4.24).

Since the LG shares in the linear motion of the nearby region, there is no dipole distortion of the nearby region in the LG frame. A dipole distortion appears in the CMB frame because of the streaming of galaxies past the stationary observer. Galaxies streaming toward the observer appear compressed to a smaller volume in redshift space by the spherical convergence of the volume element, and conversely galaxies downstream appear decompressed.

Two general strategies to deal with the redshift distortion produced (or rather removed) by the motion of the LG are described below. A third strategy is to ignore the problem, which is what the plane-parallel approximation does.

### 4.3.1. *The Distortion Operator in the Local Group Frame*

The first strategy, used by Fisher, Scharf & Lahav (1994), is to work entirely in the LG frame, and to write the LG peculiar velocity as its value in linear theory, which is given by the usual formula (4.19) for peculiar velocity, evaluated at the position of the LG, the origin, $\boldsymbol{r} = 0$:

$$\boldsymbol{v}^{\rm LG} = -\beta \left. \frac{\partial}{\partial \boldsymbol{r}} \right|_{\boldsymbol{r}=0} \nabla^{-2} \delta \ . \tag{4.43}$$

An equivalent expression is

$$\boldsymbol{v}^{\rm LG} = \beta \int \frac{\hat{\boldsymbol{r}} \delta(\boldsymbol{r})}{4\pi r^2}\, d^3 r \tag{4.44}$$

which expresses the usual fact that peculiar velocity in linear theory is proportional to peculiar gravity, which is proportional to the integrated dipole overdensity about the observer.

The value (4.43), inserted in equation (4.42), combined with the redshift distortion equation $\delta^s = \mathbf{S}\delta$ in the CMB frame, equation (4.22), yields the distortion equation for the redshift overdensity $\delta^{s{\rm LG}}$ in the LG frame

$$\delta^{s{\rm LG}} = \mathbf{S}^{\rm LG}\, \delta \tag{4.45}$$

where $\mathbf{S}^{\mathrm{LG}}$ is the linear redshift distortion operator (4.23) modified to the LG frame

$$\mathbf{S}^{\mathrm{LG}} = 1 + \beta \left[ \frac{\partial^2}{\partial r^2} + \frac{\alpha(\boldsymbol{r}) \hat{\boldsymbol{r}}}{r} \cdot \left( \frac{\partial}{\partial \boldsymbol{r}} - \left. \frac{\partial}{\partial \boldsymbol{r}} \right|_{\boldsymbol{r}=0} \right) \right] \nabla^{-2} . \tag{4.46}$$

The first $\partial/\partial\boldsymbol{r}$ in the $\alpha$ term is the gradient evaluated at the position $\boldsymbol{r}$, while $\partial/\partial\boldsymbol{r}|_{\boldsymbol{r}=0}$ is the gradient evaluated at the position of the LG. The assumption of linear theory here automatically guarantees that the peculiar velocities of galaxies in the LG frame are small compared to their distances,

$$|v - \hat{\boldsymbol{r}}.\boldsymbol{v}^{\mathrm{LG}}| \ll r , \tag{4.47}$$

unlike the corresponding condition (4.29) in the CMB frame, which had to be incorporated as an additional assumption.

The redshift distortion operator $\mathbf{S}^{\mathrm{LG}}$ in the LG frame, equation (4.46), can also be derived directly along the same lines as the derivation in §4.2 of the distortion operator $\mathbf{S}$ in the CMB frame, equation (4.23). The main difference is to replace $v \to v - \hat{\boldsymbol{r}}.\boldsymbol{v}^{\mathrm{LG}}$ throughout the derivation; for example, equation (4.30) in the CMB frame changes, in the LG frame, to

$$\delta^s(\boldsymbol{r}) = \delta(\boldsymbol{r}) - \left( \frac{\partial}{\partial r} + \frac{\alpha(\boldsymbol{r})}{r} \right) (v - \hat{\boldsymbol{r}}.\boldsymbol{v}^{\mathrm{LG}}) \tag{4.48}$$

which is Kaiser's (1987) equation (3.3).

The redshift correlation function $\xi^{s\mathrm{LG}}(r_{12}, r_1, r_2)$ in the LG frame is related to the unredshifted correlation function $\xi(r_{12})$ by

$$\xi^{s\mathrm{LG}}(r_{12}, r_1, r_2) \; = \; \mathbf{S}_1^{\mathrm{LG}} \mathbf{S}_2^{\mathrm{LG}} \, \xi(r_{12}) \tag{4.49}$$

which may be compared to the corresponding equation (4.32) in the CMB frame.

The Hermitian conjugate of the linear distortion operator $\mathbf{S}^{\mathrm{LG}}$ in the LG frame can be written in a variety of ways. One of the simpler ways, which however does not reveal the subtraction of the LG velocity as manifestly as does the expression (4.46) for the distortion operator $\mathbf{S}^{\mathrm{LG}}$ itself, is

$$\mathbf{S}^{\mathrm{LG}\dagger} = 1 + \beta \left[ \nabla^{-2} r^{-2} \frac{\partial}{\partial r} \left( \frac{\partial}{\partial r} - \frac{\alpha(\boldsymbol{r})}{r} \right) r^2 - \frac{\hat{\boldsymbol{r}}}{r^2} \cdot \left. \frac{\partial}{\partial \boldsymbol{r}} \right|_{\boldsymbol{r}=0} \nabla^{-2} \alpha(\boldsymbol{r}) r \right] , \tag{4.50}$$

in which the last term inside the square brackets on the right hand side is the term arising from the motion of the LG. This equation (4.50) may be compared to the corresponding equation (4.40) for the Hermitian conjugate $\mathbf{S}^\dagger$ of the distortion operator in the CMB frame. Equation (4.50) is used in equation (8.13), and the equation (8.14) following that one gives a rearrangement that brings out the subtraction of the LG velocity more clearly.

4.3.2. *Using the Known Value of the Local Group Motion*
The second strategy to deal with the motion of the LG is to use its known value, measured from the dipole anisotropy of the CMB radiation, corrected for the motion of the solar system about the centre of the Milky Way, and for the motion of the Milky Way within the Local Group (Yahil, Tammann & Sandage 1977). The LG peculiar velocity is (Kogut *et al.* 1993, Table 3; Lineweaver *et al.* 1996)

$$\boldsymbol{v}^{\mathrm{LG}} = 627 \pm 22\,\mathrm{km\,s^{-1}} \quad \text{towards} \quad l = 276^\circ \pm 3^\circ,\ b = 30^\circ \pm 3^\circ \tag{4.51}$$

in the constellation of Hydra.

Thus if the redshift overdensity $\delta^{s\mathrm{LG}}$ is measured in the LG frame, then it can be corrected to yield the overdensity $\delta^s$ in the CMB frame by subtracting a dipole term in accordance with equation (4.42),

$$\delta^s(\boldsymbol{r}) = \delta^{s\mathrm{LG}}(\boldsymbol{r}) - \alpha(\boldsymbol{r})\frac{\hat{\boldsymbol{r}}.\boldsymbol{v}^{\mathrm{LG}}}{r} \ . \tag{4.52}$$

If $\alpha(r)$ is a function only of depth $r$, independent of direction, then the adjustment is a pure dipole overdensity, but in the general case the adjustment involves other harmonics also. The redshift correlation function of the overdensity in the CMB frame is given by equation (4.32).

An alternative version of the second strategy is to adjust the redshift distances from the LG frame to the CMB frame using equation (4.41), $s = s^{\mathrm{LG}} + \hat{\boldsymbol{r}}.\boldsymbol{v}^{\mathrm{LG}}$, and to measure the overdensity $\delta^s$ in the resulting CMB frame. However, this alternative version is not equivalent to the original when the condition $|v| \ll r$ is violated. Closer examination of the derivation of the distortion operator $\mathbf{S}$, equation (4.23), reveals that in effect it *assumes* that any streaming motion past the observer induces a dipole in the redshift overdensity. It follows that the best procedure is to measure the overdensity in the frame where the streaming motion vanishes, which is the LG frame, and then to transform to the CMB frame by making a dipole correction to the overdensity, as previously recommended, equation (4.52).

If the overdensity $\delta^s(\boldsymbol{r})$ in the CMB frame is constructed according to the recommended procedure — measure the overdensity in the LG frame and then correct to the CMB frame by subtracting a dipole streaming term, equation (4.52) — then the distortion equation (4.22) in the CMB frame will remain correct even without the condition $|v| \ll r$, equation (4.29).

Which strategy is better for dealing with the motion of the LG, to work entirely in the LG frame as described in §4.3.1, or to make a dipole correction and to work in the CMB frame, as proposed in this subsection? It seems to be largely a matter of judgement or convenience. The difference in information content is not great. If one likes, the difference can be isolated into the single mode with the specific form $\alpha(\boldsymbol{r})\hat{\boldsymbol{r}}.\boldsymbol{v}^{\rm LG}/r$, equation (4.42), by making all other modes orthogonal to it (Tegmark *et al.* 1997). In the case of an all-sky survey with a uniform flux limit, the mode is a dipole mode. For a non-all-sky survey with a uniform flux limit, the $\alpha(\boldsymbol{r})\hat{\boldsymbol{r}}.\boldsymbol{v}^{\rm LG}/r$ mode is effectively a 'cut' dipole mode, and making observed modes orthogonal to this mode means making them orthogonal to the cut dipole.

Note that the peculiar velocity of the LG in the second strategy is not used in order to allow a measurement of $\beta$ from any direct comparison between peculiar velocity and density fields; rather the LG velocity is used to introduce a dipole redshift distortion, which combines statistically with all the other redshift distortions to yield a statistical estimate of $\beta$.

### 4.3.3. *Rocket Effect*

The fact that the peculiar motion of the observer induces a dipole overdensity in redshift space in the direction of motion leads to an effect called the **rocket effect**, first pointed out by Kaiser (1987, §2; Kaiser & Lahav 1988, p. 366). This dipole needs to be distinguished carefully from the true dipole anisotropy of the density around the observer, which in the standard gravitational instability picture is responsible for inducing the observer's peculiar velocity, equation (4.44) (Strauss *et al.* 1992b).

Consider for example an observer situated in a perfectly uniform distribution of galaxies, all with zero peculiar velocities. Suppose that the observer hypothesizes that he is actually moving at some velocity $\boldsymbol{v}$ relative to the background. This motion will induce a spurious dipole in the direction of motion, which the witless observer may interpret as the dipole pulling him.

A complementary problem arises if an observer attempts to correct the dipole around him for the reflex motion induced on the observer by that dipole. Here there is a tendency to instability at large depths, beyond the median depth of the survey, where $\alpha(\boldsymbol{r})$ goes negative (Nusser & Davis 1994).

## 4.4. REAL AND REDSHIFT SPACE SELECTION FUNCTIONS

The distortion equation (4.22) with the distortion operator $\mathbf{S}$ given by (4.23) provides a valid prediction $\delta^s(\boldsymbol{r})$ of the observed redshift overdensity $\delta^s_{\rm obs}(\boldsymbol{r})$ provided that the redshift overdensity is defined relative to the real space selection function $\bar{n}(\boldsymbol{r})$, as in equation (3.17). In the practical case however, the selection function is likely to be estimated in redshift space. This subsection examines how the real and redshift space selection functions $\bar{n}(\boldsymbol{r})$ and $\bar{n}^s(\boldsymbol{r})$ are related for the typical case of a flux-limited redshift survey.

Suppose that the redshift survey includes all galaxies brighter than some flux limit $F_{\rm lim}$, and suppose for simplicity that this limit is uniform over the angular extent of the survey. It is convenient to characterize the luminosity of a galaxy of flux $F$ and true distance $r$ by the maximum distance $r_L$ to which the galaxy could be moved and still remain above the flux limit:

$$r_L \equiv r \left( \frac{F}{F_{\rm lim}} \right)^{1/2} . \tag{4.53}$$

The limiting distance $r_L$ of a galaxy is just a measure of its luminosity $L$, since $L = 4\pi r^2 F = 4\pi r_L^2 F_{\rm lim}$. Define $\phi(r_L) d\ln r_L$ to be the luminosity function, the mean number density of galaxies in a logarithmic interval $d\ln r_L$ of limiting distances (which is the same, up to a factor, as the mean number density of galaxies in a logarithmic interval of luminosity, or in an interval of absolute magnitude). The selection function $\bar{n}(r)$ at (true, not redshift) depth $r$ in this flux-limited redshift survey is the number density of galaxies luminous enough to be seen to depth $r$, which is the luminosity function integrated over $r_L > r$

$$\bar{n}(r) = \int_r^\infty \phi(r_L)\, d\ln r_L \ . \tag{4.54}$$

All modern methods of measuring the selection function (Binggeli, Sandage & Tammann 1988; Koranyi & Strauss 1997) seek to eliminate the (otherwise dominant) uncertainty arising from density inhomogeneity by assuming that the luminosity function $\phi(r_L)$ is a universal function, independent of the local density, and estimating the luminosity function by counting relative numbers of bright and faint galaxies in identical volumes. If bright and faint galaxies at any point share the same peculiar velocity, as seems probable in the linear regime, then the measurement of the selection function in redshift space should remain unbiased by inhomogeneity. A systematic difference between the real and redshift space luminosity functions, hence selection functions, does however result from the fact that radial peculiar velocities $v$ shift galaxies not only in space, $s = r + v$, but also in luminosity, since the apparent limiting distance $s_L$ of a galaxy in redshift space, defined by

$$s_L \equiv s \left( \frac{F}{F_{\rm lim}} \right)^{1/2} = \frac{s r_L}{r} \ , \tag{4.55}$$

is $s/r$ times its true limiting distance $r_L$, equation (4.53).

Hamilton & Culhane (1996) claim that the real and redshift selection functions agree to linear order, but this is false. What is true is that the expectation values of the real and redshift selection functions, averaged over an ensemble of observers, agree to linear order, essentially because the peculiar velocity averaged over an ensemble of observers is zero everywhere, $\langle \boldsymbol{v}(\boldsymbol{r}) \rangle = 0$ (a difference between ensemble-averaged real and redshift selection functions arises at next order, since the ensemble average dispersion of peculiar velocities is non-zero). In the actual case, however, the peculiar velocity does not vanish, and the actual real and redshift selections differ to linear order.

### 4.4.1. *Turner's Method*

As a typical procedure, consider Turner's (1979) procedure for estimating the selection function. It can be shown that Turner's method, when applied over a grid of depths $r$, yields the same selection function as that of the "Stepwise Maximum Likelihood" method of Efstathiou, Ellis & Peterson (1988), although Turner's method does not yield uncertainties. Turner's procedure is to construct the selection function $\bar{n}(r)$ from

$$\frac{d \ln \bar{n}(r)}{d \ln r} = \left. \frac{\partial N(r_L, r)}{N(r, r) \, \partial \ln r_L} \right|_{r_L = r} \tag{4.56}$$

where $N(r_L, r)$ is the observed number of galaxies brighter than $r_L$ and at the same time closer than $r$. In practice, equation (4.56) is solved on a grid of depths $r$, in which case the equation becomes, for bins of logarithmic width $\Delta \ln r$,

$$\frac{\Delta \ln \bar{n}(r)}{\Delta \ln r} = \frac{N(r e^{\Delta \ln r}, r) - N(r, r)}{N(r, r)} . \tag{4.57}$$

As pointed out by Strauss, Yahil & Davis (1991), if observed fluxes are rounded into discrete bins (for example, Zwicky magnitudes are rounded mostly to 0.1 magnitudes), then the grid in $r$ must be chosen carefully to avoid a discretization bias. For a logarithmic spacing $\Delta \ln F$ of discretized fluxes, the estimate of the selection function is unbiased if the logarithmic spacing $\Delta \ln r$ of depths is taken to be an integral multiple of $\Delta \ln(F^{1/2})$; the estimate is most biased if $\Delta \ln r$ is taken to be half of $\Delta \ln(F^{1/2})$. As with all such methods, Turner's procedure leaves the overall normalization of the selection function undetermined. The normalization must be fixed as a separate step.

Suppose that Turner's method (4.56) is applied in redshift space, so that the redshift space selection function $\bar{n}^s(s)$ is estimated from

$$\frac{d \ln \bar{n}^s(s)}{d \ln s} = \left. \frac{\partial N^s(s_L, s)}{N^s(s, s) \, \partial \ln s_L} \right|_{s_L = s} \tag{4.58}$$

where $N^s(s_L, s)$ is the observed number of galaxies brighter than $s_L$ and closer than $s$ in redshift space. How is the redshift space selection function estimated by (4.58) related to the real space selection function? Let $f(r_L, \boldsymbol{r})$ denote the number density of galaxies in an interval $d\ln r_L\, d^3r$ of luminosity and position in real space, and similarly let $f^s(s_L, \boldsymbol{s})$ denote the number density of galaxies in an interval $d\ln s_L\, d^3s$ in redshift space. Conservation of galaxies implies

$$f^s(s_L, \boldsymbol{s})\, d\ln s_L\, d^3s = f(r_L, \boldsymbol{r})\, d\ln r_L\, d^3r \tag{4.59}$$

with $s = r + v$ and $s_L = r_L s/r$, which generalizes equation (4.25). The relation $s = r + v$ is valid in the CMB frame; in the LG frame, replace $v \to v - \hat{\boldsymbol{r}}.\boldsymbol{v}^{\mathrm{LG}}$. The assumption that the luminosity function $\phi(r_L)$ is a universal function, independent of position, implies that the real space number density $f(r_L, \boldsymbol{r})$ factorizes to

$$f(r_L, \boldsymbol{r}) = \phi(r_L)[1 + \delta(\boldsymbol{r})] \ . \tag{4.60}$$

Given this assumption (4.60), a derivation similar to that leading from (4.25) to (4.30) shows that, to linear order in the overdensity $\delta(\boldsymbol{r})$ and peculiar velocity $v(\boldsymbol{r})$, equation (4.59) implies

$$f^s(r_L, \boldsymbol{r}) = \phi(r_L)\left[1 + \delta(\boldsymbol{r}) - \left(\frac{\partial}{\partial r} + \frac{\alpha_\phi(r_L)}{r}\right) v(\boldsymbol{r})\right] \tag{4.61}$$

where the dimensionless quantity $\alpha_\phi(r_L) \equiv d\ln[r_L^2 \phi(r_L)]/d\ln r_L$ is similar but not identical to the quantity $\alpha(\boldsymbol{r})$ defined by equation (4.24). By definition, the number $N^s(s_L, s)$ of galaxies brighter than $s_L$ and closer than $s$ is

$$N^s(s_L, s) \equiv \int_0^s \int_A \int_{s_L}^\infty f^s(s'_L, \boldsymbol{s}')\, d\ln s'_L\, d^3s' \tag{4.62}$$

where the angular part of the spatial integral is over the solid angle $A$ of the survey. A few-step calculation using equation (4.61) in equation (4.62) leads to

$$\frac{\partial N^s(r_L, r)}{N^s(r_L, r)\, \partial \ln r_L} = \frac{d\ln \bar{n}(r_L)}{d\ln r_L} - \frac{d^2 \ln \bar{n}(r_L)}{d\ln r_L^2} \left\langle \frac{v}{r} \right\rangle_r \tag{4.63}$$

where $\bar{n}(r_L)$, the real space selection function, is the integral (4.54) of the luminosity function, and $\langle v/r \rangle_r$ is the volume-average of $v/r$ within depth $r$ over the solid angle $A$ of the survey

$$\left\langle \frac{v}{r} \right\rangle_r \equiv \frac{3}{Ar^3} \int_0^r \int_A \frac{v(\boldsymbol{r}')}{r'}\, d^3r' \ . \tag{4.64}$$

Equation (4.63) shows that the basic relation between the real and redshift space selection functions, when the latter is estimated according to Turner's procedure (4.58), is, to linear order in the overdensity and peculiar velocity,

$$\frac{d\ln[\bar{n}(r)/\bar{n}^s(r)]}{d\ln r} = \frac{d^2 \ln \bar{n}(r)}{d\ln r^2} \left\langle \frac{v}{r} \right\rangle_r \ . \tag{4.65}$$

Integrating equation (4.65) yields

$$\ln\left(\frac{\bar{n}(r)}{\bar{n}^s(r)}\right) = -\int_r^{r_{\max}} \frac{d^2 \ln \bar{n}(r')}{d \ln r'^2} \left\langle \frac{v}{r} \right\rangle_{r'} \frac{dr'}{r'} \tag{4.66}$$

where the limit $r_{\max}$ of integration is an arbitrary depth which can conveniently be taken to be, say, the maximum depth of the survey. Changing $r_{\max}$ changes $\ln[\bar{n}(r)/\bar{n}^s(r)]$ by an integration constant, the arbitrariness of which reflects the fact that Turner's method leaves the overall normalization of the selection function undetermined. Changing the order of integration transforms equation (4.66) to

$$\ln\left(\frac{\bar{n}(r)}{\bar{n}^s(r)}\right) = -\int_0^{r_{\max}}\!\!\int_A a[\max(r,r')]\,\frac{v(\boldsymbol{r}')}{r'}\,d^3r' \tag{4.67}$$

where the integration is over the volume of the survey, and the function $a(r)$ is

$$a(r) \equiv \frac{3}{A}\int_r^{r_{\max}} \frac{1}{r'^3}\frac{d^2 \ln \bar{n}(r')}{d \ln r'^2}\frac{dr'}{r'} \ . \tag{4.68}$$

In practice it would normally be better to estimate the redshift selection function $\bar{n}^{s\,\mathrm{LG}}(r)$ in the LG frame rather than in the CMB frame, since the derivation of equation (4.61) requires amongst other things that $|v| \ll r$, which is not necessarily satisfied at small depths $r$, whereas the corresponding condition $|v - \hat{\boldsymbol{r}}.\boldsymbol{v}^{\mathrm{LG}}| \ll r$ in the LG frame is automatically satisfied, at least over linear scales. Equations in the LG frame are obtained by replacing $v \rightarrow v - \hat{\boldsymbol{r}}.\boldsymbol{v}^{\mathrm{LG}}$ everywhere. In the LG frame, equations (4.66) and (4.67) become

$$\begin{aligned}\ln\left(\frac{\bar{n}(r)}{\bar{n}^{s\,\mathrm{LG}}(r)}\right) &= -\int_r^{r_{\max}} \frac{d^2 \ln \bar{n}(r')}{d \ln r'^2}\left(\left\langle \frac{v}{r}\right\rangle_{r'} - \left\langle \frac{\hat{\boldsymbol{r}}}{r}\right\rangle_{r'}.\boldsymbol{v}^{\mathrm{LG}}\right)\frac{dr'}{r'} \\ &= -\int_0^{r_{\max}}\!\!\int_A a[\max(r,r')]\,\frac{[v(\boldsymbol{r}') - \hat{\boldsymbol{r}}'.\boldsymbol{v}^{\mathrm{LG}}]}{r'}\,d^3r' \end{aligned} \tag{4.69}$$

where $\langle \hat{\boldsymbol{r}}/r\rangle_r \equiv 3/(Ar^3)\int_0^r\!\int_A (\hat{\boldsymbol{r}}'/r')d^3r'$ is defined to be the volume average of $\hat{\boldsymbol{r}}/r$ within depth $r$ over the solid angle $A$ of the survey, similarly to the definition (4.64) of $\langle v/r\rangle_r$. Note that $\langle \hat{\boldsymbol{r}}/r\rangle_r = 0$ in the case of an all-sky survey.

Just as there were two strategies to deal with the LG motion, described in §§4.3.1 and 4.3.2, so here there are two general strategies to deal with the difference between real and redshift selection functions. The two strategies are described in the next two subsections, §§4.4.2 and 4.4.3.

4.4.2. *The Distortion Operator Relative to the Redshift Selection Function* The first strategy is to define the redshift overdensity $\delta^{ss}(\boldsymbol{s})$ relative to the selection function $\bar{n}^s(\boldsymbol{s})$ in redshift space, as in equation (3.18), and to modify the linear redshift distortion operator $\mathbf{S}$ accordingly.

It is straightforward to ascertain that the overdensities $\delta^{ss}(\boldsymbol{r})$ and $\delta^s(\boldsymbol{r})$ defined respectively relative to the redshift and real space selection functions, equations (3.18) and (3.17), are related to linear order by

$$\delta^{ss}(\boldsymbol{r}) = \delta^s(\boldsymbol{r}) + \ln[\bar{n}(\boldsymbol{r})/\bar{n}^s(\boldsymbol{r})] \ . \tag{4.70}$$

Equation (4.67), together with the usual linear expression (4.31) for the radial peculiar velocity $v$, shows that, in the case of Turner's (1979) method applied to a flux-limited redshift survey, $\ln[\bar{n}(r)/\bar{n}^s(r)]$ can be written

$$\ln[\bar{n}(r)/\bar{n}^s(r)] = \int \mathbf{S}^{\bar{n}}(r, \boldsymbol{r}')\delta(\boldsymbol{r}')\, d^3r' \tag{4.71}$$

(more compactly, the right hand side of eq. [4.71] is just $\mathbf{S}^{\bar{n}}\delta$ in the notation of matrices and vectors in Hilbert space, §3.3) where the linear operator $\mathbf{S}^{\bar{n}}$ is

$$\mathbf{S}^{\bar{n}}(r, \boldsymbol{r}') = \beta\, a[\max(r, r')]\omega(\boldsymbol{r}') \frac{\partial}{r'\partial r'} \nabla'^{-2} \tag{4.72}$$

with $\omega(\boldsymbol{r}') \equiv 1$ inside, and 0 outside, the volume of the survey.

Thus the linear redshift distortion equation for the redshift overdensity $\delta^{ss}$ is

$$\delta^{ss} = \mathbf{S}^s \delta \tag{4.73}$$

where the linear distortion operator $\mathbf{S}^s$ is the sum of the linear distortion operator $\mathbf{S}$, equation (4.23), and an extra piece $\mathbf{S}^{\bar{n}}$ given, for Turner's method, by equation (4.72):

$$\mathbf{S}^s = \mathbf{S} + \mathbf{S}^{\bar{n}} \ . \tag{4.74}$$

A complete expression for the distortion operator $\mathbf{S}^s$ is

$$\mathbf{S}^s(\boldsymbol{r}, \boldsymbol{r}') = \delta_D(\boldsymbol{r} - \boldsymbol{r}') \left(1 + \beta \frac{\partial^2}{\partial r^2} \nabla^{-2}\right) + \beta \frac{\alpha^s(\boldsymbol{r}, \boldsymbol{r}')}{r'} \frac{\partial}{\partial r'} \nabla'^{-2} \tag{4.75}$$

where $\alpha^s(\boldsymbol{r}, \boldsymbol{r}')$ is

$$\alpha^s(\boldsymbol{r}, \boldsymbol{r}') \equiv \delta_D(\boldsymbol{r} - \boldsymbol{r}')\alpha(r) + a[\max(r, r')]\omega(\boldsymbol{r}') \tag{4.76}$$

with $\alpha(r)$ and $a(r)$ given by (4.24) and (4.68), and again $\omega(\boldsymbol{r}') \equiv 1$ inside, and 0 outside, the volume of the survey.

In practice, it makes more sense to apply this strategy in the LG frame than in the CMB frame, for the reason given in the paragraph containing equation (4.69). In the LG frame, equations (4.70)–(4.75) carry through, but superscripted LG appropriately. In the LG frame, the overdensities $\delta^{ss\,\mathrm{LG}}$ and $\delta^{s\,\mathrm{LG}}$ are related by

$$\delta^{ss\,\mathrm{LG}}(\boldsymbol{r}) = \delta^{s\,\mathrm{LG}}(\boldsymbol{r}) + \ln[\bar{n}(\boldsymbol{r})/\bar{n}^{s\,\mathrm{LG}}(\boldsymbol{r})] \ . \tag{4.77}$$

In the case of Turner's method applied to a flux-limited survey, equation (4.69) shows that $\ln[\bar{n}(r)/\bar{n}^{s\,\mathrm{LG}}(r)] = \int \mathbf{S}^{\bar{n}\,\mathrm{LG}}(r, \boldsymbol{r}')\delta(\boldsymbol{r}')\, d^3r'$ where the operator $\mathbf{S}^{\bar{n}\,\mathrm{LG}}$ is

$$\mathbf{S}^{\bar{n}\,\mathrm{LG}}(r, \boldsymbol{r}') = \beta\, \frac{a[\max(r, r')]\omega(\boldsymbol{r}')\hat{\boldsymbol{r}}'}{r'} \cdot \left( \frac{\partial}{\partial \boldsymbol{r}'} - \left.\frac{\partial}{\partial \boldsymbol{r}'}\right|_{\boldsymbol{r}'=0} \right) \nabla'^{-2} \ . \tag{4.78}$$

The distortion equation for the redshift overdensity $\delta^{ss\,\mathrm{LG}}$ in the LG frame is

$$\delta^{ss\,\mathrm{LG}} = \mathbf{S}^{s\,\mathrm{LG}}\delta \tag{4.79}$$

where the linear distortion operator $\mathbf{S}^{s\,\mathrm{LG}}$ is the sum of the distortion operator $\mathbf{S}^{\mathrm{LG}}$, equation (4.46), and the extra piece $\mathbf{S}^{\bar{n}\,\mathrm{LG}}$ given, for Turner's method, by equation (4.78):

$$\mathbf{S}^{s\,\mathrm{LG}} = \mathbf{S}^{\mathrm{LG}} + \mathbf{S}^{\bar{n}\,\mathrm{LG}} \ . \tag{4.80}$$

A complete expression for the distortion operator $\mathbf{S}^{s\,\mathrm{LG}}$ is

$$\begin{aligned} \mathbf{S}^{s\,\mathrm{LG}}(\boldsymbol{r}, \boldsymbol{r}') \ &= \ \delta_D(\boldsymbol{r} - \boldsymbol{r}') \left( 1 + \beta\, \frac{\partial^2}{\partial r^2} \nabla^{-2} \right) \\ &\quad + \beta\, \frac{\alpha^s(\boldsymbol{r}, \boldsymbol{r}')\hat{\boldsymbol{r}}'}{r'} \cdot \left( \frac{\partial}{\partial \boldsymbol{r}'} - \left.\frac{\partial}{\partial \boldsymbol{r}'}\right|_{\boldsymbol{r}'=0} \right) \nabla'^{-2} \end{aligned} \tag{4.81}$$

with $\alpha^s(\boldsymbol{r}, \boldsymbol{r}')$ given by equation (4.76). In evaluating $\alpha^s(\boldsymbol{r}, \boldsymbol{r}')$, the selection function $\bar{n}(r)$ which enters the definitions (4.24) of $\alpha(r)$ and (4.68) of $a(r)$ can be replaced to linear order by the measured redshift selection function $\bar{n}^{s\,\mathrm{LG}}(r)$.

Equation (4.81) is an important equation. To recapitulate, it gives the form of the linear redshift distortion operator $\mathbf{S}^{s\,\mathrm{LG}}$ for the case of a flux-limited redshift survey in which the redshift space overdensity $\delta^{ss\,\mathrm{LG}}(\boldsymbol{r})$ is defined (a) in the frame of reference of the LG, and (b) relative to the redshift space selection function $\bar{n}^{s\,\mathrm{LG}}(r)$ measured using Turner's (1979) method, equation (4.58), again in the LG frame. That is, the distortion equation (4.79) gives the theoretically predicted form of the overdensity

$\delta^{ss\,\mathrm{LG}}(\boldsymbol{r})$ in the linear regime, to be compared to the observed redshift overdensity $\delta^{ss\,\mathrm{LG}}_{\mathrm{obs}}(\boldsymbol{r})$ defined by

$$\delta^{ss\,\mathrm{LG}}_{\mathrm{obs}}(\boldsymbol{r}) \equiv \frac{n^{s\,\mathrm{LG}}(\boldsymbol{r}) - \bar{n}^{s\,\mathrm{LG}}(r)}{\bar{n}^{s\,\mathrm{LG}}(r)} \tag{4.82}$$

with $n^{s\,\mathrm{LG}}(\boldsymbol{r})$ the observed galaxy density at redshift position $\boldsymbol{r}$ in the LG frame, and $\bar{n}^{s\,\mathrm{LG}}(r)$ the selection function measured in redshift space in the LG frame by Turner's method (4.58).

The redshift correlation function $\xi^{ss}(r_{12}, r_1, r_2)$ of the overdensity $\delta^{ss}(\boldsymbol{r})$ in the linear regime is related to the unredshifted correlation function $\xi(r_{12})$ by

$$\xi^{ss}(r_{12}, r_1, r_2) = \mathbf{S}^s_1 \mathbf{S}^s_2 \, \xi(r_{12}) \ . \tag{4.83}$$

Similarly the redshift correlation function $\xi^{ss\,\mathrm{LG}}(r_{12}, r_1, r_2)$ of the overdensity $\delta^{ss\,\mathrm{LG}}(\boldsymbol{r})$ in the LG frame is

$$\xi^{ss\,\mathrm{LG}}(r_{12}, r_1, r_2) = \mathbf{S}^{s\,\mathrm{LG}}_1 \mathbf{S}^{s\,\mathrm{LG}}_2 \, \xi(r_{12}) \ . \tag{4.84}$$

### 4.4.3. *Correcting the Redshift to the Real Space Selection Function*

The second strategy is to measure the redshift space selection function $\bar{n}^s(\boldsymbol{r})$ as before, but then to correct it to the real space selection function $\bar{n}(\boldsymbol{r})$.

The correction is given to linear order by equation (4.66), for the case of a flux-limited redshift survey in which the redshift space selection function is estimated by Turner's (1979) method, equation (4.58). Actually it would normally be preferable to estimate the redshift selection function $\bar{n}^{s\,\mathrm{LG}}(r)$ in the LG frame, and to correct this to the real selection function using equation (4.69). More precisely, the correction is

$$\ln\left(\frac{\bar{n}(r)}{\bar{n}^{s\,\mathrm{LG}}(r)}\right) = -\int_r^{r_{\max}} \frac{d^2 \ln \bar{n}^{s\,\mathrm{LG}}(r')}{d \ln r'^2} \left( \left\langle \frac{v}{r} \right\rangle_{r'} - \left\langle \frac{\hat{\boldsymbol{r}}}{r} \right\rangle_{r'} . \boldsymbol{v}^{\mathrm{LG}} \right) \frac{dr'}{r'} \tag{4.85}$$

which is the same as equation (4.69) except that the redshift selection function $\bar{n}^{s\,\mathrm{LG}}(r')$ (the thing you have) replaces the real selection function $\bar{n}(r')$ (the thing you want) in the integrand on the right hand side of equation (4.85). The replacement is valid to linear order because $\bar{n}^{s\,\mathrm{LG}}(r)$ and $\bar{n}(r)$ agree to zeroth order.

The correction (4.85) involves modelling the volume-average $\langle v/r \rangle_r$, equation (4.64), of the radial peculiar velocity $v$ divided by depth $r$, as a function of depth $r$ in the survey. Note that $\langle v/r \rangle_r$ is an average of the actual $v/r$, not an ensemble average. One way to do this is to carry out a complete reconstruction of the unredshifted density and velocity field

(Yahil *et al.* 1991; Fisher *et al.* 1995b; Webster, Lahav & Fisher 1997). Complete reconstruction seems extravagant merely for the purpose of estimating $\langle v/r \rangle_r$; perhaps a simpler approximation could be devised.

It is to be noted that in an all-sky survey $\langle v/r \rangle_r$ is negative if galaxies are on average converging towards the observer, which occurs if the observer is in an overdense region of the Universe. The LG lies in such an overdense region, as is typical for galaxy-bound observers, which is an example of local bias.

In a non-all-sky survey, $\langle v/r \rangle_r$ depends in addition on the alignment of the survey with large scale flows.

However carefully $\langle v/r \rangle_r$ is modelled, there is liable to be some uncertainty in it, hence some uncertainty in the correction $\ln[\bar{n}(r)/\bar{n}^{s\,\mathrm{LG}}(r)]$, equation (4.85). The correction $\ln[\bar{n}(r)/\bar{n}^{s\,\mathrm{LG}}(r)]$ to the selection function is essentially a correction to the redshift overdensity, as in equation (4.77). Measurements of the redshift space overdensity can be immunized against uncertainty in the radial correction by discarding from the statistical analysis the uncertain radial mode or modes — possibly all radial modes, to be conservative — and by making all other observed modes orthogonal to these radial modes (Tegmark *et al.* 1997). In the case of an all-sky survey with a uniform flux limit, radial modes are monopole modes about the observer. For a non-all-sky survey with a uniform flux limit, radial modes are effectively 'cut' monopole modes, and observed modes can be immunized against radial uncertainty by making them orthogonal to the cut monopole modes.

## 5. Methods to Measure $\beta$

To date, essentially three types of method have been used to measure $\beta$ from linear redshift distortions:

1. The ratio of angle-averaged redshift space to real space power spectrum or correlation function;
2. The ratio of quadrupole to monopole moments of the redshift space power spectrum;
3. A maximum likelihood approach in which the data are the amplitudes of (many hundreds of) individual modes, and the parameters are $\beta$ and the power spectrum $P(k)$, parametrized in some way.

These methods are discussed in turn below.

### 5.1. RATIO OF REDSHIFT SPACE TO REAL SPACE ANGLE-AVERAGED POWER SPECTRA

This method was proposed in Kaiser's (1987) original paper. It assumes the plane-parallel, or distant observer, approximation, where structures are sufficiently far away that line-of-sight peculiar velocities are effectively plane-parallel, as illustrated in the left panel of Figure 1.

Kaiser's formula (4.36) predicts that in the linear regime the angle-averaged redshift power spectrum $P^s(k) \equiv \int P^s(\boldsymbol{k}) do_{\boldsymbol{k}}/(4\pi)$ [which is the same thing as the monopole redshift power $P_0^s(k)$ discussed in §5.2] is amplified by a constant factor over the unredshifted power spectrum $P(k)$:

$$\frac{P^s(k)}{P(k)} = 1 + \frac{2}{3}\beta + \frac{1}{5}\beta^2 . \tag{5.1}$$

As $\beta$ varies between 0 and 1, the amplification factor varies between 1 and 28/15, almost a factor of 2.

Equation (5.1) is valid for any wavenumber $k$ in the linear regime, so remains valid if the real and redshift space power spectra are first smoothed over any window $W(k)$ that picks out linear wavenumbers. That is, if

$$\tilde{P}(k) \equiv \int_0^\infty W(k) P(k)\, 4\pi k^2 dk/(2\pi)^3 \tag{5.2}$$

and similarly for $\tilde{P}^s(k)$, then the ratio of the smoothed redshift to real space power spectra is predicted to be

$$\frac{\tilde{P}^s}{\tilde{P}} = 1 + \frac{2}{3}\beta + \frac{1}{5}\beta^2 . \tag{5.3}$$

In particular, the ratio of the angle-averaged redshift space correlation function $\xi^s(r) \equiv \int \xi^s(\boldsymbol{r}) do_{\boldsymbol{r}}/(4\pi)$ [which is the monopole redshift correlation function $\xi_0^s(r)$] to its unredshifted counterpart $\xi(r)$ is, at linear separations $r$,

$$\frac{\xi^s(r)}{\xi(r)} = 1 + \frac{2}{3}\beta + \frac{1}{5}\beta^2 , \tag{5.4}$$

which follows from equation (5.3) with the smoothing window in (5.2) taken to be the zeroth spherical Bessel function, $W(k) = \int e^{-i\boldsymbol{k}.\boldsymbol{r}} do_{\boldsymbol{k}}/(4\pi) = j_0(kr)$.

The unredshifted correlation function $\xi(r)$ and/or the unredshifted power spectrum $P(k)$ can be measured by deprojecting the angular correlation function of a survey. This deprojection is liable to be quite noisy, unless the survey is large. Thus the method of inferring $\beta$ from the ratio of redshift to real correlation functions or power spectra is most appropriate if the redshift survey happens to be a subset of a larger angular survey. Examples of this are the Stromlo-APM survey (Loveday *et al.* 1996b), which is a 1-in-20 redshift survey of galaxies brighter than $b_J = 17.15$ in the APM angular survey, or the QDOT survey (e.g. Saunders *et al.* 1990; Rowan-Robinson *et al.* 1991; Saunders, Rowan-Robinson & Lawrence 1992; Lawrence *et al.* 1997, in preparation) which is a 1-in-6 redshift survey of galaxies brighter than 0.6 Jy from the *IRAS* Point Source Catalog.

## 5.2. RATIO OF QUADRUPOLE TO MONOPOLE HARMONICS OF THE POWER SPECTRUM

The idea of decomposing the redshift correlation function into harmonics was proposed by Hamilton (1992), and the connection to the power spectrum was clarified by Cole, Fisher & Weinberg (1994). Integral expressions for the shape of the redshift correlation function had previously been given by Kaiser (1987), Lilje & Efstathiou (1989), and McGill (1990). This method, like method 1, assumes the plane-parallel approximation.

For plane-parallel redshift distortions, the redshift power spectrum can be written as a sum of even harmonics $P^s_\ell(k)$

$$P^s(\boldsymbol{k}) = \sum_{\ell \text{ even}} \mathcal{P}_\ell(\mu_{\boldsymbol{k}}) P^s_\ell(k) \ , \quad P^s_\ell(k) \equiv (2\ell+1) \int \mathcal{P}_\ell(\mu_{\boldsymbol{k}}) P^s(\boldsymbol{k}) \, do_{\boldsymbol{k}}/4\pi \tag{5.5}$$

where $\mathcal{P}_\ell(\mu_{\boldsymbol{k}})$ are Legendre polynomials, and $\mu_{\boldsymbol{k}} \equiv \hat{\boldsymbol{z}}.\hat{\boldsymbol{k}}$ is the cosine of the angle between the wavevector $\boldsymbol{k}$ and the line of sight $\boldsymbol{z}$. The odd harmonics vanish by pair exchange symmetry, and non-zero azimuthal harmonics (corresponding to $Y_{\ell m}$'s with $m \neq 0$) vanish by symmetry about the line of sight.

In the linear regime, Kaiser's formula (4.36) shows that the redshift power spectrum reduces to a sum of monopole ($\ell = 0$), quadrupole ($\ell = 2$), and hexadecapole ($\ell = 4$) harmonics

$$P^s(\boldsymbol{k}) = \mathcal{P}_0(\mu_{\boldsymbol{k}}) P^s_0(k) + \mathcal{P}_2(\mu_{\boldsymbol{k}}) P^s_2(k) + \mathcal{P}_4(\mu_{\boldsymbol{k}}) P^s_4(k) \tag{5.6}$$

where the harmonics $P^s_\ell$ of the redshift space power spectrum are related to the unredshifted power spectrum $P(k)$ by

$$\begin{aligned} P^s_0(k) &= \left(1 + \frac{2}{3}\beta + \frac{1}{5}\beta^2\right) P(k) \\ P^s_2(k) &= \left(\frac{4}{3}\beta + \frac{4}{7}\beta^2\right) P(k) \\ P^s_4(k) &= \frac{8}{35}\beta^2 P(k) \ . \end{aligned} \tag{5.7}$$

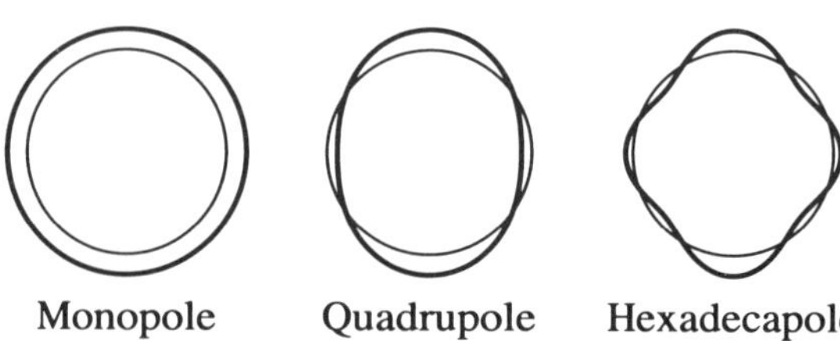


*Figure 6.* Shapes of the monopole ($\ell = 0$), quadrupole ($\ell = 2$), and hexadecapole ($\ell = 4$) harmonics (thick lines), drawn as perturbations of a sphere (thin lines). The quadrupole has positive amplitude as drawn, as appropriate for the redshift power spectrum. The factor of $i^\ell$ between the power spectrum and the correlation function, equation (5.10), introduces a minus sign in the quadrupole correlation function, which therefore appears squashed, opposite to what is drawn.

Equations (5.7) imply in particular that the ratio $P_2^s(k)/P_0^s(k)$ of quadrupole to monopole harmonics of the redshift power spectrum in the linear regime is

$$\frac{P_2^s(k)}{P_0^s(k)} = \frac{\frac{4}{3}\beta + \frac{4}{7}\beta^2}{1 + \frac{2}{3}\beta + \frac{1}{5}\beta^2} \tag{5.8}$$

which therefore provides a means of measuring $\beta$. Equations (5.7) predict that the hexadecapole harmonic $P_4^s(k)$ should be quite small, $\lesssim 0.1$ of the monopole and quadrupole harmonics for $\beta \leq 1$, and it also proves to be noisier, especially at the largest scales, so in practice the quadrupole-to-monopole ratio has been the statistic of choice for many authors.

The harmonics $\xi_\ell^s(r)$ of the redshift correlation function are defined similarly to those (5.5) of the power spectrum

$$\xi^s(\boldsymbol{r}) = \sum_{\ell \text{ even}} \mathcal{P}_\ell(\mu_{\boldsymbol{r}})\xi_\ell^s(r) \ , \quad \xi_\ell^s(r) \equiv (2\ell+1)\int \mathcal{P}_\ell(\mu_{\boldsymbol{r}})\xi^s(\boldsymbol{r})\, do_{\boldsymbol{r}}/(4\pi) \tag{5.9}$$

where $\mu_{\boldsymbol{r}} \equiv \hat{\boldsymbol{z}}.\hat{\boldsymbol{r}}$ is the cosine of the angle between the separation vector $\boldsymbol{r}$ and the line of sight $\boldsymbol{z}$. The harmonics $P_\ell^s(k)$ of the redshift power spectrum are related to the harmonics $\xi_\ell^s(r)$ of the redshift correlation function by

$$P_\ell^s(k) = i^\ell 4\pi \int_0^\infty j_\ell(kr)\xi_\ell^s(r)\, r^2 dr \ , \ \ \xi_\ell^s(r) = i^{-\ell} 4\pi \int_0^\infty j_\ell(kr) P_\ell^s(k)\, k^2 dk/(2\pi)^3 \tag{5.10}$$

where $j_\ell(kr)$ are spherical Bessel functions. For the quadrupole harmonic, $\ell = 2$, the $i^\ell$ factor in equation (5.10) introduces a negative sign between the power spectrum and the correlation function. Thus the large scale squashing effect of the redshift correlation function, a negative quadrupole, corresponds to a stretching of the redshift power spectrum along the line of sight, a positive quadrupole (Cole, Fisher & Weinberg 1994, Fig. 1).

### 5.2.1. *Smoothed Harmonics of the Redshift Power Spectrum*

In any finite survey, it is necessary to measure the power spectrum at finite resolution, given roughly by the inverse scale size of the survey. Consider then the harmonics of the redshift power spectrum smoothed over some window $W(k)$ (the following equation generalizes eq. [5.2])

$$\tilde{P}^s_\ell \equiv \int_0^\infty W(k) P^s_\ell(k)\, 4\pi k^2 dk/(2\pi)^3 \,. \tag{5.11}$$

As long as the window $W(k)$ picks out wavenumbers in the linear regime, the quadrupole-to-monopole ratio of the smoothed harmonics $\tilde{P}_\ell$ will also satisfy the relation (5.8),

$$\frac{\tilde{P}^s_2}{\tilde{P}^s_0} = \frac{\frac{4}{3}\beta + \frac{4}{7}\beta^2}{1 + \frac{2}{3}\beta + \frac{1}{5}\beta^2} \,. \tag{5.12}$$

The relation (5.10) between the harmonics of the power spectrum and the correlation function means that the smoothed harmonics $\tilde{P}^s_\ell$ of the power spectrum, equation (5.11), can also be written as smoothed harmonics of the correlation function:

$$\tilde{P}^s_\ell = \int_0^\infty W_\ell(r) \xi^s_\ell(r)\, 4\pi r^2 dr \tag{5.13}$$

where the real space smoothing window $W_\ell(r)$ is related to the Fourier window $W(k)$ by

$$W_\ell(r) = i^\ell 4\pi \int_0^\infty j_\ell(kr) W(k)\, k^2 dk/(2\pi)^3 \,. \tag{5.14}$$

The point of equation (5.13) is that it shows how smoothed redshift power spectra $\tilde{P}^s_\ell$ can be measured from quantities in real (redshift) space, which is where the data lie.

### 5.2.2. *Smoothing Windows*

Childers (1978, p. 1) terms the business of choosing smoothing windows $W(k)$ "window carpentry". One possibility, proposed by Hamilton (1995) and used in §6 below, is to use smoothing windows that are power laws times a Gaussian, $W(k) \sim k^n e^{-k^2}$ with $n$ an even integer, suitably scaled and normalized. These windows $W(k)$ have the desirable properties that:

- they are everywhere positive, so preserving the intrinsic positivity of the power spectrum;

- they vanish at zero wavenumber, $W(k) = 0$ at $k = 0$, provided that $n \geq 2$, so immunizing the measurement of power against uncertainty in the mean density (which makes a delta-function contribution to power at zero wavenumber);
- they are analytically convenient;
- they yield Gaussian convergence as a function of pair separation $r$ in the corresponding real space windows $W_\ell(r)$, equation (5.14), for harmonics $\ell \leq n$, provided that $n$ is chosen to be an even integer.

Suitably scaled, and normalized so $\int W(k)d^3k/(2\pi)^3 = 1$, these smoothing windows are

$$W(k)\frac{d^3k}{(2\pi)^3} \equiv \frac{2e^{-q^2}q^{n+2}\,dq}{\Gamma[(n+3)/2]}\ , \quad q \equiv \frac{\alpha k}{\tilde{k}}\ . \tag{5.15}$$

The wavenumber $\tilde{k}$ is an 'effective' wavenumber, the constant $\alpha$ being chosen so that $W(k)$ peaks at approximately $k \approx \tilde{k}$. In Hamilton (1995) and §6, the constant $\alpha$ is chosen so that the smoothed monopole power at wavenumber $\tilde{k}$ is equal to the unsmoothed monopole power at the same wavenumber, $\tilde{P}_0^s(\tilde{k}) = P_0^s(\tilde{k})$, for the particular case where the power spectrum is a power law of index $-1.5$, that is, for $P_0^s(k) \propto k^{-1.5}$. For the case $n = 2$ in the window (5.15), as used in Figure 7, this fixes $\alpha = 1.279$. The harmonics of the power spectrum smoothed over the window (5.15) are, according to equation (5.13), equal to the harmonics of the correlation function smoothed over corresponding windows $W_\ell(r)$ given by equation (5.14):

$$W_\ell(r) = \frac{i^\ell[(n-\ell)/2]!}{(3/2)_{(n/2)}}\, s^\ell e^{-s^2} L_{(n-\ell)/2}^{\ell+(1/2)}(s^2)\ , \quad s \equiv \frac{\tilde{k}r}{2\alpha} \tag{5.16}$$

(note that $W_0(0) = 1$) where $L_\nu^\lambda$ are Laguerre polynomials (Abramowitz & Stegun 1964) and $(3/2)_{(n/2)} = \Gamma[(n+3)/2]/\Gamma(3/2)$ is a Pochhammer symbol. The smoothing windows (5.16) happen to be $e^{-s^2/2}$ times the radial wavefunctions of a 3-dimensional simple harmonic oscillator at energy level $n$ and angular momentum $\ell$ (Landau & Lifshitz 1958, §33 Problem 4). The quantization condition that the index $(n-\ell)/2$ be a non-negative integer ensures Gaussian convergence at large separations $r$. Despite the forbidding appearance, the smoothing windows $W_\ell(r)$ are just polynomials times a Gaussian, with a rapid stable recurrence relation available to evaluate the polynomials.

It is noteworthy that in order to measure the $\ell$'th harmonic with one of the windows (5.15), it is necessary to choose $n \geq \ell$, so that the window must go to zero at least as fast as $W(k) \sim k^\ell$ as $k \to 0$.

### 5.3. MAXIMUM LIKELIHOOD

Maximum likelihood (ML) techniques offer an optimal way to estimate $\beta$ and its uncertainty from redshift distortions. The technique is not only optimal, yielding both the most probable answer along with a complete probability distribution of answers, for a given prior, but is flexible enough to admit complications, such as the radial (as opposed to plane-parallel) character of redshift distortions, which can be difficult to deal with by direct methods.

The ML technique was first applied to measure $\beta$ from linear redshift distortions by Fisher, Scharf & Lahav (1994), and a more thorough implementation was carried through by Heavens & Taylor (1995) and then by Ballinger, Heavens & Taylor (1995).

Likelihood techniques, applied according to Bayesian precepts (Loredo 1990), require a prior, that is, a precise statement of prior assumptions, including the range of models to be considered, and the prior probability distributions of the parameters to be measured. In the usual case, the prior probability of the parameters is taken to be uniform, in which case the posterior probability, the probability distribution of the parameters given the observed data (which is what you want), is proportional to the likelihood function, which is the probability of the observed data given the parameters (which is what theory tells you).

When estimating $\beta$ from linear redshift distortions, the prior is liable to include such assumptions as that the Universe is statistically homogeneous and isotropic, that the distribution of overdensity on linear scales is a multivariate Gaussian, that the power spectrum takes such and such a parametrized form (e.g. Cold Dark Matter (CDM) or one of its cousins), that observations introduce such and such additional sources of uncertainty (notably Poisson sampling noise), that linear redshift distortions conform to the standard model described in §4, and that all values of $\beta$ are a priori equally likely (which is called a uniform prior on the parameter $\beta$).

At first sight the need to state a prior seems a disadvantage compared to traditional direct approaches (where for example one measures $\beta$ from some ratio of power spectra, and one measures power spectra by, well, by measuring power spectra). With second sight however, one realizes the advantages of the ML procedure:

- Being forced to state one's prior assumptions explicitly is a virtue, not a drawback;
- If changing your prior (within plausible limits) makes much difference to the values of the quantities you are trying to estimate, then you are not learning much from the data;

- The mathematical formalism is quite general, specifying the best way to proceed even in highly complicated situations.

This having been said, it is true that the ML formalism will happily provide answers for parameters even if the prior is completely wrong. Thus the prior should, correctly, always be tested for consistency with the data (a step that to date has generally not been taken).

If the real overdensity $\boldsymbol{\delta}$ forms a multivariate Gaussian field, as may well be true on linear scales in reality, then the observed linear redshift overdensity $\boldsymbol{\delta}^s_{\rm obs}$ is also predicted to be a multivariate Gaussian (any linear combination of a Gaussian field is a Gaussian field), and the likelihood function $\mathcal{L}$ is Gaussian

$$\mathcal{L} \propto \frac{1}{|\mathbf{C}^s|^{1/2}} \exp\left(-\frac{1}{2}\boldsymbol{\delta}^{s\dagger}_{\rm obs}\mathbf{C}^{s-1}\boldsymbol{\delta}^s_{\rm obs}\right) \tag{5.17}$$

where $\mathbf{C}^s \equiv \langle \boldsymbol{\delta}^s \boldsymbol{\delta}^{s\dagger} \rangle$ is the expectation value of the survey covariance matrix, and $|\mathbf{C}^s|$ and $\mathbf{C}^{s-1}$ are its determinant and inverse. The $\boldsymbol{\delta}^s_{\rm obs}$ in the likelihood (5.17) is the observed vector of overdensities, while the covariance matrix $\mathbf{C}^s$ is the prior. The likelihood function (5.17) gives the probability of the observed data, $\boldsymbol{\delta}^s_{\rm obs}$, given the prior, $\mathbf{C}^s$. With a uniform prior on the parameters of the covariance matrix $\mathbf{C}^s$, the likelihood function becomes (up to a normalization factor) the probability of the parameters given the data.

In general, the prior covariance $\mathbf{C}^s$ is a combination of signal plus noise terms. In the simplest case, one supposes that the only source of noise in a redshift survey of galaxies is Poisson sampling noise (§§3.1.2, 3.2.2). The survey covariance $\mathbf{C}^s$ is then a sum of cosmic and Poisson sampling terms (the following equation is a repeat of eq. [3.28])

$$C^s_{ij} = \xi^s(\boldsymbol{r}_i, \boldsymbol{r}_j) + \delta_D(\boldsymbol{r}_i - \boldsymbol{r}_j)[\bar{n}(\boldsymbol{r}_i)]^{-1} \tag{5.18}$$

where $\xi^s(\boldsymbol{r}_i, \boldsymbol{r}_j)$ is the redshift space correlation function, and $\delta_D(\boldsymbol{r}_i - \boldsymbol{r}_j)$ is a Dirac delta function. For fluctuations in the linear regime, the redshift correlation function $\xi^s(\boldsymbol{r}_i, \boldsymbol{r}_j)$ is predicted to depend on the linear distortion parameter $\beta$ and the unredshifted correlation function $\xi(r)$ (or equivalently the unredshifted power spectrum $P(k)$, eq. [3.6]) in accordance with equation (4.32). The parameters to be measured are then the distortion parameter $\beta$, and the parameters of some parametrization of the unredshifted power spectrum $P(k)$.

Actually, the 'observed' redshift overdensity $\delta^s_{\rm obs}(\boldsymbol{r})$, equation (3.17), depends not only on the observed galaxy density $n^s(\boldsymbol{r})$ in redshift space, but also on the selection function $\bar{n}(\boldsymbol{r})$, which is itself estimated from the survey. The shape of the selection function (or the parameters of some parametrization thereof) would normally be measured as a separate operation, since that measurement typically involves additional assumptions about the universality of the luminosity function (see §4.4). This leaves the overall normalization of the selection function as an extra parameter to be measured from the likelihood function (5.17). In principle, uncertainty in the measurement of the shape of the selection function should be allowed for by marginalizing over it, that is, by integrating the likelihood function over the probability distribution of the parameters of the shape of the selection function. In practice, the shape of the selection function is normally so accurately measured, compared to other uncertainties, that it can be treated as a fixed, known quantity in the likelihood analysis.

In order to proceed numerically, it is necessary to pixelize the data in some fashion, so that the covariance matrix $C_{ij}$ becomes a finite matrix. In principle, the choice of pixelization is irrelevant, if the pixels are fine and many enough to represent the survey data accurately. In practice, the matrix $C_{ij}$ needs to be of tractable size, and some cunning is needed to pick a pixelization that is small but nevertheless contains virtually all the information about the parameters to be measured (Tegmark, Taylor & Heavens 1996; Tegmark *et al.* 1997). Fisher, Scharf & Lahav (1994) chose the pixels to be spherical harmonics to $\ell \leq 10$ on the sky, with 4 Gaussian pixels in the radial direction. Heavens & Taylor (1995) and Ballinger, Heavens & Taylor (1995) chose a basis of spherical waves, that is, spherical harmonics on the sky, with (weighted) spherical Bessel functions in the radial direction.

Clearly the ML technique, or methods essentially equivalent to it, is the method of choice for measuring redshift distortions, certainly on the largest scales where a small number of pixels contains all relevant information, perhaps also on smaller scales if clever compression techniques can be applied (Vogeley & Szalay 1996; Tegmark, Taylor & Heavens 1996; Tegmark *et al.* 1997). Undoubtedly there will be further development of this powerful approach in the future.

One mysterious aspect of ML results published to date is that they yield estimates of $\beta$ that appear to be systematically larger than other methods (see Table 8.1). One possible reason for the difference is the fact that the plane-parallel approximation underestimates $\beta$. However, authors who use the plane-parallel approximation have generally confined their measurements to opening angles no larger than 50°, at which point the $N$-body simulations of Cole, Fisher & Weinberg (1994, Fig. 8) indicate that the plane-parallel approximation underestimates $\beta$ by only 5%. Another possibility, suggested by Ballinger, Heavens & Taylor (1995), is that direct methods unfairly penalize larger values of $\beta$, since larger values of $\beta$ predict a larger variance in the power spectrum, which is not taken into consideration by direct methods.

The resolution of this mystery remains unclear.

## 6. Example of Measuring $\beta$ from Linear Redshift Distortions

It is helpful to present an example of the measurement of $\beta$ from linear redshift distortions, which is intended not only to illustrate how things work out in practice, but also (a) to bring out the difference between *IRAS* and optically selected galaxies when they are analysed with (essentially) the same procedure, and (b) to demonstrate the importance of nonlinearities (fingers-of-god).

The results are given first, in §6.1, and then §6.2 provides some details of the analysis.

### 6.1. RESULTS

Figure 7 shows the ratio $\tilde{P}_2^s(\tilde{k})/\tilde{P}_0^s(\tilde{k})$ of smoothed quadrupole to smoothed monopole power measured as a function of effective wavenumber $\tilde{k}$ in the *IRAS* QDOT + 1.2 Jy survey and the Stromlo-APM survey. These are the same surveys whose redshift correlation functions are plotted in Figure 3. The result for the QDOT + 1.2 Jy survey is from Hamilton (1995), who gives additional details of the analysis. The Stromlo-APM result, new to this review, follows essentially the same analysis.

On linear scales (to the right of the graphs in Figure 7) the quadrupole-to-monopole ratio is expected to go over to a constant. While there is some indication, more particularly in the *IRAS* surveys, that the ratio is indeed asymptoting to a constant, it is evident that nonlinearities have some effect even at scales as large as a half-wavelength[10] of $\pi/\tilde{k} \approx 40h^{-1}$Mpc. One should probably not believe the quadrupole-to-monopole ratio for $\pi/\tilde{k} \gtrsim 50h^{-1}$Mpc, where the uncertainties are becoming large. On small scales the nonlinear finger-of-god effect lowers the quadrupole-to-monopole ratio, so that the quadrupole power is negative below a half-wavelength of $\pi/\tilde{k} \approx 5h^{-1}$Mpc in the QDOT + 1.2 Jy survey, and below $\pi/\tilde{k} \approx 8h^{-1}$Mpc in the Stromlo-APM survey.

For the QDOT + 1.2 Jy survey, nonlinear effects appear to be adequately described by a model in which the linear distortion is modulated by a random exponentially distributed pairwise velocity dispersion (Fisher *et al.* 1994b) (see §7.2). A least squares fit to the quadrupole-to-monopole ratio over half-wavelengths 5–44 $h^{-1}$Mpc gives a one-dimensional pairwise velocity dispersion of $322\,\mathrm{km\,s^{-1}}$. Although this method of measuring the

[10]The uncertainty principle ensures that the correspondence between wavenumber and separation is inevitably uncertain. Empirically however, I find that a half-wavelength $\pi/k$ typically gives a better estimate of the physical scale of separation than for example a full wavelength $2\pi/k$ or an inverse wavenumber $1/k$.

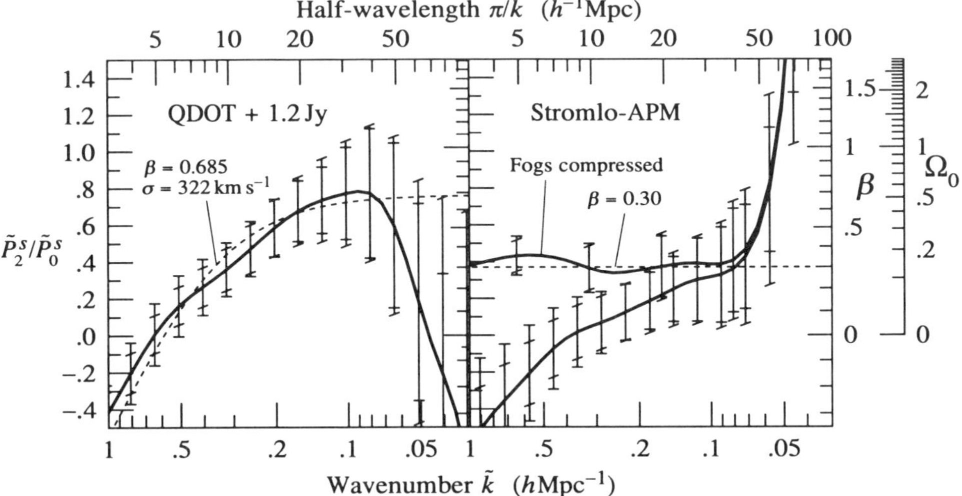


*Figure 7.* Ratio $\tilde{P}_2^s(\tilde{k})/\tilde{P}_0^s(\tilde{k})$ of smoothed quadrupole to smoothed monopole power as a function of effective wavenumber $\tilde{k}$ in: (left) the merged *IRAS* QDOT plus 1.2 Jy survey; and (right) the Stromlo-APM survey. These are the same surveys whose redshift correlation functions are illustrated in Figure 3. For the QDOT + 1.2 Jy sample in the left panel, the dashed line is a fit to a model in which the linear redshift distortion is modulated by a random exponentially distributed pairwise velocity dispersion (§7.2). For the Stromlo-APM survey in the right panel, the upper thick line shows the quadrupole-to-monopole ratio which results when fingers-of-god are compressed, and the dashed line is a straight line fit to this line. The scales on the right show the value of $\beta$ corresponding to the linear ratio $\tilde{P}_2^s/\tilde{P}_0^s$, equation (5.12), and the inferred value of $\Omega_0$, equation (1.5), in the absence of bias, $b = 1$. Beware that the smoothing window $W(k) \propto k^2 \exp[-(1.279k/\tilde{k})^2]$, equation (5.15), is broad, so the points are correlated; roughly every third error bar is uncorrelated. The horizontal and slanted bars on the ($1\sigma$) errors come from two separate ways of measuring uncertainties.

velocity dispersion is crude compared to Fisher *et al.*'s (1994b) procedure, the number is nonetheless in good agreement with Fisher *et al.*'s scale-dependent estimate (§8.2.2). As reported by Hamilton (1995), the best fitting value of the linear redshift distortion parameter in the QDOT + 1.2 Jy survey is $\beta = 0.69^{+0.21}_{-0.19}$ ($1\sigma$).

For the Stromlo-APM survey, the nonlinear effects appear large enough that simple model-fitting to the quadrupole-to-monopole ratio seems a hazardous procedure. Certainly there would be a large covariance between $\beta$ and the velocity dispersion, so that neither would be well measured. One alternative procedure is to deal with nonlinearities at a more fundamental level, by identifying individual fingers-of-god in the survey, and compressing them (Gramann, Cen & Gott 1994). Fingers can be identified using a friends-of-friends algorithm with a window elongated in the radial direction (e.g. Ramella, Geller & Huchra 1989). Figure 7 shows the quadrupole-to-monopole ratio in the Stromlo-APM survey that results from compressing fingers-of-god whose overdensities exceeded 10 when measured through elliptical windows 10 times longer in the radial than transverse direction (so the overdensity of the compressed fingers exceeded $10 \times 10 = 100$; to avoid spurious fingers in distant, sparsely sampled regions of the survey, the transverse link length was also limited to less than $5\,h^{-1}$Mpc). The result is encouraging in two ways: first, that the quadrupole-to-monopole ratio is flattened almost to a constant on all scales (at least until the uncertainties go out of control at $\pi/\tilde{k} \gtrsim 50\,h^{-1}$Mpc); and second, that the ratio is hardly changed at half-wavelengths $\pi/\tilde{k} \gtrsim 40\,h^{-1}$Mpc, suggesting that the ratio is close to linear at these scales. I choose to quote the value $\beta = 0.30^{+0.17}_{-0.15}$ ($1\sigma$) measured at a half-wavelength of $\pi/\tilde{k} = 20\,h^{-1}$Mpc. Even though the statistical uncertainty is smaller at shorter wavelengths, the systematic uncertainty associated with compressing the fingers-of-god presumably becomes larger.

That $\beta$ is smaller for optical than *IRAS* galaxies is consistent with the standard interpretation in which $\beta \approx \Omega_0^{0.6}/b$, and optical galaxies are positively biased relative to *IRAS* galaxies. The values of $\beta$ measured here indicate $b_{\mathrm{optical}}/b_{IRAS} \sim 2.3$ with an uncertainty of a factor of 2 or 3.

### 6.2. ANALYSIS

The smoothing window used in Figure 7 is the power law times Gaussian window $W(k) \propto k^2 \exp[-(1.279k/\tilde{k})^2]$ of equation (5.15), with index $n = 2$, the smallest $n$ that allows the quadrupole to be measured. This choice of $n$ yields the broadest smoothing window of the form (5.15), so the errors in the ratio $\tilde{P}_2^s(\tilde{k})/\tilde{P}_0^s(\tilde{k})$ of smoothed power spectrum harmonics are smallest for this $n$, although the results at different $\tilde{k}$ are also then most correlated.

A feature of this analysis is that, although it is the power spectrum that is being measured, all the calculations are done in real (redshift) space rather than in Fourier space. In measuring redshift distortions, it is important to disentangle the true distortion from the artificial distortion introduced by a non-uniform survey window. In real (redshift) space, the observed galaxy density is the product of the true density and the selection function. In Fourier (redshift) space, this product becomes a convolution. Thus the natural place to 'deconvolve' observations from the selection function is real space, where deconvolution reduces to division, and where the observations exist in the first place. The deconvolution procedure is 'exact', stable in practice, and admits a near minimum variance pair weighting.

Specifically, the smoothed harmonics $\tilde{P}_\ell^s(\tilde{k})$ of the redshift power spectrum are computed, in accordance with equations (5.13) and (5.9), by taking a suitably weighted sum over galaxy pairs of $W_\ell(r)$, equation (5.16) for

$n = 2$, times $\mathcal{P}_\ell(\mu_{\boldsymbol{r}})$, the $\ell$'th Legendre polynomial:

$$\tilde{P}_\ell^s = (2\ell+1) \int W_\ell(r) \mathcal{P}_\ell(\mu_{\boldsymbol{r}}) \xi^s(r, \mu_{\boldsymbol{r}})\, d^3 r \tag{6.1}$$

in which the redshift correlation function $\xi^s(r, \mu_{\boldsymbol{r}})$ at separation $r$ and cosine angle $\mu_{\boldsymbol{r}} = \hat{\boldsymbol{z}}.\hat{\boldsymbol{r}}$ to the line of sight $\boldsymbol{z}$ is estimated by Hamilton's (1993b) estimator

$$\xi^s(r, \mu_{\boldsymbol{r}}) = \frac{\langle DD \rangle \langle RR \rangle}{\langle DR \rangle^2} - 1 \tag{6.2}$$

where, following the conventional notation of the literature, $D$ signifies data, and $R$ signifies random background points, (although in practice all the background integrals here were done as integrals, not as Monte-Carlo integrals), and the angle brackets $\langle\, \rangle$ represent near-minimum-variance weighted (Hamilton 1993b, eqs. [60] & [61]) averages over pairs at separation $r$ and $\mu_{\boldsymbol{r}}$. The line of sight $\boldsymbol{z}$ is defined separately for each pair as the angular bisector of the pair, and to ensure the validity of the plane-parallel approximation, only pairs closer than 50° on the sky are retained. Poisson sampling noise is removed by excluding self-pairs (pairs consisting of a galaxy and itself).

Uncertainties in the ratio $\tilde{P}_2^s(\tilde{k})/\tilde{P}_0^s(\tilde{k})$ plotted in Figure 7 were estimated by two distinct methods, distinguished in Figure 7 by horizontal and tilted bars on the errors. Both methods take account of the correlated character of the galaxy distribution.

The first method (horizontal bars) is described and justified in detail by Hamilton (1993b, §4). The procedure is to subdivide the survey into a few hundred volume elements, measure the fluctuation in the power attributable to each volume element, and form a variance by adding up the covariances of the fluctuations between the volume elements. However, there is an integral constraint that implies that the sum of all covariances between all volume elements is exactly zero, so it is necessary to truncate the sum at some point. The procedure adopted is to order the covariances in order of increasing distance between each pair of volume elements, and to take the variance to be the maximum value of the cumulative covariance. The advantage of this method is that it makes minimal assumptions about the origin and nature of the errors. The main defect of the method is that it necessarily underestimates the variance on scales approaching the scale of the survey.

The second method (tilted bars), inspired by Feldman, Kaiser & Peacock (1994), is to take the Poisson (i.e. self-pair) variance of pairs $ij$ weighted by $[1 + \bar{n}^s(\boldsymbol{r}_i)\tilde{P}_0^s(\tilde{k})][1 + \bar{n}^s(\boldsymbol{r}_j)\tilde{P}_0^s(\tilde{k})]$. Properly, this method is valid only for Gaussian fluctuations, and at wavelengths that are less than some fraction of the scale of the survey; moreover this estimate fails to take into account the fact that the covariance of the redshift power spectrum $P^s(\boldsymbol{k})$ is different in different directions of the wavevector $\boldsymbol{k}$. However, the estimate is a useful check on the reliability of the uncertainties computed by the first method. The method may be expected to overestimate the variance somewhat on scales approaching the scale of the survey, and Figure 7 shows that it clearly underestimates the uncertainties in the nonlinear regime, where fluctuations are non-Gaussian.

## 7. Translinear Regime

Figure 7 indicates that nonlinearities affect the redshift power spectrum even at half-wavelengths as large as $\pi/k \approx 40h^{-1}$Mpc (although one should bear in mind that the influence of nonlinearity may be exaggerated in Fig. 7 because of the broad smoothing window) (see also Bromley, Warren & Zurek 1997). Studies of the ratio of redshift to real space power spectra (Fisher *et al.* 1993; Brainerd *et al.* 1996; see also: Suto & Suginohara 1991; Bahcall, Cen & Gramann 1993; Gramann, Cen & Bahcall 1993; Brainerd & Villumsen 1993, 1994) lead to the same conclusion, that nonlinearity can affect the redshift power spectrum even at quite large scales. Moreover, there is good reason to push to smaller scales, to improve the statistics. Thus a good understanding of nonlinearities would appear essential to a reliable measurement of the linear distortion parameter $\beta$. With such an understanding, it might even become possible to break the degeneracy in $\beta \approx \Omega_0^{0.6}/b$ between the cosmological density $\Omega_0$ and bias $b$.

### 7.1. LINEAR OR NONLINEAR?

Before going any further, it is worth emphasizing that *redshift distortions are more linear than they look.*

This can be seen from Figure 2 back at the beginning of this review. For example, a region that is just turning around in real space, a mildly nonlinear condition, appears to be collapsed to a sheet in redshift space. The edge of the sheet forms a caustic, a line of infinite density, a thoroughly nonlinear condition (note that only the boundary of the collapsed sheet appears at infinite density; the complete, infinite density, caustic surface of a collapsing overdensity is the envelope of the finger-of-god in Figure 1).

This suggests that theories that go a step beyond the linear regime, for example the Zel'dovich approximation (§7.3), or perturbation theory (e.g. Scoccimarro & Frieman 1996; Hivon *et al.* 1995), may prove quite successful in modelling redshift distortions in the translinear regime.

## 7.2. RANDOM ISOTROPIC PAIRWISE VELOCITY DISPERSION

The most common approach to nonlinearity has been phenomenological rather than based on any rigorous theory. The approach has been to suppose that the nonlinear redshift correlation function $\xi^s$ is the result of convolving the linear redshift correlation function $\xi^s_L$ with the line-of-sight component of a random isotropic pairwise velocity distribution $f(v)$. In the plane-parallel approximation, which should normally be adequate for small, nonlinear scales, this model gives (with $f(v)$ normalized to $\int_{-\infty}^{\infty} f(v)dv = 1$)

$$\xi^s(r_{/\!/}, r_\perp) = \int_{-\infty}^{\infty} \xi^s_L(r_{/\!/} - v, r_\perp) f(v) dv \ . \tag{7.1}$$

Note that this model does not require that the individual galaxies of a pair should separately have random isotropic velocity distributions, although some authors have found it convenient also to assume the latter (e.g. Peacock & Dodds 1994, to allow them to consider cross-correlations between different populations; or Heavens & Taylor 1995, to allow them to deal with radial rather than plane-parallel redshift distortions). Convolution in real space becomes multiplication in Fourier space, so the redshift power spectrum in this model is the product of the linear redshift power spectrum $P^s_L(\boldsymbol{k}) = (1 + \beta\mu^2_{\boldsymbol{k}})^2 P(k)$ with the line-of-sight Fourier transform $\hat{f}(k_{/\!/}) = \int_{-\infty}^{\infty} f(v) e^{ik_{/\!/}v} dv$ (note $k_{/\!/} = k\mu_{\boldsymbol{k}}$) of the velocity distribution:

$$P^s(\boldsymbol{k}) = \hat{f}(k\mu_{\boldsymbol{k}}) P^s_L(\boldsymbol{k}) = \hat{f}(k\mu_{\boldsymbol{k}})(1 + \beta\mu^2_{\boldsymbol{k}})^2 P(k) \ . \tag{7.2}$$

Those authors who have invoked this model have generally adopted either a Gaussian velocity distribution (Peacock & Dodds 1994; Heavens & Taylor 1995; Tadros & Efstathiou 1996), for which $f(v) = [(2\pi)^{1/2}\sigma]^{-1} \exp[-(v/\sigma)^2/2]$ and

$$\hat{f}(k\mu_{\boldsymbol{k}}) = \exp\Big[-(\sigma k\mu_{\boldsymbol{k}})^2/2\Big] \ ; \tag{7.3}$$

or else an exponential pairwise velocity distribution (Fisher *et al.* 1994b; Hamilton 1995; Ballinger, Peacock & Heavens 1996; Peacock 1997; Bromley, Warren & Zurek 1997; Ratcliffe *et al.* 1997; Cole, Fisher & Weinberg 1995 use an exponential single-particle distribution, equivalent to a pairwise distribution that is the convolution of two exponentials), for which $f(v) = (2^{1/2}\sigma)^{-1} \exp(-2^{1/2}|v|/\sigma)$ and

$$\hat{f}(k\mu_{\boldsymbol{k}}) = \frac{1}{1 + \frac{1}{2}(\sigma k\mu_{\boldsymbol{k}})^2} \ . \tag{7.4}$$

The velocity dispersions $\sigma$ above are, after the convention established by Peebles (1976, 1980 eq. [76.14]) and Davis & Peebles (1983), 1-dimensional pairwise velocity dispersions, which are $2^{1/2}$ larger than a random 1-dimensional single-point (galaxy) dispersion, and $3^{1/2}$ smaller than a 3-dimensional pairwise dispersion.

The exponential pairwise velocity distribution was first proposed as a fit to observations by Peebles (1976), and has held up as a good approximation both in observations (Davis & Peebles 1983, CfA1; Fisher *et al.* 1994b, 1.2 Jy survey; Marzke *et al.* 1995, CfA2 + SSRS2; Lin 1995, LCRS), and in $N$-body experiments (Fisher *et al.* 1994b, Fig. 5; Zurek *et al.* 1994, Fig. 7), although the latter reveal appreciable skewness in the distribution of pairwise infall velocities.

## 7.3. ZEL'DOVICH APPROXIMATION

The Zel'dovich approximation (e.g. Hui & Bertschinger 1996) is essentially linear theory expressed in Lagrangian space. In linear theory, particles (galaxies) move in straight lines from their initial comoving positions $\boldsymbol{r}_i$, with comoving displacements $\Delta\boldsymbol{r} = D(t)\Delta\boldsymbol{r}_L$ growing in proportion to the linear growth factor $D(t)$

$$\boldsymbol{r}(t) = \boldsymbol{r}_i + D(t)\Delta\boldsymbol{r}_L(\boldsymbol{r}_i) \tag{7.5}$$

where $\Delta\boldsymbol{r}_L = -\boldsymbol{\nabla}\nabla^{-2}\delta_L$ is the linear displacement field, with $\delta_L$ the linear overdensity, evaluated at some suitably early epoch. The Zel'dovich approximation supposes that the linear result (7.5) remains true also at nonlinear epochs. By construction, the Zel'dovich approximation satisfies the continuity equation, but it fails to satisfy the Euler equation beyond the linear regime.

Redshift distortions in the Zel'dovich approximation have been studied by Fisher & Nusser (1996), Taylor & Hamilton (1996), and Hatton & Cole (1997). All three sets of authors tested the predictions of the Zel'dovich approximations against $N$-body simulations, and all three adopted the plane-parallel approximation and chose the ratio of quadrupole to monopole harmonics of the redshift power spectrum as the test statistic. Fisher & Nusser and Hatton & Cole examined $N$-body simulations with CDM-like power spectra, while Taylor & Hamilton examined simulations with power law power spectra, $P(k) \propto k^n$ with $n = -1$, $-1.5$, and $-2$.

The conclusions can be summarized as follows. The quadrupole-to-monopole ratio $P_2^s(k)/P_0^s(k)$ as a function of $k$ can be characterized by a shape, an overall amplitude, and an overall scale $k_0$ that is conveniently taken to be the zero-crossing of the quadrupole power, $P_2^s(k_0) = 0$. Then:

- The shape of the quadrupole-to-monopole ratio depends on the shape of the power spectrum, but is insensitive to the redshift distortion pa-

rameter $\beta$. This shape is accurately predicted by the Zel'dovich approximation at least down to the zero-crossing of the quadrupole, $k \lesssim k_0$.
- The overall amplitude of the quadrupole-to-monopole ratio depends on the linear redshift distortion parameter $\beta$ according to the usual linear formula (5.8).
- The zero-crossing scale $k_0$ of the quadrupole is not reliably predicted by the Zel'dovich approximation.

Actually, Fisher & Nusser argued that the shape of the quadrupole-to-monopole ratio is a universal function that is insensitive to the shape of the power spectrum. Within the range of CDM-like power spectra this is probably a reasonable approximation. However, the investigations of Taylor & Hamilton showed that if a broader range of power spectra is considered, then there is a dependence on the shape of the power spectrum which, according to the Zel'dovich approximation, becomes marked as the spectral index $n \to -3$. There is general agreement that the shape of the quadrupole-to-monopole ratio is insensitive to $\beta$, the latter serving simply to set the overall normalization of the ratio.

The third conclusion, that the Zel'dovich approximation does not predict the zero-crossing of the quadrupole power reliably, is based mainly on the results of Taylor & Hamilton, but the conclusion seems to be consistent with the results graphed by Hatton & Cole (Fisher & Nusser offer no information on this point). It is not entirely clear what is responsible for the discrepancy in the zero-crossing.

Fisher & Nusser and Taylor & Hamilton both interpreted the agreement between the predicted Zel'dovich and measured $N$-body shape of the quadrupole-to-monopole ratio as suggesting that the departure of this ratio from a constant value in the translinear regime is caused not by random velocities in nonlinear virialized fingers-of-god, but rather by coherent infall towards collapsing clusters. This is encouraging if true, because it lends hope that redshift distortions can be modelled accurately with more rigorous theoretical treatments in the translinear regime.

## 7.4. FUTURE WORK

The shortcomings of modelling nonlinearity in redshift distortions either with random pairwise velocities, or with the Zel'dovich approximation, have been emphasized by Hatton & Cole (1997). It is clear that a more thorough and rigorous treatment of redshift distortions in the translinear regime, for example using perturbation theory (e.g. Scoccimarro & Frieman 1996), is needed.

As usual, bias complicates the problem. One clear bias is that highly nonlinear fingers-of-god are more prominent in optical and elliptical populations than in *IRAS* and spiral populations (Fig. 3; Guzzo *et al.* 1997). Long fingers in rich clusters appear stretched to large separations in redshift space, contaminating the linear and translinear regime. Perhaps this problem can be side-stepped by identifying and compressing prominent fingers (§6.1; Gramann, Cen & Gott 1994). Or perhaps estimators can be constructed that are somehow orthogonal[11] to the nonlinear fingers, as suggested by Bromley, Warren & Zurek (1997; these authors claimed to do just this, but actually their estimator, based on eq. [8.24] in §8.2.3 below, merely corrects for nonlinearity, and is not orthogonal to it).

## 8. Measurements of $\beta$ from Linear Redshift Distortions

### 8.1. COMPILATION

Table 8.1 lists measurements of the distortion parameter $\beta$ from redshift distortions in the linear regime. The table is intended to be complete up to the first half of 1997. Compare this compilation to that of Strauss & Willick (1995, Table 3, p. 412). Note that Table 8.1 does not include measurements of $\beta$ from direct measurements of peculiar velocities, which Strauss & Willick do include.

The Table is subdivided into the three principal methods described §5. However, precise procedures used by different authors differ in many respects, and the threefold categorization is not as clean as the headings might suggest. Comments on the individual measurements are given in the next subsection, §8.2.

The average and standard deviation of values listed in Table 8.1 is $\beta_{\rm optical} = 0.52 \pm 0.26$ for optical galaxies, and $\beta_{IRAS} = 0.77 \pm 0.22$ for *IRAS* selected galaxies (the quoted uncertainty is the deviation of a single measurement; the lower limit from Taylor & Hamilton was omitted). In averaging the numbers, I felt it was better to assign equal weight to all measurements, rather than to rely either on the authors' quoted uncertainties or on my own judgement of their quality. Of course the data sets — especially the *IRAS* data — are not all independent of each other, so the standard deviation reflects differences in methods as much as anything else. All the same, the dispersion of the measurements turns out to be about the same as the uncertainties that people quote, so is perhaps not unreasonable. If the difference in optical and *IRAS* $\beta$'s is attributed to bias, it implies a relative bias of $b_{\rm optical}/b_{IRAS} \approx 1.5$, consistent with the ratio $b_{\rm optical}/b_{IRAS} \approx 1.3$ measured from the square root of the ratio of their power spectra (Peacock & Dodds 1994; Peacock 1997, Fig. 2). If

[11] A nice example of what is meant here by orthogonality is the case of the angular correlation function or angular power spectrum, which is orthogonal to — i.e. completely unaffected by — redshift distortions.

TABLE 1. Values of $\beta$

| Survey | $\beta$ | Reference |
|---|---|---|
| 1. Ratio of redshift space to real space power spectra (plane-parallel) | | |
| *CfA* | $0.53 \pm 0.15$ | Fry & Gaztañaga (1994) |
| *SSRS* | $1.10 \pm 0.16$ | ... |
| *IRAS* | $0.84 \pm 0.45$ | ... |
| *IRAS* | $1.0 \pm 0.2$ | Peacock & Dodds (1994) |
| Optical | $0.77 \pm 0.15$ | ... |
| Stromlo-APM | $0.20^{+0.19}_{-0.22}$ | Baugh (1996) |
| Stromlo-APM | $0.48 \pm 0.12$ | Loveday *et al.* (1996a) |
| Stromlo-APM | $0.20 \pm 0.44$ | Tadros & Efstathiou (1996) |
| Optical | $0.40 \pm 0.12$ | Peacock (1997) |
| Durham/UKST | $0.52 \pm 0.39$ | Ratcliffe *et al.* (1997) |
| 2. Ratio of quadrupole to monopole redshift power (plane-parallel) | | |
| *IRAS* 2 Jy | $0.69^{+0.28}_{-0.24}$ | Hamilton (1993a) |
| Perseus-Pisces | $0.75 \pm 0.3$ | Bromley (1994) |
| *IRAS* 1.2 Jy | $0.45^{+0.27}_{-0.18}$ | Fisher *et al.* (1994b) |
| *IRAS* 1.2 Jy | $0.52 \pm 0.15$ | Cole, Fisher & Weinberg (1995) |
| QDOT | $0.54 \pm 0.3$ | ... |
| LCRS | $0.5 \pm 0.25$ | Lin (1995) |
| QDOT + *IRAS* 1.2 Jy | $0.69^{+0.21}_{-0.19}$ | Hamilton (1995) |
| *IRAS* 1.2 Jy | $0.6 \pm 0.2$ | Fisher & Nusser (1996) |
| QDOT + *IRAS* 1.2 Jy | $> 0.5$ (95% confidence) | Taylor & Hamilton (1996) |
| *IRAS* 1.2 Jy | $0.8^{+0.4}_{-0.3}$ | Bromley *et al.* (1997) |
| Durham/UKST | $0.48 \pm 0.11$ | Ratcliffe *et al.* (1997) |
| Stromlo-APM | $0.30^{+0.17}_{-0.15}$ | This paper, §6 |
| 3. Maximum Likelihood (radial) | | |
| *IRAS* 1.2 Jy | $0.96^{+0.20}_{-0.18}$ | Fisher, Scharf & Lahav (1994) |
| *IRAS* 1.2 Jy | $1.1 \pm 0.3$ | Heavens & Taylor (1995) |
| *IRAS* 1.2 Jy | $1.04 \pm 0.3$ | Ballinger, Heavens & Taylor (1995) |

optical galaxies are unbiased, $b_{\text{optical}} = 1$, then the inferred value of the cosmological density is $\Omega_0 = 0.33^{+0.32}_{-0.22}$. If on the other hand *IRAS* galaxies are unbiased, $b_{IRAS} = 1$, then $\Omega_0 = 0.63^{+0.35}_{-0.27}$.

## 8.2. COMMENTS ON INDIVIDUAL MEASUREMENTS

The comments in this subsection are arranged in approximately the chronological order in which the measurements were published, except that measurements by the same author or group of authors are kept together.

### 8.2.1. *Hamilton et al.*

*Hamilton (1993a)* measured $\beta$ in the *IRAS* 2 Jy survey (Strauss *et al.* 1992a) from a ratio of quadrupole-to-monopole correlation functions, weighted with pair separation in the fashion proposed by Hamilton (1992):

$$\frac{\xi_2^s(r)}{\xi_0^s(r) - \bar{\xi}_0^s(r)} = \frac{\frac{4}{3}\beta + \frac{4}{7}\beta^2}{1 + \frac{2}{3}\beta + \frac{1}{5}\beta^2} \tag{8.1}$$

where $\bar{\xi}_0^s(r) \equiv 3r^{-3} \int_0^r \xi_0^s(s) s^2 ds$. Cole, Fisher & Weinberg (1994, eq. [2.17] and Appendix B) subsequently pointed out that Hamilton's ratio (8.1) was actually just a ratio of smoothed quadrupole to smoothed monopole power, equation (5.12), in which the smoothing function is taken to be the second spherical Bessel function, $W(k) = j_2(kr)$ in equation (5.11). This smoothing window vanishes at zero wavenumber, $W(k) = 0$ at $k = 0$, immunizing the estimate against uncertainty in the mean density.

*Hamilton (1995)* and the present paper, §6 and Figure 7, use essentially the same procedure as Hamilton (1993a), but with the smoothing function taken to be a power law times a Gaussian, equation (5.15) with $n = 2$.

A principal topic of Hamilton (1995) was the presentation of evidence of a large, by the look of it systematic, difference in clustering between the near and far regions of the QDOT sample, which Hamilton suggested might arise from some systematic in the *IRAS* PSC fluxes. To date no satisfactory explanation of this problem has emerged.

*Taylor & Hamilton (1996)* studied the Zel'dovich approximation as a means of approximating redshift distortions in the mildly nonlinear regime (see §7.3). As an application, they fitted the ratio of quadrupole-to-monopole power measured by Hamilton (1995) from the *IRAS* QDOT + 1.2 Jy survey, the same data as are plotted in Figure 7. They concluded that $\beta > 0.5$ at the 95% confidence level, using a full covariance matrix estimated from the data.

### 8.2.2. *Fisher et al.*

The paper by Fisher *et al.* (1994b) was published after that of Fisher, Scharf & Lahav (1994), but the former seems to be the intellectual predecessor, so here it appears first.

*Fisher et al.'s (1994b)* paper on the *IRAS* 1.2 Jy redshift survey is especially useful because of its careful study of nonlinear effects through Cold

Dark Matter (CDM) $N$-body simulations engineered to the statistical properties of the 1.2 Jy survey. They pointed out that the redshift correlation function along the line of sight of sight axis, $\xi^s(r_{/\!/}, 0)$ with $r_\perp = 0$, gives a clear signature of the functional form of the pairwise velocity distribution, and find that an isotropic exponential pairwise velocity distribution fits the *IRAS* data well — indeed better than a more detailed anisotropic model that they tried.

Fisher *et al.* measured the linear distortion parameter $\beta$ from the statistic

$$\beta = \frac{\partial \ln \bar{\xi}(r)}{\partial \ln a^2} = \frac{3v_{12}(r)[1+\xi(r)]}{2r\bar{\xi}(r)} \tag{8.2}$$

where $v_{12}(r)$ is the mean pairwise infall velocity at separation $r$, and $\bar{\xi}(r) \equiv 3r^{-3}\int_0^r \xi(s)s^2 ds$. The correlation functions $\xi(r)$ and $\bar{\xi}(r)$ in equation (8.2) are in real space, not redshift space, and Fisher *et al.* adopted the power law fit $\xi(r) = (r/3.76\, h^{-1}\text{Mpc})^{1.66}$ to the *IRAS* real space correlation function estimated by Fisher *et al.* (1994a) from the angular correlation function. The last equality in (8.2) is just the pair conservation equation (Peebles 1980, eq. [71.7]), true in both linear and nonlinear regimes, while the first equality in (8.2) is true in the linear regime. Fisher *et al.* showed that the statistic (8.2) gave a reliable measure of $\beta$ at separations $r \gtrsim 10\, h^{-1}\text{Mpc}$ in three CDM simulations, one flat and unbiased, one flat and biased ($b = 1.6$), and one open and unbiased.

To determine the infall velocity $v_{12}(r)$ in the estimator (8.2), they modelled the pairwise velocity distribution as an isotropic exponential with mean infall velocity $v_{12}(r)$ and dispersion $\sigma(r)$, the shape (but not amplitude) of these two velocities as a function of separation $r$ being determined in essence by the CDM simulations, but scaled and adjusted to match the power law fit to the *IRAS* real space correlation function. To determine the overall amplitudes of the infall velocity $v_{12}(r)$ and dispersion $\sigma(r)$, they performed least squares fits to two functions that they measured from the data at separations up to $16\, h^{-1}\text{Mpc}$: the quadrupole-to-monopole ratio of correlation functions $\xi_2^s(r)/\xi_0^s(r)$; and the line-of-sight correlation function $\xi^s(r_{/\!/}, r_\perp)$ with $r_\perp < 2\, h^{-1}\text{Mpc}$. In carrying out the fits, they used a full covariance matrix estimated from the simulations. The measured pairwise velocity dispersion (with scale-dependence fixed by their procedure) was $\sigma(r) = 317^{+40}_{-49}\,\text{km}\,\text{s}^{-1}$ at a separation of $r = 1\, h^{-1}\text{Mpc}$, rising to $\approx 400\,\text{km}\,\text{s}^{-1}$ at $r = 4\, h^{-1}\text{Mpc}$, then declining again to $\approx 350\, h^{-1}\text{Mpc}$ at $r = 10\, h^{-1}\text{Mpc}$ (Fisher *et al.*, Fig. 3). The final result from the estimator (8.2) was $\beta = 0.45^{+0.27}_{-0.18}$.

*Fisher, Scharf & Lahav (1994)* were the first to study radial redshift distortions without using the plane-parallel approximation, and the first to apply a maximum likelihood (ML) formalism to measure $\beta$ (albeit with a rather crude choice of radial density modes). They expanded the *IRAS* 1.2 Jy density field in spherical harmonics about the observer (us), windowing the density in the radial direction with Gaussian windows centred at four depths, 38, 58, 78, and 98 $h^{-1}$Mpc, each with a dispersion of 8 $h^{-1}$Mpc. To make spherical harmonics uncorrelated requires full sky coverage, and the 1.2 Jy survey has the asset that it already covers almost all the sky at galactic latitudes $|b| > 5°$. While it is possible to take into account the coupling of harmonics induced by incomplete sky coverage, Fisher *et al.* chose instead to complete the sky coverage of the 1.2 Jy survey by interpolating over the plane of the Milky Way in a way that smoothly continued structure. They worked in the Local Group (LG) frame (§4.3.1), and the equations below do likewise.

In the course of their analysis, Fisher *et al.* did a cute trick, which is worth pointing out here. Let $a$ be the redshift space galaxy density $n^s(\boldsymbol{s})$ in the LG frame, smoothed over some window $f(\boldsymbol{s})$ (superscripts LG's on $\boldsymbol{s}^{\rm LG}$ and $n^{\rm LG}$ are omitted for brevity):

$$\begin{aligned} a &\equiv \int f(\boldsymbol{s}) n^s(\boldsymbol{s})\, d^3 s = \int f(\boldsymbol{s}) n(\boldsymbol{r})\, d^3 r \\ &\approx \int \left[ f(\boldsymbol{r}) + \left( v - \hat{\boldsymbol{r}}.\boldsymbol{v}^{\rm LG} \right) \frac{\partial f(\boldsymbol{r})}{\partial r} \right] \bar{n}(\boldsymbol{r}) [1 + \delta(\boldsymbol{r})]\, d^3 r \ . \end{aligned} \tag{8.3}$$

The second equality in equation (8.3) is true because galaxies are conserved, equation (4.25), and the last approximation, just a Taylor expansion of $f(\boldsymbol{s})$, is valid for windows $f(\boldsymbol{s})$ that are sufficiently smooth in the radial direction. Fisher *et al.* emphasized the point that the selection function $\bar{n}(\boldsymbol{r})$ in a flux-limited survey is in fact a function of the true distance $r$ to a galaxy, not of its redshift distance $s$ (see footnote[12]). Equation (8.3), which is essentially an integration by parts, recasts the effect of redshift distortions into the window $f(\boldsymbol{r})$ rather than the selection function and the density, which seems neat.

Actually, the redshift distortion is still there in the last expression of equation (8.3), lurking in the peculiar velocity $v - \hat{\boldsymbol{r}}.\boldsymbol{v}^{\rm LG}$. In the linear regime, the velocities $v$ and $\boldsymbol{v}^{\rm LG}$ are given as usual by equations (4.31) and (4.43), and equation (8.3) reduces to

$$a = \int \bar{n}(\boldsymbol{r}) \left[ f(\boldsymbol{r}) - \beta \frac{\partial f(\boldsymbol{r})}{\partial r} \hat{\boldsymbol{r}} \cdot \left( \frac{\partial}{\partial \boldsymbol{r}} - \left. \frac{\partial}{\partial \boldsymbol{r}} \right|_{\boldsymbol{r}=0} \right) \nabla^{-2} \right] \delta(\boldsymbol{r})\, d^3 r \ , \tag{8.4}$$

[12] It remains necessary nonetheless to estimate the real space selection function somehow, in order to evaluate the quantities (8.7) and (8.8). In practice Fisher *et al.* used the selection function $\bar{n}^s(\boldsymbol{r})$ estimated in redshift space, and discarded all monopole ($\ell = 0$) modes. This is fine for an all-sky survey — see the final paragraph of §4.4.3.

where it has been assumed that $\int f(\boldsymbol{r})\bar{n}(\boldsymbol{r})d^3r = 0$. The latter is true for all the windows considered by Fisher *et al.*, and is a desirable feature in any case to immunize measurements against uncertainty in the mean density.

Fisher *et al.* took the spherical transform of the redshift space density folded through Gaussian radial windows, which corresponds to choosing windows $f_{\ell m}(\boldsymbol{r})$ in equation (8.3) (the subscript $\ell m$ is appended to distinguish the windows) of the form

$$f_{\ell m}(\boldsymbol{r}) \equiv f(r) Y_{\ell m}(\hat{\boldsymbol{r}}) \tag{8.5}$$

with $f(r) \propto e^{-[(r-r_0)/\sigma]^2/2}$. They showed from equation (8.4) that, for an all-sky survey, the resulting spherical transform $a_{\ell m}$ could be written (Fisher *et al.* eq. [8])

$$a_{\ell m} = \int_0^\infty \left[\hat{w}_\ell(k) + \beta \hat{w}_\ell^C(k)\right]^* \hat{\delta}_{\ell m}(k)\, k^2 dk/(2\pi)^3 \tag{8.6}$$

where $\hat{\delta}_{\ell m}(k)$ is the spherical transform (3.7) of the unredshifted overdensity, $\hat{w}_\ell(k)$ is

$$\hat{w}_\ell(k) = i^\ell 4\pi \int_0^\infty j_\ell(kr) f(r) \bar{n}(r)\, r^2 dr\ , \tag{8.7}$$

and $\hat{w}_\ell^C(k)$ is a correction term embodying the redshift distortions:

$$\hat{w}_\ell^C(k) = i^\ell 4\pi \int_0^\infty \left[j'_\ell(kr) - \frac{1}{3}\delta^K_{\ell 1}\right] \frac{df(r)}{kdr} \bar{n}(r)\, r^2 dr\ , \tag{8.8}$$

with $j'_\ell(x) \equiv dj_\ell(x)/dx$ the derivative of the $\ell$'th spherical Bessel function. The Kronecker delta term $\frac{1}{3}\delta^K_{\ell 1}$, which subtracts from the dipole ($\ell = 1$) term $j'_1(kr)$ its value at $r = 0$, is the term that arises from the motion of the LG.

It is important in equation (8.6) that the survey covers the entire sky. If the survey were not all-sky, then the functions $\hat{w}_\ell(k)$ in equation (8.6) would be replaced by matrices $\hat{w}^{\ell' m'}_{\ell m}(k)$, which act by matrix multiplication on the spherical transform $\hat{\delta}_{\ell' m'}(k)$ of the overdensity. Only in the case of an all-sky survey does the matrix $\hat{w}^{\ell' m'}_{\ell m}(k)$ become diagonal; the quantities $\hat{w}_\ell(k)$ in equation (8.6) are then the eigenvalues of the matrix.

For an all-sky survey, statistical isotropy implies that the $a_{\ell m}$ are uncorrelated, with variances from equation (8.6)

$$\langle |a_{\ell m}|^2 \rangle = \int \left|\hat{w}_\ell(k) + \beta \hat{w}_\ell^C(k)\right|^2 P(k)\, k^2 dk/(2\pi)^3 + N_{\ell m} \tag{8.9}$$

where $N_{\ell m}$ is the Poisson sampling noise, which is just the (expected) self-galaxy contribution to $\langle |a_{\ell m}|^2 \rangle$. Fisher *et al.* used four separate Gaussian radial windows. Here there are also covariances $\langle a^i_{\ell m} a^{j*}_{\ell m} \rangle$ between the different radial windows $i$ and $j$, but still only covariances with the same $\ell m$ are non-zero. To determine $\beta$, Fisher *et al.* combined the observed amplitudes $a^i_{\ell m}$ measured in the 1.2 Jy survey for $\ell = 1$ to 10 into a likelihood function, adopting as prior an unredshifted power spectrum $P(k)$ with fixed shape corresponding to a CDM model with $\Gamma = 0.2$ (Efstathiou, Bond & White 1992), and fixed amplitude $\sigma_8 = 0.69 \pm 0.04$, as determined by Fisher *et al.* (1993). The final result was $\beta = 0.96^{+0.20}_{-0.18}$, which is the value quoted in Table 8.1.

Fisher *et al.* also quote (in proof) a result reported by Fisher (1993), in which the likelihood analysis is done with the shape parameter $\Gamma$ as well as $\beta$ treated as a free parameter, with the result $\beta = 0.94 \pm 0.17$ and $\Gamma = 0.17 \pm 0.05$. The uncertainty here is the conditional $1\sigma$ uncertainty, which comes from projecting the $\Delta \ln \mathcal{L} = -0.5$ contour on to the parameter axes. If the amplitude $\sigma_8$ is also treated as a free parameter, then $\beta = 0.47 \pm 0.25$, $\Gamma = 0.15 \pm 0.05$, and $\sigma_8 = 0.81 \pm 0.06$. There is a strong anti-correlation between $\beta$ and $\sigma_8$, and Fisher (1993) argues that the more reliable value of $\beta$ is probably the higher one obtained when $\sigma_8$ is set to the value measured from the real space correlation function.

It is instructive to derive a more general version of equation (8.6). Let $w(\boldsymbol{r})$ denote the product of a (complex-valued, in general) window $f(\boldsymbol{r})$ (not necessarily of the form [8.5]) with the selection function $\bar{n}(\boldsymbol{r})$ (not necessarily all-sky)

$$w^*(\boldsymbol{r}) \equiv f(\boldsymbol{r})\bar{n}(\boldsymbol{r}) \ . \tag{8.10}$$

Assume, following Fisher *et al.*, that $\int w(\boldsymbol{r}) d^3r = 0$, always a good idea in any case to protect against uncertainty in the mean density. Then the windowed overdensity $a$, equation (8.3) is, for linear distortions,

$$a = \int w^*(\boldsymbol{r})\delta^s(\boldsymbol{r})\, d^3r = \int w^*(\boldsymbol{r})\mathbf{S}^{\mathrm{LG}}\delta(\boldsymbol{r})\, d^3r = \int \left[\mathbf{S}^{\mathrm{LG}\dagger} w(\boldsymbol{r})\right]^* \delta(\boldsymbol{r})\, d^3r \tag{8.11}$$

where $\mathbf{S}^{\mathrm{LG}}$ is the linear redshift distortion operator in the LG frame, equation (4.46), and $\mathbf{S}^{\mathrm{LG}\dagger}$ is its Hermitian conjugate, equation (4.50). It follows that the windowed density $a$ is

$$a = \int \left[w(\boldsymbol{r}) + \beta w^C(\boldsymbol{r})\right]^* \delta(\boldsymbol{r})\, d^3r \tag{8.12}$$

where $w^C(\boldsymbol{r})$ is the result of the distortion term (the part proportional to $\beta$) in $\mathbf{S}^{\mathrm{LG}\dagger}$, equation (4.50), acting on the window $w(\boldsymbol{r})$:

$$w^C(\boldsymbol{r}) = \left[\nabla^{-2} r^{-2} \frac{\partial}{\partial r}\left(\frac{\partial}{\partial r} - \frac{\alpha(\boldsymbol{r})}{r}\right) r^2 - \frac{\hat{\boldsymbol{r}}}{r^2} \cdot \frac{\partial}{\partial \boldsymbol{r}}\bigg|_{\boldsymbol{r}=0} \nabla^{-2} \alpha(\boldsymbol{r}) r\right] w(\boldsymbol{r}) \ . \tag{8.13}$$

The correction term (8.13) can be rewritten in a variety of ways, including one of the same ilk as Fisher *et al.*'s expression, equation (8.8),

$$w^{C*}(\boldsymbol{r}) = \int \frac{\hat{\boldsymbol{r}}'}{4\pi} \cdot \left( \frac{\boldsymbol{r} - \boldsymbol{r}'}{|\boldsymbol{r} - \boldsymbol{r}'|^3} - \frac{\boldsymbol{r}}{r^3} \right) \frac{\partial f(\boldsymbol{r}')}{\partial r'} \bar{n}(\boldsymbol{r}')\, d^3 r' \ . \tag{8.14}$$

The $\boldsymbol{r}/r^3$ term in the integrand, which subtracts from $(\boldsymbol{r} - \boldsymbol{r}')/|\boldsymbol{r} - \boldsymbol{r}'|^3$ its value at $\boldsymbol{r}' = 0$, is the term that arises from the motion of the LG. Recasting the integral (8.12) in spherical transform space, equations (3.7), (3.9), and (3.33), gives

$$a = \int_0^\infty \sum_{\ell m} \Big[ w_{\ell m}(k) + \beta w^C_{\ell m}(k) \Big]^* \hat{\delta}_{\ell m}(k)\, k^2 dk/(2\pi)^3 \tag{8.15}$$

which generalizes equation (8.6). Equation (8.13) reveals the nature of the correction term $w^C$, and it is readily confirmed that Fisher *et al.*'s result (8.8) is regained if the window is chosen to be $w^*(\boldsymbol{r}) = f(r) Y_{\ell m}(\hat{\boldsymbol{r}}) \bar{n}(r)$ in an all-sky survey, where the selection function $\bar{n}(r)$ is the same in all directions over the sky.

The same derivation can be written compactly in terms of vectors and matrices in Hilbert space (see §3.3)

$$a = \boldsymbol{w}^\dagger \boldsymbol{\delta}^s = \boldsymbol{w}^\dagger \mathrm{S}^{\mathrm{LG}} \boldsymbol{\delta} = (\mathrm{S}^{\mathrm{LG}\dagger} \boldsymbol{w})^\dagger \boldsymbol{\delta} = (\boldsymbol{w} + \beta \boldsymbol{w}^C)^\dagger \boldsymbol{\delta} \ . \tag{8.16}$$

Here $\boldsymbol{w}$ is a weighting function, a vector, $\mathrm{S}^{\mathrm{LG}}$ is the linear distortion operator (4.46) in the LG frame, a matrix, and $\boldsymbol{\delta}$ is the vector of (unredshifted) overdensities. Equation (8.16) generalizes easily to the case of an array of quantities: the scalar $a$ in equation (8.16) simply becomes a vector $\boldsymbol{a}$, and the weighting vector $\boldsymbol{w}$ becomes a matrix. The covariance matrix of such an array is

$$\langle \boldsymbol{a}\, \boldsymbol{a}^\dagger \rangle = (\boldsymbol{w} + \beta \boldsymbol{w}^C)^\dagger \boldsymbol{\xi} \, (\boldsymbol{w} + \beta \boldsymbol{w}^C) + \mathrm{N} \tag{8.17}$$

where $\boldsymbol{\xi}$ is the (unredshifted) power spectrum matrix, and $\mathbf{N}$ is the Poisson sampling noise matrix, the self-galaxy contribution to $\langle \boldsymbol{a}\, \boldsymbol{a}^\dagger \rangle$.

### 8.2.3. *Bromley et al.*

*Bromley (1994)* proposed the idea of measuring $\beta$ from a ratio of variances of the redshift space density windowed through anisotropic sampling functions. Consider the statistic $\tilde{\delta}^s$ which is the value of the density filtered through some window (sampling function) $w(\boldsymbol{r})$ randomly positioned in the redshift survey

$$\tilde{\delta}^s \equiv \int w(\boldsymbol{r}) \delta^s(\boldsymbol{r})\, d^3 r = \int w(\boldsymbol{r}) \frac{n^s(\boldsymbol{r})}{\bar{n}^s(\boldsymbol{r})}\, d^3 r \tag{8.18}$$

where the last equality is true as long as the window is chosen to have vanishing volume integral, $\int w(\boldsymbol{r}) d^3r = 0$, as was true for the windows considered by Bromley. The real space integral in (8.18) can be rewritten as a Fourier space integral

$$\tilde{\delta}^s = \int \hat{w}^*(\boldsymbol{k}) \hat{\delta}^s(\boldsymbol{k})\, d^3k/(2\pi)^3 \tag{8.19}$$

where $\hat{w}(\boldsymbol{k})$ and $\hat{\delta}^s(\boldsymbol{k})$ are the Fourier transforms of the sampling function and overdensity. It follows that the expected shot-noise-subtracted variance $\tilde{P}^s$ of the filtered density is equal to the redshift space power spectrum $P^s(\boldsymbol{k})$ folded with the power spectrum $|\hat{w}(\boldsymbol{k})|^2$ of the sampling function

$$\tilde{P}^s \equiv \langle(\tilde{\delta}^s)^2\rangle - N = \int |\hat{w}(\boldsymbol{k})|^2 P^s(\boldsymbol{k})\, d^3k/(2\pi)^3 \tag{8.20}$$

where $N$ is the shot noise, the self-pair contribution to $\langle(\tilde{\delta}^s)^2\rangle$. Shifting the window $w(\boldsymbol{r})$ to another random position in the survey simply changes the phase of $\hat{w}(\boldsymbol{k})$, which leaves the variance (8.20) unaltered. Bromley now proposed to choose windows $w(\boldsymbol{r})$ that were the gradient along some unit direction $\boldsymbol{n}$ of some spherically symmetric function $R(r)$, equivalent in Fourier space to $i\boldsymbol{n}.\boldsymbol{k}$ times its Fourier transform $\hat{R}(k) = \int R(r) e^{i\boldsymbol{k}.\boldsymbol{r}} d^3r$:

$$w(\boldsymbol{r}) = \boldsymbol{n}.\boldsymbol{\nabla} R(r) \ , \quad \hat{w}(\boldsymbol{k}) = i\boldsymbol{n}.\boldsymbol{k}\hat{R}(k) \ . \tag{8.21}$$

If the unit directions $\boldsymbol{n}$ are chosen to be either parallel to or perpendicular to the line of sight to the window, then the ratio of parallel to perpendicular variances is (Bromley eq. [9]; see also Gramann, Cen & Gott 1994, eq.[10], which however contains an error)

$$\frac{\tilde{P}^s_{/\!/}}{\tilde{P}^s_{\perp}} = \frac{\int \mu^2_{\boldsymbol{k}} W(k) P^s(\boldsymbol{k})\, d^3k}{\frac{1}{2}\int (1-\mu^2_{\boldsymbol{k}}) W(k) P^s(\boldsymbol{k})\, d^3k} = \frac{1 + \frac{6}{5}\beta + \frac{3}{7}\beta^2}{1 + \frac{2}{5}\beta + \frac{3}{35}\beta^2} \tag{8.22}$$

where the last equality follows from Kaiser's equation (4.36), and the smoothing window is $W(k) = k^2 |\hat{R}(k)|^2$. If the parallel and perpendicular variances are measured from many random samplings in a redshift survey, then their ratio $\tilde{P}^s_{/\!/}/\tilde{P}^s_{\perp}$ should yield a measure of $\beta$.

Bromley applied this idea to the Giovanelli & Haynes (1991, and references therein; Wegner, Haynes & Giovanelli 1993) redshift survey of the Pisces-Perseus region, which is complete to blue magnitude $m_B = 15.7$ and contains over 4000 galaxies. He used a quartic window $R = [1 - (r/\lambda)^2]^2$, with $\lambda = 14$–$17\,h^{-1}$Mpc. The value $\beta = 0.75 \pm 0.3$ listed in Table 8.1 is measured from Bromley's Figure 2 at $\lambda = 16\,h^{-1}$Mpc, a length scale that he states gave the best results in CDM simulations of the procedure.

It is useful to relate Bromley's statistic (8.22) to the harmonics of the power spectrum. Both numerator and denominator are sums of monopole and quadrupole harmonics of the power spectrum filtered through the window $W(k) = k^2|\hat{R}(k)|^2$, equation (5.11),

$$\frac{\tilde{P}^s_{/\!/}}{\tilde{P}^s_{\perp}} = \frac{\tilde{P}^s_0 + \frac{2}{5}\tilde{P}^s_2}{\tilde{P}^s_0 - \frac{1}{5}\tilde{P}^s_2} \ . \tag{8.23}$$

Thus the procedure is effectively equivalent to measuring a quadrupole-to-monopole ratio of smoothed power spectra.

*Bromley, Warren & Zurek (1997)* developed Bromley's (1994) procedure further, using a neat trick to correct for nonlinearity. The paper is notable also for reporting redshift distortions from two large (17 million particle) high resolution CDM simulations with $\Omega = 1$ and $H_0 = 50\,\mathrm{km/s/Mpc}$, one $125\,h^{-1}$Mpc on a side, the other $500\,h^{-1}$Mpc on a side. The 1-dimensional pairwise velocity dispersion of the simulations was $1100\,\mathrm{km\,s^{-1}}$ and $850\,\mathrm{km\,s^{-1}}$ (at a separation of $1\,h^{-1}$Mpc) respectively for the mass and the haloes ('galaxies'), considerably higher than observations. However, Bromley *et al.* were able to engineer a set of haloes with velocity dispersion $384\,h^{-1}$Mpc and clustering properties consistent with those of the *IRAS* 1.2 Jy sample (Brainerd *et al.* 1996) by the device of eliminating all objects that were within 3 Abell radii of the highest density peaks and that had peculiar velocities greater than $750\,\mathrm{km\,s^{-1}}$. Such culling is, as they say, 'rather contrived', but it is true that *IRAS* galaxies are conspicuous by their absence from rich clusters such as the Coma cluster, and it is possible that nature pulls a similar trick.

Bromley *et al.*'s idea for nonlinearity was to modify the anisotropy of sampling functions in a way designed to cancel nonlinearity. Suppose that $\hat{w}_L(\boldsymbol{k})$ are a set of 'linear' sampling functions, and suppose that nonlinearity can be modelled with a random isotropic pairwise velocity dispersion $f(v)$, as in §7.2. Then the sampling functions can be 'corrected' for nonlinearity by dividing them in Fourier space by the square root of the line-of-sight Fourier transform $\hat{f}(k_{/\!/})$ (note $k_{/\!/} = k\mu_{\boldsymbol{k}}$) of the velocity dispersion:

$$\hat{w}(\boldsymbol{k}) = \frac{\hat{w}_L(\boldsymbol{k})}{\hat{f}(k\mu_{\boldsymbol{k}})^{1/2}} \ . \tag{8.24}$$

Bromley *et al.* assumed the exponential model for the pairwise velocity distribution, equation (7.4), which appears to give a good description of pairwise velocities in the 1.2 Jy survey (Fisher *et al.* 1994b). With this choice, and with the linear sampling functions chosen as before, equation (8.21), to be the parallel and perpendicular gradients of a spherically symmetric function, the expected shot-noise-subtracted variances (8.20) of the filtered densities are equal to the redshift space power spectrum folded through the functions

$$\begin{aligned} |\hat{w}_{/\!/}(\boldsymbol{k})|^2 &= \mu_{\boldsymbol{k}}^2 \left(1 + \frac{1}{2}\sigma^2 k^2 \mu_{\boldsymbol{k}}^2\right) k^2 |\hat{R}(k)|^2 \\ |\hat{w}_{\perp}(\boldsymbol{k})|^2 &= \frac{1}{2}(1 - \mu_{\boldsymbol{k}}^2)\left(1 + \frac{1}{2}\sigma^2 k^2 \mu_{\boldsymbol{k}}^2\right) k^2 |\hat{R}(k)|^2 . \end{aligned} \tag{8.25}$$

The factor $(1 + \frac{1}{2}\sigma^2 k^2 \mu_{\boldsymbol{k}}^2)$ 'corrects' the redshift power spectrum for nonlinearity, equations (7.2) and (7.4), so the resulting ratio $\tilde{P}^s_{/\!/}/\tilde{P}^s_{\perp}$ of parallel to perpendicular variances should be equal to the linear ratio (8.22). Once again, as in (8.23), it is instructive to express this ratio in terms of harmonics of the smoothed power spectrum:

$$\frac{\tilde{P}^s_{/\!/}}{\tilde{P}^s_{\perp}} = \frac{\tilde{P}^s_0 + \frac{2}{5}\tilde{P}^s_2 + \sigma^2(\frac{3}{10}\tilde{P}^{sNL}_0 + \frac{6}{35}\tilde{P}^{sNL}_2 + \frac{4}{105}\tilde{P}^{sNL}_4)}{\tilde{P}^s_0 - \frac{1}{5}\tilde{P}^s_2 + \sigma^2(\frac{1}{10}\tilde{P}^{sNL}_0 + \frac{1}{70}\tilde{P}^{sNL}_2 - \frac{2}{105}\tilde{P}^{sNL}_4)} \tag{8.26}$$

where $\tilde{P}^{sNL}_\ell$ are the harmonics of the power spectrum smoothed over the window $k^2 W(k) = k^4 |\hat{R}(k)|^2$ instead of $W(k)$ (cf. eq. [5.11]),

$$\tilde{P}^{sNL}_\ell = \int_0^\infty W(k) P^s_\ell(k)\, 4\pi k^4 dk/(2\pi)^3 . \tag{8.27}$$

The nonlinear correction is equivalent to replacing 'linear' estimators of the monopole and quadrupole harmonics of the redshift power spectrum by 'nonlinear' estimators

$$\begin{aligned} \tilde{P}^s_0 &\to \tilde{P}^s_0 + \sigma^2 \left(\frac{1}{6}\tilde{P}^{sNL}_0 + \frac{1}{15}\tilde{P}^{sNL}_2\right) \\ \tilde{P}^s_2 &\to \tilde{P}^s_2 + \sigma^2 \left(\frac{1}{3}\tilde{P}^{sNL}_0 + \frac{11}{42}\tilde{P}^{sNL}_2 + \frac{2}{21}\tilde{P}^{sNL}_4\right) . \end{aligned} \tag{8.28}$$

Two comments can be made about the nonlinear correction. Firstly, it involves incorporating judicious quantities of monopole, quadrupole, and hexadecapole terms into the linear estimators of the parallel and perpendicular (or monopole and quadrupole) power, which seems like a simple and stable procedure. Secondly, the nonlinear correction terms involve the power spectrum multiplied by an extra factor of $k^2$, equation (8.27), so that the correction becomes larger further into the nonlinear regime, at larger $k$, as one might expect.

Bromley *et al.* applied their procedure both to their $N$-body simulations, and to the *IRAS* 1.2 Jy survey. For the spherical function $R(r)$, they adopted a Gaussian $R(r) = e^{-(r/\lambda)^2/2}$, in place of Bromley's (1994) quartic. As it happens, the resulting (linear) smoothing window $W(k) =$

$k^2|\hat{R}(k)|^2 \propto k^2 e^{-(k\lambda)^2}$ coincides with that of Hamilton (1995), which is the same smoothing window used in Figure 7, and the nonlinear smoothing window $k^2 W(k) \propto k^4 e^{-(k\lambda)^2}$ also belongs to the class of windows specified in equation (5.15).

The simulations gave good agreement with expectation. For the 1.2 Jy survey, Bromley *et al.* assumed a velocity dispersion of $\sigma = 320\,\mathrm{km\,s^{-1}}$, which is a reasonable approximation to the (scale-dependent) velocity dispersion measured by Fisher *et al.* (1994b). The final result was $\beta = 0.8^{+0.4}_{-0.3}$.

### 8.2.4. *Cole, Fisher et al.*

*Cole, Fisher & Weinberg (1994, 1995)* wrote a pair of papers, in the first of which they described and tested with $N$-body simulations a procedure to measure the redshift power spectrum in the plane-parallel approximation, and in the second of which they applied their procedure to measure $\beta$ from the quadrupole-to-monopole ratio of power in the *IRAS* 1.2 Jy and QDOT surveys.

To ensure the validity of the plane-parallel approximation, they first windowed the surveys through sets of spherical bowler hat windows $w(|\boldsymbol{r} - \boldsymbol{r}_c|)$ (a bowler is a top hat convolved with a Gaussian), centred at random positions $\boldsymbol{r}_c$ sufficiently far from the observer that the windows subtended an opening angle no greater than 50°. Cole *et al.* (1994, Fig. 8) showed that at this opening angle and using their procedure, the plane-parallel approximation causes $\beta$ to be underestimated by approximately 5%. They defined the line of sight to each window as the direction to its centre $\boldsymbol{r}_c$. In the plane-parallel approximation, the shot-noise-subtracted power spectrum $\tilde{P}^s(k, \mu_{\boldsymbol{k}})$ measured through a window $w(|\boldsymbol{r} - \boldsymbol{r}_c|)$ (normalized to $\int w(r) d^3 r = 1$) is the convolution of the true redshift power spectrum $P^s(k, \mu_{\boldsymbol{k}})$ with the power spectrum $|\hat{w}(k)|^2$ of the window (Cole *et al.* 1994, eq. [3.4]),

$$\tilde{P}^s(k, \mu_{\boldsymbol{k}}) = \int |\hat{w}(|\boldsymbol{k} - \boldsymbol{k}'|)|^2 \, P^s(k', \mu_{\boldsymbol{k}'}) \, d^3 k'/(2\pi)^3 \, . \tag{8.29}$$

It follows (Cole *et al.* 1994, eq. [3.5]) that the harmonics $\tilde{P}^s_\ell(k)$ of the power spectrum measured through the window are equal to the harmonics $P^s_\ell(k)$ of the actual power spectrum convolved with the harmonics $W_\ell(k, k')$ of the power spectrum of the window:

$$\tilde{P}^s_\ell(k) = \int_0^\infty W_\ell(k, k') P^s_\ell(k') \, 4\pi k'^2 dk'/(2\pi)^3 \tag{8.30}$$

where

$$|\hat{w}(|\boldsymbol{k}' - \boldsymbol{k}|)|^2 = \sum_\ell (2\ell + 1) \mathcal{P}_\ell(\hat{\boldsymbol{k}}.\hat{\boldsymbol{k}}') W_\ell(k, k') \, . \tag{8.31}$$

The spherical symmetry of the window $w(r)$ is essential here to ensure that the $\ell$'th harmonic of the true power spectrum maps only to the $\ell$'th harmonic of the observed power spectrum. The fact that the kernels $W_\ell(k, k')$ are different for different harmonics $\ell$ means that the ratio $\tilde{P}_2^s(k)/\tilde{P}_0^s(k)$ of measured quadrupole-to-monopole harmonics no longer satisfies the simple formula (5.8). However, Cole *et al.* (1994, Table 1 and Fig. 5) showed that the measured ratio $\tilde{P}_2^s(k)/\tilde{P}_0^s(k)$ is equal to the true ratio $P_2^s(k)/P_0^s(k)$ multiplied by a correction factor $\approx 0.7$ which, although appreciably different from unity, is nevertheless insensitive to the shape of the power spectrum. In practice they adopted correction factors appropriate for the $\Gamma = 0.25$ CDM spectrum of Efstathiou, Bond & White (1992).

The Fourier transform $\hat{w}(k)$ of the bowler hat window is the product of the first spherical Bessel function $j_1(kR_{\text{sph}})$ with a Gaussian, where $R_{\text{sph}}$ is the radius of the top hat. Cole *et al.* used only wavenumbers $k$ for which $kR_{\text{sph}}$ are zeros of $j_1(kR_{\text{sph}})$ (in practice they used only the first two zeros, at $kR_{\text{sph}} = 4.49$ and $7.72$). With this choice, the integrand in equation (8.29) vanishes at $\boldsymbol{k}' = 0$, which eliminates leakage from power at zero wavevector, so immunizing the measurement of power against uncertainty in the mean density (which makes a delta-function contribution to power at zero wavevector). This clever trick was used first by Fisher *et al.* (1993).

Cole *et al.* (1995) combined their measurements of the redshift power spectrum $\tilde{P}^s(k, \mu_{\boldsymbol{k}})$ from bowler windows at different depths using an inverse-variance weighting, the variance of the power from each window being estimated in the manner of Feldman, Kaiser & Peacock (1994). They fitted the averaged $\tilde{P}^s(k, \mu_{\boldsymbol{k}})$ to a sum of harmonics, from which they formed the quadrupole-to-monopole ratio $\tilde{P}_2^s(k)/\tilde{P}_0^s(k)$, which they divided by the aforesaid correction factor $\approx 0.7$ to obtain the true ratio $P_2^s(k)/P_0^s(k)$. To allow for nonlinearity, they adopted the model in which the linear redshift distortion is modulated by a random one-point (not pairwise) exponential velocity dispersion. To arrive at a final value of $\beta$ and the velocity dispersion $\sigma$ in each of the 1.2 Jy and QDOT surveys, they carried out a least squares fit to the quadrupole-to-monopole ratios as a function of wavenumber $k$, using a full covariance matrix estimated from an ensemble of mock catalogues constructed from $N$-body simulations.

*Fisher & Nusser (1996)* studied the Zel'dovich approximation as a means of approximating redshift distortions in the mildly nonlinear regime (see §7.3). As an application, they fitted the ratio of quadrupole-to-monopole power measured by Cole, Fisher & Weinberg (1995) from the *IRAS* 1.2 Jy survey, finding $\beta = 0.6 \pm 0.2$.

### 8.2.5. *Fry & Gaztañaga*

*Fry & Gaztañaga (1994)* compared redshift to real space correlation functions in each of three redshift surveys, the first Center for Astrophysics (CfA1) survey (Huchra *et al.* 1983), the Southern Sky Redshift Survey (SSRS) (da Costa *et al.* 1991), and the *IRAS* 2 Jy survey (Strauss *et al.* 1992a). Their procedure was: measure the volume-averaged correlation functions $\bar{\xi} = V^{-2} \int_V \xi(r_{12}) d^3r_1 d^3r_2$, in spherical volumes $V$ for redshift space, and conical volumes $V$ for real space; find the best power law fits to $\xi^s(r)$ and $\xi(r)$ that reproduce the behaviour of the volume-averaged $\bar{\xi}$ in each case; infer $\beta$ from the ratio $\xi^s(r)/\xi(r)$ of the fitted power law correlation functions at the largest separations probed, $r \sim 5$–$10\,h^{-1}$Mpc; repeat this for several volume-limited subsamples, and adopt an average $\beta$. The resulting values of $\beta$ seem surprisingly large for such modest separations. However, given the indirectness of the procedure for measuring large scale power, and some lack of rigour in the error analysis, one might be inclined to take the numbers with a pinch of salt.

### 8.2.6. *Peacock et al.*

*Peacock & Dodds' (1994)* principal goal was not so much to measure $\beta$ as to reconstruct the linear power spectrum of mass fluctuations from a compilation of available observational evidence. A feature of the paper was the derivation of empirical analytic formulae relating linear and nonlinear power spectra, using a procedure inspired by one proposed by Hamilton *et al.* (1991). Updated versions of Peacock & Dodds' formulae are given by Peacock & Dodds (1996).

The value of $\beta$ measured by Peacock & Dodds (1994) followed from a comparison of redshift space to real space power spectra. For the real space power spectrum they adopted the APM power spectrum of Baugh & Efstathiou (1993), while for redshift space power spectra they considered power spectra from the *IRAS* QDOT survey (Feldman, Kaiser & Peacock 1994), the Stromlo-APM survey (Loveday *et al.* 1992), and the CfA2 survey (Vogeley *et al.* 1992). To combine the various observed power spectra into a single canonical linear power spectrum, they introduced linear bias factors $b$, equation (4.15), corrected the galaxy power spectra for nonlinear evolution, and modelled nonlinearity in the redshift distortions with a random Gaussianly distributed velocity dispersion. To break the degeneracy between $\Omega_0$ and bias $b$ that occurs in $\beta \approx \Omega_0^{0.6}/b$, they brought into consideration the power spectra of Abell clusters (Peacock & West 1992) and of radio galaxies (Peacock & Nicholson 1991), which they assumed to be biased but (unlike galaxies) unaffected by nonlinear evolution. With these assumptions they found $b_{\rm optical}/b_{IRAS} = 1.3$ and $\beta_{IRAS} = 1.0 \pm 0.2$, which imply also $\beta_{\rm optical} = 0.77 \pm 0.15$. The absolute values of the bias factors were less well determined than their relative values; the best models had $b_{IRAS} \approx 0.8$.

*Peacock (1997)* carried out an improved version of Peacock & Dodds' (1994) analysis, arriving at the notably lower value of $\beta_{\rm optical} = 0.40 \pm 0.12$ from optical data alone. As in the earlier paper, for optical data Peacock used the real space APM power spectrum from Baugh & Efstathiou (1993; also 1994), and redshift space power spectra from the Stromlo-APM survey (Loveday *et al.* 1992) and the CfA2 survey (Vogeley *et al.* 1992). The reason for the lower value of $\beta$ was that Peacock increased the real space APM power spectrum from Baugh & Efstathiou (1993) by a factor of 1.25 in order to match it to the amplitude of the real space correlation function of Stromlo-APM, as measured by Loveday *et al.* (1995) using the cross-correlation technique of Saunders *et al.* (1992). This adjustment reduces $\beta_{\rm optical}$ from 0.77 to 0.40, in accordance with equation (5.1).

Peacock carried out a separate analysis for *IRAS* data, but here the results were less conclusive. He used a real space *IRAS* power spectrum obtained by transforming the cross-correlation function between the 1-in-6 QDOT survey and its parent QIGC catalogue, measured by Saunders, Rowan-Robinson & Lawrence (1992), while for the redshift power spectrum he adopted Tadros & Efstathiou's (1995) power spectrum of the combined *IRAS* 1.2 Jy and QDOT surveys, in place of the Feldman *et al.* (1994) QDOT power spectrum. The redshift power spectrum of the combined *IRAS* 1.2 Jy and QDOT surveys is lower than that of the QDOT survey thanks in large part to a region in Hercules, which Tadros & Efstathiou showed produces an anomalously large upward excursion in the QDOT power spectrum (cf. Hamilton 1995; Oliver *et al.* 1996). Tadros & Efstathiou's downwardly revised redshift power spectrum lay below Saunders *et al.*'s real space power spectrum at wavenumbers $k < 0.05\,h\,{\rm Mpc}^{-1}$, nominally suggesting a negative $\beta$, and causing Peacock to discard those points. Such machinations do not inspire confidence, and Peacock avoided drawing any definitive conclusion.

Besides improved data, Peacock applied various theoretical improvements: he used an improved treatment of nonlinear evolution (Peacock & Dodds 1996); the ratio of optical and *IRAS* real space power spectra indicated some scale dependence in bias, which he fitted with a two-parameter power law model (Mann, Peacock & Heavens 1997) in place of the one-parameter linear bias model; he abandoned the use of Abell cluster and radio galaxy power spectra; and he used an exponential rather than Gaussian pairwise velocity distribution to model nonlinearity in redshift distortions.

In the latter half of the paper, Peacock went on to discuss bias, $\Omega_0$, and the shape of the reconstructed, linearized power spectrum, adducing a variety of arguments to conclude that a low bias, low $\Omega_0$ model is preferred over a high bias, high $\Omega_0$ model.

### 8.2.7. *Heavens, Taylor, et al.*

*Heavens & Taylor (1995)* were the first to apply a full-blown maximum likelihood procedure to measure $\beta$, in a manner designed to retain as much information as possible about $\beta$ in the linear regime. They took the *IRAS* 1.2 Jy survey as the data set, conservatively cut to galactic latitude $|b| > 10°$. They chose to work in a basis of spherical waves, and in the frame of the Local Group (LG). First, they defined the unredshifted overdensity modes $\delta_{n\ell m}$ by the discrete spherical transform (the definition below is the complex conjugate of Heavens & Taylor's definition, but it conforms to the convention of the present review, eq. [3.8], which follows that of Peebles 1980, §46)

$$\delta_{n\ell m} \equiv c_{n\ell} \int_{r<r_{\max}} j_\ell(k_{n\ell} r) Y_{\ell m}(\hat{\boldsymbol{r}}) \delta(\boldsymbol{r})\, d^3 r \ , \quad \delta(\boldsymbol{r}) = \sum_{n\ell m} c_{n\ell} j_\ell(k_{n\ell} r) Y^*_{\ell m}(\hat{\boldsymbol{r}}) \delta_{n\ell m} \tag{8.32}$$

where the integration is over a finite sphere of radius $r_{\max} = 200\, h^{-1}$Mpc, the $c_{n\ell}$ are normalization constants, and the wavenumbers $k_{n\ell}$ are chosen to satisfy the boundary condition $j'_\ell(k_{n\ell} r_{\max}) = 0$, which ensures that the predicted linear peculiar velocity field vanishes on the boundary of the sphere (see Fisher *et al.* 1995b for a discussion of boundary conditions). Next, they defined observed, redshifted modes $D_{n\ell m}$ by

$$D_{n\ell m} \equiv c_{n\ell} \int_{r<r_{\max}} j_\ell(k_{n\ell} r) Y_{\ell m}(\hat{\boldsymbol{r}}) w(r) \bar{n}(\boldsymbol{r}) \delta^s(\boldsymbol{r})\, d^3 r \tag{8.33}$$

which, besides being in redshift space, involve an additional factor $w(r)\bar{n}(\boldsymbol{r})$ in the integrand compared to (8.32). The radial weighting function $w(r)$ (which in practice depended also on $n\ell$) is a near-minimum-variance weighting which helps to reduce the noise in each mode, hence to increase the information content of each mode, and hence to reduce the number of modes required. They derived approximate expressions for the weightings $w_P(r)$ and $w_\beta(r)$ that optimize each mode $D_{n\ell m}$ for measuring respectively the normalization of the power spectrum $P$ and the distortion parameter $\beta$. What it means to optimize modes for the measurement of particular parameters is clarified by Tegmark, Taylor & Heavens (1997) and Tegmark *et al.* (1997). Heavens & Taylor showed that their modes $D_\mu$ (abbreviating $n\ell m = \mu$) are linearly related to the unredshifted overdensity modes $\delta_\mu$ by

$$D_\mu = \sum_\nu (\Phi_{\mu\nu} + \beta V_{\mu\nu}) \delta_\nu \ . \tag{8.34}$$

The matrix $\Phi_{\mu\nu} + \beta V_{\mu\nu}$ can be recognized as the representation in spherical transform space of the operator $w(r)\bar{n}(\boldsymbol{r})\mathbf{S}^{\rm LG}$, where $\mathbf{S}^{\rm LG}$ is the redshift distortion operator (4.46) in the LG frame[13]. Heavens & Taylor introduced one further refinement, to model nonlinearity as a random Gaussian velocity field (cf. §7.2), which has the effect of modifying the matrix $\Phi_{\mu\nu} + \beta V_{\mu\nu}$ in equation (8.34) by pre-multiplying it by another matrix. The expected covariance matrix of the modes is then

$$C_{\mu\nu} \equiv \langle D_\mu D_\nu^* \rangle = \sum_\alpha (\Phi_{\mu\alpha} + \beta V_{\mu\alpha})(\Phi_{\nu\alpha} + \beta V_{\nu\alpha})^* P(k_\alpha) + N_{\mu\nu} \quad (8.35)$$

where $N_{\mu\nu}$ is the Poisson sampling noise, which is just the (expected) self-galaxy contribution to $\langle D_\mu D_\nu^* \rangle$. In all, Heavens and Taylor's covariance matrix $C_{\mu\nu}$ took into account finite sky coverage, linear redshift distortions, nonlinearity modelled as a random Gaussian velocity distribution, and Poisson sampling noise.

The prior covariance $C_{\mu\nu}$, equation (8.35), depends on the unredshifted power spectrum $P(k)$ and on the distortion parameter $\beta$. Heavens & Taylor adopted as prior the shape of the power spectrum measured by Peacock & Dodds (1994), but retained its overall amplitude $\sigma_8^2$ and the distortion parameter $\beta$ as parameters to be measured from the likelihood analysis. They assumed a Gaussian likelihood function, equation (5.17), and included observed modes $D_{n\ell m}$ with $k_{n\ell} < 0.1\,h\mathrm{Mpc}^{-1}$, a total of 604 modes, with maximum values $\ell = 17$ and $n = 6$. The prior covariance $C_{\mu\nu}$, equation (8.35), was computed including modes $\delta_{n\ell m}$ up to $\ell = 30$ and $n = 20$. The final result of the likelihood analysis was $\beta = 1.1 \pm 0.3$ . The uncertainty here is the conditional $1\sigma$ uncertainty, which comes from projecting the $\Delta \ln \mathcal{L} = -0.5$ contour on to the parameter axes. Heavens (1997, private communication) states that a more conservative estimate of the uncertainty in $\beta$ is the marginal uncertainty, $\pm 0.5$ ($1\sigma$), which results from integrating the likelihood over the probability distribution of the amplitude $\sigma_8$. Heavens & Taylor commented that the maximum likelihood value of $\beta$ dropped to 0.5 when the maximum $k_{n\ell}$ was increased from 0.1 to $0.12\,h\mathrm{Mpc}^{-1}$, which increases the number of modes from 604 to 1048. They attributed the change in $\beta$ to the onset of nonlinearity, but it seems likely that statistical fluctuation rather than nonlinearity would cause so dramatic change in $\beta$ for so small a change in the cutoff wavenumber.

[13]Actually they used the distortion operator $\mathbf{S}$ appropriate to the CMB frame, equation (4.23), rather than the distortion operator $\mathbf{S}^{\mathrm{LG}}$ in the LG frame, equation (4.46), and they dropped the dipole ($\ell = 1$) modes from the likelihood analysis. This is not quite correct, although the error should be minor for a near all-sky survey. They used a selection function, their equation (54), from Fisher, private communication; whether this was a real or redshift selection function is unclear. If a redshift selection function, then strictly the distortion operator $\mathbf{S}^{s\,\mathrm{LG}}$ relative to the redshift space selection function should be used, as discussed in §4.4.2. Again, Heavens & Taylor dropped the monopole ($\ell = 0$) modes from the likelihood analysis, so whatever the case, any error should be minor for this near all-sky survey.

*Ballinger, Heavens & Taylor (1995)* took Heavens & Taylor's procedure a step further by parametrizing the power spectrum by its amplitude in six logarithmically spaced bins, rather than just an overall normalization. Otherwise the analysis proceeded essentially exactly as in Heavens & Taylor, with a similar result, $\beta = 1.04 \pm 0.3$.

Tegmark, Taylor & Heavens (1997) report that the 604 modes used by Heavens & Taylor (1995) can be compressed to as few as 60 signal-to-noise eigenmodes (so called Karhunen-Loève modes) of $\beta$, without appreciable loss of accuracy in $\beta$. This compression step, which involves solving an eigenvalue equation in the covariance matrix $C_{\mu\nu}$, does not obviate the necessity to compute the covariance matrix at full size, but it can reduce the work involved in computing likelihood contours for the parameters.

8.2.8. *Lin*

*Lin (1995)* examined redshift distortions, and measured the pairwise velocity dispersion, in the Las Campanas Redshift Survey (LCRS) (Shectman *et al.* 1996), which contains 23697 galaxies in six narrow slices each $1\overset{\circ}{.}5$ in declination and 80° in right ascension, three each in the north and south galactic caps. Lin measured $\beta = 0.5 \pm 0.25$ using Hamilton's (1992) estimator of the quadrupole-to-monopole power, equation (8.1).

8.2.9. *Baugh*

*Baugh (1996)* compared the real space correlation function $\xi(r)$ of the APM survey obtained by Fourier transforming the power spectrum derived by Baugh & Efstathiou (1993), to the redshift space correlation function $\xi^s(r)$ of the Stromlo-APM survey from Loveday *et al.* (1995). Uncertainty in the APM real space power spectrum arises from uncertainty in cosmological geometry (i.e. in $\Omega_0$), and in the evolution of the power spectrum with redshift, parametrized as $P(k) \propto (1+z)^{-\alpha}$ at fixed comoving wavenumber $k$ (so $\alpha = 2$ for linear growth in an $\Omega = 1$ Universe). Baugh (his Table 2) quoted three values of $\beta$, two assuming no evolution, $\alpha = 0$ and $\Omega_0 = 1$ or 0.2, giving respectively $\beta = 0.61^{+0.20}_{-0.23}$ or $0.45^{+0.21}_{-0.23}$, and one value with evolution, $\alpha = 2$ and $\Omega_0 = 1$, giving $\beta = 0.20^{+0.19}_{-0.22}$. Evidently the value of $\beta$ is quite sensitive to the assumed geometry and evolution. Observations suggest quite strong evolution. For example, Le Fèvre *et al.* (1996) find from the Canada-France Redshift Survey (CFRS) an evolutionary index[14] $\epsilon = 0.4 \pm 1.1$, corresponding to $\alpha = 1.7 \pm 1.1$. Table 8.1 in this review therefore lists Baugh's value $\beta = 0.20^{+0.19}_{-0.22}$ with evolution. Presumably the uncertainty arising from evolution could be resolved in a manner similar

[14]Le Fèvre *et al.* define evolution with respect to proper, not comoving coordinates: their index $\epsilon$ is related to Baugh's index $\alpha$ by $3 + \epsilon = \alpha + \gamma$, where $\gamma \approx 1.7$ is the logarithmic slope of the correlation function.

to Peacock (1997) — the APM power spectrum could be normalized so as to agree with the amplitude of the real space correlation function of Stromlo-APM deduced by Loveday *et al.* (1995).

8.2.10. *Loveday et al.*

*Loveday et al. (1996a)* examined the Stromlo-APM redshift survey (Loveday *et al.* 1996b), a 1-in-20 survey of galaxies brighter $b_J = 17.15$ in the Automatic Plate Measuring (APM) survey (Maddox *et al.* 1990a,b, 1996). They considered both of the first two methods described in §5. They measured the real space correlation function $\xi(r)$ from the projected cross-correlation between the Stromlo-APM survey and its parent APM survey, using the procedure of Saunders, Rowan-Robinson & Lawrence (1992), as described by Loveday *et al.* (1995). They measured the monopole, quadrupole, and hexadecapole harmonics $\xi^s_\ell(r)$ of the redshift space correlation function using the procedure of Hamilton (1993b).

They tested the different methods with mock catalogues constructed from two ensembles of $N$-body simulations described by Croft & Efstathiou (1994), 10 flat CDM simulations with a cosmological constant, $\Omega_M = 0.2$ and $\Omega_\Lambda = 0.8$, and 9 flat Mixed Dark Matter simulations, $\Omega_{\rm CDM} = 0.7$ and $\Omega_\nu = 0.3$ ($\nu$ for neutrinos denotes Hot Dark Matter). Loveday *et al.* found from the simulations that nonlinearities affect the angle-averaged (monopole) correlation function $\xi^s(r)$ up to separations $r \sim 10\,h^{-1}$Mpc, its volume integral $J^s(r) \equiv \int_0^r \xi^s(s)s^2 ds$ to separations $r \sim 20\,h^{-1}$Mpc, and the quadrupole and hexadecapole harmonics $\xi^s_2(r)$ and $\xi^s_4(r)$ to scales $r \sim 30\,h^{-1}$Mpc. They concluded that the first method, the ratio of redshift to real space correlation functions (or its integrals) is likely to yield more reliable measures of $\beta$. In practice the ratio $J^s(r)/J(r)$ of integrated correlation functions proved considerably less noisy than $\xi^s(r)/\xi(r)$, so the former was their statistic of choice.

For the Stromlo-APM survey, Loveday *et al.* found $\beta = 0.48 \pm 0.12$ from the ratio $J^s(r)/J(r)$ at its most accurately measured point, a separation of $r = 17.8\,h^{-1}$Mpc. This is also the point where the observed ratio $J^s(r)/J(r)$ reached its maximum value, so they also concluded more conservatively that the data favoured $\beta \lesssim 0.6$. They also considered Hamilton's (1992) quadrupole-to-monopole ratio, equation (8.1), but were unable to draw any conclusions from this estimator. The conclusion that nonlinearities have a large effect on the quadrupole-to-monopole ratio in the Stromlo-APM survey agrees with the analysis of the present review, §6.

8.2.11. *Tadros & Efstathiou*

*Tadros & Efstathiou (1996)* carried out an analysis similar to Baugh (1996), §8.2.9, except that they worked with the power spectrum instead of the correlation function. Tadros & Efstathiou computed the redshift space power spectrum $P^s(k)$ of the Stromlo-APM survey, and compared this to the real space power spectrum $P(k)$ of the APM survey computed by Baugh & Efstathiou (1993). Table 8.1 quotes the 'Flux Limited' (misprinted 'Flux Weighted') result from Tadros & Efstathiou's Table 2, since this weighting is expected to (and in fact did) give lower variance than the 'Volume Limited' weighting. As with Baugh (1996), systematic uncertainty in the real space APM power spectrum arises from uncertainty in its rate of evolution with redshift, and again Tadros & Efstathiou gave values both without evolution, $\beta = 0.74 \pm 0.48$ for $\alpha = 0$, and with evolution, $\beta = 0.20 \pm 0.44$ for $\alpha = 1.3$. Table 8.1 quotes the value with evolution.

### 8.2.12. *Ratcliffe et al.*

*Ratcliffe et al. (1997)* studied the Durham/UKST (UK Schmidt Telescope) redshift survey, which is a $> 75\%$ complete 1-in-3 redshift survey of galaxies brighter than $b_J \approx 17$ in the Edinburgh/Durham Southern Galaxy Catalogue (Collins, Heydon-Dumbleton & MacGillivray 1988; Collins, Nichol & Lumsden 1992). The survey contains $\approx$ 2500 galaxies over a region $\sim 20^\circ \times 70^\circ$ centred on the South Galactic Pole.

Ratcliffe *et al.* used both of the first two methods described in §5. For the first method, they measured both the angle-averaged redshift correlation function $\xi^s(r)$ and the real space correlation function $\xi(r)$ from the survey, inferring the latter from the projected redshift correlation function using Saunders *et al.*'s (1992) procedure. The resulting ratio $\xi^s(r)/\xi(r)$ of redshift to real correlation functions was quite noisy, so they considered instead the ratio $J^s(r)/J(r)$ of the integrated correlation functions $J(r) \equiv \int_0^r \xi(s)s^2 ds$, finding $\beta = 0.52 \pm 0.39$ at a separation $r = 20\,h^{-1}$Mpc.

For the second method, Ratcliffe *et al.* used Hamilton's (1992) estimator of the quadrupole-to-monopole power, equation (8.1). Combining results over separations $r \approx$ 10–20 $h^{-1}$Mpc gave $\beta = 0.48 \pm 0.11$.

Ratcliffe *et al.* tested their procedures extensively with two sets of CDM $N$-body simulations, one set an $\Omega_M = 1$ biased $b = 1.6$ model, the other a spatially flat $\Omega_M = 0.2$, $\Omega_\Lambda = 0.8$ unbiased model with a cosmological constant. These simulations were not matched in every respect to the observations — in particular the simulations had a considerably larger small scale velocity dispersion. On the assumption of an exponential pairwise velocity distribution, the 1-dimensional pairwise velocity dispersions were measured to be $980 \pm 22\,\mathrm{km\,s^{-1}}$ and $835 \pm 60\,\mathrm{km\,s^{-1}}$ respectively for the $\Omega_M = 1$ and 0.2 models, compared to $416 \pm 36\,\mathrm{km\,s^{-1}}$ for the Durham/UKST survey.

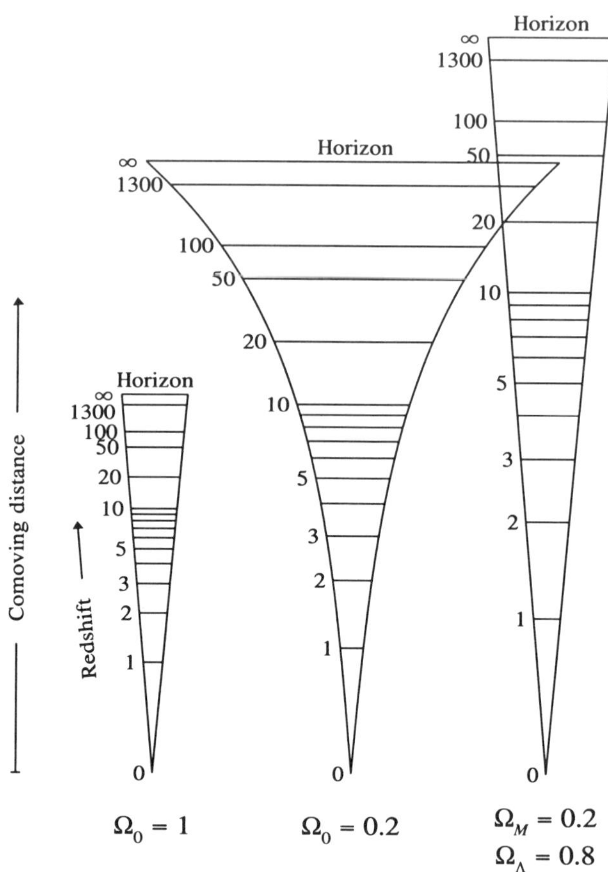


*Figure 8.* In this diagram, each wedge represents a cone of fixed opening angle, with the observer (us) at the point of the cone, at zero redshift. The wedges show the relation between physical sizes, namely the comoving distances in the radial (vertical) and transverse (horizontal) directions, and observable quantities, namely redshift and angular separation, in three different cosmological models: (left) flat matter-dominated Universe; (middle) open matter-dominated Universe with $\Omega_0 = 0.2$; (right) flat Universe with a cosmological constant, $\Omega_\Lambda = 0.8$.

## 9. Cosmological Redshift Distortions

The relation between redshift and radial comoving distance, and between angle and transverse comoving distance, is different for different cosmological models, as illustrated in Figure 8. The differences produce a cosmological redshift distortion that is zero at zero redshift, but that becomes more marked at higher redshift.

The idea of using cosmological redshift distortions to measure cosmological parameters, notably the cosmological constant, was first proposed by

Alcock & Paczyński (1979). More recently, several authors have considered the combined effect of distortions from peculiar velocities and from cosmology, both to see how cosmological distortion affects the measurement of $\beta$, and to assess the prospect of measuring cosmological parameters from upcoming redshift surveys (Ballinger, Peacock & Heavens 1996; Matsubara & Suto 1996; Nakamura, Matsubara & Suto 1997; de Laix & Starkman 1997).

Alcock & Paczyński (1979) argued, and subsequent authors have concurred, that the largest cosmological distortion is a $\Lambda$-squashing, a squashing in the radial direction that occurs if there is a large cosmological constant. By comparison, the differences in redshift distortions between low and high $\Omega_M$ models in the absence of a cosmological constant are relatively small.

The differences between redshift distortions predicted by models with and without a cosmological constant at first increase with redshift, but then saturate at $z \approx 1$ for physically interesting models, those with $\Omega_M \gtrsim 0.1$ (Ballinger, Peacock & Heavens 1996; Matsubara & Suto 1996). Even at these redshifts, the cosmological distortion is not easy to distinguish from redshift distortions caused by peculiar velocities. This degeneracy could in principle be resolved because the cosmological and peculiar velocity signals evolve differently with redshift, but in practice the uncertain evolution of bias muddies the issue (Ballinger, Peacock & Heavens 1996). Nakamura, Matsubara & Suto (1997) emphasize that cosmological redshift distortions will affect the linear distortion parameter $\beta$ at the 10–20% level in the Sloan Digital Sky Survey (SDSS), which aims to go to a median depth of $z \approx 0.1$. de Laix & Starkman (1997) conclude that the SDSS will not provide a clean signal of cosmological parameters from redshift distortions in the linear regime.

## Acknowledgements

This work was supported by NSF grant AST93-19977 and by NASA Astrophysical Theory Grant NAG 5-2797. I thank Jon Loveday and George Efstathiou for providing a copy of the Stromlo-APM survey in advance of publication.

## References

Abramowitz, M. and Stegun, I.A. (1968) *Handbook of Mathematical Functions*. Dover, New York.

Alcock, C. and Paczyński, B. (1979) An Evolution Free Test for Non-Zero Cosmological Constant, *Nature*, **281**, 358–359.

Bahcall, N.A., Cen, R. and Gramann, M. (1993) Redshift Space Clustering of Galaxies and Cold Dark Matter Model, *Ap. J.*, **408**, L77–L80.

Ballinger, W.E., Heavens, A.F. and Taylor, A.N. (1995) The Real-Space Power Spectrum of *IRAS* Galaxies on Large Scales and the Redshift Distortion, *MNRAS*, **276**, L59–L63.

Ballinger, W.E., Peacock, J.A. and Heavens, A.F. (1996) Measuring the Cosmological Constant with Redshift Distortions, *MNRAS*, **282**, 877–888.

Bardeen, J., Bond, J.R., Kaiser, N. and Szalay, A. (1986) The Statistics of Peaks of Gaussian Random Fields, *Ap. J.*, **304**, 15–61.

Baugh, C.M. (1996) The Real-Space Correlation Function Measured from the APM Survey, *MNRAS*, **280**, 267–275.

Baugh, C.M. and Efstathiou, G. (1993) The Three-Dimensional Power Spectrum Measured from the APM Galaxy Survey – I. Use of the Angular Correlation Function, *MNRAS*, **265**, 145–156.

Baugh, C.M. and Efstathiou, G. (1994) The Three-Dimensional Power Spectrum Measured from the APM Galaxy Survey – II. Use of the Two-Dimensional Power Spectrum, *MNRAS*, **267**, 323–332.

Bennett, C.L., Banday, A.J., Górski, K.M., Hinshaw, G., Jackson, P., Keegstra, P., Kogut, A., Smoot, G.F., Wilkinson, D.T. and Wright, E.L. (1996) Four-Year *COBE* DMR Cosmic Microwave Background Observations: Maps and Basic Results, *Ap. J.*, **464**, L1–L4.

Binggeli, B., Sandage, A. and Tammann, G.A. (1988) The Luminosity Function of Galaxies, *Ann. Rev. Astr. Ap.*, **26**, 509–560.

Blackman, R.B. and Tukey, J.W. (1959) *The Measurement of Power Spectra from the Point of View of Communications Engineering.* Dover, New York.

Brainerd, T.G., Bromley, B.C., Warren, M.S. and Zurek, W.H. (1996) Velocity Dispersion and the Redshift-Space Power Spectrum, *Ap. J.*, **464**, L103–L106.

Brainerd, T.G. and Villumsen, J.V. (1993) The Redshift Space Power Spectrum of Galaxies in a Cold Dark Matter Universe, *Ap. J.*, **415**, L67–L70.

Brainerd, T.G. and Villumsen, J.V. (1994) The Power Spectra of Dark Halos in a Cold Dark Matter Universe, *Ap. J.*, **425**, 403–417.

Bromley, B.C. (1994) Sampling Functions for Measuring the Cosmic Mass Density, *Ap. J.*, **423**, L81–L84.

Bromley, B.C., Warren, M.S. and Zurek, W.H. (1997) Estimating $\Omega$ from Galaxy Redshifts: Linear Flow Distortions and Nonlinear Clustering, *Ap. J.*, **475**, 414–420.

Childers, D.G. (1978) *Modern Spectrum Analysis.* IEEE Press, New York.

Cole, S., Fisher, K.B. and Weinberg, D.H. (1994) Fourier Analysis of Redshift-Space Distortions and the Determination of $\Omega$, *MNRAS*, **267**, 785–799.

Cole, S., Fisher, K.B. and Weinberg, D.H. (1995) Constraints on $\Omega$ from the *IRAS* Redshift Surveys, *MNRAS*, **275**, 515–526.

Colless, M. and Dunn, A.M. (1996) Structure and Dynamics of the Coma Cluster, *Ap. J.*, **458**, 435–454.

Collins, C.A., Heydon-Dumbleton, N.H and MacGillivray, H.T. (1988) The Edinburgh/Durham Southern Galaxy Catalogue – I. First Results on the Galaxy Angular Correlation Function, *MNRAS*, **236**, 7P–12P.

Collins, C.A., Nichol, R.C. and Lumsden, S.L. (1992) The Edinburgh/Durham Southern Galaxy Catalogue – III. $w(\theta)$ from the Full Survey, *MNRAS*, **254**, 295–300.

Croft, R.A.C. and Efstathiou, G. (1994) Peculiar Velocities of Galaxy Clusters: A Comparison of Mixed Dark Matter and Low-Density Cold Dark Matter, *MNRAS*, **268**, L23–L26.

Davis, M., Miller, A. and White, S.D.M. (1997) A Galaxy-Weighted Measure of the Relative Peculiar Velocity Dispersion, *Ap. J.*, in press (astro-ph/9705224).

Davis, M. and Peebles, P.J.E. (1983) A Survey of Galaxy Redshifts. V. The Two-Point Position and Velocity Correlations, *Ap. J.*, **267**, 465–482.

Dekel, A. (1994) Dynamics of Cosmic Flows, *Ann. Rev. Astr. Ap.*, **32**, 371–418.

de Laix, A.A. and Starkman, G. (1997) Sensitivity of Redshift Distortion Measurements to Cosmological Parameters, preprint (astro-ph/9707008).

de Lapparent, V., Geller, M.J. and Huchra, J.P. (1986) A Slice of the Universe, *Ap. J.*,

**302**, L1–L5.
Dell'Antonio, I.P., Geller, M.J. and Bothun, G.D. (1996) Peculiar Velocities for Galaxies in the Great Wall. II. Analysis, *A. J.*, **112**, 1780–1793.
Efstathiou, G., Bond, J.R. and White, S.D.M. (1992) *COBE* Background Radiation Anisotropies and Large-Scale Structure in the Universe, *MNRAS*, **258**, 1P–6P.
Efstathiou, G., Ellis, R.S. and Peterson, B.A. (1988) Analysis of a Complete Galaxy Redshift Survey – II. The Field-Galaxy Luminosity Function, *MNRAS*, **232**, 431–461.
Feldman, H.A., Kaiser, N. and Peacock, J.A. (1994) Power Spectrum Analysis of Three-Dimensional Redshift Surveys, *Ap. J.*, **426**, 23–37.
Fisher, K.B. (1993) A Spherical Harmonic Approach to Redshift Distortion: Implications for $\Omega_0$ and the Power Spectrum, *Cosmic Velocity Fields*, Proc. 9th IAP Astrophysics Meeting, ed. Bouchet, F.R. and Lachièze-Rey, M. Editions Frontières, pp. 177–185.
Fisher, K.B., Davis, M., Strauss, M.A., Yahil, A. and Huchra, J.P. (1993) The Power Spectrum of *IRAS* Galaxies, *Ap. J.*, **402**, 42–57.
Fisher, K.B., Davis, M., Strauss, M.A., Yahil, A. and Huchra, J.P. (1994a) Clustering in the 1.2-Jy *IRAS* Galaxy Redshift Survey – I. The Redshift and Real Space Correlation Functions, *MNRAS*, **266**, 50–64.
Fisher, K.B., Davis, M., Strauss, M.A., Yahil, A. and Huchra, J.P. (1994b) Clustering in the 1.2-Jy *IRAS* Galaxy Redshift Survey – II. Redshift Distortions and $\xi(r_p, \pi)$, *MNRAS*, **267**, 927–948.
Fisher, K.B., Huchra, J.P., Strauss, M.A., Davis, M., Yahil, A. and Schlegel, D. (1995a) The *IRAS* 1.2 Jy Survey: Redshift Data, *Ap. J. Supplement*, **100**, 69–103.
Fisher, K.B., Lahav, O., Hoffman, Y., Lynden-Bell, D. and Zaroubi, S. (1995b) Wiener Reconstruction of Density, Velocity and Potential Fields from All-Sky Galaxy Redshift Surveys, *MNRAS*, **272**, 885–908.
Fisher, K.B. and Nusser, A. (1996) The Nonlinear Redshift Space Power Spectrum: Omega from Redshift Surveys, *MNRAS*, **279**, L1–L5.
Fisher, K.B., Scharf, C.A. and Lahav, O. (1994) A Spherical Harmonic Approach to Redshift Distortion and a Measurement of $\Omega_0$ from the 1.2 Jy *IRAS* Redshift Survey, *MNRAS*, **266**, 219–226.
Fry, J.N. and Gaztañaga, E. (1993) Biasing and Hierarchical Statistics in Large-Scale Structure, *Ap. J.*, **413**, 447–452.
Fry, J.N. and Gaztañaga, E. (1994) Redshift Distortions of Galaxy Correlation Functions, *Ap. J.*, **425**, 1–13.
Giovanelli, R. and Haynes, M.P. (1993) Redshift Surveys of Galaxies, *Ann. Rev. Astr. Ap.*, **29**, 499–541.
Górski, K.M. (1997) Cosmic Microwave Background Anisotropy in the *COBE* DMR 4-Yr Sky Maps, *Microwave Background Anisotropies*, Proc. XXXIst Rencontres de Moriond, to appear (astro-ph/9701191).
Gramann, M., Cen, R. and Bahcall, N.A. (1993) Clustering of Galaxies in Redshift Space: Velocity Distortion of the Power Spectrum and Correlation Function, *Ap. J.*, **419**, 440–450.
Gramann, M., Cen, R. and Gott, J.R. III. (1994) Recovering the Real Density Field of Galaxies from Redshift Space, *Ap. J.*, **425**, 382–391.
Guzzo, L., Strauss, M.A., Fisher, K.B., Giovanelli, R. and Haynes, M.P. (1997) Redshift-Space Distortions and the Real-Space Clustering of Different Galaxy Types, *Ap. J.*, **489**, in press (astro-ph/9706150).
Hamilton, A.J.S. (1992) Measuring Omega and the Real Correlation Function from the Redshift Correlation Function, *Ap. J.*, **385**, L5–L8.
Hamilton, A.J.S. (1993a) Omega from the Anisotropy of the Redshift Correlation Function in the IRAS 2 Jansky Survey, *Ap. J.*, **406**, L47–L50.
Hamilton, A.J.S. (1993b) Towards Better Ways to Measure the Galaxy Correlation Function, *Ap. J.*, **417**, 19–35.
Hamilton, A.J.S. (1995) Redshift Distortions and Omega in *IRAS* Surveys, *Clustering in*

*the Universe*, Proc. XVth Rencontres de Moriond, ed. Maurogordato, S., Balkowski, C., Tao, C. and Trân Thanh Vân, J. Editions Frontières, pp. 143–155.
Hamilton, A.J.S. (1997) Towards Optimal Measurement of Power Spectra – II. A Basis of Positive, Compact, Statistically Orthogonal Kernels, *MNRAS*, **289**, 295–304.
Hamilton, A.J.S. and Culhane, M. (1996) Spherical Redshift Distortions, *MNRAS*, **278**, 73–86.
Hamilton, A.J.S., Kumar, P., Lu, E. and Matthews, A. (1991) Reconstructing the Primordial Spectrum of Fluctuations of the Universe from the Observed Nonlinear Clustering of Galaxies, *Ap. J.*, **374**, L1–L4; Erratum in *Ap. J.*, **442**, L73.
Hatton, S.J. and Cole, S. (1997) Modelling the Redshift-Space Distortion of Galaxy Clustering, *MNRAS*, submitted (astro-ph/9707186).
Heavens, A.F. and Taylor, A.N. (1995) A Spherical Harmonic Analysis of Redshift Space, *MNRAS*, **275**, 483–497.
Hivon, E., Bouchet, F.R., Colombi, S. and Juszkiewicz, R. (1995) Redshift Distortions of Clustering: A Lagrangian Approach, *Astron. & Astrophys.*, **298**, 643–660.
Hubble, E. (1929) A Relation between Distance and Radial Velocity among Extra-Galactic Nebulae, *Proc. NAS*, **15**, 168–173.
Huchra, J., Davis, M., Latham, D. and Tonry, J. (1983) A Survey of Galaxy Redshifts. IV. The Data, *Ap. J. Supplement*, **52**, 89–119.
Hui, L. and Bertschinger, E. (1996) Local Approximations to the Gravitational Collapse of Cold Matter, *Ap. J.*, **471**, 1–12.
Kaiser, N. (1984) On the Spatial Correlations of Abell Clusters, *Ap. J.*, **284**, L9–L12.
Kaiser, N. (1987) Clustering in Real Space and in Redshift Space, *MNRAS*, **227**, 1–21.
Kaiser, N. and Lahav, O. (1988) Theoretical Implications of Cosmological Dipoles, *Large-Scale Motions in the Universe*, ed. Rubin, V.C. and Coyne, G.V. Princeton University Press, Princeton, pp. 339–384.
Kogut, A., *et al.* (1993) Dipole Anisotropy in the *COBE* Differential Microwave Radiometers First-Year Sky Maps, *Ap. J.*, **419**, 1–6.
Kogut, A., Banday, A.J., Bennett, C.L., Górski, K.M., Hinshaw, G., Smoot, G.F. and Wright, E.L. (1996) Tests for Non-Gaussian Statistics in the DMR Four-Year Sky Maps, *Ap. J.*, **464**, L29–L33.
Koranyi, D.M. and Strauss, M.A. (1997) Testing the Hubble Law with the *IRAS* 1.2 Jy Redshift Survey, *Ap. J.*, **477**, 36–46.
Lahav, O., Lilje, P.B., Primack, J.R. and Rees, M.J. (1991) Dynamical Effects of the Cosmological Constant, *MNRAS*, **251**, 128–136.
Landau, L.D. and Lifshitz, E.M. (1958) *Quantum Mechanics.* Pergamon Press, Oxford.
Lawrence A., Rowan-Robinson M., Crawford J., Parry I., Xia X.-Y., Ellis R.S., Frenk C.S., Saunders W., Efstathiou G. and Kaiser N. (1997) in preparation (QDOT redshift survey).
Le Fèvre, O., Hudon, D., Lilly, S.J., Crampton, D., Hammer, F. and Tresse, L. (1996) The Canada-France Redshift Survey. VIII. Evolution of the Clustering of Galaxies from $z \sim 1$, *Ap. J.*, **461**, 534–545.
Lightman, A.P. and Schechter, P.L. (1990) The $\Omega$ Dependence of Peculiar Velocities Induced by Spherical Density Perturbations, *Ap. J. Supplement*, **74**, 831–832.
Lilje, P.B. and Efstathiou, G. (1989) Gravitationally Induced Velocity Fields in the Universe – I. Correlation Functions, *MNRAS*, **236**, 851–864.
Lin, H. (1995) Redshift-Space Distortions in the Las Campanas Redshift Survey, *The Las Campanas Redshift Survey*, PhD thesis, Ch. IV, available at http://manaslu.astro.utoronto.ca/~lin/Dissertations/Dissertations.html.
Lineweaver, C.H., Tenorio, L., Smoot, G.F., Keegstra, P., Banday, A.J. and Lubin, P. (1996) The Dipole Observed in the *COBE* DMR 4 Year Data, *Ap. J.*, **470**, 38–42.
Loredo, T.J. (1990) From Laplace to Supernova SN1987A: Bayesian Inference in Astrophysics, *Maximum Entropy and Bayesian Methods*, Proc. 9th MaxEnt Workshop, Dartmouth, ed. Fougère, P.F. Kluwer, Dordrecht, pp. 81–142.
Loveday, J., Efstathiou, G., Maddox, S.J. and Peterson, B.A. (1996a) The Stromlo-APM

Redshift Survey. III. Redshift Space Distortions, Omega and Bias, *Ap. J.*, **468**, 1–16.
Loveday, J., Efstathiou, G., Peterson, B.A. and Maddox, S.J. (1992) Large-Scale Structure in the Universe: Results from the Stromlo-APM Redshift Survey, *Ap. J.*, **400**, L43–L46.
Loveday, J., Maddox, S.J., Efstathiou, G. and Peterson, B.A. (1995) The Stromlo-APM Redshift Survey. II. Variation of Galaxy Clustering with Morphology and Luminosity, *Ap. J.*, **442**, 457–468.
Loveday, J., Peterson, B.A., Maddox, S.J. and Efstathiou, G. (1996b) The Stromlo-APM Redshift Survey. IV. The Redshift Catalog, *Ap. J. Supplement*, **107**, 201–214.
Maddox, S.J, Efstathiou, G. and Sutherland, W.J. (1990b) The APM Galaxy Survey – II. Photometric Corrections, *MNRAS*, **246**, 433–457.
Maddox, S.J, Efstathiou, G. and Sutherland, W.J. (1996) The APM Galaxy Survey – III. An Analysis of Systematic Errors in the Angular Correlation Function and Cosmological Implications, *MNRAS*, **283**, 1227–1263.
Maddox, S.J, Sutherland, W., Efstathiou, G. and Loveday, J. (1990a) The APM Galaxy Survey – I. APM Measurements and Star-Galaxy Separation, *MNRAS*, **243**, 692–712.
Mann, R.G., Peacock, J.A. and Heavens, A.F. (1997) Eulerian Bias and the Galaxy Density Field, *MNRAS*, in press (astro-ph/9702228).
Marzke, R.O., Geller, M.J., da Costa, L.N. and Huchra, J.P. (1995) Pairwise Velocities of Galaxies in the CfA and SSRS2 Redshift Surveys, *A. J.*, **110**, 477–501.
Matsubara, T. and Suto, Y. (1996) Cosmological Redshift-Space Distortion of Correlation Functions as a Probe of the Density Parameter and the Cosmological Constant, *Ap. J.*, **470**, L1–L5.
McGill, C. (1990) The Redshift Projection – I. Caustics and Correlation Functions, *MNRAS*, **242**, 428–438.
Nakamura, T.T., Matsubara, T. and Suto, Y. (1997) Cosmological Redshift Distortion: Deceleration, Bias and Density Parameters from Future Redshift Surveys of Galaxies, *Ap. J.*, submitted (astro-ph/9706034).
Nusser, A. and Davis, M. (1994) On the Prediction of Velocity Fields from Redshift Space Galaxy Samples, *Ap. J.*, **421**, L1–L4.
Oliver, S.J., Rowan-Robinson, M., Broadhurst, T.J., McMahon, R.G., Saunders, W., Taylor, A., Lawrence, A., Lonsdale, C.J., Hacking, P. and Conrow, T. (1996) Large-Scale Structure in a New Deep *IRAS* Galaxy Redshift Survey, *MNRAS*, **280**, 673–688.
Ostriker, J.P. and Cowie, L.L. (1981) Galaxy Formation in an Intergalactic Medium Dominated by Explosions, *Ap. J.*, **243**, L127–L131.
Padmanabhan, T. (1993) *Structure Formation in the Universe.* Cambridge University Press, Cambridge.
Peacock, J.A. (1997) The Evolution of Galaxy Clustering, *MNRAS*, **284**, 885–898.
Peacock, J.A. and Dodds, S.J. (1994) Reconstructing the Linear Power Spectrum of Cosmological Mass Fluctuations, *MNRAS*, **267**, 1020–1034.
Peacock, J.A. and Dodds, S.J. (1996) Nonlinear Evolution of Cosmological Power Spectra, *MNRAS*, **280**, L19–L26.
Peacock, J.A. and Nicholson, D. (1991) The Large-Scale Clustering of Radio Galaxies, *MNRAS*, **253**, 307–319.
Peacock, J.A. and West, M.J. (1992) The Power Spectrum of Abell Clusters Correlations, *MNRAS*, **259**, 494–504.
Peebles P.J.E. (1973) Statistical Analysis of Catalogs of Extragalactic Objects. I. Theory, *Ap. J.*, **185**, 413–440.
Peebles P.J.E. (1976) A Cosmic Virial Theorem, *Ap. Space Sci.*, **45**, 3–19.
Peebles, P.J.E. (1980) *The Large Scale Structure of the Universe.* Princeton University Press, Princeton.
Praton, E.A., Melott, A.L. and McKee, M.Q. (1997) The Bull's-Eye Effect: Are Galaxy Walls Observationally Enhanced?, *Ap. J.*, **479**, L15–L18.
Ramella, M., Geller, M.J. and Huchra, J.P. (1992) The Distribution of Galaxies within the "Great Wall", *Ap. J.*, **384**, 396–403.

Ramella, M., Geller, M.J. and Huchra, J.P. (1993) Groups of Galaxies in the Center for Astrophysics Redshift Survey, *Ap. J.*, **344**, 57–74.

Ratcliffe, A., Shanks, T., Fong, R. and Parker, Q.A. (1997) The Durham/UKST Galaxy Redshift Survey IV. Redshift Space Distortions in the 2-Point Correlation Function, *MNRAS*, submitted (astro-ph/9702228).

Regős, E. and Geller, M.J. (1989) Infall Patterns around Rich Clusters of Galaxies, *A. J.*, **98**, 755–765.

Rowan-Robinson M., Saunders W., Lawrence A., Leech K.J. (1991) The QMW *IRAS* Galaxy Catalogue: A Highly Complete and Reliable *IRAS* 60-$\mu$m Galaxy Catalogue, *MNRAS*, **253**, 485–495.

Sargent, W.L.W. and Turner, E.L. (1977) A Statistical Method for Determining the Cosmological Density Parameter from the Redshifts of a Complete Sample of Galaxies, *Ap. J.*, **212**, L3–L7.

Saunders, W., Rowan-Robinson, M. and Lawrence, A. (1992) The Spatial Clustering of *IRAS* Galaxies on Small and Intermediate Scales, *MNRAS*, **258**, 134–146.

Saunders, W., Rowan-Robinson, M., Lawrence, A., Efstathiou, G., Kaiser, N., Ellis, R.S. and Frenk, C.S. (1990) The 60-$\mu$m and Far-Infrared Luminosity Functions of *IRAS* galaxies, *MNRAS*, **242**, 318–337.

Scoccimarro, R. and Frieman, J.A. (1996) Loop Corrections in Nonlinear Cosmological Perturbation Theory. II. Two-Point Statistics and Self-Similarity. *Ap. J.*, **473**, 620–644.

Shectman, S.A., Landy, S.D., Oemler, A., Tucker, D.L., Lin, H., Kirshner, R.P. and Schechter, P.L. (1996) The Las Campanas Redshift Survey, *Ap. J.*, **470**, 172–188.

Stirling, A.J. and Peacock, J.A. (1996) Power Correlations in Cosmology: Limits on Primordial Non-Gaussian Density Fields, *MNRAS*, **283**, L99–L104.

Strauss, M.A. (1997) Recent Advances in Redshift Surveys of the Local Universe, *Structure Formation in the Universe*, ed. Dekel, A. and Ostriker, J.P., to appear (astro-ph/9610033).

Strauss, M.A., Huchra, J.P., Davis, M., Yahil, A., Fisher, K.B., and Tonry, J. (1992a) A Redshift Survey of *IRAS* Galaxies. VII. The Infrared and Redshift Data for the 1.936 Jansky Sample, *Ap. J. Supplement*, **83**, 29–63.

Strauss, M.A. and Willick, J.A. (1995) The Density and Peculiar Velocity Fields of Nearby Galaxies, *Physics Reports*, **261**, 271–431.

Strauss, M.A., Yahil, A. and Davis, M. (1991) On the Derivation of Selection Functions from Redshift Survey Data, *Pub. Astr. Soc. Pacific*, **103**, 1012–1019.

Strauss, M.A., Yahil, A., Davis, M., Huchra, J.P. and Fisher, K.B. (1992b) A Redshift Survey of *IRAS* Galaxies. V. The Acceleration on the Local Group, *Ap. J.*, **397**, 395–419.

Suto, Y. and Suginohara, T. (1991) Redshift-Space Correlation Functions in the Cold Dark Matter Scenario, *Ap. J.*, **370**, L15–L18.

Tadros, H. and Efstathiou, G. (1995) The Power Spectrum of *IRAS* galaxies, *MNRAS*, **276**, L45–L50.

Tadros, H. and Efstathiou, G. (1996) Power Spectrum Analysis of the Stromlo-APM Redshift Survey, *MNRAS*, **282**, 1381–1396.

Taylor, A.N. and Hamilton, A.J.S. (1996) Nonlinear Cosmological Power Spectra in Real and Redshift-Space, *MNRAS*, **282**, 767–778.

Tegmark, M. and Bromley, B.C. (1995) Real-Space Cosmic Fields from Redshift-Space Distributions: A Green's Function Approach, *Ap. J.*, **453**, 533–540.

Tegmark, M., Hamilton, A.J.S, Strauss, M.A., Vogeley, M.S. and Szalay, A.S. (1997) Measuring the Galaxy Power Spectrum with Future Redshift Surveys, *Ap. J.*, submitted (astro-ph/9708020).

Tegmark, M., Taylor, A.N. and Heavens, A.F. (1997) Karhunen-Loève Eigenvalue Problems in Cosmology: How Should We Tackle Large Data Sets?, *Ap. J.*, **480**, 22–35.

Turner, E.L. (1979) Statistics of the Hubble Diagram. II. The Form of the Luminosity Function and Density Variations with Application to Quasars, *Ap. J.*, **231**, 645–652.

Vogeley, M.S., Park, C., Geller, M. and Huchra, J.P. (1992) Large-Scale Clustering of Galaxies in the CfA Redshift Survey, *Ap. J.*, **391**, L5–L8.
Vogeley, M.S. and Szalay, A.S. (1996) Eigenmode Analysis of Galaxy Redshift Surveys. I. Theory and Methods, *Ap. J.*, **465**, 34–53.
Webster, M., Lahav, O. and Fisher, K.B. (1997) Wiener Reconstruction of the *IRAS* 1.2-Jy Galaxy Redshift Survey: Cosmographical Implications, *MNRAS*, **287**, 425–444.
Wegner, G., Haynes, M.P. and Giovanelli, R. (1993) A Survey of the Pisces-Perseus Supercluster. V. The Declination Strip +33°.5 to +39°.5 and the Main Supercluster Ridge, *A. J.*, **105**, 1251–1270.
Yahil, A., Strauss, M.A., Davis, M. and Huchra, J.P. (1991) A Redshift Survey of *IRAS* Galaxies. II. Methods for Determining Self-Consistent Velocity and Density Fields, *Ap. J.*, **372**, 380–393.
Yahil, A., Tammann, G.A. and Sandage, A. (1977) The Local Group: The Solar Motion Relative to its Centroid, *Ap. J.*, **217**, 903–915.
Yu J.T. and Peebles P.J.E. (1969) Superclusters of Galaxies?, *Ap. J.*, **158**, 103–113.
Zaroubi, S. and Hoffman, Y. (1996) Clustering in Redshift Space: Linear Theory, *Ap. J.*, **462**, 25–31.
Zurek, W.H., Quinn, P.J., Salmon, J.K. and Warren, M.S. (1994) Large-Scale Structure after *COBE*: Peculiar Velocities and Correlations of Cold Dark Matter Halos, *Ap. J.*, **431**, 559–568.

# THE SLOAN DIGITAL SKY SURVEY

A.S. SZALAY

*Department of Physics and Astronomy*
*The Johns Hopkins University, Baltimore*

*(for the SDSS Collaboration)*

## 1. Introduction

Astronomy is about to undergo a major paradigm shift, with data sets becoming larger, and more homogeneous, for the first time designed in the top-down fashion. In a few years it may be much easier to "dial-up" a part of the sky, when we need a rapid observation than wait for several months to access a (sometimes quite small) telescope. With several projects in multiple wavelengths under way, like the SDSS, 2MASS, GSC-2, POSS2, ROSAT, FIRST and DENIS projects, each surveying a large fraction of the sky, the concept of having a "Digital Sky", with multiple, TB size databases interoperating in a seamless fashion is no longer an outlandish idea. More and more catalogs will be added and linked to the existing ones, query engines will become more sophisticated, and astronomers will have to be just as familiar with mining data as with observing on telescopes.

The Sloan Digital Sky Survey, hereafter the SDSS, is a project to digitally map about 1/2 of the Northern sky in five filter bands from UV to the near IR, and is expected to detect over 200 million objects in this area. Simultaneously, redshifts will be measured for the brightest 1 million galaxies. The SDSS will revolutionize the field of astronomy, increasing the amount of information available to researchers by several orders of magnitude. The resultant archive that will be used for scientific research will be large (exceeding several Terabytes) and complex: textual information, derived parameters, multi-band images, and spectra. The catalog will allow astronomers to study the evolution of the universe in greater detail and is intended to serve as the standard reference for the next several decades.



*D. Hamilton (ed.), The Evolving Universe, 277–289.*

*Figure 1.* The view of the SDSS 2.5m telescope from the East.

## 2. The Sloan Digital Sky Survey

The Sloan Digital Sky Survey (SDSS) is a collaboration between the University of Chicago, Princeton University, the Johns Hopkins University, the University of Washington, Fermi National Accelerator Laboratory, the Japanese Promotion Group, the United States Naval Observatory, and the Institute for Advanced Study, Princeton, with additional funding provided by the Alfred P. Sloan Foundation and the National Science Foundation. In order to perform the observations, a dedicated 2.5 meter Ritchey-Chretien telescope was constructed at Apache Point, New Mexico, USA. This telescope is designed to have a large, flat focal plane which provides a 3 field of view. This design results from an attempt to balance the areal coverage of the instrument against the detector's pixel resolution.

The survey has two main components: a photometric survey, and a spectroscopic survey. The photometric survey is produced by drift scan imaging of 10,000 square degrees centered on the North Galactic Cap using five broad-band filters that range from the ultra-violet to the infra-red. The photometric imaging will use a CCD array that consists of 30 2K x 2K imaging CCDs, 22 2K x 400 astrometric CCDs, and 2 2K x 400 Focus CCDs. The data rate from this camera will exceed 8 Megabytes per second, and the total amount of raw data will exceed 40TB.

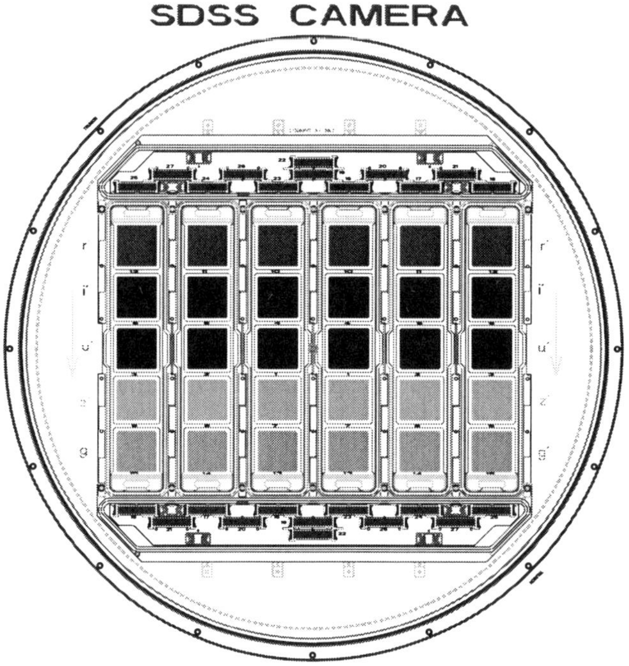


*Figure 2.* Front view of the SDSS camera assembly. This diagram shows the camera as it would be seen with the front cover and shutters removed, showing the 30 photometric and 24 astrometric/focus CCDs and their associated dewars and kinematic supports.

A considerable care has gone into selecting the filter system. The SDSS has designed a filter system similar to ugriz which should transform to and from it with little difficulty, in which the g , r and i filters are as wide as practicable consistent with keeping the overlap small. If we wish the u band to be a "good" one which is almost entirely contained between the Balmer jump and the atmospheric cutoff, it cannot be significantly wider than Thuan-Gunn u . We call the new system u' g' r' i' z' . The response curves and sensitivity data with the coatings and CCDs we will use are shown in Figure 3.

The spectroscopic survey will target over a million objects chosen from the photometric survey in an attempt to produce a statistically uniform sample. This survey will utilize two multi-fiber medium resolution spectrographs, with a total of 640 optical fibers, of 3 seconds of arc in diameter, that provide spectral coverage from 3900 - 9200 Å. The telescope will gather

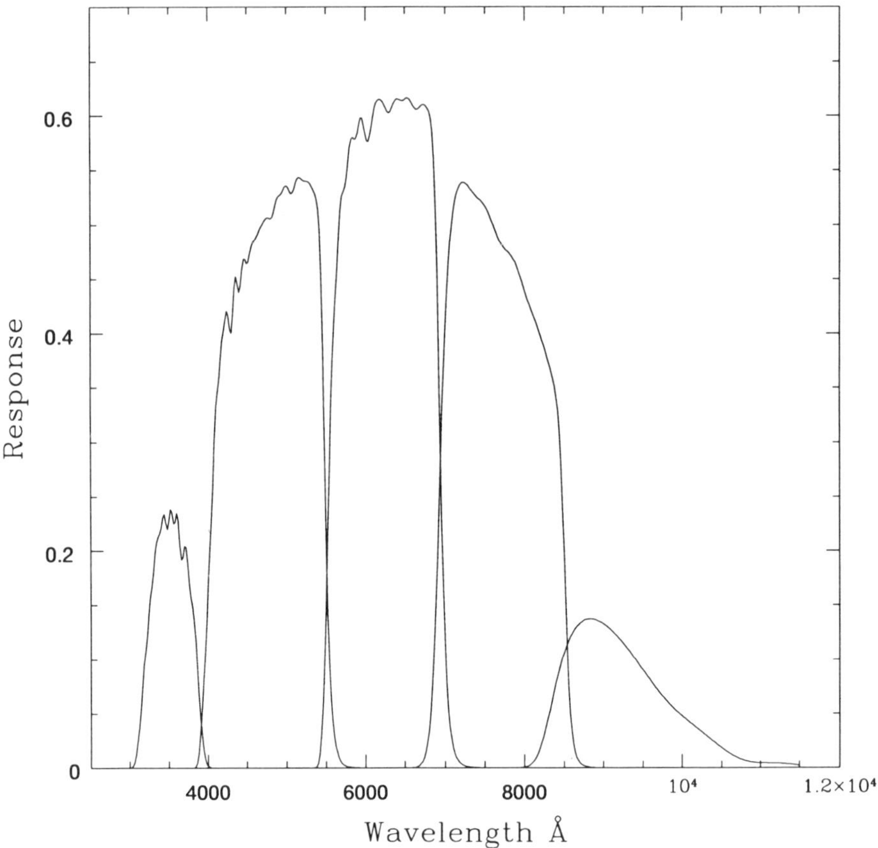


*Figure 3.* The SDSS system response curves. The responses are shown without atmospheric extinction. The curves represent expected total quantum efficiencies of the camera plus telescope on the sky.

about 5000 galaxy spectra in one night. The total number of spectra known to astronomers today is about 60,000 - only 12 days of SDSS data! Whenever the Northern Galactic cap is not accessible from the telescope site, a complementary survey will repeatedly image several areas in the Southern Galactic cap to study fainter objects and identify any variable sources.

### 2.1. THE DATA PRODUCTS

The SDSS will create four main data sets: a photometric catalog, a spectroscopic catalog, images, and spectra. The photometric catalog is expected to contain one hundred million galaxies, one hundred million stars, and

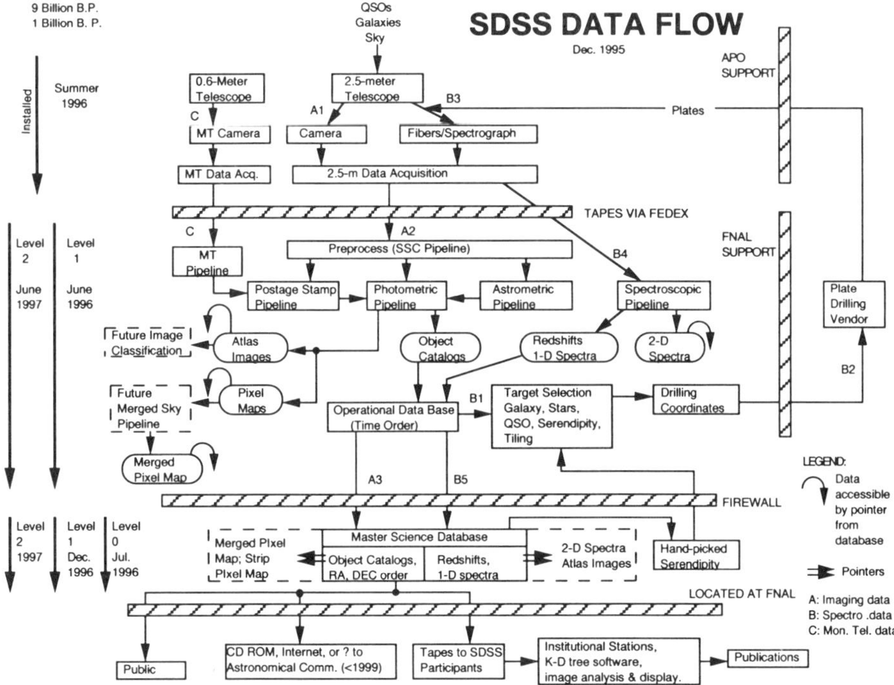


*Figure 4.* A diagram of the data reduction software for the SDSS project. Due to the large amount of data, a fully automated reduction system is necessary. The software is organized into pipelines which communicate via the Operational Database (ODB).

one million quasars, with magnitudes, profiles, and observational information recorded in the archive. The anticipated size of this product is about 250GB. Each detected object will also have an associated image cutout ("atlas image") for each of the five filters, adding up to about 700GB. The spectroscopic catalog will contain identified emission and absorption lines, and one dimensional spectra for one million galaxies, one hundred thousand stars, one hundred thousand quasars, and about ten thousand clusters, totaling about 50GB. In addition, derived custom catalogs may be included, such as a photometric cluster catalog, or QSO absorption line catalog. Thus the amount of tracked information in these products is about 1TB.

The collaboration will release the data to the public after a period of thorough verification. The actual distribution method is still under discussion. This public archive is expected to remain the standard reference catalog for the next several decades, presenting additional design and legacy problems. Furthermore, the design of the SDSS science archive must allow for the archive to grow beyond the actual completion of the survey. As the reference astronomical data set, each subsequent astronomical survey will want to cross-identify its objects with the SDSS catalog, requiring that the

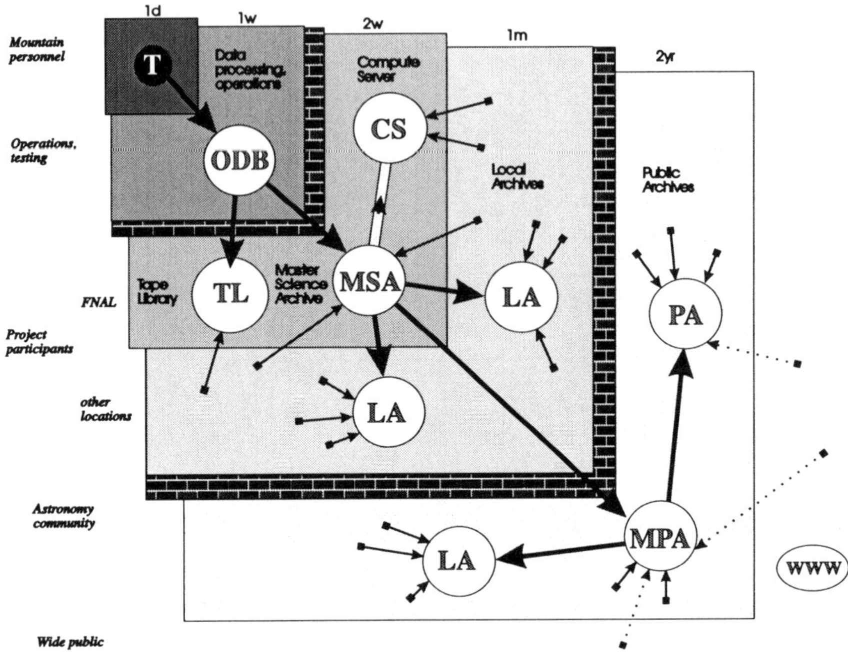


*Figure 5.* A conceptual data-flow diagram of the SDSS data. The data is taken at the telescope (T), and is shipped on tapes to FNAL, where it is processed within one week, and ingested into the operational archive (ODB), protected by a firewall, accessible only the personnel working on the data processing. Within two weeks, data will be transferred into the Master Science Archive (MSA). From there data will be replicated to local archives (LA) within another two weeks. The data gets into the public domain (MPA, LA) after two years of science verification, and recalibration, if necessary. These servers will provide data for the astronomy community. We will also provide a WWW based access for the wide public, to be defined in the near future.

archive, or at least a part of it be dynamic.

## 3. The SDSS Archives

The survey archive is split into two orthogonal functionalities and the corresponding distinct components: an *operational archive*, where the raw data is reduced and mission critical information is stored; and the *science archive*, where calibrated data is available to the collaboration for analysis and is optimized for such queries. In the operational archive. data is reduced, but uncalibrated, since the calibration data is not necessarily taken at the same time as the observations. Calibrations will be provided on the fly, via method functions, and several versions will be accessible. The Science Archive will contain only calibrated data, reorganized for efficient science use, using as much data clustering as possible. If a major revision of the calibrations is necessary, the Science Archive and its replications will have to be regenerated from the ODB. At any time the Master Science Archive will be the standard to be verified against.

High level requirements for the archive included (a) easy transfer of the binary database image from one architecture to another, (b) easy multi-platform availability and interoperability, (c) easy maintainance for future operating systems and platforms. In order to satisfy these, the Science Archive employs a three-tiered architecture: the user interface, the query support component, and the data warehouse. This distributed approach provides maximum flexibility, while maintaining portability, by isolating hardware specific features. The data warehouse, where most of the low-level I/O access happens is based upon an OODBMS (Objectivity/DB), where the porting issues depend mainly on the database vendor. Objectivity's database image is binary compatible, can be copied between different platforms, since the architecture is encoded on every page in the archive.

### 3.1. ACCESSING TERABYTES OF DATA

Today's approaches to accessing astronomical data do not scale into the Terabyte regime — brute force does not work! Assume a hypothetical 500 GB data set. The most popular data access technique today is the World Wide Web. Most universities can receive data at the bandwidth of about 15 kbytes/sec. The transfer time for this data set would be 1 year! If the data is residing locally within the building (access via Ethernet at 1 Mbytes/sec), the transfer time drops to 1 week. If the astronomer is logged on to the machine which contains the data, all of it on hard disk, then with SCSI bandwidth it still takes 1 day to scan through the data. Even faster hardware cannot support hundreds of "brute force" queries per day. With Terabyte catalogs, even small custom datasets are in the 10 GB range, thus a high level data management is needed.

## 4. Survey Goals

Both the photometric and the redshift survey will cover a larger fraction of the sky, a slightly elliptical region in the North, centered close to the NGP, and three strips at -10, 0 and +15 declinations. The length of these stripes is determined by extinction.

The design of the SDSS redshift survey is crucial to its effectiveness as a tool for studying large scale structure. The goal of this survey is not to

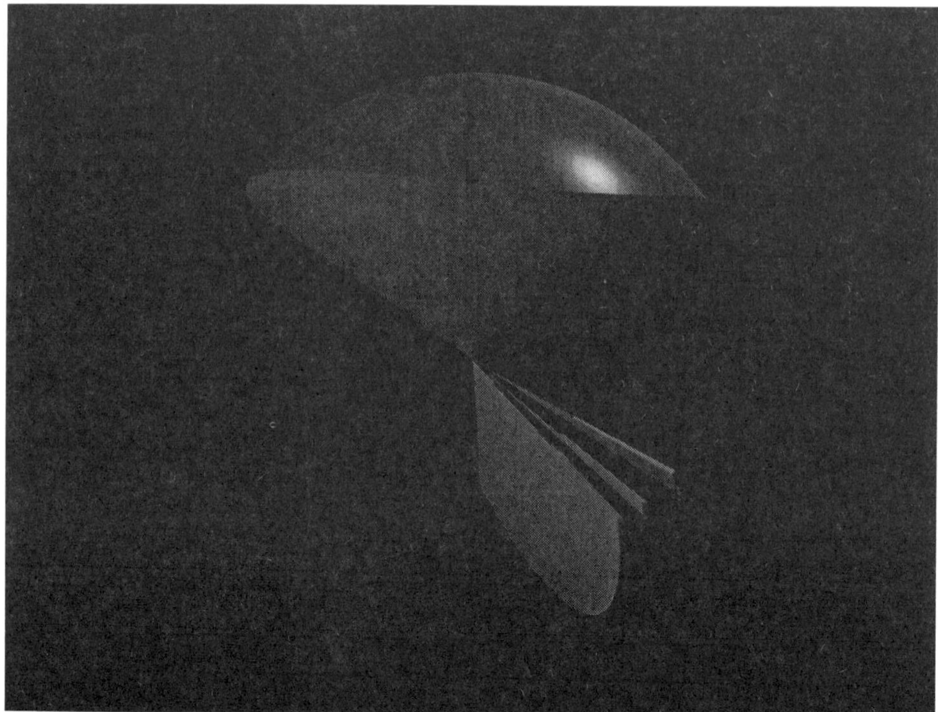

*Figure 6.* A ray-traced visualization of the volume covered by the SDSS survey: the upper cone is the Northern survey, while the three narrow fans below are the Southern stripes.

measure a single statistical property of the distribution of galaxies, such as the power spectrum of density fluctuations, or the maximum size of non-linear structures. Instead, we desire an accurate description of the galaxy distribution over a wide range of physical scales, in order to address the questions listed above. To meet this goal, we plan to do a wide-angle, deep redshift survey which fully samples the galaxy distribution.

Figure 7. displays a view of a simulated SDSS spectroscopic sample, drawn from a large N-body simulation of a low-density, cold dark matter Universe. For these simulations, the spectroscopic sample consisted of all galaxies with apparent magnitude $r' < 17.55$ and apparent half-light diameter greater than 2 arcseconds, selecting just under one million galaxies over the survey area. Since the time that this simulation was done, the details of the galaxy selection criteria have changed; see below, but the qualitative nature of this figure is not sensitively dependent on these details.

This figure shows the distribution of galaxies in redshift space in a 6 deg by 130 deg slice along the survey equator, which contains about 6% of the spectroscopic sample. The median depth of the sample is $300h^{-1}$ Mpc

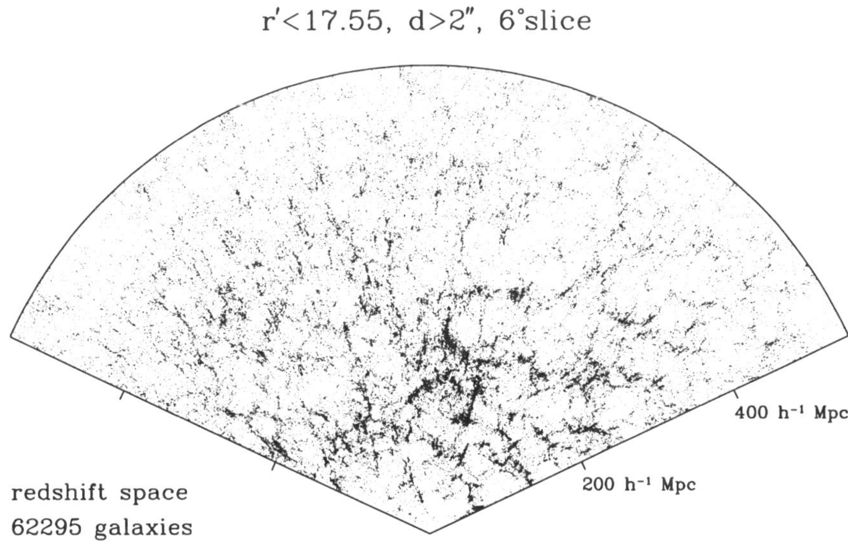


*Figure 7.* The redshift-space distribution of galaxies. The objects lie in a slice 6deg thick along the survey equator. Galaxies are plotted at the distance indicated by their redshift, hence the appearance of clusters as "fingers-of-God." This slice contains roughly 6% of the galaxy redshift sample. (D. Weinberg)

, and there is useful information on the density field to $600h^{-1}$ Mpc, with a few galaxies as far away as $1000h^{-1}$ Mpc.

## 5. Large Scale Structure and Survey Strategy

A few examples from the recent history of redshift surveys illustrate how the detection of structure depends on both the survey geometry and the sampling rate. Observing in narrow pencil beams in the direction of the constellation Bootes, Kirshner *et al.* (1981) detected a $60h^{-1}$ Mpc diameter void in the galaxy distribution. The geometry and sampling of the Kirshner *et al.* survey was well suited to detection of a single void, but it could not answer the question of how common such structures are. De Lapparent, Geller and Huchra (1986, hereafter CfA2) measured redshifts of a magnitude-limited sample of galaxies in a narrow strip on the sky, and found that (1) large voids fill most of the volume of space and (2) these voids are surrounded by thin, dense, coherent structures. Demonstration of the first result was possible because of the geometry and depth of the survey, demonstration of the second because of the dense sampling. There were hints of this structure in the shallower survey of the same region by

*Figure 8.* A schematic three-dimensional comparison of window functions. The window function is the Fourier transform of the survey volume; it can be thought of as the point-spread function for measurements of the galaxy power spectrum. The narrower the window function, the better the sensitivity to large scale clustering. Surveys with highly anisotropic shapes (slices, pencil beams) have highly anisotropic window functions; as a result, estimates of large scale clustering can be polluted by aliased power from small scales. The window functions of five surveys are shown here: lower left is QDOT, upper left is CfA2, upper right is LC, lower right is the BEKS survey, and the small dot in the center is the SDSS.

Huchra et al. (1983), but the deeper CfA2 sample was required to reveal the structure clearly. Sparse samples of the galaxy distribution, *e.g.*, the QDOT (Saunders *et al.* 1991) survey of one in six IRAS galaxies, have provided useful statistical measures of low-order clustering on large scales, but sparse surveys have less power to detect coherent overdense and underdense regions. Perhaps because of this limitation, members of the QDOT team have elected to follow up their one-in-six survey by obtaining a complete (one-in-one) survey of IRAS galaxies to the same limiting flux, now nearing completion.

In order to encompass a fair sample of the Universe, a redshift survey must sample a very large volume. We do not know a priori how large such a volume must be, but existing observations provide the minimum constraints. Our picture of the geometry and size of existing structures became clearer as the CfA2 survey was extended over a larger solid angle (*e.g.*, Vogeley *et al.* 1994). The largest of the structures found is a dense region measuring $150h^{-1}$ Mpc by $50h^{-1}$ Mpc – the "Great Wall" (Geller and Huchra 1989). Giovanelli *et al.* (1986) find a similar coherent structure in a

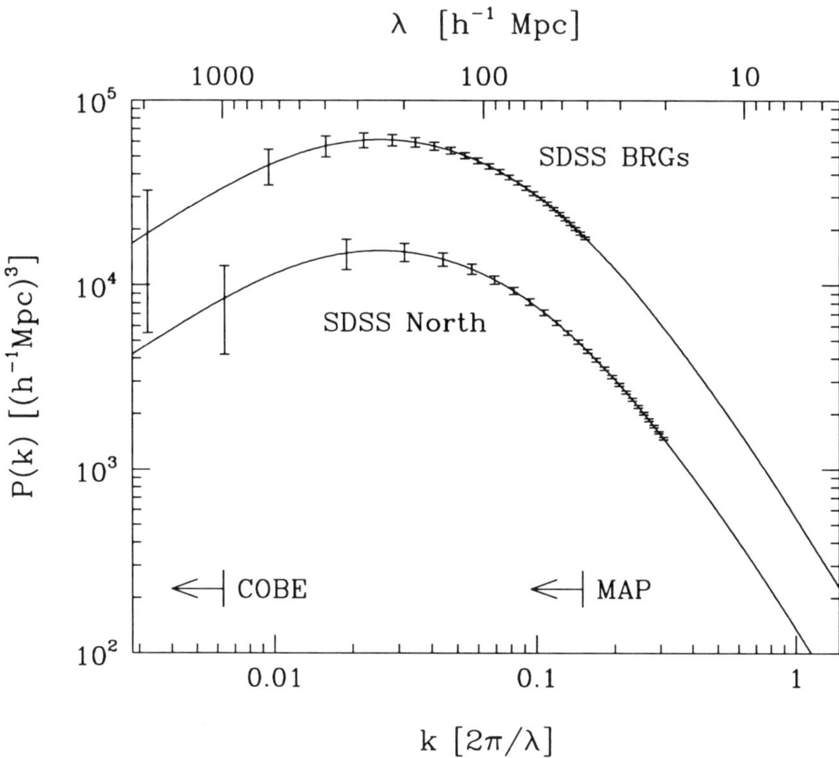


*Figure 9.* Model power spectra. The power spectra for the SDSS Northern galaxy survey and for the luminous red galaxies (BRGs) are shown. The latter are assumed to be biased by a factor of two with respect to the galaxies. The individual points and their error bars are statistically independent. On small scales, the errors are smaller than the smallest bars. Also shown are the smallest comoving wavelength scales accessible to COBE and to the upcoming Microwave Anisotropy Probe (MAP).

complete redshift survey of the Perseus-Pisces region, and da Costa *et al.* (1988) find another such "wall" in the Southern hemisphere (cf. Santiago et al. 1996). However, none of these surveys covers a large enough volume to investigate the frequency of these structures. In quantitative terms, the observed power spectrum of galaxy density fluctuations continues to rise on scales up to $> 100h^{-1}$ Mpc (Vogeley *et al.* 1992; Loveday *et al.* 1992; Fisher *et al.* 1993; Feldman *et al.* 1994; Park *et al.* 1994; Baugh and Efstathiou 1993; Peacock and Dodds 1994; da Costa *et al.* 1994; Landy *et al.* 1996). The observed power spectrum may be influenced by these few, largest, nearby structures. Thus a fair sample clearly must include many volumes with scale $100h^{-1}$ Mpc .

The sampling rate and geometry of a survey can strongly influence the largest structures seen (Szalay *et al.* 1991). A very deep pencil-beam survey by Broadhurst *et al.* (1990) showed remarkable clustering signatures on a scale of $128h^{-1}$ Mpc . First results from the redshift survey of Shectman *et al.* (1996, hereafter LC for the Las Campanas survey), has found evidence for an excess of power on similar scales (Landy *et al.* 1996). The LC survey covers a two narrow wedges in the Southern sky; it is deeper than CfA2 but less deep than the BEKS survey. While different survey geometries are appropriate for elucidating different features of large scale structure, a deep survey with a large opening angle and full sampling of the galaxy population is the only way to get a complete picture of the galaxy distribution. Coverage of the whole sky is extremely helpful if one wants to compare velocity and density fields in the nearby Universe, and redshift surveys of infrared-selected galaxies (Strauss *et al.* 1992; Fisher *et al.* 1995; Lawrence *et al.* 1995) are invaluable for this purpose. However, these surveys are much shallower than the SDSS, with almost two orders of magnitude fewer galaxies, so they are not nearly as powerful for statistical analyses of clustering.

## 6. Summary

We are in the middle of designing and constructing an extremely ambitious project, aiming to provide a useful tool for almost all astronomers in the world. This endavour would not have been possible ten years ago, and even now we are pushing the limits of technology. We hope that our efforts will be successful, and the result will substantially change the way scientists do astronomy today. Having 200 million objects at our fingertips will undoubtedly lead to new major discoveries, and the spectroscopic followup of even a fraction of our fainter objects occupy astronomers for decades. The day when we have a “Digital Sky” at our desktop may be nearer than most astronomers think. Given the enormous public interest in astronomy, we hope that the resulting archive will also provide a challenge and inspiration to thousands of interested high-school students, and a lot of fun for the web-surfing public.

## 7. Acknowledgments

This material is based upon the Grey Book, the SDSS proposal to the National Science Foundation. The author would like to acknowledge helpful discussions with the SDSS participants, and emphasize the heroic efforts of the collaboration on making this data set the highest quality possible.

## References

Baugh, C. M., and Efstathiou, G. 1993, MNRAS 265, 145.
Broadhurst, T.J., Ellis, R.S., Koo, D.C., and Szalay, A.S. 1990, Nature 343, 726.
da Costa, L. N., Pellegrini, P., Sargent, W., Tonry, J., Davis, M., Meiksin, A., Latham D., Menzies, J., and Coulson, I. 1988, ApJ 327, 544.
da Costa, L. N., Vogeley, M. S., Geller, M. J., Huchra, J. P., and Park, C. 1994, ApJL 437, 1.
de Lapparent, V., Geller, M.J., and Huchra, J.P. 1986, ApJL 302, 1.
Feldman, H., Kaiser, N., and Peacock, J. 1994, ApJ 426, 23.
Fisher, K.B., Davis, M., Strauss, M.A., Yahil, A., and Huchra, J.P. 1993, ApJ 402, 42.
Fisher, K. B., Huchra, J. P., Davis, M., Strauss, M. A., Yahil, A., and Schlegel, D. 1995, ApJSuppl, 100, 69.
Geller, M.J., and Huchra, J.P. 1989, Science 246, 897.
Giovanelli, R., Haynes, M.P., and Chincarini, G. 1986, ApJ 300, 77.
Huchra, J., Davis, M., Latham, D., and Tonry, J. 1983, ApJSuppl 52, 89.
Kirshner, R.P., Oemler, A., Schechter, P.L., and Shectman, S.A. 1981, ApJL 248, 57.
Landy, D.S., Shectman, S.A., Lin, H., Kirshner, R.P., Oemler, A.A., and Tucker, D. 1996, ApJL 456, 1.
Loveday, J., Efstathiou, G., Peterson, B., A., and Maddox, S. J. 1992, ApJ 400, 43.
Park, C., Vogeley, M.S., Geller, M.J., and Huchra, J.P. 1994, ApJ 431, 569.
Peacock, J. A., and Dodds, S. J. 1994, MNRAS 267, 1020.
Santiago, B. X., Strauss, M. A., Lahav, O., Davis, M., Dressler, A., and Huchra, J. P. 1996, ApJ 461, 38.
Saunders, W., Frenk, C. S., Rowan-Robinson, M., Efstathiou, G., Lawrence, A., Kaiser, N., Ellis, R. S., Crawford, J., Xia, X.-Y., and Parry, I. 1991, Nature 349, 32.
Shectman, S.A., Landy, S.D., Oemler, A., Tucker, D.L., Lin, H., Kirshner, R.P., and Schechter, P.L. 1996, ApJ 470, 172.
Strauss, M.A. and Willick, J.A. 1995, Physics Reports, 261, 271.
Szalay, A.S., Broadhurst, T.J., Ellman, N., Koo, D.C., and Ellis, R. 1991, Proc. Natl. Acad. Sci., 90, 4858.
Vogeley, M.S., Park, C., Geller, M.J., Huchra, J.P., 1992, ApJL 391, 5.
Vogeley, M.S., Park, C., Geller, M.J., Huchra, J.P., and Gott, J.R. 1994, ApJ 420, 525.

# THE LARGE-SCALE DISTRIBUTION OF BARYONS AT HIGH REDSHIFT

MICHAEL RAUCH
*Astronomy Department 105-24*
*California Institute of Technology*
*Pasadena, CA 91125, USA*

**Abstract.**
Recent advances in high resolution spectroscopy together with the increasingly realistic cosmological simulations now available provide us with detailed insights into the nature of the high redshift universe. Observations of the Lyman $\alpha$ forest can be used to constrain the properties of the numerical models, with the aim of understanding the cosmic baryon distribution and measuring cosmological parameters. We discuss some aspects of the large-scale distribution and the physical state of baryons at high redshift, and describe a new measurement of $\Omega_{baryon}$ from observations of cosmic gas in absorption.

## 1. Introduction

The past four years have seen remarkable, technical progress in two seemingly disjoint areas of astronomy: using QSO absorption spectroscopy with the first of the new generation of large optical telescopes (Keck) (*e.g.*, Tytler *et al.* 1995, Cowie *et al.* 1995, Lu *et al.* 1996, Rauch *et al.* 1996, Churchill, Steidel & Vogt 1996) and UV sensitive space telescopes (HST, HUT) it has become possible to probe the nature of baryonic matter at redshifts from zero to almost five at great detail. Individual QSO spectra allow us to measure the temperature, kinematics, ionization state, and metal abundances of the gas along the line of sight, while spatial correlations between absorption features in adjacent QSO beams yield information about the large-scale structure at high $z$. At about the same time, high resolution hydrodynamic models of structure formation have become capable of simulating typical ($\sim$ average density) regions of the universe (*e.g.*, Cen *et al.*



*D. Hamilton (ed.), The Evolving Universe,* 291–302.

1994; Petitjean, Muecket & Kates 1995; Zhang, Anninos & Norman 1995; Hernquist *et al.* 1996; Miralda-Escudé *et al.* 1996; Haehnelt, Steinmetz, & Rauch 1996). These simulations help interpreting the physical environment giving rise to the observed spectral pattern. Conversely, the observational data can be used to constrain the cosmological parameter space occupied by these models. Matching simulated spectra with observed ones we are now close to performing genuine "measurements" of several quantities relevant to cosmology and structure formation which are beyond reach of traditional, emission-based astronomy.

Here we will be concerned in a rather general sense with the fate of baryons at high redshift (*i.e.*, $2 < z < 5$). We will briefly consider their observational signature, large-scale distribution, their physical properties, and their contribution to the closure density of the universe.

## 2. Observational signature of the general baryon distribution

The high sensitivity of resonance line absorption spectroscopy allows us to detect baryons with a sort of "peripheral vision." Rather than looking at conspicuous places like galaxies (the baryonic inventory of which is highly processed and poorly understood), we can observe matter in absorption against a background QSO to measure the baryon density in randomly selected, typical regions of the universe where the gas density is not far from the universal average. Looking at such gaseous regions in HI Lyman $\alpha$ absorption at $z \sim 3$ has the advantage of sampling structures on scales which have yet to break away from the expansion of the universe and to collapse (assuming that the gas dynamics at this epoch is dominated by gravitational structure formation). The thermal and ionization history of such low density, unprocessed gas is relatively simple and can be modelled *ab initio*, using appropriate initial conditions for the density and radiation field.

This approach provides a relatively sensitive measure of the cosmic matter distribution because, fortuitously, the transition from the *linear density regime* of gravitational collapse on scales of several hundred kpc coincides roughly with the *linear part of the curve of growth* for the ubiquitous Lyman $\alpha$ absorption line, as shown in Fig. 1.

Of the four simulated Ly$\alpha$ absorption lines in that Figure (at redshift $\sim 3$), the weakest line with an HI column density N(HI) $= 10^{12}$ cm$^{-2}$ illustrates an easily attainable detection limit for a large telescope. Between this value and $10^{14}$ - $10^{15}$ cm$^{-2}$, where Ly$\alpha$ enters saturation, the column density of HI can be measured very precisely. The overdensities causing absorption in this column density range extend from the underdense case ($\delta < 1$) at N(HI) $= 10^{12}$ to $\delta \sim 10 - 20$ at the transition to saturation, *i.e.*,

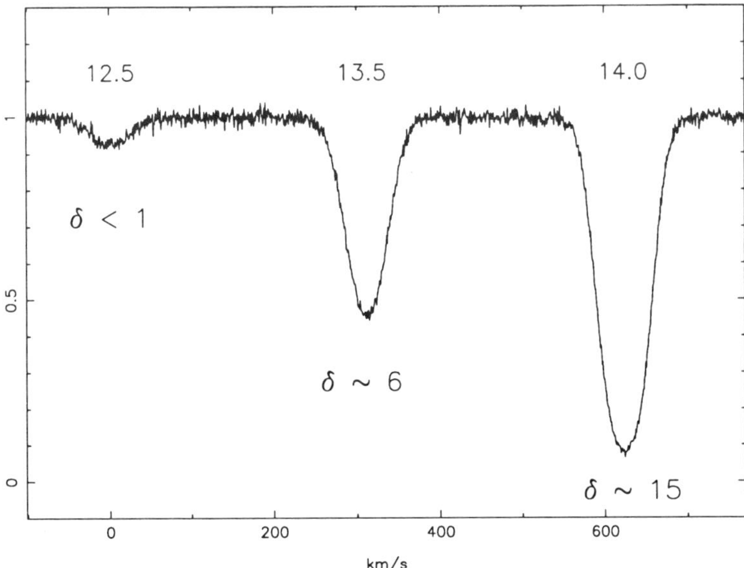


*Figure 1.* Illustration of the linear part of the curve of growth. Three artifical HI Lyman $\alpha$ lines with Doppler parameter b = 30 kms$^{-1}$ are shown. The logarithmic column densities and corresponding typical baryonic overdensities are also shown. The signal to noise ratio is 80 per pixel, typical of Keck spectra of bright QSOs.

they conveniently include the density range where structures are supposed to turn around ($\bar{\delta} \sim 5.5$ for a spherical top-hat perturbation).

## 3. The large-scale distribution of baryons

Assigning a baryonic density to the gas clouds with a given HI column density requires a knowledge of the ionization state of the gas. Given the low density and the presence of an ionizing radiation background there will be a large ionization correction, with by far most of the gas being ionized. To estimate the ionized fraction for a given radiation background intensity we need to know the gas density or, with the column density being measured, the size of the absorbing structure.

It has been known for several years (*e.g.*, Petitjean *et al.* 1993, Meiksin & Madau 1993, Shapiro, Giroux, & Babul 1994) that the number of baryons in low and intermediate column density (N(HI) $< 10^{16}$ cm$^{-2}$) Lyman $\alpha$ forest "clouds" could in principle be very large, dominating the baryon content of the universe. The piece missing from a determination of $\Omega_b$ was the lack of useful constraints on the density/size of the Ly$\alpha$ absorbing clouds. Recent measurements of the transverse size of the absorbers, investigating common absorption features in multiple lines of sight (Smette *et al.* 1993,

1995, Bechtold *et al.* 1994, Dinshaw *et al.* 1994, 1995, Crotts *et al.* 1994, Fang *et al.* 1996) have supplied this important piece of evidence, in that the Lyman $\alpha$ absorbers were found to be structures coherent on scales of several hundred kpc. From that it follows immediately that the baryon density in the Lyman $\alpha$ forest "clouds" has to be high: the larger the structures, the lower the gas density (for a given column density) and the larger the ionized gas fraction and thus the hidden amount of baryons. It turns out that in order to reconcile the observed sizes (which really are *transverse* sizes on the sky) with the standard assumptions about the temperature $T$, the ionizing background intensity $I$, and the baryon density $\Omega_b$ expected from nucleosynthesis calculations (Walker *et al.* 1991), the transverse sizes $L$ need to be much larger than the sizes $D$ along the line of sight (Rauch & Haehnelt 1995). Otherwise the Ly$\alpha$ forest, to produce the observed amount of absorption, would need to contain more baryons than available in total. This condition can be satisfied by flattened structures with typical axis ratios

$$D/L \leq 0.056 h_{75}^{-2} f_{Ly\alpha}^{2} \left(\frac{T}{3\times10^{4}}\right)^{-0.7} \left(\frac{I}{10^{-21}}\right)^{-1} \left(\frac{L}{1 h_{75}^{-1}\mathrm{Mpc}}\right)^{-1}, \quad (1)$$

where $f_{Ly\alpha}$ is the fraction of all baryons in intermediate and low column density absorption systems (N(HI)$\sim 10^{13} - 10^{15}$ cm$^{-2}$). Thus, even if all baryons at high $z$ were in low column density Lyman $\alpha$ clouds, these objects would need to be significantly flattened. For $f_{Ly\alpha}$ less than 1, this condition becomes even more severe. Obviously, this argument is independent on the cosmological model; however, for this reason, it cannot be very precise because the actual density field and to a lesser degree the temperature does dependent on the structure formation model adopted. For a quantitative picture we need to look to models including both, the gas dynamics and the cosmological background.

Throughout the past decade, most of the theoretical work on intergalactic gas has been based on the cold dark matter (CDM) scenario or its variants. This line of modelling started with the suggestion (by Rees 1986, and Ikeuchi 1986) that Lyman $\alpha$ forest absorbers are caused by gas accumulated in mini-CDM-halos, which subsequently failed to develop into galaxies. This model has several advantages over the older picture of gas clouds pressure-confined by an ambient, hotter intercloud medium: it treats the absorbers as by-products of a general structure formation scenario; it provides a confining agent (dark matter gravity), an order of magnitude stronger than the barely sufficient self-gravity (Sargent *et al.* 1980) of the baryons; and it avoids a number of difficulties inherent in the pressure-confined cloud model. Aside from the difficulty of accomodating a hot intercloud medium (Barcons, Fabian & Rees 1991), the pressure-confined cloud model suffered from the problem, that, with its homogeneous clouds occupying only a narrow density range the wide range of column densities observed must be due to a very large range of sizes and cloud masses (Williger & Babul 1992). The problem is solved in the dark matter dominated mini-halo picture, where the density gradients within an individual halo can produce the power law distribution function of column densities even if Lyman $\alpha$ clouds were a class of objects identical in size and mass (Rees 1986, Milgrom 1986). More realistic configurations with gas falling into and partially re-expanding from the CDM potential wells were studied by Bond, Szalay, and Silk (1987) and Meiksin (1994); non-spherical geometries by Charlton, Salpeter, and Hogan (1993). McGill (1990), Bi, Börner, and Chu (1992), Miralda-Escudé & Rees (1993), and Reisenegger & Miralda-Escudé (1995) have advocated a more general picture, where density and velocity caustics produced by gravitational collapse may be contributing to the Lyman $\alpha$ forest, such that there is no strict distinction between fluctuations in the general density field and discrete, gaseous objects.

Three-dimensional gas dynamics simulations of CDM or related cosmogonies (see references in the introduction) have confirmed many features of the analytical work adding new information on the geometry, temporal evolution, spatial correlations and the velocity field. The Lyman $\alpha$ forest in a hierarchical model appears to be caused by a coherent network of filaments, sheets, and knots of gas, in which more spherical, higher density condensations, galaxies or mini-halos, are embedded. The fraction of baryons in Lyman $\alpha$ clouds is indeed very high, with (even by $z \sim 2$) of order 80% of all baryons in low column density clouds, dominated by the range ($14 < \log N < 15.5$), (Miralda-Escudé *et al.* 1996, Hernquist *et al.* 1996), not in virialized galaxies, or damped Lyman $\alpha$ systems. The spatial arrangement of this baryonic reservoir is illustrated by Fig. 2. Column density contours at a level as high as $10^{14}$ cm$^{-2}$ are stretching continuously over many hundreds of kpcs. Obviously, the *bulk of the baryons is sitting in the filaments*, without having collapsed into galaxies yet.

The properties of this intermediate column density gas are gradually becoming known: From detections of CIV absorption by Tytler *et al.* (1995) and Cowie *et al.* (1995) we know that much of the gas at these column densities is already metal-enriched, albeit at at low level (perhaps $10^{-2.5}$ $\odot$), and probably with a composition favoring the $\alpha$ elements. Thus, the filaments may have formed from a gas phase already partly enriched by stellar nucleosynthesis (population III); alternatively, supernovae of type II may have blown the enriched gas from the currently forming galaxies back into space, from where it is being re-accreted. High resolution spectroscopy also shows that the CIV absorbing gas is at temperatures of order $4 \times 10^4$ $K$, with the individual absorption components having rather little bulk motion (typically $< 10$ kms$^{-1}$) (Rauch *et al.* 1996), in agreement with the hypothesis that many of these absorption systems indeed arise in clumps in the moderately overdense, filamentary structure outside the virial radii of galaxies (Rauch, Haehnelt, & Steinmetz 1997).

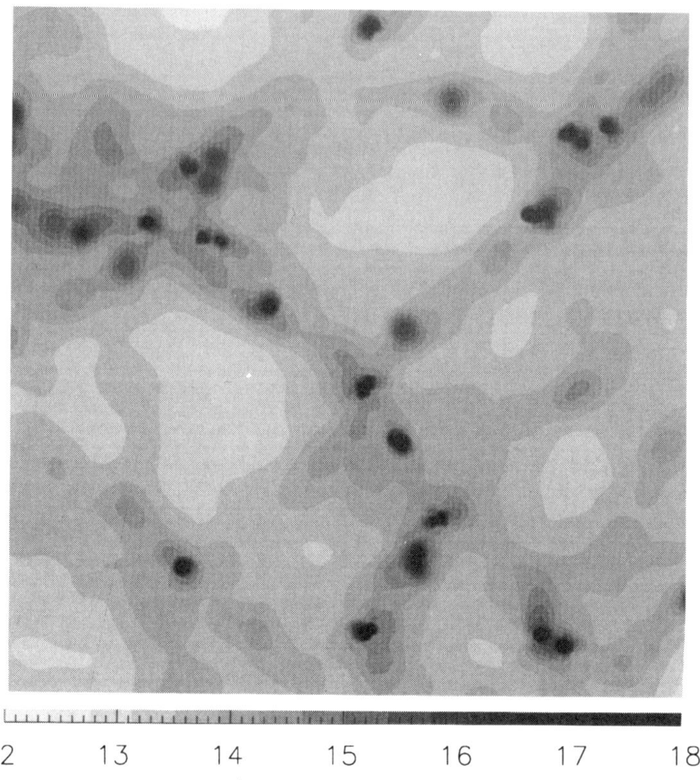


*Figure 2.* The projected HI column density distribution in a cube of size 700 kpc (proper), at $z$=3.1, from a standard CDM SPH simulation by M. Steinmetz. The contours refer to logN(HI).

## 4. The baryon content of the universe

The hydrodynamic simulations are certainly illustrative and have furthered the interpretation of QSO absorption spectra immensely, but they can also be used for quantitative measurements, *e.g.*, by comparing large sets of simulated QSO spectra with observed ones. The problem here is to find observables which can be measured at a reasonable precision, and at the same time correspond in a unique way to the ingredients of the simulated model.

One of the most easily measurable quantities is the distribution of pixel intensities $I = e^{-\tau}$ (or alternatively, flux decrements $D = 1 - I$, or optical depths $\tau$), *i.e.*, the amount of light per unit velocity absorbed by Ly$\alpha$ of intervening HI clouds from the beam of a QSO. The optical depth $\tau$ is a measure of the distribution of the neutral hydrogen in real and velocity space, $\tau \propto \mathrm{d}N_{HI}/\mathrm{d}v$, where $\mathrm{d}N_{HI}$ is the neutral hydrogen column density spread out over velocity interval $\mathrm{d}v$. The simulations predict density $\rho$, temperature $T$, peculiar velocity $v_{pec}$, and thus the ionization state of the gas as a function of the cosmological model, with the ionizing radiation background, and the total $\Omega_b$ in the universe as free parameters. Over a wide range of realistic gas densities the gas is highly ionized, and photoionization dominates the ionization equilibrium. The optical depth for absorption (neglecting here for simplicity thermal line broadening, which amounts to a further convolution of $\tau$ with a Maxwellian thermal velocity distribution) is then proportional to

$$\tau \propto \frac{(\Omega_b H_0^2)^2}{\Gamma\, H(z)} (1+z)^6 \,\alpha(T) \left(\frac{\rho}{\bar{\rho}}\right)^2 \left(1 + \frac{dv_{pec}}{H(z)dr}\right)^{-1} , \tag{2}$$

where $\Gamma$ is the photoionization probability per second, $\alpha(T)$ the recombination coefficient, $H(z)$ the Hubble constant, $\rho$ the gas density, and $dv_{pec}/dr$ the gradient of the peculiar velocity along the line of sight.

For a given temperature, the optical depth then scales with the ratio of the total density squared, divided by the ionization rate due to background radiation, $\Gamma$, *i.e.*, $\tau \propto \Omega_b^2 h^4/(H(z)\Gamma)$. Comparing observed and predicted $\tau$ distributions we can measure this quantity, as a function of redshift. If we are able to obtain an independent estimate of one of the parameters, *e.g.*, the ionizing flux, we can determine the other, *e.g.*, $\Omega_b h^{3/2}$ or *vs.*

We have observed a sample of seven QSOs with the Keck telescope, and compared the observed distribution of flux decrements $D = 1 - e^{-\tau}$ with the predicted distribution from the hydro-simulation of a CDM+$\Lambda$ universe by Cen *et al.* (1994), and a standard CDM universe from the SPH simulation by Hernquist *et al.* (1996). The measurement consists of applying a suitable global scaling to the simulated optical depths such that the mean flux decrements agree between simulation and observation for each of three redshift bins,

$$\overline{D_{obs}}(z) = \overline{D_{sim}}(z), \qquad z = 2, 3, 4. \tag{3}$$

Observed and simulated cumulative flux decrement distributions are shown in Fig. 3. The continuum level is at $D = 0$, the zero level (no transmitted light) at $D = 1$. One remarkable feature is the rapid decrease of the mean absorption with redshift. In the hierarchical models this is mostly due to the expansion of the universe which decreases the gas density and the neutral fraction. Excellent agreement with the observed distribution is attained for both models, after scaling the optical depth globally. It is worth pondering that this is the result of a one-parameter fit which is equivalent to adjusting the area under the curve giving the cumulative distribution. The agreement between the shapes of the distribution is an independent bit of information, telling us that these models do indeed produce a Ly$\alpha$ forest with a realistic intensity distribution. The Eulerian $\Lambda$CDM model is doing slightly better than the SPH SCDM simulation, in that the latter has more absorption at low column densities, leading to a steeper slope than observed. This may reflect differences in the way the heating during reionization was incorporated.

Having obtained the scalefactors by which the optical depths in the simulations have to be multiplied in order to get the same mean absorption, we can derive a lower limit on $\Omega_b$. For an independent estimate of the radiation background intensity we adopt a temporally constant UV background with $\Gamma > 7 \times 10^{-13}$ s$^{-1}$. This value corresponds to $J_\nu \approx 1.6 - 2.5 \times 10^{-22}$ (depending on the spectral shape). It is a *lower limit* in that it includes only the radiation background from QSOs alone, as estimated on the basis of the QSO luminosity function). The corresponding lower limits for $\Omega_b h^2$ at $< z > = 2$, 3, and 4 are shown in Figure 4.

The region between the dashed lines is the 95% confidence area for the "low D/H" value (Tytler, Fan & Burles 1996) as derived by Hata et al. (1996). The dotted lines show the corresponding limits for the "high D/H" value (*e.g.*, Rugers & Hogan 1996). The dash-dotted line denotes the value from primordial nucleosynthesis calculations based on solar system abundances of the light elements (after Hata *et al.* 1996). The $\Omega_b h^2$ of the SCDM simulation is lower by $\sim 30\%$ than the $\Lambda$CDM value, a difference probably mostly due to the different initial temperatures (a higher gas temperature means a lower recombination coefficient and so a smaller neutral fraction, requiring a larger $\Omega_b$).

Our measurement supports a high $\Omega_{baryon}$ (low D/H) universe, still consistent with the upper range permitted by the solar system light element abundances, but at variance with the "high D/H" values, at least for the cosmological models applied here. This confirms the conclusions Tytler, Fan & Burles (1996) have drawn from their D/H measurements. Apparently, the "high D/H" measurements (D/H $\sim$ a few$\times 10^{-4}$) may be upper limits, rather than actual detections of deuterium. Alternatively, D/H may not be a good measure of the global baryon density. From the firm lower limit on $\Gamma$ due to QSOs alone we can determine a lower limit for the amount of baryons in the universe:

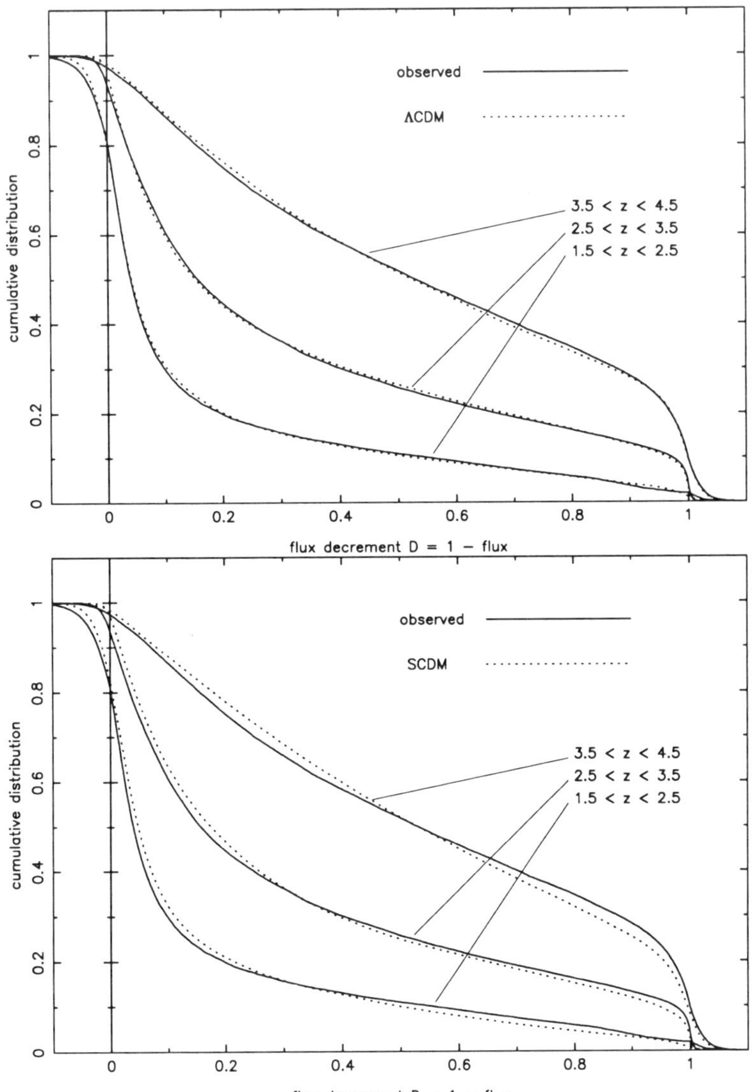


*Figure 3.* top: Comparison between observed (solid line) and simulated (best fit: dotted line) cumulative flux decrement distribution for the CDM+$\Lambda$ model of Cen *et al.* 1994. bottom: Same statistic for the standard CDM SPH simulation of Hernquist *et al.* 1996)

$$\Omega_{baryon}h^2 > 0.017$$

Remaining uncertainties stem from the assumed intensity of the ionizing radiation background, and the temperature the lower column density gas has attained by the time reionization is complete. Furthermore, the data sample is still relatively small, especially at lower redshifts. The finite reso-

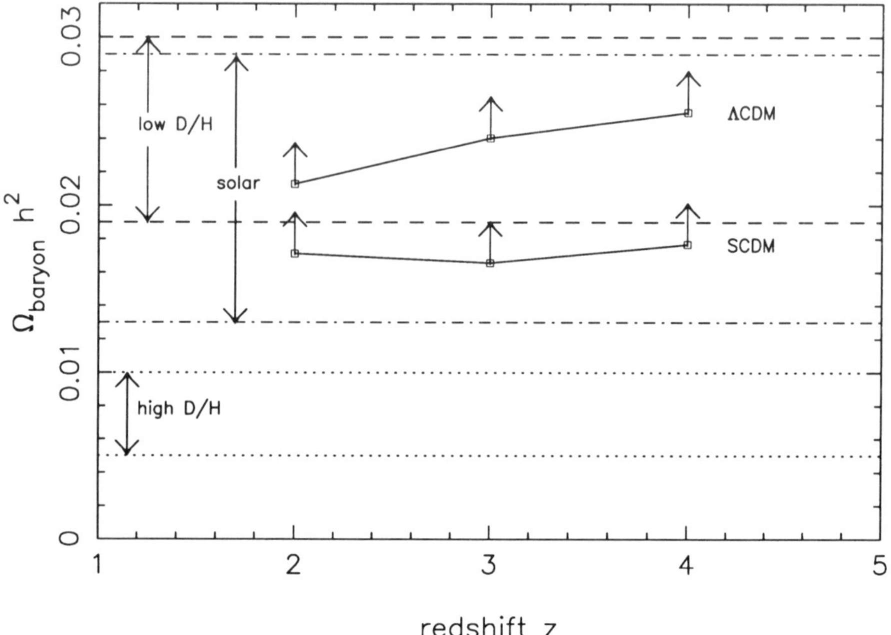


*Figure 4.* Lower limits to $\Omega_b h^2$ assuming a constant minimum contribution to the ionizing flux as given in the text.

lution of the simulations could possibly lead to the simulated gas being less clumpy than in reality. In that case we may be overestimating the $\Omega_b$ necessary to reproduce the observations. However, from the small variations seen between common absorption features in close lines of sight to gravitationally lensed QSOs (Smette *et al.* 1993,1995) it appears that such structure, if present, does not show up in Ly$\alpha$ forest absorption. Weinberg *et al.* (1997) have recently put forward general arguments why in a hierarchical model (where the low column density clouds are bulk motion broadened) the $\Omega_b$ in the Ly$\alpha$ forest has to be large, so unless the structure formation model is wrong, our result should hold. Further (probably smaller) uncertainties may enter through the dependence of the optical depth distribution on the cosmological model, an effect currently under investigation for various other scenarios. Although the precise cosmological model may be more difficult to ascertain with this method, the degree of agreement between observed and simulated flux distribution functions achieved by manipulating only one free parameter may be considered as strong evidence in favor of a hierarchical structure formation scenario.

## 5. Conclusions

We have described some new results on the distribution of baryons at high redshift, made possible by new theoretical methods and progress with high resolution QSO absorption line spectroscopy. If current theoretical ideas are correct, and we are living in a hierarchical universe, most of the baryons are (at $z > 2$) in the form of highly ionized gas in a filamentary network, not incorporated in galaxies. The temperature of this gas is typically $4 \times 10^4$ $K$, and many random lines of sight through it show metal-enrichment at a level of $10^{-2.5}\odot$. We have attempted to measure the baryon content comparing the mean absorption between actual and simulated QSO spectra from a Eulerian $\Lambda$CDM and an SPH standard CDM model. We obtain a lower limit to $\Omega_b$, in agreement with the value predicted by the low "D/H" values recently measured. The good agreement between observed and simulated flux distribution functions favors a hierarchical structure formation scenario. The method presented here is currently the most direct way for measuring the universal baryonic $\Omega$, being independent of primordial nucleosynthesis. Baryons are counted in the reservoir where most of them actually reside.

Results on the flux distribution and $\Omega_b$ presented here are based on work done in an ongoing collaboration with Jordi Miralda-Escudé, Wal Sargent, Tom Barlow, David Weinberg, Lars Hernquist, Neal Katz, Renyue Cen, and Jeremiah Ostriker.

## Acknowledgements

I would like to thank the organizers of this workshop for inviting me to a very pleasant meeting, and to my collaborators for allowing me to present this work prior to publication. I am also grateful to NASA for support through grant HF-01075.01-94A from the Space Telescope Science Institute.

## References

Barcons, X., Fabian, A.C., Rees, M.J., 1991, *Nature*, 350, 685
Bi, H.G., Börner, G., Chu, Y., 1992, *A & A*, 266, 1
Bechtold, J., Crotts, A.P.E., Duncan, R.C., Fang, Y.H., 1994, *ApJ*, 437, L83
Bond, J.R., Szalay, A.S., Silk, J., 19888, *ApJ*, 324, 627
Cen, R., Miralda-Escudé, J., Ostriker, J.P., Rauch, M., 1994, *ApJ*,437, L9
Charlton, J.C., Salpeter, E.E., Hogan, C.J., 1993, *ApJ*, 402, 493
Churchill, C.W., Steidel, C.C., Vogt, S.S., 1996, *ApJ*, 471, 164
Cowie L.L., Songaila A., Kim T.-S., Hu E. 1995, *AJ*,109, 1522
Crotts, A.P.E., Bechtold, J., Fang, Y., Duncan, R.C., 1994, *ApJ* 437, 79
Dinshaw, N., Impey, C.D., Foltz, C.B., Weymann, R.J., Chaffee, F.H., 1994, *ApJ*, 437, L87
Dinshaw, N., Foltz, C.B., Impey, C.D., Weymann, R.J., Morris, S.L., 1995, *Nature*, 373, 223

Fang, Y.H., Duncan, R.C., Crotts, A.P.S., Bechtold, J. 1996, *ApJ*, 462, 77
Haardt F., Madau P., 1996, *ApJ*, 461, 20 (HM)
Haehnelt, M.G., Steinmetz, M., Rauch, M., 1996, *ApJL*,465, L95
Hata, N., Steigman, G., Bludman, S., Langacker, P., 1997, *Phys. Rev. D*, 55, 540
Hernquist L., Katz N., Weinberg D.H., Miralda-Escudé J., 1996,*ApJ*,457, L5
Lu L., Sargent W.L.W., Womble W.S., Takada-Hidai M., 1996, *ApJ*, 472, 509
McGill, C., 1990, *MNRAS*, 242, 544
Meiksin, A., Madau, P., 1993, *ApJ*, 412, 34
Meiksin, A., 1994, *ApJ*, 431, 109
Milgrom, M., 1988, *A & A*, 202. L9
Miralda-Escudé, J., Rees, M.J., 1993, *MNRAS*, 260, 617
Miralda-Escudé J., Cen R., Ostriker J.P., Rauch M., 1996, *ApJ*, 471, 582
Petitjean, P., Webb, J.K., Rauch, M., Carswell, R.F., Lanzetta, K., 1993, *MNRAS*,262, 499
Petitjean, P., Mücket, J.P., Kates, R.E., 1995, *A& A*, 295, L9
Rauch M., Haehnelt M.G., 1995, *MNRAS*, 275, L76
Rauch M., Sargent, W.L.W., Womble, D.S., Barlow, T.A., 1996, *ApJL*,467, L5,
Rauch M., Haehnelt M.G., Steinmetz M., 1996, *ApJ* , 481, 601
Rees, M.J., 1986, *MNRAS*. 218, 25
Reisenegger A., Miralda-Escudé J., 1995, *ApJ*, 449, 476
Rugers M., Hogan C.J., 1996, *AJ*, 111, 2135
Sargent, W.L.W., Young, P.J., Boksenberg, A., Tytler, D., 1980, *ApJSupp.*, 42, 41
Shapiro, P.R., Giroux, M.L., Babul, A., 1994, *ApJ* 427, 25
Smette, A., Surdej, J., Shaver, P. A., Foltz, C. B., Chaffee, F. H., Weymann, R. J., Williams, R. E., & Magain, P. 1992, ApJ, 389, 39
Smette, A., Robertson, J. G., Shaver, P. A., Reimers, D., Wisotzki, L., & Köhler, Th. 1995, A&AS, 113, 199
Tytler D., Fan X.-M., Burles S., Cottrell L., Davis C., Kirkman D., Zuo L., 1995, in *QSO Absorption Lines*, Proc. ESO Workshop, ed. G.Meylan (Heidelberg: Springer), p. 289.
Tytler D., Fan, X.M, Burles, S. 1996, *Nature*, 381, 207
Walker, T.P., Steigman, G., Schramm, D.N., Olive, K.A., Kang, H.S, 1991, *ApJ*, 376, 51
Williger, G.M., Babul, A., 1992, *ApJ*, 399, 385
Weinberg D.H., Miralda-Escudé J., Hernquist L., Katz N., 1997, subm. (astro-ph/9701012)
Zhang Y., Anninos P., Norman M.L., 1995, *ApJ*, 453, L57

# SIMULATIONS OF THE LARGE-SCALE STRUCTURE FORMATION & THE EVOLUTION OF THE LYMAN-$\alpha$ FOREST COMPARISON WITH OBSERVATIONS

J.P. MÜCKET AND R. RIEDIGER
*Astrophysikalisches Institut Potsdam*
*An der Sternwarte 16, D-14482 Potsdam, Germany*

AND

P. PETITJEAN
*Institut d'Astrophysique de Paris - CNRS*
*98bis Boulevard Arago, F-75014 Paris, France*
*UA CNRS 173- DAEC, Observatoire de Paris-Meudon*
*F-92192 Meudon Principal Cedex, France*

## 1. Introduction

The study of the Ly$\alpha$ absorbers seen in quasar spectra is a very sensitive way to probe the baryonic material in the whole redshift range $0 < z < 5$. In particular the evolution of the spatial distribution of the Ly$\alpha$ clouds offers the unique chance to obtain direct information about the structure formation processes throughout a wide redshift range. During recent years considerable success has been obtained in modeling the distribution of the neutral hydrogen gas in the context of the evolution of large-scale structure (Cen *et al.* 1994, Petitjean *et al.* 1995, Mücket *et al.* 1996, Hernquist *et al.* 1996, Miralda-Escudé *et al.* 1996, Zhang *et al.* 1996, Bi & Davidsen 1996) By means of simulations many of the characteristics of the Ly$\alpha$ forest could be reproduced. One of the outstanding problems is the relation between the Lyman-$\alpha$ forest and galaxies. Although conclusions are uncertain, it seems that at least the strongest lines in the Lyman-$\alpha$ forest at low redshift are anyhow associated with galaxies (Lanzetta *et al.* 1995, Le Brun *et al.* 1996) since the density of the intergalactic gas is expected to be higher in the vicinity of the galactic potential wells. The case for the weak line to be associated with galaxies is less clear. Indeed observations of the line of sight to 3C273 that are the most sensitive to the presence of weak lines



*D. Hamilton (ed.), The Evolving Universe,* 303–314.

(Morris *et al.* 1991, Bahcall *et al.* 1991) indicate the presence of a large number of these lines and no clear association with galaxies is seen (Morris *et al.* 1993). Moreover, Stocke *et al.* (1995) have detected weak absorption lines located in regions devoided of galaxies.

Recent observations have shown that at redshift $z \sim 3$, CIV is found in 90% of the clouds with $N(\text{HI}) > 10^{15}$ cm$^{-2}$ and in about 50% of the clouds with $3 \times 10^{14}$ cm$^{-2}$ $< N(\text{HI}) < 10^{15}$ cm$^{-2}$ (Songaila & Cowie 1996, Cowie *et al.* 1995).

Following the simulations the Lyman-$\alpha$ absorption line properties can be understood if the gas traces the development of structures in the Universe. In this picture, part of the gas is located inside filaments where star formation can occur very early in small halos that subsequently merge to build-up a so-called galaxy (Haehnelt *et al.* 1996). This gas contains metals. The remaining part of the gas has very low metallicity or no metals and either is loosely associated with the filaments and has $N(\text{HI}) \geq 10^{14}$ cm$^{-2}$ or is located in the underdense regions and has $N(\text{HI}) \leq 10^{14}$ cm$^{-2}$.

To clarify this issue, we have refined our simulations introducing a distinction between shocked and unshocked particles leading in the result, as will be shown, to two different populations of Lyman-$\alpha$ clouds.

## 2. Simulations

We described in Mücket *et al.* (1996) the main characteristics of a pseudo-hydrodynamic code using the particle-mesh (PM) code developed by Kates *et al.* (1991) including temperature evolution of the gas component associated with the dark matter and effects of photoionisation. The simulations used $128^3$ particles on a $256^3$ grid. We carefully analyzed the length scales to be assigned to the simulation box and to the cells. Taking into account the restrictions due to scales of nonlinearity at $z = 0$ and recent observations which give estimates for the cloud sizes the simulations were carried out using a box size of 12.8 Mpc which corresponds to a co-moving cell size of 50 kpc. The baryonic mass is assumed to be proportional to the dark matter mass inside a cell. This assumption has been shown by detailed hydro-simulations to be valid for the low-density regime characteristic of the Lyman-$\alpha$ forest (see Miralda-Escudé *et al.* 1996, Hernquist *et al.* 1996). We adopt a value for the Hubble parameter $H_0 = 50$ km Mpc$^{-1}$ s$^{-1}$ and $\Omega_b = 0.05$ throughout.

The photoionizing UV background flux responsible for the photoionisation is calculated by the code self-consistently and is assumed to be homogeneous and isotropic throughout the simulation box. The ionizing spectrum is modelled as $J_\nu \propto J_0 \, \nu^{-1}$ where $J_0 = J_{-21} \cdot 10^{-21}$ erg cm$^{-2}$ s$^{-1}$ Hz$^{-1}$ sr$^{-1}$ is the ionizing flux at 13.6 eV which depends on redshift, *i.e.*, $J_{-21} = f(z)$.

The UV background flux is assumed to be determined by star formation processes in collapsing cool and dense regions. The variation of the flux intensity at redshift $z$ is related to the rate $\Delta m(T_4 < 0.5; z)$ at which the baryonic material cools below $T_4 = 0.5$ (with $T = T_4 \cdot 10^4$ K) in the simulation and to the expansion of the Universe:

$$
\begin{aligned}
f(z) = \; & C_{\rm cool}\, \Delta m(T_4 < 0.5; z) \\
& + f(z + \Delta z) \left( \frac{1+z}{1+z+\Delta z} \right)^4 ,
\end{aligned} \tag{1}
$$

where $C_{\rm cool}$ is a factor of proportionality. To avoid overcooling (*e.g.*, Blanchard *et al.* 1992), we assume that the cool gas is transformed into stars with an efficiency $\varepsilon$ of about 8%. The remainder of the gas is reheated to temperatures above 50 000 K. The characteristic time period for those processes is of order $t_* \approx 10^8$ years. This procedure provide simulation results that are independent of the time step.

Whereas the time dependence of the UV flux is almost not affected by the value of $t_*$ within a reasonable range the actual value of the efficiency parameter has considerable influence on the flux intensity at small redshifts (see also Miralda-Escudé *et al.* 1996). Namely in this scenario it determines the amount of gas still available for star formation.

During the simulation we distinguish between two different populations of clouds: Part of matter particles are involved in shell-crossing processes (shocks). In those cases the procedure for temperature assignment proposed by Kates *et al.* (1991) can be applied:

(1) A particle is labeled "shocked" if the Jacobi determinant of the transformation from Lagrange (particle) to Euler (grid) coordinates becomes negative.

(2) If the particle enters a shocked region, defined as entering a cell on a coarse grid of cell size $2l_c$ containing at least one shocked particle. In either case, we attempt to define a local velocity field $\mathbf{U}$ on the coarse grid, and we assign to the gas associated with the particle a temperature according to $kT = \mu_{\rm H} m_{\rm H} (\mathbf{v} - \mathbf{U})^2/3$, where $\mathbf{v}$ is the particle velocity, $\mu_{\rm H}$ the molecular weight and $m_{\rm H}$ the mass of the hydrogen atom.

We assume ionization equilibrium which is a good approximation when the gas is fully ionized (Duncan *et al.* 1989). The temperature is determined using the time dependent equations (see Mücket *et al.* 1996). The shocked particles are mostly found in big halos and elongated, filamentary structures (regions of enhanced density). The remainder of the gas is unshocked. It is found in the surroundings of the structures formed by shocked particles but mostly in the voids delineated by these structures. The gas here is photo-ionized by the background flux and assumed in thermal equilibrium.

We have monitored all results for both populations separately. The modeling of the cloud distribution along a full line of sight up to a fictitious QSO at redshift $z = 5$ is described in detail in Mücket *et al.* (1996).

## 3. Results

The simulation must reproduce the evolution with time of the number of Lyman-$\alpha$ absorption lines. Due to blending, the comparison between simulations and observations is not simple. A first constrain is thus to reproduce the evolution of the average Lyman-$\alpha$ decrement. As emphasized by Miralda-Escudé *et al.* (1996), this number depends on the mean HI density directly related to $J_0^{-1}(\Omega_b h^2)^2$ where $\Omega_b$ is the baryon density and $J_0 = J_{-21} \cdot 10^{-21}$ erg cm$^{-2}$ s$^{-1}$ Hz$^{-1}$ sr$^{-1}$ is the ionizing flux at 13.6 eV. In our simulations, the baryon density is given a value 0.05 ($\Omega_b h^2 = 0.0125$) and the evolution with redshift of the ionizing flux is computed assuming that its variation is related to the amount of gas that collapses in the simulation at any time (see Section 2). The only free parameter is thus the normalization of $J_{-21}(z)$. A value of $J_{-21}(z_0) = 0.1$ at $z_0 \sim 3$ (see Fig. 1) fits the decrement evolution quite well (see Fig. 2).

It must be noticed, that the redshift range considered here is large ($0 < z < 5$) and the flux is significantly changing. It is thus necessary and important to calculate the variations consistently in order to be able to discuss the evolution of the Lyman-$\alpha$ forest over the whole redshift range.

Our result for the flux normalisation is consistent with the findings by Hernquist *et al.* (1996) and Miralda-Escudé *et al.* (1996) using hydro simulations over a much smaller redshift range.

Fitting the Lyman-$\alpha$ forest is ususally done using Voigt profile deblending procedures. Even though blending is a severe limitation for this analysis at $z > 2.5$, it is interesting to compare the observed number of lines with the simulated number of clouds that is perfectly defined. For $N$(HI) $> 10^{14}$ cm$^{-2}$, the lines are not numerous enough for blending to be a problem. It can be seen from Fig. 3a that the evolution of the total number of strong lines is well reproduced. Data are taken from Lu *et al.* (1991), Petitjean *et al.* (1993) and Bahcall *et al.* (1993).

If the number of lines per unit redshift is approximated by a power-law ($dn/dz \propto (1+z)^\gamma$), we find $\gamma \approx 2.6$ for $1.5 < z < 5$ and $\gamma \approx 0.6$ for $0 < z < 1.5$. The simulated number is also consistent with observation at $z > 4$ (Williger *et al.* 1994).

Fig. 3b shows the contributions of the two populations of clouds with $N$(HI) $> 10^{14}$ cm$^{-2}$, shocked (solid line) and unshocked (dash-dotted line). It is apparent that the dominant population is different before and after $z \sim 3$. At high redshift, most of the lines arise in unshocked particles

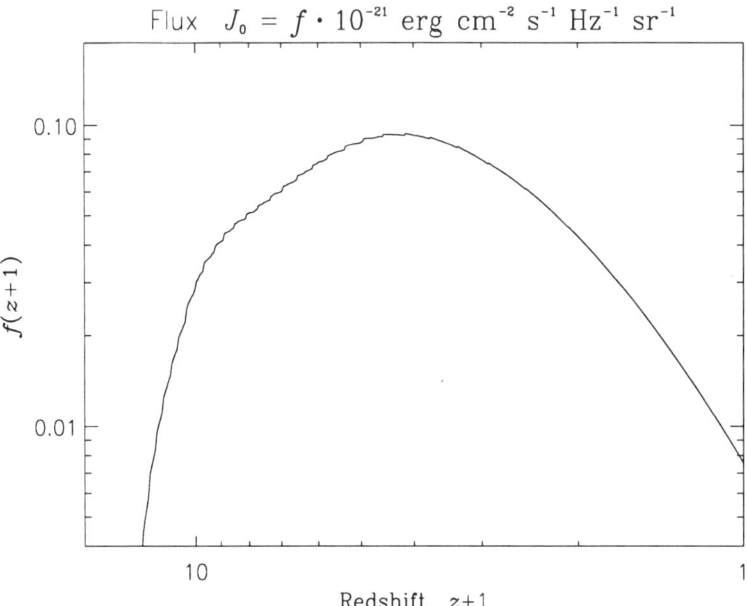


*Figure 1.* The UV-background flux computed in the course of our simulations.

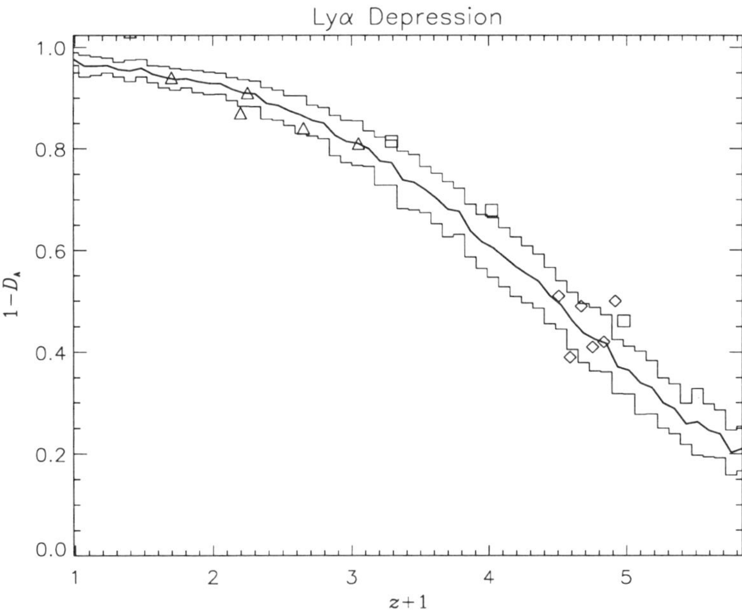


*Figure 2.* Lyman-$\alpha$ continuum depression versus redshift. Observed data are from: diamonds, Lu *et al.* (1996); triangles, a compilation by Jenkins & Ostriker (1991) at low redshifts; squares: Rauch *et al.* (1997).

whereas at low redshift, most of the gas is condensed in filamentary structures (see Petitjean *et al.* 1995).

Let us assume that most of the unshocked clouds are co-expanding and optically thin. If we also assume the flux to be nearly constant throughout the considered redshift range, then the evolution of the column density for each cloud is $\propto J_0^{-1} n_{\rm H}^2 l_c \propto (1+z)^5$, *i.e.*, the column density of such clouds is a rapidly decreasing function of time. Therefore, at high redshift, a similar behavior is expected for the number density of the clouds providing a column density threshold is given. That might explain the very steep slope found for the number density evolution of the unshocked clouds as shown in Fig. 3b.

The number density of lines with $N(\text{H}\,\text{I}) > 10^{12}$ cm$^{-2}$ is about constant over the redshift range $1 < z < 5$ (see Fig. 5 and decreases slowly at lower redshift. Low density gas is found in regions delineated by filamentary and sheet-like structures at high redshift. This gas slowly disappears. The total number density of lines stays constant because the high column density gas has column density decreasing with time. Such a difference in the evolution of the number density of weak and strong lines, although to be confirmed, has been noticed in intermediate resolution data (Bechtold 1994). Fig. 8 gives a visual impression of the distributions of clouds having column densities within the range $10^{14}$ cm$^{-2}$ $< N(\text{H}\,\text{I} < 10^{14}$ cm$^{-2}$ at different redshifts. Most of the clouds seen at $z = 3$ are distributed in sheet-like and filamentary structures are not longer detectable at this column density range at $z = 1$ or even $z = 0$. For the fixed density threshold $N(\text{H}\,\text{I} \approx 10^{14}$ cm$^{-2}$ the detectable structures in the HI distribution are transformed from sheet-like to filamentary and eventually to single clumpy structures.

Fig. 4 shows an example for a simulated full spectrum along a single line of sight at a resolution similar to Keck HIRES observations. The absorption features are calculated from the parameters obtained for each cell in the simulation. The line number density is calculated as in Mücket *et al.* (1996). The resulting H I column density distribution is given in Fig. 6 (solid line) together with the observed points taken from Hu *et al.* (1996). The agreement is good for column densities $N(\text{H}\,\text{I}) \geq 10^{14}$ cm$^{-2}$. In this range the distribution is dominated by the contribution from shocked particles (dotted line). For lower column densities the distribution is dominated by the contribution from the unshocked particles (dashed line) and is too large by a factor of two compared to observations. However this factor is well within the uncertainty of both observational and simulated data. The discrepancy could result from the limited resolution of simulations. A flattening is apparent toward lower column densities. The comparison of the column density distributions at different redshifts shows that the flattening happens at larger $N$ for higher redshifts in agreement with the predictions

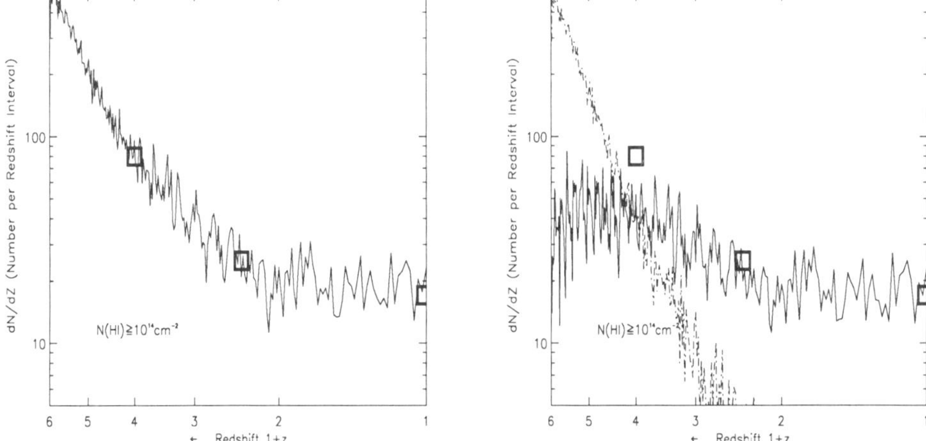


*Figure 3.* (a) Number density $\mathrm{d}n/\mathrm{d}z$ of clouds with column density $\log N(\mathrm{H\,I}) > 14$ versus redshift $z$. (b) As in (a). The number density of lines drawn from the shocked and unshocked populations are plotted as full and dash-dotted lines respectively

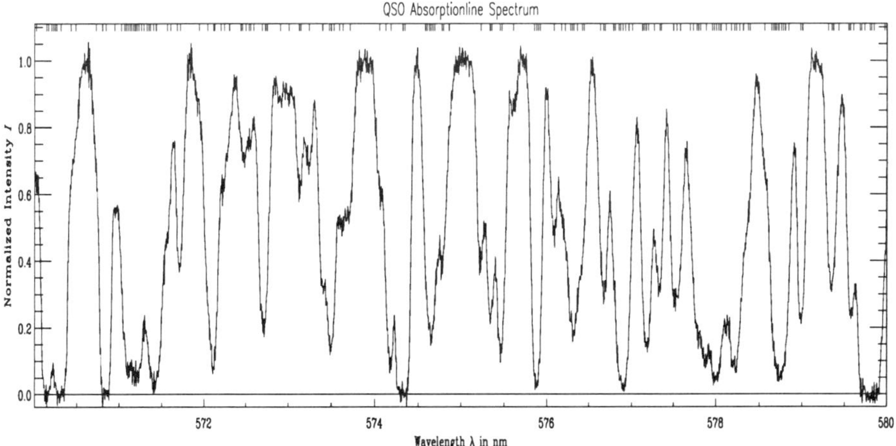


*Figure 4.* Part of a synthetic spectrum derived from simulation data along a single line of sight.

by Mücket *et al.* 1996 (cp. Eq. 7). So the onset of the flattening at small column densities is related to the resolution of the simulations. On the other hand, the behavior is similar to that recently observed at redshifts $z \approx 3 \ldots 4$ that is usually attributed to the blending of weak lines with stronger ones.

The lines are considered broadened by thermal and turbulent motions. The thermal broadening is derived from the temperature. The turbulent broadening is estimated as the root mean square of peculiar velocities in

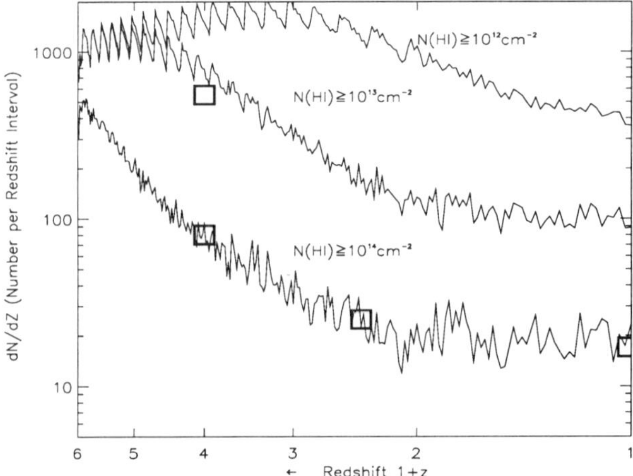


*Figure 5.* Number density of lines versus redshift for different column densities thresholds $\log N(\mathrm{H\,I}) > 12, 13, 14$.

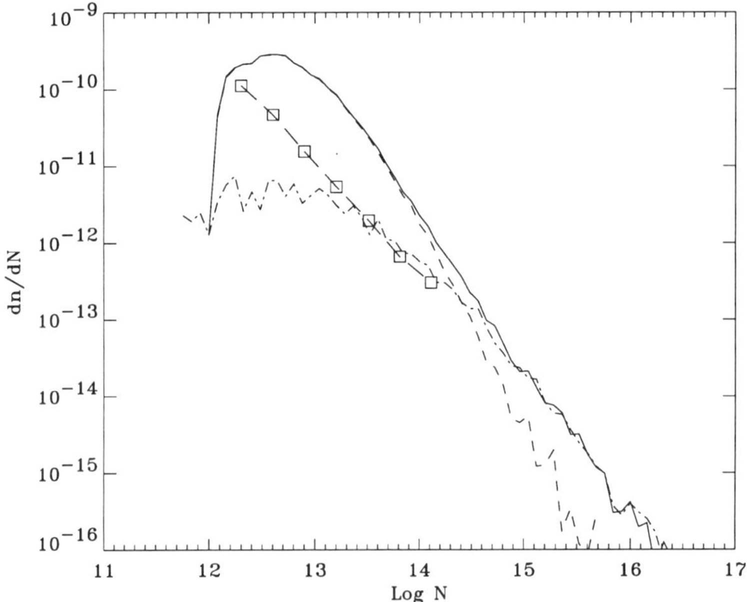


*Figure 6.* H I column density distribution at $z = 3$ for the shocked particles (dotted line), the unshocked particles (dashed line) and all the particules together (solid line). Observational data points from Hu *et al.* (1995) and Petitjean *et al.* (1993) are indicated by square symbols.

one cell. The resulting distribution together with data from Lu *et al.* (1996) is shown in Fig. 7. There is a lack of small Doppler parameters in both

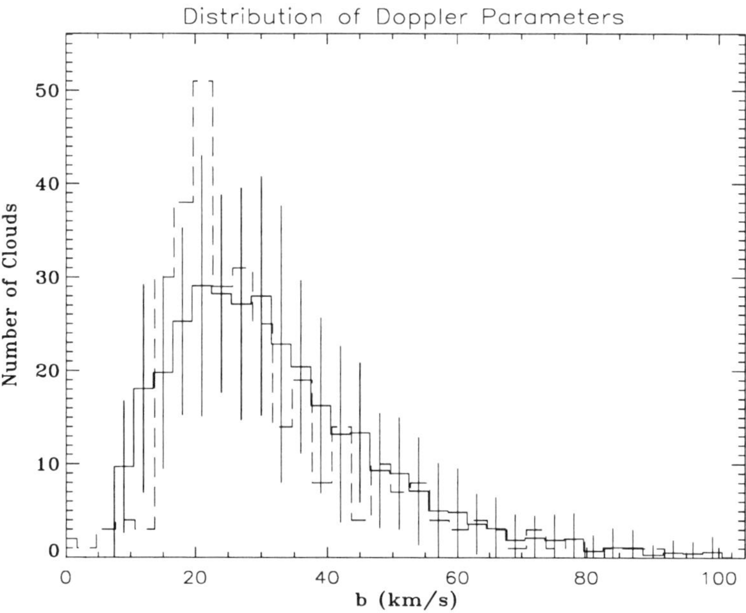


*Figure 7.* Doppler parameter distribution (solid line). Results from Lu *et al.* (1996) are overplotted as a dashed line.

the observed and and simulated data. The mean temperature of the gas beeing 20 000 K, this means that broadening of the lines is dominated by turbulent motions. Note that this procedure does not apply to unshocked particles since the number of particles within one cell is too small to define turbulent motions this way. A weak dependence of the Doppler parameter distribution with redshift is observed in the simulations: The maximum of the distribution is marginally shifted to higher Doppler parameters with decreasing redshift.

## 4. Conclusion

We have refined our simulations (Mücket *et al.* 1996) introducing a distinction between shocked particles predominantly found within dense structures such as filaments and unshocked particles which populate underdense regions.

At redshifts $z < 2.5$ the number density of lines arising from unshocked particles decreases very rapidly for H I column densities larger than $10^{14}$ $cm^{-2}$. Therefore at low redshifts, the main contribution for this column density threshold comes from shocked particles. As mentioned above early star formation probably occurs in the filamentary structures and most of the Lyman-$\alpha$ clouds with column densities $\log N(\mathrm{H\,I}) > 14$ are expected to

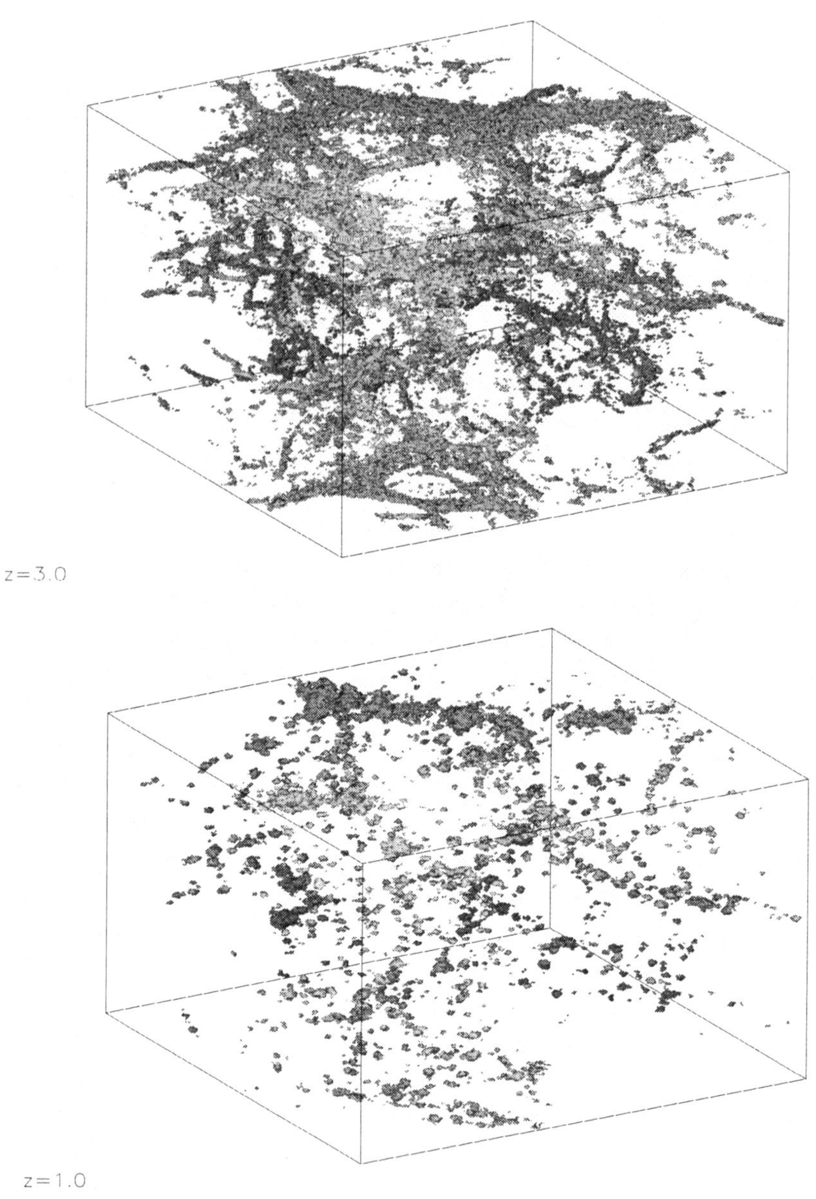


*Figure 8.* The distribution of HI clouds having column densities within the range $14 < \log_{10}(NHI/cm^{-2}) < 16$ at the epochs z=3 and z=1. The box length is 12.8 Mpc.

contain metals. In addition a large fraction of those clouds at low redshift should somehow be correlated with galaxies (Petitjean *et al.* 1995).

The number density of lines with $\log N(\mathrm{H\,I}) > 12$ remains about constant over the redshift range $5 > z > 1$ and decreases slowly at lower redshift. This is the result of both decreasing ionizing flux and mean hydrogen density. The number density of such weak lines is predicted to be larger than 200 per unit redshift at $z \sim 0$. This prediction can be tested along the line of sight to 3C273 with the new instrumentation to be installed on the Hubble Space Telescope.

Determination of realistic Doppler parameters for clouds in the underdense regions needs consideration of sub-cellular effects. The internal velocity and temperature distribution for one-particle clouds is thus beyond the limit of our simulations. However the contribution of the shocked particles alone yields a Doppler parameter distribution in reasonable agreement with observations.

Our simulation yields a consistent determination of the evolution of the UV flux intensity over the whole redshift range $0 < z < 5$ (see Fig. 1). The normalisation is obtained by fitting the Lyman-$\alpha$ decrement; a value of $J_{-21}(z_0) = 0.1$ at $z_0 \sim 3$ fits the decrement evolution quite well (see Fig. 2) for $\Omega_b h^2 \sim 0.0125$.

## References


Bahcall J.N., Bergeron J., Boksenberg A., *et al.*, 1993, *ApJS* **87**, 1
Bahcall J.N., Januzzi B.T., Schneider D.P., *et al.*, 1991, *ApJL* **377**, 5
Bechtold J., 1994, *ApJS* **91**, 1
Bi HongGuang, Davidsen A.F., 1996, astro-ph/9611062
Blanchard A., Valls-Gabaud D., Mamon G.A., 1992, A&A **264**, 365
Cen R., Miralda-Escudé J., Ostriker J.P., Rauch M., 1994, *ApJL* **437**, L9
Cowie L.L., Songaila A., Kim Tae-Sun, Hu E.M., 1995, *AJ* **109**, 1522
Duncan R.C., Ostriker J.P., Bajtlik S., 1989, *ApJ* **345**, 39
Haehnelt M.G., Rauch M., Steinmetz M., 1996, *MNRAS* (in press)
Hernquist L., Katz N., Weinberg D.H., Miralda-Escudé J., 1996, *ApJ* **457**, L51
Hu E.M., Kim Tae-Sun, Cowie L.L., Songaila A., Rauch M., 1995,*AJ* **110**, 1526
Jenkins E.B., Ostriker J.P., 1991, *ApJ* **376**, 33
Kates R.E., Kotok E.V., Klypin A.A., 1991, *A&A* **243**, 295
Lanzetta K.M., Bowen D.V., Tytler D., Webb J.K., 1995, *ApJ* **442**, 538
Le Brun V., Bergeron J., Boissé P., 1996, *A&A* **306**, 691
Lu L., 1991, *ApJ* **379**, 99
Lu L., Sargent W.L.W., Womble D.S., Takada-Hidai M., 1996, *ApJ* **472**, 509
Miralda-Escudeé J., Cen R., Ostriker J.P., Rauch M., 1996, *ApJ* **471**, 582
Morris S.L., Weymann R.J., Dressler A., *et al.*, 1993, *ApJ* **419**, 524
Morris S.L., Weymann R.J., Savage B.D., Gilliland R.L., 1991, *ApJL* **377**, 21
Mücket J.P., Petitjean P., Kates R., Riediger R., 1996, *A&A* **308**, 17
Petitjean P., Bergeron J., 1994, *A&A* **283**, 759
Petitjean P., Mücket J., Kates R.E., 1995, *A&A* **295**, L9
Petitjean P., Webb J.K., Rauch M., *et al.*, 1993, *MNRAS* **262**, 499
Rauch M., Miralda-Escude J., Sargent W.L.W., Barlow T.A., Weinberg D.H., Hernquist

L., Katz N., Cen R., Ostriker J.P., 1997, astro-ph/9612245
Songaila A., Cowie L.L., 1996, astro-ph/9605102
Stocke J.T., Shull J.M., Penton S., *et al.*, 1995, *ApJ* **451**, 24
Williger G.M., Baldwin J.A., Carswell R.F., *et al.*, 1994, *ApJ* **428**, 574
Zhang Yu, Meiksin A., Anninos P., Norman M.L., 1996, astro-ph/9601130

# PROBING SMALL- AND LARGE-SCALE STRUCTURE AT HIGH REDSHIFT WITH THE TPCF OF CIV ABSORPTION LINES

MARTIN G. HAEHNELT
*Max-Planck-Institut für Astrophysik*
*Karl-Schwarzschild-Straße 1*
*85740 Garching*
*Germany*

## 1. Introduction

While at low redshift metal absorption systems have been convincingly demonstrated to arise in the haloes of rather normal galaxies [1] [2] as suggested by Bahcall & Spitzer [3], much less is known about the nature of metal absorption system at $z > 2$ [4] [5] [6]. Recently, numerical simulations performed with GRAPESPH [7] and CLOUDY [8] and assuming a cold dark matter cosmogony were used to show that the prominent complex CIV absorption features observed at high redshift can be well reproduced by the absorbing properties of regions in which galaxies form by hierarchical merging [9] (see Navarro & Steinmetz [10] for more numerical details of the simulation and Rauch, Haehnelt & Steinmetz [11] for a detailed description of their metal absorption properties). The two-point correlation function (TPCF) is one of the standard tools to study the clustering properties of the matter distribution and the TPCF of galaxies and galaxy clusters has been instrumental in establishing the gravitational instability picture and hierarchical models of structure formation. Results on the TPCF of QSO absorption systems, however, have been inconclusive and contradictory. While metal absorption systems at high redshift show strong correlations on scales of several hundred km s$^{-1}$and are still weakly correlated on scales of several thousand km s$^{-1}$[4][12], only very weak or no correlations were reported for Ly$\alpha$ absorption systems even on very small scales [13][14][15]. Cristiani *et al.* [16], however, were able to demonstrate a relation between correlation strength and Ly$\alpha$ column density. Here some



*D. Hamilton (ed.), The Evolving Universe,* 315–319.

results concerning the correlation properties of CIV absorption systems in our numerical simulations are presented.

## 2. The two-point correlation function of QSO absorption systems

The TPCF of QSO absorption systems in velocity space can be approximately related to that of the matter distribution in real space [16][17]

$$\xi_{\rm abs}(v) \approx b^2_{\rm abs-mass} \int_{r_{\rm cl}}^{\infty} \xi_{\rm mass}(r)\, f(v)\, H(z)\, {\rm d}r, \tag{1}$$

where $b_{\rm abs-mass}$ is the bias factor relating absorber and mass correlation function in real space, $r_{\rm cl}$ is a measure of the cloud size, $f(v)$ is the distribution of peculiar velocities of the absorption systems and $H(z)$ is the Hubble constant. Equation (1) shows that there are three basic problems in predicting $\xi_{\rm abs}$:

- It is not clear to which extent absorption systems trace the mass distribution and $b_{\rm abs-mass}$ might depend on column density in a complicated way.
- $\xi_{\rm mass}(r)$ generally diverges towards small radii and for small velocities the TPCF will depend strongly on the absorber "size".
- Little is known about the functional form and amplitude of the peculiar velocity distribution of absorption systems.

Numerical simulation of the large-scale gas distribution suggest that the traditional picture of absorbing clouds with well defined spatial boundaries is not adequate and that much of the absorption is due to large-scale density fluctuations in a warm photoionized intergalactic medium [18] [19] [20][21]. These fluctuations are better characterized by a coherence length which depends strongly on column density than by a definite size. Recent observations of close quasar pairs suggest that the coherence length of intermediate column density absorbers ($\log N \sim 14 - 15$) is of the order of several hundred kpc and approaches or exceeds the scale where the expected correlation strength is unity [22][23]. The large column density dependent coherence length of the absorbers is a plausible explanation for the weak and column-density dependent correlation strength which is found for Ly$\alpha$ absorption systems. As discussed above our numerical simulations have shown that the high-redshift building blocks of present-day galaxies predicted by hierarchical structure formation scenarios are associated with multiple metal absorption features which resemble those typical for observed strong CIV systems [9][11]. It seems therefore worthwhile to compare the absorber TPCF at high redshift directly to that of galaxies at $z = 0$. Assuming that the TPCF of galaxy progenitors has the same power-law form as that of present-day galaxies and assuming a Gaussian distribution of peculiar velocities with dispersion $\sigma$ the absorber TPCF can be written as

$$\xi_{abs}(v) \approx b^2_{\mathrm{abs-gal}} \frac{1}{\sqrt{2\pi}} \int_{r_{\mathrm{cl}}}^{\infty} (1+z)^{-(3+\epsilon)} \left(\frac{r}{r_0}\right)^{-1.8}$$

$$\times \left\{ \exp\left[-\frac{(Hr-v)^2}{2\sigma^2}\right] + \exp\left[-\frac{(Hr+v)^2}{2\sigma^2}\right] \right\} \frac{H(z)\,\mathrm{d}r}{\sigma}, \quad (2)$$

where $r_0 \approx 5.5h^{-1}$Mpc is the correlation length of present-day galaxies and $\epsilon$ characterizes the evolution of the correlation amplitude with redshift as usual ($\epsilon = 0$ for stable clustering and $\epsilon = 0.8$ for the linear regime). To get a feeling for the expected correlation strength we consider the limit of small absorber size and negligible peculiar velocities and set $z = 3$, then

$$\xi_{abs}(v) \approx b^2_{\mathrm{abs-gal}} \left(\frac{1}{4}\right)^{\epsilon} \left(\frac{v}{400\mathrm{km\,s^{-1}}}\right)^{-1.8}. \quad (3)$$

## 3. Observed and simulated CIV TPCF

Figure 1 shows the TPCF for two different simulation boxes at redshift three. The box investigated in the left panel contains three galaxies with circular velocities of order $v_c \sim 100\,\mathrm{km\,s^{-1}}$at redshift zero while the box investigated in the right panel contains one galaxy with $v_c \sim 200\,\mathrm{km\,s^{-1}}$. The velocities of the CIV systems were obtained by Voigt profile fits to significant continuum depressions. The absolute normalization is from the observed TPCF for three QSOs ($\langle z \rangle$=2.78) [12]. Similar to the observed TPCF, the correlation function of CIV lines in the simulated spectra exhibits a narrow peak at the origin (the lowest velocity bin is incomplete) and a long tail out to velocities of 500-600 $\mathrm{km\,s^{-1}}$ and beyond. The correlation strength is of order unity at about 400 $\mathrm{km\,s^{-1}}$.

## 4. Conclusions

The observed TPCF of CIV absorbers is about that expected from evolving the present-day galaxy TPCF backwards. This suggests that the bias factor $b_{\mathrm{abs-gal}}$ is not much different from unity. There is reasonable agreement between observed and simulated TPCF. The rather strong variation between different simulation boxes which differ mainly in the depth of the potential wells of the objects contained suggests, however, that the effects of finite size and peculiar velocity significantly affect the absorber TPCF on scales below 400 $\mathrm{km\,s^{-1}}$. Simulations of the kind discussed here allow to study these effects in detail. Once these effects are understood the TPCF of CIV

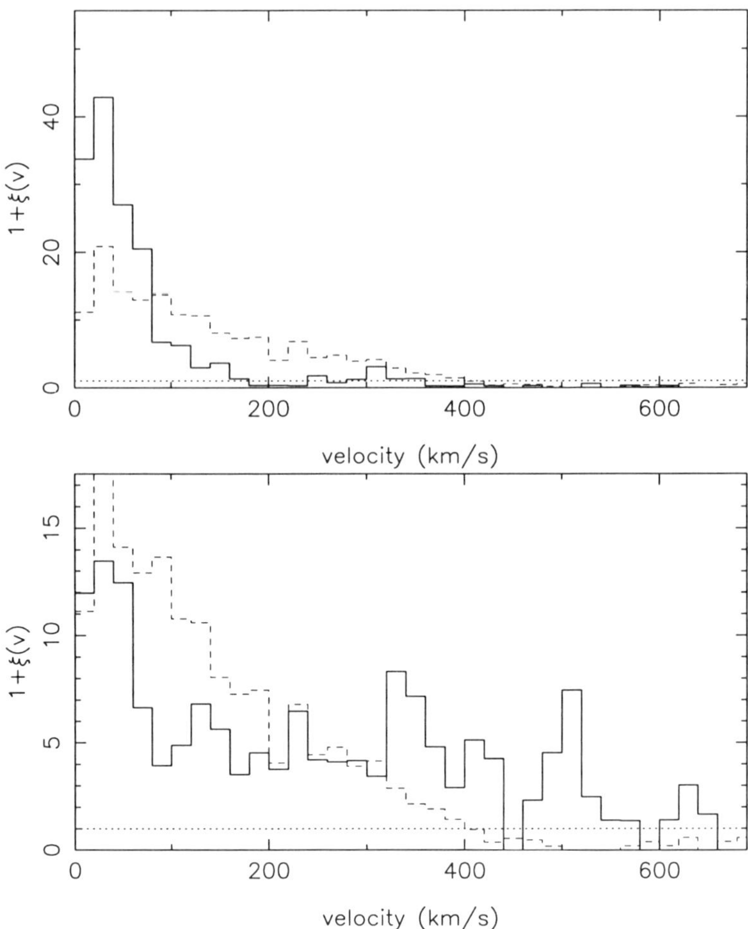


*Figure 1.* Upper panel: Two-point correlation function for the CIV systems detected in a simulation box at $z$=3.07 (to contain three galaxies with $v_c \sim 100\,\mathrm{km\,s^{-1}}$at $z = 0$). The observed TPCF is show by the dashed curve. Lower panel: Same but for a simulation box containing a single galaxy with $v_c \sim 200$ $\mathrm{km\,s^{-1}}$ by $z = 0$. Note the different scales of the axes.

absorbers should become a useful tool to study the redshift evolution of the clustering of galaxies and eventually also that of the mass.

## References


1. Bergeron J., Boissé P., 1991, A&A, 243, 344
2. Steidel C.C., 1995, in *QSO Absorption Lines*, Proc. ESO Workshop, ed. G.Meylan (Heidelberg: Springer), p. 139.
3. Bahcall J.N., Spitzer L., 1969, ApJ, 156, L63
4. Sargent W.L.W., Boksenberg A., Steidel C.C., 1988, ApJS, 68, 539

5. Petitjean P., Bergeron J.A., 1994, A&A, 283, 759
6. Aragon-Salamanca A., Ellis R.S., Schwartzenberg J.-M., Bergeron J.A., 1994, ApJ, 421, 27
7. Steinmetz M., 1996, MNRAS, 278, 1005
8. Ferland G.J., 1993, University of Kentucky Department of Physics and Astronomy Internal Report
9. Haehnelt M.G., Steinmetz M., Rauch M., 1996, ApJ, 465, L95
10. Navarro J., Steinmetz M., 1997, ApJ, submitted
11. Rauch M., Haehnelt M.G., Steinmetz M., 1997, ApJ, in press
12. Rauch M., Sargent W.L.W., Womble D.S., Barlow T.A., 1996, ApJ, 467, L5
13. Sargent W.L.W., Young P.J., Boksenberg A., Tytler D., 1980, ApJS, 68, 539
14. Rauch M., Carswell R.G., Chaffee F.H., Foltz C.B., Webb J.K., Weyman R.J. Bechthold J. Green R.G., 1992, ApJ, 390, 387
15. Cristiani S., D'Odorico S., Fontana A., Giallongo E., Savaglio S., 1995, MNRAS, 273, 1016
16. Cristiani S., D'Odorico S., Fontana A., Giallongo E., Savaglio S., 1997, MNRAS, 285, 209
17. Heisler J., Hogan C.J., White S.D.M.,1989, ApJ, 347, 52
18. Cen R., Miralda-Escudé J., Ostriker J.P., Rauch M., 1994, ApJ, 437, L9
19. Petitjean, P., Mücket, J.P., Kates, R.E., 1995, A&A, 295, L9
20. Hernquist L., Katz N., Weinberg D.H., Miralda-Escudé J., 1996, ApJ,457, L51
21. Miralda-Escudé, J., Cen, R., Ostriker, J.P., Rauch, M., 1996, ApJ, 471, 582
22. Dinshaw N., Impey, C.D., Foltz C.B., Weymann R.J., Chaffee F.H., 1994, ApJ, 437, L87
23. Bechtold J., Crotts A.P.S., Duncan R.S., Fang Y., 1994, ApJ, 437, L83

# LINES AND LENSES ON THE HUBBLE DEEP FIELD

R. D. BLANDFORD
*130-33 Caltech*
*Pasadena*
*CA91125*
*USA*

## 1. Introduction

In this contribution I describe progress on and preliminary results from a redshift survey on the Hubble Deep Field (HDF) that is being carried out using the Low Resolution Imaging Spectrograph on the Keck Telescope. The primary goal of this research (which involves the combined effort and cooperation of several research groups) is to assemble a uniform redshift sample for about 500 galaxies and use it to probe the spatial distribution, the birth rate and the evolution of field galaxies. There are various secondary goals, one of which - the investigation of strong and weak gravitational lensing - I will also discuss here.

## 2. The Hubble Deep Field

The Hubble Deep Field was a special initiative, supported under Director's discretionary time, in which a "blank" field was observed for thirty complete orbits in each of four carefully chosen filters with central wavelengths 300, 450, 606 and 814 nm (Williams *et al.* 1996). In total there are approximately 4 sq. arcminutes of high quality imaging area (plus $\sim$ 40 sq armin of "flanking" fields observed to a shallower depth on the three WF/PC2 chips. Over 2,500 faint galaxies can be counted down to $B \sim 30$, corresponding to just over $10^{11}$ sources on the sky. They are relatively blue and appear to be unusually compact relative to local galaxies. There have been several parallel observations from the ground, notably in the infrared (*e.g.* , Hogg *et al.* 1997) and radio (Kellermann *et al.* 1997) bands. Observations and study of the HDF have greatly enhanced our understanding of a variety of aspects of faint galaxies including especially the $z \gtrsim 3$ galaxy population (Steidel



*D. Hamilton (ed.), The Evolving Universe,* 321–332.

*et al.* 1996), evolution of the luminosity function, and the high incidence (roughly 40 percent) of morphologically "irregular" galaxies (Abraham *et al.* 1997).

## 3. Deep Redshift Survey

The HDF was chosen specifically to avoid known clusters, Galactic absorption (like high latitude cirrus) and quasars. As such, it ought to be repesentative of the field galaxy population. It offers an excellent opportunity to carry out a pencil beam redshift survey that will allow us to combine deep imaging and spectroscopy in unprecedented detail. This acquisition of redshifts has been carried out exclusively using the Keck 10 m telescopes by several cooperating teams (Cohen *et al.* 1996b, Phillips *et al.* 1997 and references therein). To date, over 300 redshifts have been acquired and the combined efforts have already produced a sample that is complete on the HDF to $R = 23$. This number should almost double by the end of the current observing season. One more season will be necessary to complete these observations to the practical limit imposed by the current capabilites of 10 m class, ground-based telescopes. The present sample selection is still quite heterogeneous, but we aspire to completeness in terms of observation to depths of $R \sim 24$, $K \sim 20$ on the HDF and $R \sim 23$, $K \sim 19$ within a circle of radius $\sim 4'$ centered on the HDF. (We remark in passing that absolute competeness, *per se* is no longer necessary provided that one's sampling is unbiased and the selection function is properly understood. Nonetheless, it turns out that near completeness is what gives this survey its unique diagnostic powers and distinguishes it from the extensive wide field surveys of, for example, Lilly *et al.* 1995)

I will concentrate upon the intermediate redshift sample with $z \lesssim 1.5$, which are selected preferentially with red or infrared photometry and will ignore the important high redshift galaxies that are identified using the "UV dropout" technique. For reference, our galaxies have a median redshift of $z \sim 0.6$ at which distance, $1'' \equiv 4h^{-1}$ kpc in an Einstein-De Sitter universe. If galaxies have halos with diameters $\gtrsim 100h^{-1}$ kpc, then they will overlap and each one will cover ten percent of a WF/PC2 chip, far more than their optical images. When this survey was initiated, it was far from clear how complete it would be. What has transpired is that it has proven easier than anticipated to acquire redshifts of bright galaxies with $z \lesssim 1.3$. This is because the [OII] emission line $\lambda 3727$ is far more common in the past than now, evidence for a declining rate of star formation. The concern that this raises is that many of the redshifts for fainter galaxies are really based upon only one sure feature. As far as can be seen from scrutiny of the derived 4000Å break and Balmer lines, there is internal consistency in this identification, but it is still possible that a small minority of the redshifts presented are incorrect.

As the data are still being analyzed, it would be misleading to present a partial redshift list here. Suffice it to say that that when plotted in a plot of the rest luminosity in the $B$ band, scaled by the local value of $L_B^*$, against $V$, the comoving volume surveyed out to the measured galaxy redshift, the observed galaxies seem to be drawn from a total luminosity function that is fairly similar to what is observed locally. In particular, there are relatively few galaxies with $L_B \gtrsim 4L_B^*$ consistent with passive stellar evolution. For a Schechter luminosity function, with $\alpha \sim -1$, we expect the accessible region on this plot with $L_B \lesssim 0.3L_B^*$ to be uniformly populated; the current sample shows a mild relative excess of low luminosity galaxies, as is expected from the integrated source counts on the HDF and other studies. An extensive maximum likelihood analysis is underway (Hogg 1997, in preparation).

## 4. Redshift Peaks and Voids

One feature of the survey that is less sensitive to incompleteness and spectroscopic selection effects is the fine scale redshift distribution. Having determined that there are not dramatic, evolutionary changes in the luminosity function, the overall redshift distribution is prescribed. Furthermore, we know that local galaxies are clustered and that this non-uniformity can be described by a two-point correlation function with a correlation length $\xi_0 \sim 8h^{-1}$ Mpc, much larger than the size of even the flanking fields. What the two-point correlation function does not and cannot tell us about is the anisotropy in this clustering. We already have strong visual indications that this is very important on both larger and smaller scales. From the work of, *e.g.* , de Lapparent, Geller & Huchra 1986, Landy *et al.* 1996, we have a strong visual impression that galaxies, in general, are preferentially located on surfaces that bound giant voids of size $\sim 50 - 100h^{-1}$ Mpc. Is also clear that these surfaces are not uniformly dense in galaxies. Filaments and nodes are clearly discernible as superclusters and rich clusters and the majority of the remaining objects are located in "groups" some of which are quite compact (*e.g.* , Hickson 1993). The shapes, sizes and connectivity of these features are still not well-determined. (I should emphasize that it is very important to explore these issues empirically, independent of theoretical cosmogony, not because there is necessarily anything wrong with the current suite of impressive CDM simulations, *e.g.* , White, Carlberg, these proceedings, but simply because, historically, extragalactic astronomy has always surprized us and useful, quantitive measures of large-scale structure are quite elusive and are just as likely to be suggested by direct observations as simulations.)

Pencil beam surveys can attack these aspects of the problem in three separate ways. The first is to determine the average, filling factors of the galaxy-forming surfaces. A single pencil beam penetrates $\sim 1.7h^{-1}$ Gpc of comoving distance out to $z \sim 1$ equivalent to $\sim 10 - 20$ real or incipient voids, which is enough for simple averaging. Secondly, we can explore the lateral coherence of the sheets at intermediate redshifts by drilling auxiliary pencils at angular offsets of up to several degrees. In this way it should be possible to distinguish galaxy distributions where the $\sim$ 100 Mpc structure is a relatively insignificant manifestation of large-scale power from cosmologies where the $\sim$ 100 Mpc scale is an intrinsic, outer scale imprinted on the expanding universe at the time of recombination and, perhaps, associated with the acoustic peak. Finally, and this is really all that can be discussed now, we can probe the small scale structure of the redshift features (Cohen *et al.* 1996ab). A cautionary word on pencil beam surveys is perhaps in order. As has been demonstrated, (*e.g.* , Broadhurst *et al.* 1990, Feldman, Kaiser & Peacock 1993), it is possible to be seriously misled by the analysis of a single pencil beam unless this is done in the context of a secure understanding of the overall two-point correlation function derived from larger-scale surveys.

On the basis of the current redshift sample, it is apparent that roughly 40 percent of bright galaxies are concentrated within $\sim$ 5 significant redshift featuress. We assess the significance of these features using the following procedure:

1. Smooth the observed galaxy redshift to obtain an average galaxy distribution function.
2. Construct histograms of the observed redshift distribution assuming all possible bin widths and phases. We use $\ln(1+z)$ as the distance measure as this corresponds to uniform local, radial velocity differences.
3. Identify those choices exhibiting unusually tall bins.
4. Repeat this procedure with the simulated distribution many times and determine how often features like those under investigation recur. This frequency is an estimate of the *a posteriori* probability of the features and a measure of their relative significance.

As a somewhat arbitrary rule, a relative frequency of one percent can be adopted as a threshold for significance of a feature.

If we now inspect the individual redshift features, we can measure the local velocity dispersions and find that they range from 170 - 560 km $s^{-1}$. These are surprisingly small. (Note how important it is to measure individual redshifts with accuracies $\delta z \sim 0.001$ as is done here but is not always the case in larger-scale redshift surveys.) The next task is to explore the local angular correlation. The galaxies in individual features are plotted and the hypothesis that they are drawn from a uniform distribution on the sky tested using a two dimensional Kolmogorov-Smirnov test (*cf.* Peacock 1983). What is found is that this hypothesis can be rejected for some though not all features. This is not unexpected on the basis of there being a correlation length at redshift $z \sim 0.5 - 1$ somewhat smaller than the local value $\xi_0$. However, any quantitative deductions will need significantly more data.

There seem to be two complementary, dynamical interpretations of these redshift features. The first, and simplest, is that we are looking at locally self-gravitating, virialized groups where there has been time for the galaxies to cross the group many times and for the velocity distribution to isotropize. This is not incompatible with these groups lying within large sheet-like structures; indeed, we anticpate that sheets should become gravitationally unstable in just this fashion. Alternatively, we may be looking at an earlier, linear stage in the development of structure when the flow represents a breaking wave. If we just consider a single, one-dimensional wave mode, using the Zel'dovich approximation, we expect the near side of a density enhancement to recede with a speed in excess of the Hubble velocity appropriate to its location and for the motion of the far side to be retarded. Galaxies occupying a relatively large volume in configuration space can then be quite overdense in velocity space. As the perturbation grows, the flow can actually reverse so that there is a high velocity maximum on the near side and a low redshift minimum on the far side. Indeed, in special cases, the redshift distribution within an individual feature can become bimodal. The next stage finds shells crossing and this probably marks the onset of violent relaxation and randomization of the peculiar velocities. At present there seem to be no direct observational grounds for preferring one of these two limiting descriptions or, quite reasonably, some intermediate model. However, it is quite possible that future observations may provide just this discrimination.

## 5. Strong Gravitational Lensing

Theoretical and observational studies have converged upon a rough rule of thumb, namely that the incidence of multiple imaging of cosmologically distant sources by intervening galaxies is about 0.002 *per source* with redshift $z_s \gtrsim 1.5$ (*e.g.* , Myers 1996). It appears that individual deflector galaxies dominate the cross section, (although we are not yet completely confident that exotic lenses, like intergalactic massive black holes or topological defects, are unimportant.) Assuming that the number of source galaxies is $\sim 2500$, this tells us that there should be very roughly $2500 \times 0.002 = 5$ strong galaxies on the HDF. However, in order to compute more accurately the observationally more interesting sky density of lenses, we must allow for the fact that there is "no multiplication without magnification". In a flux-limited survey, strongly lensed sources are intrinsically fainter than the single sources with which they are being compared and, consequently, they are more numerous by a factor depends upon the slope of the source counts at the flux limit of the sample. This adds a factor $\sim 2$. The observed incidence of strong lensing of high redshift sources also depends upon the redshift distribution of the deflectors. It is dominated by galaxies with redshifts $z_d \sim 0.5$ but is fairly insensitive to the world model. Obviously, sources with redshifts $z_s \lesssim z_d$ will not be multiply imaged. We can use the incidence of strong lensing, at least in principle, to probe the source redshift distribution.

Armed with this realization, Hogg *et al.* (1996) scrutinized the HDF image, soon after it was released, looking for the most promising candidates on morphological grounds. Specifically, they sought image pairs and quads with similar colors arranged with respect to a brighter galaxy in a manner consistent with standard lensing geometries. They combined visual searches with semi-automatic routines similar to those used for the CLASS survey of radio sources. They produced a ranked list of roughly twelve candidates deemed suitable for further study. Most of these are really too faint for spectroscopy, but a subsample of the most promising and brightest cases has been investigated.

### 5.1. INDIVIDUAL CASES

#### 5.1.1. *J123652+621227*

This source was the most intriguing candidate of Hogg *et al.* (1996). The proposed lens is a $I = 24$ red elliptical galaxy. It is surrounded by 16 galaxies within 5" that can be partitioned into diametrically opposed pairs of similar color plus a tangential arc. However, the separation of the images was much larger than would have been expected for a normal elliptical galaxy at the redshift $z \sim 1$ indicated by the galaxy color. The most suggestive putative lens feature in this object is the $V = 26.1$ arc, for which a short, faint tentative $V = 27.4$ "counter-arc" has been identified. Now, Zepf *et al.* (1997) have carried out spectroscopy using a slit covering these two objects and argue that the spectra are sufficiently dissimilar that they are unlikely to come from the same source. Of course, this does not rule out the possibility that the arc is "naked" (where the cusp associated with the tangential caustic protudes beyond the radial caustic so that there is actually no counter-image formed.) The evidence now points to J123652+621227 not being a strong gravitational lens.

### 5.1.2. *J123656+621221*

In this case, the proposed lens is a $I$=24 red elliptical galaxy and the image is a single, blue tangential arc located only 0.8" from the galaxy. If it can be demonstrated that the blue arc has a substantially larger redshift than the elliptical then multiple imaging is almost inevitable. Cohen (private communication) has measured a tentative redshift of $z = 0.483$ for the elliptical. More integration is needed before this value can be treated as secure. Even less confidence can be attached to the best guess redshift for the arc of Zepf *et al.* (1997) which is $z = 1.02$. However, if it can be confirmed then this would be a *bona fide* gravitational lens.

### 5.1.3. *J123649+621322*

In this case, we have a $R = 24$ blue galaxy accompanied by a single red arc. Spectroscopy by Cohen (private communication) has furnished two candidate redshifts for the blue galaxy. However, even if one of these is confirmed, it will be necessary to obtain a redshift for the arc as well in order to claim this as a secure gravitational lens. This is a challenging observation.

In summary, there is, so far, no secure gravitational lens on the HDF whereas there ought to be $\sim$ 10. It is probably going to be necessary to scrutinize the images of potential deflector galaxies with much more care to find them. If we continue to fail to find these strong lenses, then we may have to conclude that the redshift distribution of the faint source population is heavily concentrated to low redshift despite good evidence to the contrary from studies of clusters (Kneib *et al.* 1996) and weak lensing (see below).

## 6. Weak Lensing by Large-Scale Structure

In recent years, there has been a resurgence of interest in weak lensing, where the deflector galaxies just make small distortions to the shapes of the source galaxies. These distortions are undetectable in individual galaxies because we do not know their intrinsic shapes, but they can be measured statistically at the level of a few percent. There are two separate contributors to this distortion, large-scale structure and the closest galaxies along the line of sight, specifically those whose halos cover the source. The HDF provides an ideal testbed for these effects because the galaxy images are of far higher quality than can routinely be obtained in much larger surveys using ground-based telescopes. It can therefore be used to calibrate the imaging of these telescopes and quantify the degradation caused by atmospheric distortion, sky noise, and telescope optics.

As light propagates through a Newtonian gravitational potential, $\phi$, its path will be deflected, through an angle $\sim \phi/c^2$. If this potential measures the characteristic fluctuation on some scale $L$, then the effect of these fluctuations will add stochastically, so that the total deflection will be $\sim N^{1/2}\phi/c^2$, where $N \sim H_0L/c$ is the number of fluctuations along the line of sight. Cosmological fluctuations have a strength $\phi \sim 3 \times 10^{-5}c^2$ independent of lengthscale and epoch for $L \gtrsim 10h^{-1}$ Mpc. This implies that $N \sim 300$ and individual rays are deflected through an angle $5 \times 10^{-4} \sim 2'$ - about the size of the HDF. However, this is quite undetectable. We can only measure differential effects and the induced ellipticity in an individual image is, to order of magnitude, $\epsilon \sim N^{3/2}\phi/c^2 \sim 0.1$. (A more careful calculation reduces this estimate by a factor $\sim 3$, although it could be larger if there is more power in the potential fluctuations than conventionally assumed on scales $L \lesssim 10h^{-1}$ Mpc.) This will be made manifest as a mean polarization of the images of all the $z \gtrsim 1$ galaxies. To measure this effect with a signal to noise ratio of unity requires observing $\sim (\epsilon/\epsilon_0)^{-2} \sim 3000$ images (where $\epsilon_0 \sim 0.4$ is the *rms* intrinsic ellipticity, *cf.* , Blandford *et al.* 1994). The HDF with no more than $\sim 2000$ high redshift images ought not to exhibit any correlated signal.

In order to test this, we prepared a galaxy catalog soon after the HDF images were generally released. (This catalog agrees in all essentials with the official catalog subsequently released by Williams *et al.* 1996.) It contained just over 2,500 galaxies within an area of $\sim 4$ sq. arcmins. These sources all have at least five edgewise continguous pixels in excess of $2.5\sigma$ in the F606W filter. Positions and colors are measured in all four filters. Second moments were measured weighting by the square root of the F606W intensity and used to compute the ellipticity, though the results were not sensitive to this choice. In carrying out weak gravitational lensing studies, it is necessary to take precautions to eliminate systematic errors. For example, in the case of weak lensing by large-scale structure, trailing of the images can produce a spurious signal. Fortunately, there are sufficient stars present that it is possible to remove this distortion across the field.

In a search for polarization due to large-scale structure, we measured the mean polarization in circles of radius 35" centered on each of the three chips. We found that there was no statistically significant mean polarization in these fields, taken collectively or singly, consistent with our expectation (*cf.* Mould *et al.* 1996). The $2\sigma$ upper limit was a polarization of 0.03.

## 7. Galaxy-Galaxy Lensing

As first discussed by Webster (1983) and Tyson *et al.* (1984), individual lens galaxies at modest redshift, ($z_d \sim 0.5$), are expected to produce a weak tangential stretching of the images of background source galaxies of significantly greater redshift, ($z_s \sim 1-3$). If we make a very simple model of the potential of the lens galaxies, (presumably dominated by dark matter), as a truncated isothermal sphere then, in round numbers, the Einstein ring subtends an angular radius $\theta_c \sim 1''$ whereas the halo is expected to extend out to $\theta_H \sim 10''$. The cross section for weak lensing by the halo is therefore roughly a hundred times that for strong lensing by the central regions of the galaxy. If we measure the halo density by its (roughly constant) circular velocity, then the expected ellipticity of a background galaxy separated by an angle $\theta$ from the galaxy center is

$$\epsilon \sim \frac{2\pi V_c^2}{\theta c^2}; \qquad \theta < \theta_H \tag{1}$$

$$\sim \frac{2\pi \theta_H V_c^2}{\theta^2 c^2}. \tag{2}$$

If we now concentrate on sources with impact parameters close to the halo radii of intervening galaxies, we find that the expected tangential ellipticity will be $\epsilon \sim 0.1$. This will be contributed by all the lens galaxies within $\theta \sim \theta_H$ and, to a pretty good approximation, the signals will add linearly. The noise in this measurement is the intrinsic galaxy ellipticity which has a value $\sim 0.3$. Now the number of lens galaxies within angle $\theta$ of a given source increases $\propto \theta^2$ whereas the signal decreases $\propto \theta^{-1}$ for $\theta < \theta_H$ and $\propto \theta^{-2}$ for $\theta > \theta_H$. Therefore the signal to noise ratio is maximized if we include all lenses within a few $\theta_H$ of the source and ignore those beyond $\theta_H$.

A positive measurement of this effect has previously been reported by Brainerd, Blandford & Smail(1996) who analyzed a $9.6' \times 9.6'$ red image with average seeing $\sim 0.8''$, taken with the COSMIC imager on the 5m Hale reflector. The detected galaxies were divided into 511 sources with $23 < r < 24$ and 439 defelectors with $r < 23$. Source-deflector separations in the range $5'' < \theta < 34''$ were included. A polarization signal $\epsilon \sim 0.01$ was measured which is consistent with a simple model of the deflector galaxies in which the circular velocity of a $L^*$ galaxy is $\sim 220$ km s$^{-1}$ and the physical radius of the halo of an $L^*$ galaxy is $\sim 50h^{-1}$ kpc. (This last quantity is not well constrained by the measurement.) In addition, Griffiths *et al.* (1996) reported detecting this field using HST images drawn from the MDS survey. They were able to distinguish ellipticals and spirals and found that the effect was much stronger for the former. Turning to the HDF itself. Dell Antonio & Tyson(1996) reported a detection of galaxy-galaxy lensing for pairs of galaxies with $\theta < 5''$, consistent with an $L^*$ circular velocity of $\sim 260$ km s$^{-1}$.

Each of these studies considered sources and deflectors in a blind, pair-wise manner. In practices this dilutes the signal by weighting all pairs equally including those where the "source" redshift is less than that of the deflector. It is apparent that if we have additional information, in particular the redshift distribution of the deflector galaxies, then we should be able to measure a much larger signal to noise ratio using from a given set of imaging data. In addition, these studies assumed a simple form of the source redshift distribution and focused on describing the average mass distribution in the deflector galaxies. In the present study, we restrict our deflector population to those galaxies with known redshifts, and, consequently, well-determined luminosities.

The deflector field comprises over 200 galaxies of which 70 are on the HDF and the remainder are on the flanking area. This ensures that the average distribution of deflector galaxies about the source galaxies is isotropic on the sky. This removes a potentially serious source of systematic error. A subset of the galaxies have nine wavelength photometry and most of the remainder have six wavelength photometry. We can, therefore, estimate their rest frame $B$ magnitudes and thereby obtain estimate of $L_B/L_B^*$. We then adopt the deflector galaxy model of Brainerd *et al.* (1996) in which the circular velocity and halo radius are given by

$$V_c = 220(L/L^*)^{1/4}\mathrm{kms}^{-1} \quad (3)$$

$$r_H = 50h^{-1}(L/L^*)^{1/2}\mathrm{kpc}. \quad (4)$$

We also adopt a specific world model, an Einstein-de Sitter cosmology with age 13 Gyr, although our results turn out to be pretty insensitive to this choice. Using this model, it is possible to compute the polarization field across the whole of the HDF as a function of source redshift $z_s$, $p(X, Y, z_s)$, where $X, Y$ are angular coordinates on the sky.

There are two potentially serious sources of systematic error in galaxy-galaxy lensing. The first is that when the source is too close to the deflector, the isophotes can overlap so that galaxies will appear to be elongated radially. The second is that some designated source galaxies will actually be companions of the deflector and be subject to Newtonian tidal forces that may stretch them out in a tail stretched out along the orbit. Both of these effects can be minimized by excluding source-deflector pairs with $\theta < 5''$, which we do.

The total source sample that we use comprises $\sim$ 1000 galaxies each covering 16 or more pixels. The magnitude distribution extends down to $V \sim 29$ and the typical galaxy ellipticity if $\epsilon \sim 0.2$. For each galaxy, we compute the complex orientation, $\chi$ and then correlate with the predicted polarization field for different assumptions about the source redshift distribution. The complex correlation coefficient $C =< p\chi^* >$ can be formed, averaging over the whole source sample. Its expectation is real, the imaginary part measuring the correlation of the source images with the polarization field after rotation through $\pi/4$. For the limiting case when all the source galaxies are assumed to lie at large redshift, we measure $C = 0.0008 + 0.0000i$. The expected error in $C$ is $\sim 0.0003$ and so this is a $\sim 3\sigma$ detection of the polarization signal. However, it is only a third of the predicted value. On this basis, we suggest that perhaps half of the source galaxies have redshifts $z_s < 1$. A fuller discussion of this analysis and related matters will be presented in Blandford *et al.* 1997 (in preparation).

## 8. Discussion

Perhaps the major role of gravitational lensing studies on the HDF is to help understand the empirical redshift distribution of the faintest galaxies. Indeed, this is one of the few ways one can imagine solving this problem. To date the results are consistent with a larger fraction of them beign at redshifts $z_d \gtrsim 1.5$, though this conclusion is far from compelling. Although many more individual strong lensing candidates remain to be studied, it is already apparent that the frequency of lensing is less than anticipated on the presumption that the all of these faint galaxies have large redshifts, $z_s \sim 3$. Similarly, the measured strength of the weak lensing signal is substantially smaller than predicted under the same hypothesis. It is clear that, in order to advance, a much larger sample of faint galaxies is required and we must probably turn to deep imaging photometry surveys from the ground for this. However the HDF sample still provides the best means of calibrating these surveys. It is, therefore, of considerable importance to observe this same field from the ground to comparable depth under conditions of excellent seeing before embarking on larger-scale surveys.

## Acknowledgements

I thank Tereasa Brainerd, Chris Fassnacht, David Hogg, Tomislav Kundić, Sangeeta Malhotra, Mike Pahre and especially Judith Cohen, who has led the Caltech portion of the redshift survey discussed here, for their collaboration. I thank Don Hamilton for financial assistance to attend this meeting and his patience. Support by STScI/NASA grant AR-06337.12-94A and NSF grant AST 95-29170 is also greatly appreciated.

## References

Blandford, R. D., Saust, A., Brainerd, T. G. & Villumsen, J. V. 1991 MNRAS 251 600
Brainerd, T. G., Blandford, R. D. & Smail, I. R. 1996 ApJ 466 623
Broadhurst, T. J., Ellis, R. S., Koo, D. C. & Szalay, A. S. 1990 Nature 343 722
Cohen, J. G., Hogg, D. W., Pahre, M. A., Blandford, R. D. 1996a ApJ 462 L9
Cohen, J. G. *et al.* 1996b ApJ 471 L5
de Lapparent, V., Geller, M. & Huchra, J. P. 1996 ApJ 302 L1

Dell Antonio, I. P. & Tyson, J. A. 1996 ApJ 43 L17
Feldman, H., Kaiser, N. & Peacock, J. A. 1993 ApJ 426 239
Fomalont, E. B., Kellermann, K. I., Richard, E. A. & Partridge, R. B.
Griffiths, R. E., Casertano, S., Im, M. & Ratnatunga, K. U. 1996 MNRAS 282 1159
Hickson, P. 1993 Ap Lett 29 1
Hogg, D. W., Blandford, R. D., Kundíc, T., Fassnacht, C. D. & Malhotra, R. 1996 ApJ 467 L73
Hogg, D. W. *et al.* 1997 AJ 113 474
Kneib, J.-P., Ellis, R. S., Smail, I. R., Couch, W. J., & Sharples, R. M. 1996 ApJ 471 643
Landy, S. J., Shectman, S., Lin, H., Kirshner, R. P., Oemler, A., & Tucker, D. 1996 ApJ 456 L1
Lilly, S. J., Tresse, L., Hammer, F., Crampton, D. & Le Févre, O. 1995 ApJ 455 108
Mould, J. R., Blandford, R. D., Villumsen, J. V. Brainerd, T. G. & Smail, I. R. 1994 MNRAS 271 31
Myers, S. T. 1996 *Astrophysical Applications of Gravitational Lensing* ed. C. S. Kochanek & J. N. Hewitt Kluwer:Dordrecht
Peacock, J. A. 1983 MNRAS 202 615
Phillips, A. C. *et al.* 1997 ApJ (in press)
Steidel, C. C. 1996 ApJ 462 L17
Tyson, J. A., Valdes, F., Jarvis, J. F. & Mill, J. R. 1984 281 L59
Webster, R. L. 1983 unpublished thesis, University of Cambridge
Williams, R. E. *et al.* 1996 AJ112 1335
Zepf, S. E., Moustakis, L. A. & Davis, M. 1997 ApJ 474 L1

# INTERPRETING HIGH-REDSHIFT GALAXIES IN THE HUBBLE DEEP FIELD

RACHEL S. SOMERVILLE, J. R. PRIMACK, AND S.M. FABER
*University of California, Santa Cruz*

**Abstract.**
Recent observations of "Lyman-break" galaxies in the Hubble Deep Field imply surprisingly large comoving number densities of bright galaxies at redshifts of $z \sim 3$ — up to several times the density of $L \geq L_*$ galaxies today (Lowenthal *et al.* 1997). We present the predictions of semi-analytic hierarchical galaxy formation models for the comoving number densities of bright galaxies at high redshift in a variety of CDM-based cosmologies. Our results suggest that a significant fraction of the Lyman-break galaxies may be low-mass starbursts, rather than the massive progenitors of present day bright galaxies. If this scenario is correct, we predict that the NICMOS observations of the Lyman-break galaxies will yield a broad distribution of $(I - K$ colors.

The Hubble Deep Field (HDF) (Williams *et al.* 1996) has provided us with an unprecedented window onto the high redshift Universe. The techniques developed by Steidel *et al.* (1996b (S96b)) allow high redshift candidates to be selected from the HDF using broadband colors ("Lyman-break" or "U/B dropout" galaxies). It has now been confirmed spectroscopically that these galaxies are in fact at or consistent with being at high redshift ($\bar{z} \sim 3$) (Steidel *et al.* 1996a (S96a); (S96b); Lowenthal *et al.* 1997 (L97)). These observations potentially provide a very strong constraint on cosmological models and theories of galaxy formation. However, any comparison with theory is difficult because we do not know the masses or virial velocities of the galaxies.

The semi-analytic approach to modeling galaxy formation (Kauffmann, White & Guiderdoni 1993 (KWG93); Cole *et al.* 1994) provides a means of testing the predictions of a given galaxy formation scenario within the context of specific cosmological models. The models discussed here are sim-



*D. Hamilton (ed.), The Evolving Universe,* 333–337.

ilar to those presented in KWG93 and developed in subsequent papers. We have developed our own code independently and tested our results against the output of the Kauffmann code. The details of our models will be given in Somerville (1997), Somerville, Primack & Faber (1997 (SPF97)), and Somerville & Primack (1997). A brief description of the approach follows.

Using Press-Schechter theory we construct a Monte Carlo realization of the merging histories of the dark matter halos. We then estimate the amount of gas that can collapse and cool in each time step, assuming a spherical infall model and radiative cooling (White & Frenk 1991). The star formation rate and the rate of gas reheating by supernova feedback are modeled by simple formulae depending on the circular velocity ($V_c$) of the halo and the dynamical time of the disk. In this way we keep track of the mass of stars, cold gas, and hot gas at each timestep as new gas cools, new stars are formed, and cold gas is reheated and returned to the halo. Assuming an initial mass function (IMF), the masses and ages of stars are convolved with the stellar population models of Bruzual & Charlot (GISSEL95: 1993; Charlot, Worthey & Bressan 1996) to obtain luminosities and colors in any desired bands. The free parameters in the models are set by requiring that a halo with circular velocity $V_c = 220$ km/s should *on average* form a central galaxy with the luminosity ($M_B \sim -20$) and cold gas mass ($\sim 10^{10} M_{\odot}$) of the Milky Way, and a satellite distribution similar to that of the Local Group. For the calculation shown here, we have assumed a Miller-Scalo IMF (Miller & Scalo 1979).

| Model | $t_0$ | $h$ | $\Omega_0$ | $\Omega_b$ | $\Omega_\nu$ | $\Omega_\Lambda$ | $n$ | $N_\nu$ | $\sigma_8$ |
|---|---|---|---|---|---|---|---|---|---|
| SCDM | 13.0 | 0.50 | 1.0 | 0.0 | 0.0 | 0.0 | 1.0 | 0 | 0.667 |
| TCDM | 14.5 | 0.45 | 1.0 | 0.1 | 0.0 | 0.0 | 0.9 | 0 | 0.732 |
| CHDM-2$\nu$ | 13.0 | 0.50 | 1.0 | 0.1 | 0.2 | 0.0 | 1.0 | 2 | 0.719 |
| OCDM | 14.7 | 0.60 | 0.5 | 0.069 | 0.0 | 0.0 | 1.0 | 0 | 0.773 |
| t$\Lambda$CDM | 14.5 | 0.60 | 0.4 | 0.035 | 0.0 | 0.6 | 0.9 | 0 | 0.878 |

TABLE 1. Model Cosmological Parameters

Figure 1 shows the comoving number density of galaxies as a function of redshift. The model cosmological parameters are given in Table 1. All the cosmologies are COBE normalized (except SCDM) and have been chosen to give the observed number densty of clusters (see Gross *et al.* 1997). One can immediately see that none of the models does very well at reproducing the *relative numbers* of high and low redshift galaxies, especially if the higher value of L97 is correct. With current normalizations, the $\Omega = 1$ models tend to overproduce galaxies at $z = 0$, while the $\Omega < 1$ models underproduce Lyman-break galaxies.

It has been suggested in several papers (S96a,b; Giavalisco et al. 1996) that the Lyman-break galaxies are the *massive* progenitors of today's $L \geq L_*$ galaxies. The large equivalent widths of saturated interstellar UV absorption lines are tentatively interpreted as implying virial velocities on the order of 250 km/s or higher. In this scenario, the galaxies are formed by the gradual funneling of gas into the deep potential well at the center of a massive dark matter halo, fueling star formation at a moderate rate over a time scale on the order of Gyrs. The implication is that each Lyman-break galaxy would evolve into a bright $L \geq L_*$ galaxy by the present day.

The galaxies that we have identified in the models as corresponding to the Lyman-break galaxies have properties that are consistent with the "massive progenitor" scenario. The "Lyman-break" galaxies in our models are in halos with circular velocities between 240 to 400 km/s and have moderate star formation rates (2-10 $M_{\odot}\mathrm{yr}^{-1}$). There is a strong correlation between the "Lyman-break" galaxies and the $L \geq L_*$ galaxies at $z = 0$ in the models. We think it is likely that *some* of the observed Lyman-break galaxies do fit this description. However, the models seem to be telling us that if *all* of the Lyman-break galaxies were to evolve into bright galaxies by $z = 0$, we would observe far more $L \geq L_*$ galaxies today. This is not surprising, since in the "massive progenitor" scenario, the Lyman-break galaxies are located in the centers of deep potential wells and will continue to accrete gas, form stars, and grow brighter for quite some time, even in a low-$\Omega$ cosmology. Therefore, in light of the higher number densities of Lyman-break galaxies suggested by the recent observations, we think it likely that a significant fraction of the Lyman-break galaxies must be different sorts of objects.

An alternative possibility is that some of the observed Lyman-break galaxies are actually lower-mass ($\sim 10^{10} M_{\odot}$) galaxies in the process of an intense starburst. The star formation rates derived from the $UV$ flux (3-9 $M_{\odot}\mathrm{yr}^{-1}$) would then imply mass-to-light ratios of about $M/L \sim 0.1$, similar to what is observed in local starburst galaxies (Guzmán *et al.* 1996). It is plausible that galaxy collisions or interactions could trigger these sorts of bursts (Mihos & Hernquist 1994). From a simple mean free path argument (supported by the simulations of Makino & Hut (1996)), random collisions between galaxies orbiting within a common dark matter halo would be expected to occur at a rate of approximately several per Gyr in a typical Local Group sized halo at $z \sim 3$. These small galaxies do not have a significant supply of new gas and could then either fade and end up as faint dwarf galaxies, or merge to form the extended Population II spheroids of large galaxies (Trager *et al.* 1997). Contrary to the "massive progenitor" scenario, there would then be no simple relationship between the number density of the Lyman-break galaxies and that of bright galaxies today. The effects of these random collisions and the resulting starbursts have not been taken into account in the existing models, and we believe that this could substantially increase the number density of "Lyman-break" galaxies at high redshift without overproducing bright galaxies at low redshift.

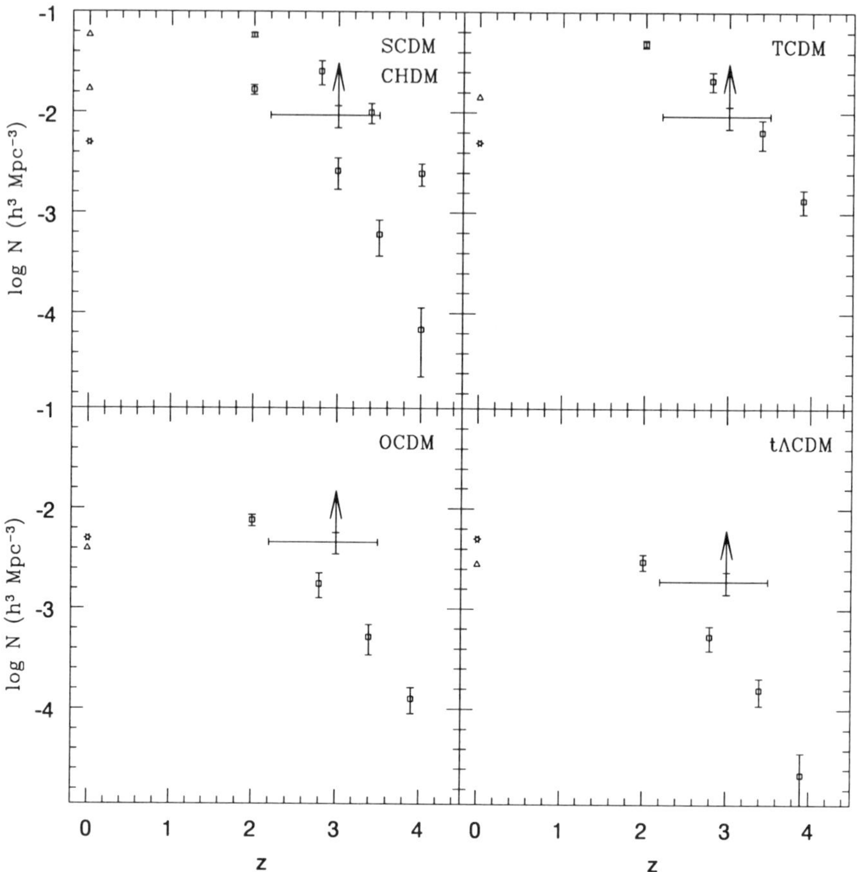


*Figure 1.* Comoving number density of galaxies as a function of redshift. The squares indicate the density of galaxies brighter than $(V_{606} + I_{814})/2 = 25.5$ (the criteria used in selecting the L97 sample) predicted by the models. The horizontal line indicates the number density implied by the L97 sample of galaxies with confirmed redshifts in the range indicated ($2.23 \leq z \leq 3.5$). The tip of the vertical arrow indicates the number density if all of the galaxies meeting the $U$ or $B$ dropout criteria (as defined by L97) are in fact in this redshift interval. The error bars on the model predictions reflect the scatter over ten different Monte Carlo realizations. The star at $z = 0$ is the comoving number density of galaxies brighter than $L_*$, obtained by integrating the luminosity function from the Las Campanas Redshift Survey (Lin *et al.* 1996). The triangle is the same quantity (in the $R$ band) for the models.

Deep near-IR images, such as those planned by the NICMOS team with HST, will probe the rest-frame visual part of the spectra. If our predictions are correct, these observations will yield a relatively broad distribution of $(I - K)$ colors, reflecting the fact that some of the Lyman-break galaxies (the ones that fit the "massive progenitor" interpretation) have a significant evolved stellar population whereas others (the low-mass starbursts) will have most of their light in the UV or B part of the spectrum. We give quantitative predictions for the distribution of $(I - K)$ colors from the collision-induced burst scenario described above in a forthcoming paper (SPF97).

There are many uncertainties in these calculations that need to be investigated. The IMF and extinction due to dust are substantial sources of uncertainty. Assuming a standard Scalo (1986) IMF instead of the Miller-Scalo IMF used here results in number densities that are a factor of 10 smaller at $z = 3$. This is because the Miller-Scalo IMF has more high mass stars, which are bright in the UV. There is currently no evidence that rules out an even more "top-heavy" IMF at high redshift or in different environments. Including the extinction due to dust results in a reduction in the $z \sim 3$ number densities in the models of about a factor of 2-3. Despite these uncertainties, the distribution of $(I - K)$ colors from NICMOS will help to clarify the interpretation of the high redshift galaxies.

## References

Bruzual, A.G., & Charlot, S. 1993, ApJ, 405, 538
Charlot, S., Worthey, G., & Bressan, A. 1996, ApJ, 457, 625
Cole, S., Aragón-Salamanca, A., Frenk, C.S., Navarro, J.F., & Zepf, S.E. 1994, MNRAS, 271, 781 (C94)
Giavalisco, M., Steidel, C.C., & Macchetto, D. 1996 ApJ, 470, 189 (G96)
Gross, M., Somerville, R.S., Primack, J.R., Holtzman, J., & Klypin, A. 1997, in prep.
Guzmán *et al.* 1996, ApJ, 460, L5
Kauffmann, G., White, S.D.M., & Guiderdoni, B. 1993, MNRAS, 264, 201
Lin *et al.* 1996, ApJ, 464, 60
Lowenthal, J.D., Koo, D.C., Guzmán, R., Gallego, J., Phillips, A.C., Faber, S.M., Vogt, N.P., & Illingworth, G.D. 1997, ApJ, in press (L97)
Makino, J., & Hut, P. 1996, astro-ph 9608159
Mihos, J.C., & Hernquist, L. 1994, ApJ, 425, L13
Miller, G.E. & Scalo J. M. 1979, ApJS, 41, 513
Scalo, J.M., 1986, Fundam. Cosmic Phys., 11, 1
Somerville, R.S., 1997, PhD dissertation
Somerville, R.S., Primack, J.R., & Faber, S.M., in prep.
Somerville, R.S. & Primack, J.R., in prep.
Steidel, C.C., Giavalisco, M., Dickinson, M., & Adelberger, K.L. 1996a, ApJ, in press
Steidel, C.C., Giavalisco, M., Pettini, M., Dickinson, M., & Adelberger, K.L. 1996b, ApJ, 462, 17L (S96b)
Trager, S.C., Faber, S.M., Dressler, A. & Oemler, A. 1997, ApJ, in press
White, S.D.M., & Frenk, C.S. 1991, ApJ, 379, 52
Williams, R.E. *et al.* 1996, AJ, 112, 1335

# THE 3-POINT CORRELATION FUNCTION OF THE COSMIC MATTER DISTRIBUTION

G. BÖRNER
*MPI für Astrophysik,*
*Karl-Schwarzschild-Str. 1, 85740 Garching, Germany*

AND

Y. JING
*MPI für Astrophysik,*
*Karl-Schwarzschild-Str. 1, 85740 Garching, Germany*
*Present address: RESCEV, Univ. of Tokyo, Bunkyo-ku, Tokyo*

## 1. Introduction

The most widely used statistical measure of the clustering properties of galaxies is the two-point correlation function $\xi$. Observations give a simple power-law from $\xi(r) \propto r^{-1.8}$ for a wide range of separations $r$ of the galaxy pair, 0.1 Mpc $\leq r <$ 10 Mpc. Theoretical models can reproduce the power law shape, but the amplitude of $\xi$ needs some readjustment to agree with the observational results. This discrepancy is ascribed to a "bias": the dark matter concentrations of the models are distributed differently from the real galaxies. Further investigation of such concepts must make use of higher-order correlation functions, since $\xi(r)$ quantifies only some gross features of the clustering. Determination of the correlation functions of higher order requires much better observational data, than those needed for the 2-point correlation function. The large survyes of galaxies, such as the LCRS and Sloan survey which have become available, respectively, will be available in the near future, will produce data of sufficient precision. In this contribution we want to discuss some properties of the 3-point correlation functions, as they appear in numerical and analytic calculations, and in the data.

The 3-point correlation function can be defined as the third central moment for the distribution of objects considered. Approximation of the distribution by a continuous function $\varrho(\vec{x})$ gives

$$\zeta(r, s, |\vec{r} - \vec{s}|) = \langle [\varrho(\vec{x} + \vec{r}) - \langle\varrho\rangle][\varrho(\vec{x} + \vec{s}) - \langle\varrho\rangle][\varrho(\vec{x}) - \langle\varrho\rangle] \rangle \langle\varrho\rangle^{-3} \quad .$$



*D. Hamilton (ed.), The Evolving Universe,* 339–342.

This immediately leads to

$$\langle \varrho(\vec{x}+\vec{r})\, \varrho(\vec{x}+\vec{s})\, \varrho(\vec{x})\rangle = n^3[1+\xi(r)+\xi(s)+\xi(|\vec{r}-\vec{s}|)+\zeta\rangle \quad ,$$

with $\langle \varrho \rangle = n$, the mean number density of objects (Peebles 1980). The 3-point correlation function thus contains additional information about the distribution of galaxies, clusters, or dark matter halos. In addition it is an important quantity, when we want to describe the dynamics of the system by the behaviour of the joint distributions in position and peculiar velocity. The BBGKY hierarchy, often used in this approach (Peebles 1980), is a hierarchy of equations, where the first equation describes how the probability distribution in the peculiar velocity of a randomly chosen particle is affected by the perturbations of neighbours; this is a relation between the one- and two-point correlation functions. The second equation treats the influence of perturbations in the abundance and motions of pairs of particles; this relates the two- and three-point correlation functions. To close the hierarchy at this stage, a model for the three-point correlation function must be employed. A possible ansatz is to express $\zeta$ in terms of two-point correlation functions:

$$\zeta(r,s,|\vec{r}-\vec{s}|) = Q[\xi(r)\,\xi(s)+\xi(s)\,\xi(|\vec{r}-\vec{s}|)+\xi(r)\,\xi(|\vec{r}-\vec{s}|)] \quad .$$

The popular "hierarchical" model assumes that $Q$ is a constant. We want to test this assumption, by determining $Q$ in various ways through the relation above.

## 2. Q from galaxy clusters in cosmological models

We have used large $P^3M$ N-body simulations to study the three-point correlation function of clusters in two models. The first model (LCDM) is a low-density , flat model with $\Omega_0 = 0.3$, $\Omega_\Lambda = 0.7$, $H_0 = 75$km s$^{-1}$ Mpc$^{-1}$, and $\sigma_8 = 1$; the Harrison-Zel'dovich form is taken for the primordial power spectrum. The second model has $\Omega_0 = 1$, $\sigma_8 = 0.8$, and a linear power spectrum extrapolated from observations (Jing, Börner, Valdarnini 1995). For both models $10^6$ particles are used in each simulation, and 8 realizations are run for each model. About $\lesssim 3200$ clusters are identified in each simulation.

We describe the sides of the triangle $(\vec{x}, \vec{x}+\vec{r}, \vec{x}+\vec{s})$ for $r < s < |\vec{r}-\vec{s}|$, by the variables
$r, u = \frac{s}{r}, \quad v = \frac{|\vec{r}-\vec{s}|-s}{r}$ (Peebles 1980); $u$ and $v$ characterize the shape, and $r$ the size of a triangle.

The function $Q(r,u,v)$ is determined from the simulations by evaluating $\xi(r, ru, ru+rv)$ and the hierarchical sum

$$\xi(r)\,\xi(ru)+\xi(ru)\,\xi(ru+rv)+\xi(ru+rv)\,\xi(r) \quad .$$

$Q$ depends only weakly on $r$ and $u$, but very strongly on $v$. $Q(r,u,v)$ is about 0.2 at $v = 0.1$ and 1.8 at $v = 0.9$, in clear contradiction to the hierarchical ansatz $Q =$ const. This dependence on triangle shapes can be approximated quite well by a function

$$Q(r,u,v) = A10^{1.3v^2} \quad ,$$

with $A \approx 0.14$. This fit works for both theoretical models considered. This $v$-dependence has not been found in the angular correlation functions of the Abell cluster catalogue. The reason may be that in a projected catalogue of clusters constructed from our simulations the $v$-dependence of the corresponding function is much weaker than for the $Q$ of the 3D case. It would be important to look for this variation in redshift samples of rich clusters.

## 3. Three-point correlation function in the quasilinear regime

A similar strong dependence of the normalized 3-point correlation function $Q$ on triangle shape has been found in the quasilinear regime, *i.e.*, for separations $r$ between $5h^{-1}$Mpc and $10h^{-1}$Mpc.

We have calculated the 3-point correlation function for realistic power spectra in a second-order Eulerian perturbation (SEPT) approach, and then also from numerical N-body simulations (Jing & Börner, 1997). We did this for two cold dark matter models (CDM) with $\Gamma = 0.5$ and $\Gamma = 0.225$ respectively ($\Gamma = \Omega h$). the first is known as the standard CDM model (SCDM), while the second is a low density CDM model (LCDM). We also have considered a mixed dark matter model (MDM) with $\Omega_{\mathrm{CDM}} = 0.6$, one species of neutrinos $\Omega_\nu = 0.3$, baryon density $\Omega_B = 0.1$, and $h = 0.5$. With SEPT we hav shown that $Q(r,u,v)$ depends sensitively on the shape of the linear power spectrum. For scale free power spectra $P(k) \propto k^n$ with index $-3 < n < 0$, $Q$ is an increasing function of $v$, and the larger $n$ (*i.e.*, the flatter the spectrum) the more rapidly $Q$ increases with $v$. For a fixed configuration of a triangle the SCDM model shows the stro! ngest and the MDM model the weakest variation with $v$. Our N-body results agree qualitatively with the prediction of SEPT, but quantitatively they show less variation of $Q$ with $v$ in the SCDM model, and more variations in the other two models, compared to the SEPT prediction. Thus the three-point correlation function is less powerful as a discriminating statistics between these models, than suggested by the SEPT results. On the other hand, the robust dependence of $Q$ on $v$ should exist in the spatial distribution of galaxies, if the galaxies trace the underlying mass. For a constant bias factor $b$, $Q$ is $\propto 1/b$, and therefore the $v$-dependence will eventually put constraints on the bias.

## 4. Further results

In the strongly non-linear regime ($\xi \gg 1$), $Q$ is a decrasing function of triangle sizes ($r$ and $u$), but $Q$ is independent of $v$. Taking values of $Q$ estimated from N-body simulations by averaging over all triplets with the smallest side equal to $r$, and with the second smallest side less than $4r$, we find that $Q$ decreases from about 2.8 at $r = 0.1h^{-1}$Mpc to 1.4 at $r = 1h^{-1}$Mpc to 0.6 at $r = 3h^{-1}$Mpc. We have carried out this analysis for several different models, and it seems that $Q$ depends only on the shape $\Gamma = \Omega h$ of the linear power spetrum and not on details of the cosmological models. This conclusion is only preliminary, because we have not investigated the parameter dependence in great detail (Jing, Mo, Börner 1997).

We have also analyzed the Las Campanas Redshift Survey, a recently released redshift sample (Shectman et al 1996). Limiting our analysis to $-23 < M_{\rm LCRC} < -18$ and to 10 000 km s$^{-1}$ $< v_{rec} <$ 45 000 km s$^{-1}$ we have to deal with 19 558 galaxies. This sample is large enough to allow a determination of the three-point correlation function. Our preliminary result for the redshift space $Q(s, u, v)$ is that the value is about 0.6 at $r \approx 1h^{-1}$Mpc.

$Q$ rises with $v$ in the quasilinear regime. There is, of course, a difference between $Q$ in redshift space and in real space, because the peculiar velocities lead to an apparent reduction of the clustering of galaxies at small scales. Therefore at small scales the value of $Q$ will be reduced in redshift space.

To account for the redshift distortions, we calculate the projected three-point correlation function. Roughly, the result is $Q(proj) \simeq 1.5$, when $r_p$ is very small ($r_p < 0.5h^{-1}$Mpc), but $Q$ decreases with increasing $r_p$ at larger scales.

Since $Q(r, u, v)$ does not show the hierarchical form in models, we must work out a careful and detailed comparison. This work is in progress.

*Acknowledgements.* YPJ thanks the Alexander-von-Humboldt-Foundation for a Research Fellowship, and the Max-Planck-Institut für Astrophysik for financial support. Both authors acknowledge support by SFB375.

## References

Jing, Y.P., Börner, G., 1997, *Astron. Ap.*, **318**, 667
Jing, Y.P., Börner, G., Valdarnini, R. (1995) *MNRAS*, **277**, 630
Jing, Y.P., Mo, H.J., Börner, G (1997) (in preparation)
Peebles, P.J.E. (1980) 'The Large-Scale Structure of the Universe' (Princeton Univ. Press)
Shectman, S.A., et al (1996) *Ap.J.*, **470**, 172

# ANALYTIC APPROXIMATIONS TO GALAXY CLUSTERING

H.J. MO
*Max-Planck-Institut für Astrophysik*
*Karl-Schwarzschild-Str. 1, 85748 Garching, Germany*

**Abstract.** We summarize some recent progress in constructing analytic approximations to the galaxy clustering. We show that successful models can be constructed for the clustering of both dark matter and dark matter haloes. Our understanding of galaxy clustering and galaxy biasing can be greatly enhanced by these models.

## 1. Introduction

The large-scale structure of the Universe is believed to have developed from small perturbations (usually assumed to be Gaussian) of the matter density field by gravitational instabilities. Under these assumptions the clustering pattern and velocity field observed today are determined by the initial conditions via the perturbation power spectrum $[P(k)]$ and the cosmological parameters such as $\Omega_0$, the cosmic density parameter. It is, therefore, possible to derive constraints on model parameters from the observed density and velocity distributions of galaxies.

There are two fundamental problems to be addressed; first, since the galaxy distribution may be biased relative to the mass density field, we need to understand such bias before making meaningful comparisons between models and observations. Second, even if the observed galaxy distribution traces the matter distribution, we still need to understand how the observed distribution is related to a cosmogonical model. This is by no means trivial, because the clustering pattern and velocity field observed today are nonlinear. N-body simulations are usually invoked to find a solution to these problems. However, such simulations are limited both in resolution and in dynamical range, and can be difficult to interpret. Our understanding of the underlying physics can be greatly enhanced by simple physical models and the analytic approximations they provide.



*D. Hamilton (ed.), The Evolving Universe, 343–353.*

In this article we summarize some of our recent progress in connection to the problems mentioned above. We show that successful semi-analytic models can be constructed for the clustering of both dark matter and dark matter haloes. Such models can enhance significantly our understanding of both the nonlinear evolution of galaxy clustering and galaxy biasing.

## 2. Cosmogonies

The models present below are for CDM-like cosmogonies. The cosmology is described by the cosmological matter density ($\Omega_0$), the cosmological constant ($\lambda_0$) and the Hubble constant ($H_0 = 100h\,\mathrm{km\,s^{-1}\,Mpc^{-1}}$). The initial power spectrum is

$$P(k) \propto kT^2(k), \tag{1}$$

$$T(k) = \frac{\ln(1+2.34q)}{2.34q}\left[1+3.89q+(16.1q)^2+(5.46q)^3+(6.71q)^4\right]^{-1/4}, \tag{2}$$

where $q \equiv k/(\Gamma h\,\mathrm{Mpc}^{-1})$, $\Gamma \equiv \Omega_0 h$ (Bardeen *et al.* 1986). The RMS mass fluctuation in top-hat windows with radius $R$, $\sigma(R)$, is defined by

$$\sigma^2(R) = \int_0^\infty \frac{dk}{k}\Delta^2(k)W^2(kR), \tag{3}$$

where $W(x)$ is the Fourier transform of the top-hat window function, and

$$\Delta^2(k) = (1/2\pi^2)k^3P(k) \tag{4}$$

is the power variance. We normalize $P(k)$ by specifying $\sigma_8 \equiv \sigma(8\,h^{-1}\mathrm{Mpc})$.

## 3. Low-Order Statistics of Dark Matter Distribution

### 3.1. MASS CORRELATION FUNCTION

The *evolved* two-point correlation function $\xi(r)$ is related to the *evolved* power variance $\Delta_E^2(k)$ by

$$\xi(r) = \int_0^\infty \frac{dk}{k}\Delta_E^2(k)\frac{\sin kr}{kr}. \tag{5}$$

Thus, in order to get $\xi(r)$ we need an expression for $\Delta_E(k)$. Following the original argument of Hamilton *et al.* (1991), Jain, Mo & White (1995), Padmanabhan *et al.* (1996) and Peacock & Dodds (1996, hereafter PD) have obtained fitting formulae which relate $\Delta_E$ to $P(k)$ for a given cosmological model. The latest version of such a fitting formula is given in PD. The solid curve in Fig.1 shows the prediction of such a formula for $\xi(r)$ in the

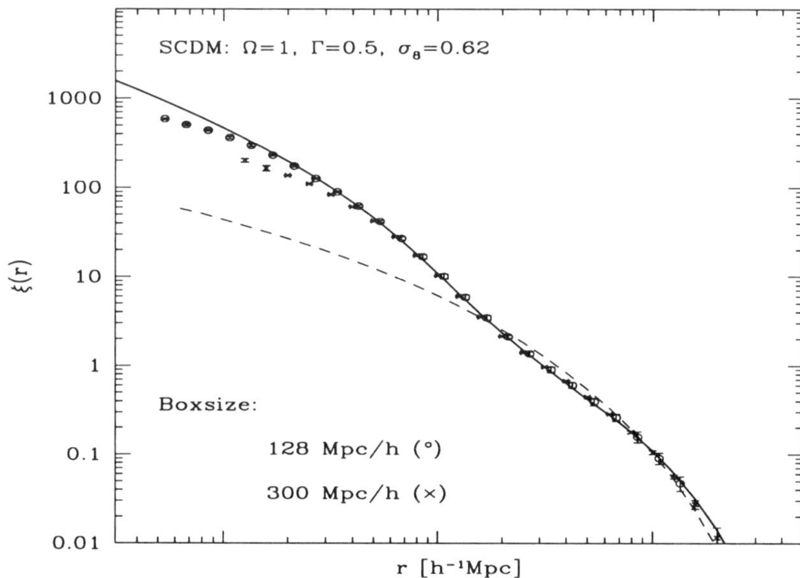


*Figure 1.* Predicted two-point correlation function of mass (solid curve) compared with simulation results (symbols). The dashed curve is given by the linear power spectrum.

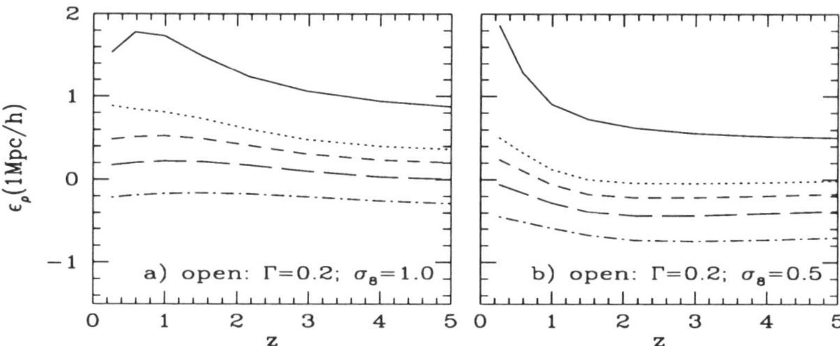


*Figure 2.* The redshift-evolution parameter of the mass correlation function as a function of redshift, for open cosmologies with $\Omega_0 = 0.1$, 0.2, 0.3, 0.5 and 1 (from bottom up).

standard cold dark matter (SCDM) model with $\Omega_0 = 1$, $\lambda_0 = 0$, $h = 0.5$ and $\sigma_8 = 0.62$ (see Mo, Jing & Börner 1997, hereafter MJB, for details). Comparing it with the N-body results, we see clearly that the fitting formula works well. This is true for various other cosmogonic models studied.

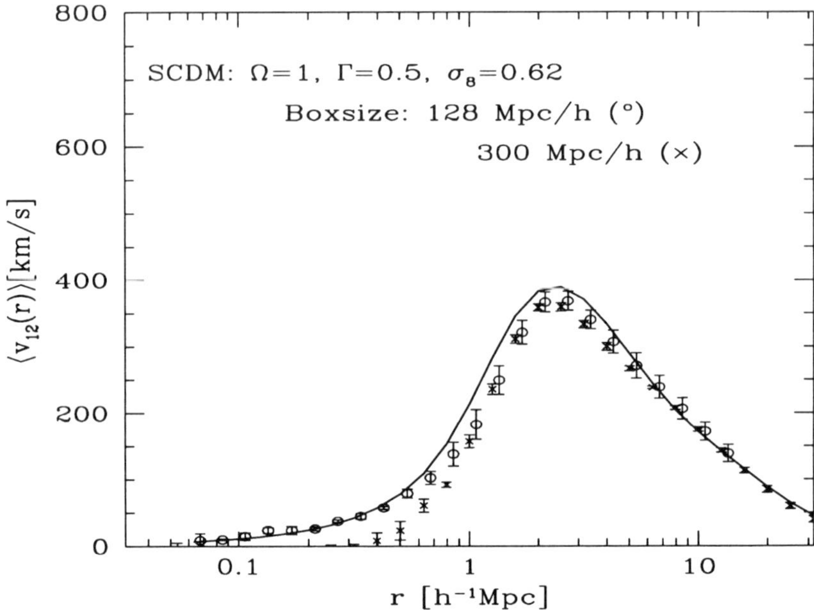


*Figure 3.* Mean pairwise peculiar velocity of dark matter particles. Curve is the model prediction; symbols are the simulation results.

### 3.2. REDSHIFT EVOLUTION OF MASS CORRELATION FUNCTION

The redshift evolution of the two-point correlation function is usually parametrized by the form

$$\xi(r, z) = \xi(r, 0)(1 + z)^{-(3+\epsilon_\rho)}, \tag{6}$$

where $\xi(r, z)$ is the amplitude of the two-point correlation function at *physical* radius $r$ at redshift $z$, $\epsilon_\rho$ is a parameter to describe the time evolution. If $\xi(r) \propto r^{-\gamma}$, then $\epsilon_\rho = \gamma - 1$ for the linear growth in an Einstein-de Sitter universe, $\epsilon_\rho = 3 - \gamma$ for clustering patterns fixed in comoving space, and $\epsilon_\rho = 0$ for stable clustering (*i.e.*, clustering patterns fixed in proper coordinates). Given the model in §3.1, it is straightforward to obtain $\epsilon_\rho$. Fig. 2 shows $\epsilon_\rho$ as a function of $z$ for open cosmologies with various $\Omega_0$. The value of $\epsilon_\rho$ at $z$ is obtained by fitting $\xi(r = 1h^{-1}\mathrm{Mpc}, z')$ in the redshift interval, $z' = 0 \to z$. The evolution is more rapid (*i.e.*, $\epsilon_\rho$ larger) for universes with larger $\Omega_0$, since linear structures grow faster in a high-$\Omega_0$ universe. For high $\Omega_0$, the evolution is faster at $z \sim 1$, because of nonlinear evolution. At late time when clustering on $r \sim 1h^{-1}\mathrm{Mpc}$ becomes stable, the evolution becomes slower.

### 3.3. MEAN PAIRWISE PECULIAR VELOCITIES

From the pair conservation equation (Peebles 1980, §71), the ensemble (pair weighted) average of the pairwise peculiar velocity $\langle v_{12}(r)\rangle \equiv \langle[\mathbf{v}(\mathbf{x})-\mathbf{v}(\mathbf{x}+\mathbf{r})]\cdot\hat{\mathbf{r}}\rangle$ can be written as

$$\frac{\langle v_{12}(r)\rangle}{H(a)r} = -\frac{1}{3}\frac{1}{[1+\xi(y,a)]}\frac{\partial\overline{\xi}(y,a)}{\partial\ln a}, \tag{7}$$

where $r$ is the proper, and $y$ the comoving, separation between the pairs; $H$ is the Hubble's constant at expansion factor $a$; $\overline{\xi}(y,a) \equiv (3/y^3)\int_0^y y^2 dy \xi(y,a)$. Thus, to obtain $\langle v_{12}(r)\rangle$, we need to work out $\partial\Delta_E(k,a)/\partial a$. This can be done directly from the fitting formula of $\Delta_E(k,a)$ (see MJB). Fig. 3 shows the comparison between the model prediction and the simulation results for the SCDM model. The agreement between the two is remarkably good, and this is true for many other cosmogonic models studied.

### 3.4. COSMIC ENERGY EQUATION

The (density weighted) mean square peculiar velocity of mass particles $\langle v_1^2\rangle$ is related to the two-point correlation function by the cosmic energy equation:

$$\frac{d}{da}a^2\langle v_1^2\rangle = 4\pi G\overline{\rho}a^3\frac{\partial I_2(a)}{\partial\ln a}, \tag{8}$$

where $\overline{\rho}$ is the mean density of the universe, and

$$I_2(a) \equiv \int_0^\infty y dy \xi(y,a) = \int_0^\infty \frac{dk}{k}\frac{\Delta_E^2(k,a)}{k^2}. \tag{9}$$

Integrating eqn. (8) once, we have

$$\langle v_1^2\rangle = \frac{3}{2}\Omega(a)H^2(a)a^2 I_2(a)\left[1 - \frac{1}{aI_2(a)}\int_0^a I_2(a)da\right]. \tag{10}$$

In the linear case, $I_2(a) \propto D^2(a)$, where $D(a) = ag(a)$ [$g(a)$ is the linear growth factor; $a_0 = 1$] and

$$\langle v_1^2\rangle = \frac{3}{2}\Omega(a)H^2(a)a^2 I_2(a)\left[1 - \frac{1}{aD^2(a)}\int_0^a D^2(a)da\right]. \tag{11}$$

We found that eqn.(11) is a good approximation (to an error of $< 10\%$) to eqn.(10) for all realistic cases. Thus, for a given cosmogonic model, we can easily obtain $\langle v_1^2\rangle$. For various cosmogonies, the model predictions fit the simulation results to an accuracy better than 10 percent (see MJB).

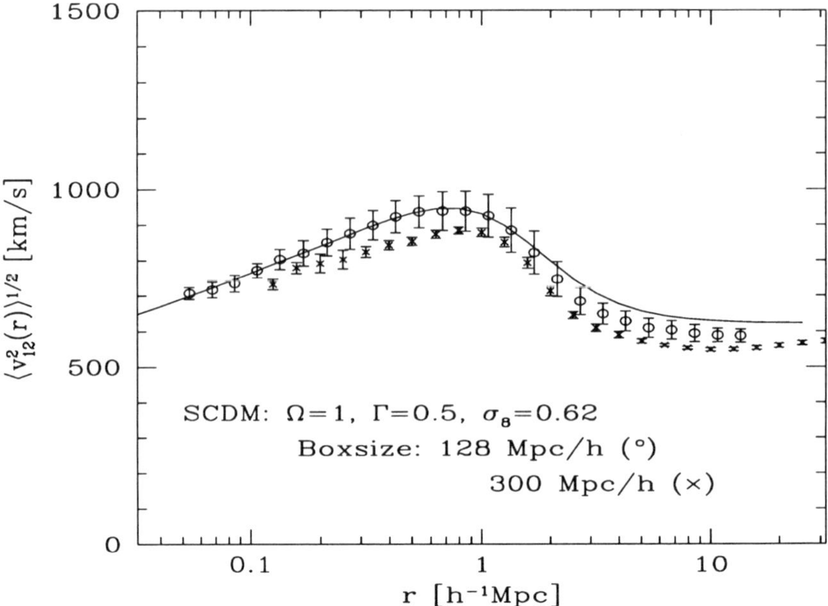


*Figure 4.* Pairwise peculiar velocity dispersion of dark matter particles. Curve is the model prediction; symbols are the simulation results.

## 3.5. PAIRWISE PECULIAR VELOCITY DISPERSION

The relative velocity dispersion of particle pairs of separation $r$ is defined as $\langle[\mathbf{v}(\mathbf{x})-\mathbf{v}(\mathbf{x}+\mathbf{r})]^2\rangle^{1/2}$. In Fig. 4 we show (by symbols) the dispersion of the pairwise peculiar velocities projected along the separations of particle pairs $[\langle v_{12}^2(r)\rangle^{1/2}]$ in the N-body simulations. The main features of $\langle v_{12}^2(r)\rangle^{1/2}$ are (a) monotonic rise at small $r$; (b) saturation at large $r$; (c) a maximum at medium $r$. As shown in MJB, these features can all be explained by physical arguments.

Based on the N-body results we make the following ansatz for $\langle v_{12}^2(r)\rangle$:

$$\langle v_{12}^2(r)\rangle^{1/2} = \Omega^{0.5} H r_c \phi(r/r_c), \tag{12}$$

where $\phi(x)$ is a universal function and $r_c$ is a nonlinear scale. We choose $r_c$ to be the virial radius of $M^*$ haloes, $r_v^*$. For a given power spectrum, the linear radius of $M^*$ haloes, $r_0^*$, is given by $\sigma(r_0^*) = 1$. The relation between $r_v^*$ and $r_0^*$ in different cosmological models can be obtained analytically, as discussed in detail in MJB. We approximate the functional form of $\phi(x)$ by:

$$\phi(x) = \frac{\phi_\infty(1+Bx^{-\beta}) + Ax^{-\alpha}}{1+V[g(a)]^{0.35}[\Omega(a)]^{0.2}x^{-(\alpha+\kappa)}}, \tag{13}$$

where $\phi_\infty = \sqrt{\frac{2}{3}}\langle v_1^2\rangle^{1/2}/(Hr_c\Omega^{0.5})$; $A$, $B$, $V$, $\alpha > \beta > 0$, and $\kappa > 0$ are constant. $\langle v_{12}^2(r)\rangle$ is forced to have the large separation asymptotic value for uncorrelated pairs. For $r \to 0$, $\langle v_{12}^2(r)\rangle^{1/2} \propto x^\kappa$ so that it increases with $r$ as a power law. The solid curve in Fig. 4 shows the prediction of our fitting formula with

$$A = 58.67;\;\; B = -0.3770;\;\; V = 4.434; \tag{14}$$

$$\alpha = 2.25;\;\; \beta = 1.90;\;\; \kappa = 0.15. \tag{15}$$

The fit to the simulation data is reasonably good.

## 4. Spatial Clustering of Dark Matter Haloes

So far we have discussed the clustering properties of dark matter. To compare model predictions with the observed galaxy distribution, we also need to understand how the galaxy distribution is related to the dark matter distribution. The bias of the galaxy distribution relative to the mass distribution can be obtained once we know how galaxies form in the mass density field. Nevertheless, some progress can still be made before the details of galaxy formation is understood. In the standard scenario of galaxy formation, a gravitationally dominant dissipationless component of dark matter is assumed to aggregate into dark matter clumps (dark matter haloes), galaxies then form by the cooling and condensation of gas within these dark haloes (*e.g.*, Kauffmann 1997). It is, therefore, important to approach the problem of galaxy biasing by first understanding how dark matter haloes are distributed relative to the mass. In the following we show that simple analytic models for such relations can be constructed (see Mo & White 1996, MW).

### 4.1. BIAS RELATION

We define dark matter haloes as virialized clumps of dark matter. In the spherical collapse model, a dark matter halo is characterized by its mass, $M$, and the redshift, $z$, when it is assembled. To describe the relation between halo and mass distributions we define a bias relation,

$$\delta_h(R) = b(R, \delta, M, z)\delta(R), \tag{16}$$

where $\delta(R) = [\rho(R) - \overline{\rho}]/\overline{\rho}$ is the overdensity of matter in a sphere of radius $R$, $\delta_h(R)$ is the same overdensity for dark matter haloes. Suppose the conditional density of dark matter haloes within a sphere of radius $R$ is $n_h(M, z|\delta, R)$. (This is the number density of dark matter haloes within a

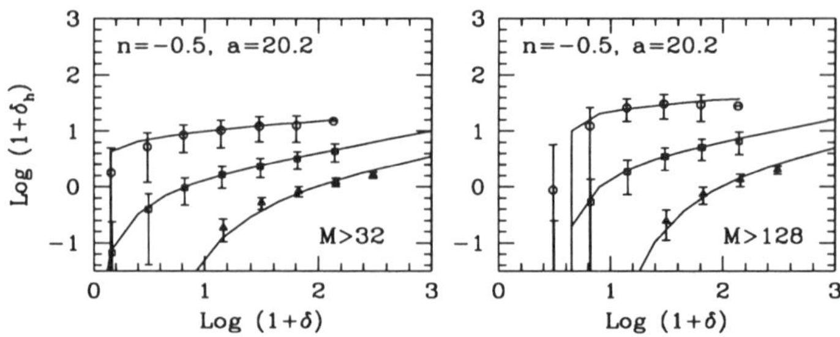


*Figure 5.* Bias relation $\delta_h(\delta)$ in a scale free model with $n = -0.5$. Results are shown for spheres with radii $R/L = 0.02$, 0.05 and 0.13 ($L$: the side of the simulation box). The results for $R/L = 0.05$ and 0.13 are shifted by 1 and 2 decades along the horizontal axis.

sphere of radius $R$, given that the mean mass overdensity within this sphere is $\delta$.) The bias relation can then be written as

$$\delta_h(M, z|\delta, R) = \frac{n_h(M, z|\delta, R)}{\overline{n}_h(M, z)} - 1, \tag{17}$$

where $\overline{n}_h(M, z)$ is the mean number density of haloes in the universe, given by the Press-Schechter formalism (Press & Schechter 1974, PS):

$$\overline{n}_h(M, z)dM = -\sqrt{\frac{2}{\pi}}\frac{\overline{\rho}}{M}\frac{\delta_z}{\sigma^2(r)}\frac{d\sigma}{dM}\exp\left[-\frac{\delta_z^2}{2\sigma^2(r)}\right]dM, \tag{18}$$

where $M = (4\pi/3)\overline{\rho}r^3$, $\delta_z = \delta_c D(a_0)/D(a)$, $\delta_c \approx 1.686$. The conditional density, $n_h(M, z|\delta, R)$, can be obtained by an extension of the PS formalism (*e.g.*, Bower 1991):

$$\begin{aligned} n_h(M, z|\delta_0, R_0)dM &= -\sqrt{\frac{2}{\pi}}\frac{\overline{\rho}}{M}\frac{(\delta_z - \delta_0)\sigma(r)}{[\sigma^2(r) - \sigma^2(R_0)]^{3/2}}\frac{d\sigma(r)}{dM} \\ &\times \exp\left\{-\frac{(\delta_z - \delta_0)^2}{2[\sigma^2(r) - \sigma^2(R_0)]}\right\}dM, \end{aligned} \tag{19}$$

where $\delta_0$ is the linear mass overdensity of spherical regions with Lagrangian radius $R_0$. Assuming spherical collapse, we have $R_0 = [1 + \delta(R)]^{1/3}R$, and $\delta_0$ is determined by $\delta(R)$ and $R$ through the spherical collapse model (see MW for details). With these the bias relation is fixed. Fig. 5 shows the bias relation for haloes in a scale free model with $P(k) \propto k^{-0.5}$. It is clear that our simple model works reasonably well.

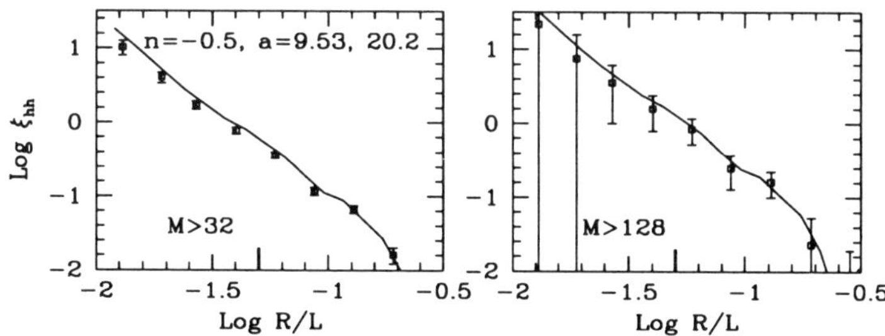


*Figure 6.* Predicted halo-halo two-point correlation functions (solid curves) compared to those from simulations (symbols).

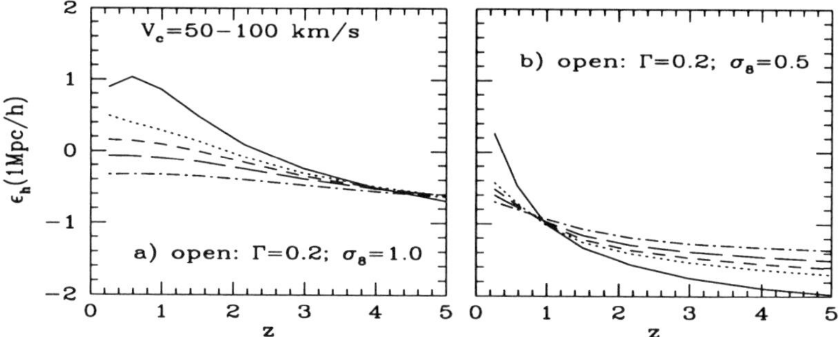


*Figure 7.* The redshift-evolution parameter of the halo-halo correlation function for haloes selected according to the first selection rule (see text). Results are shown for open cosmologies with $\Omega_0 = 0.1$, 0.2, 0.3, 0.5 and 1 (from dot-dashed to solid curves).

## 4.2. TWO-POINT CORRELATION FUNCTION OF HALOES

When $\delta \ll 1$ and $r \ll R_0$, the bias relation can be written as

$$\delta_h(R; M, z) = b(M, z)\delta(R); \quad b(M, z) = 1 + \frac{1}{\delta_z}\left[\frac{\delta_z^2}{\sigma^2(r)} - 1\right]. \qquad (20)$$

Thus the bias factor $b$ depends only on $M$ and $z$. Under the assumption of linear bias,

$$\xi_h(r) = b^2(M, z)\xi(r), \qquad (21)$$

where $\xi(r)$ is the two-point correlation function of mass. Fig. 6 shows the two-point correlation functions for haloes in a scale free model with $P(k) \propto k^{-0.5}$. The simple model works reasonably well.

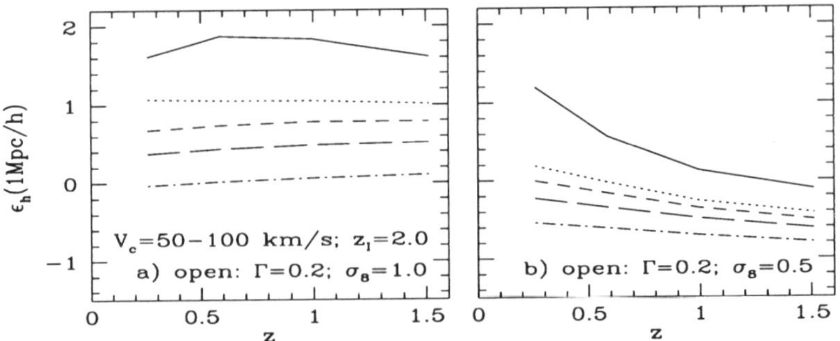


*Figure 8.* The same as Fig. 7 for haloes selected according to the second selection rule.

## 4.3. REDSHIFT EVOLUTION OF HALO CORRELATION FUNCTION

As in §3.2, we parametrize the redshift evolution of $\xi_h$ by

$$\xi_h(r,z) = \xi_h(r,0)(1+z)^{-(3+\epsilon_h)}. \tag{22}$$

Unlike the mass correlation function $\xi$, the evolution of $\xi_h$ depends on how dark matter haloes are selected. We consider two different selections. In the first, haloes are selected at the same time when the correlation function is calculated. This case is relevant for galaxies, if galaxies merge as fast as their dark haloes. In the second, haloes are selected at an earliear epoch than when their correlation function is calculated. This case is relevant for galaxies, if they remain distinct after their haloes merge. Fig. 7 shows $\epsilon_h$ as a function of $z$ in open cosmologies with two choices of $\sigma_8$. Here haloes are selected according to the first selection. As before, the value of $\epsilon_h$ at $z$ is obtained by fitting $\xi_h(r = 1h^{-1}\mathrm{Mpc}, z')$ in the redshift interval $z' = 0 \to z$. In all cases, the halo correlation function evolves less rapidly than the mass correlation function (see Fig. 2), because haloes with fixed circular velocities are more biased at higher redshifts. The evolution is also less rapid for a lower $\sigma_8$, because of the higher degree of bias involved. Fig. 8 shows the same results for haloes selected according to the second selection. In this case, the evolution of $\xi_h$ is faster than that of $\xi$, because the haloes (with $V_c = 50 - 100\,\mathrm{km\,s^{-1}}$) are antibiased relative to the mass at the time of selection ($z = 2$). For massive haloes which are (*positively*) biased relative to the mass, the evolution of their correlation function is slower than that of the mass. The evolution is also slower for a lower $\sigma_8$, because of the increased degree of bias (or decreased degree of antibias). The results in this subsection suggest that one needs to be very cautious when interpreting the redshift evolution of galaxy correlation function. Without knowing in detail what population the observed galaxies are, it is difficult to infer the time evolution of the mass correlation from the correlation functions of these galaxies.

## 5. Discussion

The models presented above show how the low order statistics of the density and velocity distributions are determined by cosmogonies. Thus they can be used to construct statistical measures of the density and peculiar velocity fields to constrain cosmogonic models by observations. In particular, the models can help in the reconstruction of cosmogonic parameters from measurements in redshift space. There are also other applications. MJB discuss the dependence of the small scale pairwise peculiar velocity dispersion on the presence (or absence) of rich clusters of galaxies. Mo, Jing & White (1996) use the model to study the correlation of galaxy clusters. Mo, Jing & White (1997) extend the model to high-order correlations of dark matter haloes and density peaks.

## References

Bardeen J., Bond J.R., Kaiser N., Szalay A.S., 1986, ApJ, 304, 15
Bower R.J., 1991, MNRAS, 248, 332
Hamilton A.J.S., Kumar P., Lu E., Matthews A., 1991, ApJ, 374, L1
Jain B., Mo H.J., White S.D.M., 1995, MNRAS, 276, L25
Kauffmann G., 1997, private communication.
Mo H.J., Jing Y.P., Börner G., 1997, MNRAS, in press (MJB)
Mo H.J., Jing Y.P., White S.D.M., 1996, MNRAS, 282, 1096
Mo H.J., Jing Y.P., White S.D.M., 1997, MNRAS, 284, 189
Mo H.J., White S.D.M., 1996, MNRAS, 282, 347 (MW)
Padmanabhan T., Cen R., Ostriker J.P., Summers F.J., 1996, preprint
Peacock J.A., Dodds S.J., 1996, MNRAS, 280, L19 (PD)
Peebles P.J.E., 1980, The Large-Scale Structure of the Universe, Princeton Univ. Press
Press W.H. & Schechter P., 1974, ApJ, 187, 425 (PS)

# HIGH REDSHIFT OBJECTS IN THE COSMIC STRING MODEL

R. MOESSNER
*Max-Planck-Institut für Astrophysik*
*Karl-Schwarzschild-Str. 1, 85740 Garching, Germany*

AND

R. BRANDENBERGER
*Physics Department, Brown University*
*Providence, RI 02912, USA*

**Abstract.** We study the accretion of hot dark matter onto moving cosmic string loops, using an adaptation of the Zeldovich approximation to HDM. We show that a large number of nonlinear objects of mass greater than $10^{12} M_{\odot}$, which could be the hosts of high redshift quasars, are formed by a redshift of $z = 4$.

## 1. Introduction

Topological defect theories provide an alternative to inflation for explaining the origin of the primordial fluctuations which have grown into present-day structures through gravitational instability. Cosmic strings are one-dimensional topological defects possibly formed in a phase transition in the early Universe.

The observed presence of massive nonlinear structures at high redshifts of 3 to 4 has provided stringent constraints on models of structure formation containing hot dark matter (HDM), due to the large thermal velocities of the HDM particles, which prevent the growth of perturbations on small scales. The scenario of structure formation with cosmic strings may still be viable even if the dark matter in the universe is hot, because cosmic strings, which provide the seeds for the density perturbations, survive the neutrino free-streaming (Vilenkin & Shafi 1983). Here we present work on



*D. Hamilton (ed.), The Evolving Universe, 355–360.*

the constraints imposed on the cosmic string theory by the abundance of high redshift quasars (Moessner & Brandenberger 1996).

According to recent results from the Palomar grism survey by Schmidt *et al.* (1995), the quasar luminosity function peaks in the redshift interval $z \epsilon [1.7, 2.7]$ and declines at higher redshifts. Quasars (QSO) are generally assumed to be powered by accretion onto black holes. It is possible to estimate the mass of the host galaxy of the quasar as a function of its luminosity, assuming that the quasar luminosity corresponds to the Eddington luminosity of the black hole. For a quasar of absolute blue magnitude $M_B = -26$ and lifetime of $t_Q = 10^8 yrs$, the host galaxy mass can be estimated as (Subramanian & Padmanabhan 1994)

$$M_G = c_1 10^{12} M_\odot , \tag{1}$$

where $c_1$ is a constant which contains some uncertainties of the model, the best estimate for $c_1$ is about 1. Models of structure formation have to pass the test of producing enough early objects of sufficiently large mass to host the observed quasars.

From the absorption lines in the quasar spectra, the amount of matter $\Omega_g$ in bound neutral gas can be estimated. For damped Ly-$\alpha$ systems, recent observational results (Storrie-Lombardi *et al.* 1996) are that for $z \epsilon [1,3]$

$$\Omega_g(z) > 10^{-3} . \tag{2}$$

## 2. Cosmic Strings and Structure Formation

Cosmic strings (Vilenkin & Shellard 1994) are one-dimensional topological defects which might have been formed in a phase transition in the very early Universe. The mass per unit length $\mu$ of the strings is given by the symmetry breaking scale. The network of cosmic strings is formed in a phase transition in the early universe and then quickly evolves towards a scaling solution. The scaling solution can be pictured as having a fixed number $N$ of long strings per Hubble volume at any given time, and a distribution $n(l,t)$ of small loops, loops being formed with a radius which is a constant fraction of the horizon, $R_f(t) = \alpha t$, and length $l = \beta R$. Numerical simulations (Albrecht & Turok 1989; Allen & Shellard 1990; Bennett & Bouchet 1988) indicate that $\alpha \leq 10^{-2}$, $\beta \simeq 10$, and $N \sim 10$. From these values it follows that – unless $\nu$ is extremely large – most of the mass of the string network resides in long strings. The long strings, therefore, are most important for structure formation. They accrete matter in the form of wakes behind them as they move through space.

However, for HDM, the first nonlinearities about wakes resulting from long strings without any small-scale structure form only at late times, at a

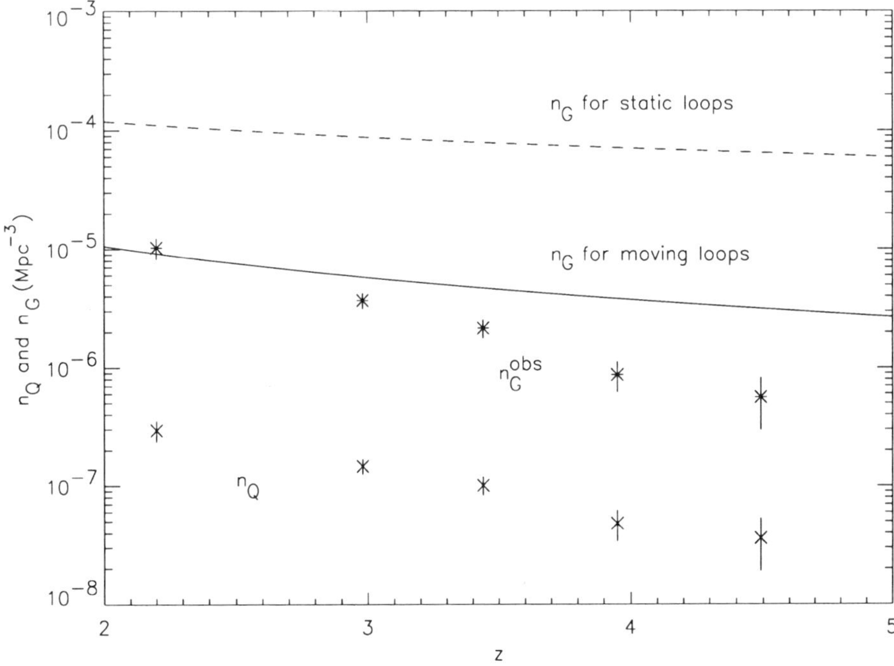


*Figure 1.* Comparison of the number density of host galaxies $n_G^{obs}$ ('*' marks) inferred from the observed number density of quasars brigher than $M_B = -26$ from the Schmidt *et al.* (1991,1995) survey('x' marks) with the number density $n_G$ of protogalaxies of mass greater than $10^{12} M_\odot$ predicted in the cosmic string theory with HDM, for the parameters discussed in the text, and $h_{50} = 1$. The horizontal axis is the redshift.

redshift of (Perivolaropoulos *et al.* 1990) about 1 for $G\mu \simeq 10^{-6}$, which is obtained when normalizing the model to COBE. Thus a different mechanism is required in order to explain the origin of high redshift objects. Here we investigate the mechanism of accretion of HDM onto loops, since in earlier work (Brandenberger *et al.* 1987) it has been shown that for HDM accreted onto static loops, the nonlinear structure around it grows from inside out, and that the first nonlinearities form early on.

## 3. Number density of quasar host galaxies

We use a simple modification of the Zeldovich approximation to HDM (Perivolaropoulos *et al.* 1990), to study the accretion of HDM onto moving cosmic string loops. For HDM, free-streaming prevents the growth of perturbations on scales $q$ below the free-streaming length, which is the mean distance traveled by neutrinos in one expansion time. We determine the mass that has gone nonlinear about a string loop as the rest mass inside of the shell which is "turning around". For any time $t$ we can thereby calculate $M(l)$, and determine a mass $M'(t)$ such that for all masses $M > M'(t)$ accreted by any loop so far, accretion has not been affected by free-streaming.

It can be shown (Brandenberger & Shellard 1989) that in an HDM model string loops accrete matter independently, at least before the large-scale structure in wakes turns nonlinear at a redshift of about one, *i.e.*, later than the times of interest in this paper. Hence, the number density $n(l, t)$ of loops of length $l$ can be combined with the mass $M(l)$ accreted by an individual loop to give the mass function $n(M, t)$. Here, $n(M, t)dM$ is the number density of objects with mass in the interval between $M$ and $M+dM$ at time $t$. This in turn determines $n_G(> M_G, t)$, the comoving density in objects of mass $M > M_G$. The functional form of $M(l)$ changes at the mass $M'$ defined in the previous paragraph, and therefore the functional forms of the mass function $n(M)$ will be different above and below $M'$, and will be denoted by $n_C$ and $n_H$ respectively. Thus

$$n_G(> M_G, t) = z^{-3}(t) \left[ \int_{M_G}^{M'(t)} dM n_H(M) + \int_{M'(t)}^{M_2(t)} dM n_C(M) \right] . \quad (3)$$

Note the upper mass cutoff $M_2(t)$ introduced, beyond which our approximation of loops as point masses breaks down. The integral in Eq. 3 is dominated at $M'(t)$ , and can be approximated as

$$n_G(> M_G, t) \sim n_C(M', t) M' z^{-3}(t) \simeq \frac{1}{5} \nu (\alpha\beta)^2 \frac{\mu^3}{M'(t)^3} z^{-3}(t) . \quad (4)$$

In Figure 1, the comoving number density of objects massive enough to be able to host quasars, $n_G(> M_G, t)$, is plotted for $G\mu = 10^{-6}$, $v_{eq} = 0.1$, and the values of the parameters $\alpha = 10^{-2}$ and $\beta = 10$. The upper curve is for static loops, the lower one for initial loop velocities of $v_i = 0.25$. The effect of loop motion is to decrease the density of these massive objects, as discussed in the next paragraph. We have to compare our results for $n_G(> M_G, t)$ with the number density of host galaxies, $n_G^{\rm obs}$, inferred from the observed quasar abundance $n_Q$. Since we assumed a finite quasar lifetime $t_Q$, the number of host galaxies $n_G^{\rm obs}$ is larger than $n_Q$ by a factor of $t_H/t_Q$, where $t_H$ is the Hubble time at time $t$,

$$n_G^{\rm obs} = \frac{t_H}{t_Q} n_Q . \quad (5)$$

This is also plotted in Fig. 1 .

The formalism presented above can be adapted to moving instead of static loops (Bertschinger 1987; Moessner & Brandenberger 1996). Loop motion changes the geometry of the accreted object, it becomes more elongated and the transverse scale going nonlinear becomes smaller than for loops at rest. Because of this, free-streaming of the neutrinos can hinder the growth of perturbations for a longer time. The smallest mass accreted at time $t$ which has been unaffected by free-streaming, $M'(t)$, is increased compared to the case of static loops, thereby decreasing $n_G(t)$ according to Eq. 4.

The result for the fraction of nonlinear mass accreted by moving loops is

$$\Omega_{nl}(t) \sim 10^{-2} \nu \alpha_{-2} (G\mu)_6^2 v_i^{-1} z(t)^{-2} h_{50}^4 , \tag{6}$$

where $v_i$ is the initial comoving velocity of the loop, $(G\mu)_6$ is defined as $(G\mu)/10^{-6}$ and $\alpha_{-2} \equiv \alpha/0.01$.

The values of $n_G(t)$ and $\Omega_{nl}$ are suppressed beyond the results plotted in Fig. 1 and Eq. 6 for redshifts larger than $z_{max} = 3h_{50}^2$ (for $v_i = 0.25$ and $v_{eq} = 0.1$). This is because beyond $z_{max}$ the condition $M' < M_2(t)$ is not satisfied, and only the tail of the loop ensemble with velocities smaller than the mean velocity $v_i$ accrete matter effectively.

## 4. Discussion

The loop accretion mechanism is able to generate nonlinear objects much earlier than the time cosmic string wakes start becoming nonlinear. For the values $\nu = 1$ and $\alpha_{-2} = 1$ which are indicated by recent cosmic string evolution simulations (Albrecht & Turok 1989; Allen & Shellard 1990; Bennett & Bouchet 1988), we conclude that the loop accretion mechanism produces enough large mass protogalaxies to explain the observed abundance of $z \leq 4$ quasars (see Figure 1). Note that the amplitude of the predicted protogalaxy density curves depends sensitively on the parameters of the cosmic string scaling solution which are still poorly determined.

In the form of Eq. 6, the condition for the cosmic string loop accretion mechanism to be able to explain the data of the damped Ly-$\alpha$ systems is also satisfied. However, Eq. 6 refers to the value of $\Omega$ in baryonic matter. The corresponding constraint on the total matter collapsed in structures associated with damped Ly-$\alpha$ systems is $\Omega_{\mathrm{DL}}(z < 3) > f_b^{-1} 10^{-3}$ where $f_b$ is the local fraction of the mass in baryons. From Eq. 5 it follows that the above constraint is only marginally satisfied, and this only if the local baryon fraction $f_b$ exceeds the average value for the whole Universe of about $f_b = 0.1$.

Here we have only reported on the mechanism of forming early nonlinear objects through accretion onto string loops. Another mechanism is through small-scale structure on the long strings leading to the formation of filaments rather than wakes, which has recently been investigated (Zanchin *et al.* 1996). It was found that this could be the most effective mechanism, and for the maximal possible amount of small-scale structure, $\Omega_{nl} \sim 1$ can be reached already at a redshift of 5. It is clearly important to determine the amount of small-scale structure present on strings.

## Acknowledgements

We are grateful to Martin Hähnelt and Houjun Mo for useful discussions.

## References

Albrecht, A. and Turok, N. 1989, *Phys. Rev.*, **D40**, 973
Allen, B. and Shellard, E.P.S. 1990, *Phys. Rev. Lett*, **64**, 119
Bennett, D. and Bouchet, F. 1988, *Phys. Rev. Lett.*, **60**, 257
Bertschinger, E. 1987, *Ap. J.*, **316**, 489
Brandenberger, R., N. Kaiser and Turok, N. 1987 *Phys. Rev.*, **D36**, 2242
Brandenberger, R. and Shellard, E.P.S. 1989, *Phys. Rev.*, **D40**, 2542
Moessner, R. and Brandenberger, R. 1996, *MNRAS*, **280**, 797
Perivolaropoulos, L., Brandenberger, R., and Stebbins, A. 1990, *Phys. Rev.*, **D41**, 1764
Schmidt, M., Schneider, D., and Gunn, J. 1991, in 'The space distribution of quasars', ed. D. Crampton (ASP, San Francisco, 1991), p. 109
Schmidt, M., Schneider, D. and Gunn, J. 1995, *A. J.*, **110**, 68
Storrie-Lombardi, L., McMahon, R., and Irwin, M. 1996, *MNRAS*, **283**, L79
Subramanian, K. and Padmanabhan, T. 1994, 'Constraints on the models for structure formation from the abundance of damped Lyman alpha systems', IUCAA preprint, astro-ph/9402006
Vilenkin, A. and Shafi, Q. 1983, *Phys. Rev. Lett.*, **51**, 1716
Vilenkin, A. and Shellard, E.P.S. 1994 *'Cosmic strings and other topological defects'* (Cambridge Univ. Press, Cambridge)
Zanchin, V., Lima, J.A.S., and Brandenberger, R. 1996, *Phys. Rev.* **D54**, 7129

# ANALYZING REDSHIFT SURVEYS TO MEASURE THE POWER SPECTRUM ON LARGE SCALES

MAX TEGMARK
*Institute for Advanced Study*
*Princeton, NJ 08540; max@ias.edu*

*Max-Planck-Institut für Physik*
*Föhringer Ring 6, D-80805 München*

AND

*Max-Planck-Institut für Astrophysik*
*Karl-Schwarzschild-Str. 1, D-85740 Garching*

**Abstract.** Upcoming large redshift surveys potentially allow precision measurements of the galaxy power spectrum. To accurately measure $P(k)$ on the largest scales, comparable to the depth of the survey, it is crucial that finite volume effects are accurately corrected for in the data analysis. Here we derive analytic expressions for the one such effect that has not previously been worked out exactly: that of the so-called integral constraint. We also show that for data analysis methods based on counts in cells, multiple constraints can be included via simple matrix operations, thereby rendering the results less sensitive to galactic extinction and mis-estimates of the shape of the radial selection function.

## 1. Introduction

Observational data on galaxy clustering are rapidly increasing in both quantity and quality, which brings new challenges when it comes to data analysis. As to quantity, redshifts have been published for a few thousand galaxies 15 years ago. Today the number is and ongoing projects such as the AAT 2dF Survey and the Sloan Digital Sky Survey (SDSS) will raise it to $10^6$ within a few years. Comprehensive reviews of past redshift surveys are given by *e.g.*, Efstathiou (1994), Vogeley (1995), Strauss & Willick (1995) and Strauss (1996), the last also including a detailed description of 2dF and



*D. Hamilton (ed.), The Evolving Universe,* 361–370.

SDSS. As to quality, more accurate and uniform photometric selection criteria (enabled by *e.g.*, the well-calibrated 5-band photometry of the SDSS) reduce potential systematic errors.

This increased data quality makes it desirable to avoid approximations in the data analysis process and to use methods that can constrain cosmological quantities as accurately as possible, without bias. Here we will focus on how to correct for the finite volume of a survey. As is well known, this causes the measured power spectrum to be a convolution of the true power spectrum with some window function which depends on the survey geometry and the data analysis method used. Exact expressions have been derived (see *e.g.*, Feldman, Kaiser & Peacock 1994, hereafter "FKP") for the window function and its normalization for the case where the number density of galaxies is assumed to be known *a priori*, but the more realistic case where the mean galaxy density is determined from the survey itself has thus far only been treated approximately (Peacock & Nicholson 1991; Park *et al.* 1994). The main purpose of this paper is to derive exact expressions for this important correction.

The methods for power spectrum estimation that have been proposed in the literature fall into two categories:

1. Direct Fourier methods
2. Pixelized methods

The direct Fourier methods make use of the exact position of each galaxy, whereas the other methods start by "pixelizing" the data set (by computing counts in cells or expansion coefficients for some set of functions), thereby reducing the problem to manipulating large vectors and matrices. In Section 2, we will derive the finite-volume correction for direct Fourier methods. The corresponding correction for pixelized methods is given in Section 3.

## 2. Finite Volume Correction for Direct Fourier Methods

### 2.1. THE POWER SPECTRUM ESTIMATION PROBLEM

It is customary (see *e.g.*, FKP) to model the observed galaxy distribution as a 3D Poisson process $n(\mathbf{r}) = \sum_i \delta(\mathbf{r} - \mathbf{r}_i)$ with intensity $\lambda(\mathbf{r}) = \bar{n}(\mathbf{r})[1 + \delta_r(\mathbf{r})]$. The function $\bar{n}$ is the selection function of the galaxy survey under consideration, *i.e.*,, $\bar{n}(\mathbf{r})dV$ is the expected (not the observed) number of galaxies in a volume $dV$ about $\mathbf{r}$. The density fluctuations $\delta_r$ are modeled as a homogeneous and isotropic (but not necessarily Gaussian) random field with power spectrum $P(k)$, and the power spectrum estimation problem is to estimate $P(k)$ given a realization of $n(\mathbf{r})$.

## 2.2. THE DIRECT FOURIER APPROACH

Due to space limitations, the method summary below is very brief, and the interested reader is referred to FKP and Tegmark (1995, hereafter "T95") for more detailed introductions to the various methods.

All direct Fourier methods not involving random numbers[1] are specified by choosing a weight function $\psi(\mathbf{r})$ in real space and a set of weights $w_i$ in Fourier space, as defined below. They all involve the following two steps:

1. At a grid of points $\mathbf{k}_i$ in Fourier space, fluctuation amplitudes are estimated by

$$\begin{aligned}\widehat{F}(\mathbf{k}_i) &\equiv \int\left[\frac{n(\mathbf{r})}{\bar{n}(\mathbf{r})}-1\right]\psi(\mathbf{r})e^{-i\mathbf{k}_i\cdot\mathbf{r}}d^3r\\ &= \sum_j \frac{\psi(\mathbf{r}_j)}{\bar{n}(\mathbf{r}_j)}e^{-i\mathbf{k}_i\cdot\mathbf{r}_j}-\widehat{\psi}(\mathbf{k}_i).\end{aligned}\tag{1}$$

   (Here and throughout, hats denote Fourier transforms.)

2. The power $P$ at some given $k$-value, say $k_*$, is estimated by squaring these fluctuation amplitudes, subtracting off their shot noise bias, rescaling them to correct for the integral constraint, and averaging them with some weights $w_i$ that add up to unity:

$$\tilde{P}(k_*)\equiv\sum_i w_i\left[\frac{|\widehat{F}(\mathbf{k}_i)|^2-\sigma_s^2(\mathbf{k}_i)}{N(\mathbf{k}_i)}\right].\tag{2}$$

As we will show in Section 2.6, the new and exact expressions for the shot noise and integral constraint corrections (when $\bar{n}$ is normalized so that $\widehat{F}(\mathbf{0})=0$) are

$$\sigma_s^2(\mathbf{k}) = \left(1+\left|\frac{\widehat{\psi}(\mathbf{k})}{\widehat{\psi}(\mathbf{0})}\right|^2\right)c_s(0)-2\,\mathrm{Re}\left\{\frac{\widehat{\psi}(\mathbf{k})^*}{\widehat{\psi}(\mathbf{0})^*}c_s(\mathbf{k})\right\},\tag{3}$$

$$N(\mathbf{k}) = \left(1+\left|\frac{\widehat{\psi}(\mathbf{k})}{\widehat{\psi}(\mathbf{0})}\right|^2\right)f(0)-2\,\mathrm{Re}\left\{\frac{\widehat{\psi}(\mathbf{k})^*}{\widehat{\psi}(\mathbf{0})^*}f(\mathbf{k})\right\},\tag{4}$$

where the functions $c_s$ and $f$ are defined by

$$c_s(\mathbf{k}) \equiv \int\frac{\psi(\mathbf{r})^2}{\bar{n}(\mathbf{r})}e^{-i\mathbf{k}\cdot\mathbf{r}}d^3r,\tag{5}$$

$$f(\mathbf{k}) \equiv \int\psi(\mathbf{r})^2e^{-i\mathbf{k}\cdot\mathbf{r}}d^3r.\tag{6}$$

[1]Including a random mock survey as in equation (2.1.3) in FKP can never give minimal error bars, since inclusion of random numbers will always increase the variance of the estimator.

If the survey is volume limited, then $\bar{n}$ is independent of $\mathbf{r}$, $c_s(\mathbf{k}) = f(\mathbf{k})/\bar{n}$, and $\sigma_s^2(\mathbf{k})/N(\mathbf{k}) = 1/\bar{n}$.

### 2.3. WEIGHTING THE GALAXIES

Four different choices of the galaxy weighting function $\psi$ have appeared in the literature:

$$\psi(\mathbf{r}) = \begin{cases} 1 \text{ inside survey volume} \\ 0 \text{ outside survey volume} \end{cases} \tag{7}$$

$$\psi(\mathbf{r}) = \bar{n}(\mathbf{r}) \tag{8}$$

$$\psi(\mathbf{r}) = \frac{\bar{n}(\mathbf{r})}{1+\bar{n}(\mathbf{r})P} \tag{9}$$

$$\psi(\mathbf{r}) = \text{eigenfunction of } \left[\nabla^2 - \frac{\gamma}{\bar{n}(\mathbf{r})}\right]. \tag{10}$$

The first choice, *i.e.*,, weighing all galaxies in a survey volume equally, was employed by *e.g.*, Fisher *et al.* (1993). The second choice was used for *e.g.*, the APM survey (Baugh & Efstathiou 1994) — since redshifts were not measured, the radial galaxy weighting by default became the selection function (moreover, modes could of course only be computed in the directions perpendicular to the line of sight). The third choice is that advocated by Feldman, Kaiser & Peacock (1994, hereafter FKP), where $P$ denotes an *a priori* guess as to the power in the band under consideration, and minimizes the variance in the limit when $k^{-1} \ll$ the depth of the survey. The fourth choice corresponds to the method of T95, and gives the narrowest window function for a given variance (the constant $\gamma$ determines the tradeoff).

### 2.4. WEIGHTING THE FOURIER MODES

As to the weights in Fourier space, $w_i$, a common choice (*e.g.*, FKP) is to perform a straight average of all modes in a spherical shell with its radius centered on $k_*$, although when the survey volume is anisotropic, a weighted average giving smaller error bars can generally be obtained by solving a quadratic programming problem (T95).

### 2.5. WINDOW FUNCTIONS

The expectation value of a power estimate $\tilde{P}$ that has been corrected for the shot noise bias and the integral constraint can always be written as

$$\langle\tilde{P}\rangle = \int W(k)P(k)dk, \tag{11}$$

where the function $W$, known as the *window function*, has the property that

$$\int_0^\infty W(k)dk = 1. \tag{12}$$

We can therefore think of $\tilde{P}(k_*)$ as measuring a weighted average of the true power spectrum, with $W$ specifying the weights (for most methods, but not all, these weights are strictly non-negative as well). The window function for a general direct Fourier method is derived in Section 2.6, and is found to be

$$\mathbf{W}(k) \propto \sum_i w_i \int |\widehat{\psi}_i(\mathbf{k})|^2 k^2 d\Omega_k, \tag{13}$$

where $\psi_i$ is given by equation (18) and the angular $\mathbf{k}$-integral is carried out over a spherical shell of radius $k$. In the limit where $k^{-1} \ll L$, where $L$ is the smallest survey dimension, the 3D window function simplifies to $\mathbf{W}(k) \propto \sum_i w_i \int |\widehat{\psi}(\mathbf{k}-\mathbf{k}_i)|^2 k^2 d\Omega_k$.

## 2.6. DERIVATION OF THE INTEGRAL CONSTRAINT CORRECTION

If we knew the selection function $\bar{n}(\mathbf{r})$ *a priori*, before counting the galaxies in our survey, we would be able to probe the power on the largest scales. For modes of wavelength much larger than the survey volume, this would essentially correspond to counting the difference between the observed and expected number of galaxies in our sample. Of course, we do not know $\bar{n}$ *a priori*, so our most accurate way of normalizing the selection function is by using the galaxies in the survey itself. When $\bar{n}$ is normalized in this way, naive application of equation (1) will give the artifact $\widehat{F}(\mathbf{k}) \to 0$ as $\mathbf{k} \to 0$ because fluctuations on the scale of the survey are forced to zero by definition (Peacock & Nicholson 1991).

Let us assume that we know the shape of the selection function but not its normalization. To reflect this, we write

$$\bar{n}(\mathbf{r}) = \eta \bar{n}_0(\mathbf{r}), \tag{14}$$

where $\bar{n}_0$ is our guess as to the shape and $\eta$ is an unknown normalization constant. If we had used $\bar{n}_0$ in place of the correct $\bar{n}$ in equation (1), we would in general not obtain the desired result $\langle \widehat{F}(\mathbf{k}_i) \rangle = 0$ but rather $\langle \widehat{F}(\mathbf{k}_i) \rangle = (\eta - 1)\widehat{\psi}(\mathbf{k})$, which would enter equation (2) as a systematic positive power bias. It is the need to eliminate this problem that forces us to impose an integral constraint. Let $\widehat{\eta}$ denote our estimate of $\eta$. We will choose $\widehat{\eta}$ so that this bias vanishes, *i.e.*,, so that the integral constraint

$$\int \left[ \frac{n(\mathbf{r})}{\widehat{\eta}\bar{n}_0(\mathbf{r})} - 1 \right] \psi(\mathbf{r}) d^3r = 0 \tag{15}$$

holds, or explicitly,

$$\widehat{\eta} \equiv \frac{1}{\widehat{\psi}(\mathbf{0})} \int \frac{n(\mathbf{r})}{\bar{n}_0(\mathbf{r})} \psi(\mathbf{r}) d^3 r = \frac{1}{\widehat{\psi}(\mathbf{0})} \sum_j \frac{\psi(\mathbf{r}_j)}{\bar{n}_0(\mathbf{r}_j)}. \tag{16}$$

This is an unbiased estimator of the density normalization, since $\langle \widehat{\eta} \rangle = \eta$, the true value. Substituting $\bar{n}(\mathbf{r}) = \widehat{\eta} \bar{n}_0(\mathbf{r})$ and equation (16) into equation (1), we obtain

$$\begin{aligned} \widehat{F}(\mathbf{k}_i) &= \frac{1}{\widehat{\eta}} \left[ \int \frac{n(\mathbf{r})}{\bar{n}_0(\mathbf{r})} e^{-i\mathbf{k}_i \cdot \mathbf{r}} \psi(\mathbf{r}) d^3 r - \widehat{\psi}(\mathbf{k}_i) \widehat{\eta} \right] \\ &= \frac{1}{\widehat{\eta}} \left[ \int \frac{n(\mathbf{r})}{\bar{n}_0(\mathbf{r})} e^{-i\mathbf{k}_i \cdot \mathbf{r}} \psi(\mathbf{r}) d^3 r - \frac{\widehat{\psi}(\mathbf{k}_i)}{\widehat{\psi}(\mathbf{0})} \int \frac{n(\mathbf{r})}{\bar{n}_0(\mathbf{r})} \psi(\mathbf{r}) d^3 r \right] \\ &= \frac{\eta}{\widehat{\eta}} \int \frac{n(\mathbf{r})}{\bar{n}(\mathbf{r})} \psi_i(\mathbf{r}) d^3 r \approx \int \frac{n(\mathbf{r})}{\bar{n}(\mathbf{r})} \psi_i(\mathbf{r}) d^3 r, \end{aligned} \tag{17}$$

where the function $\psi_i$ is defined by

$$\psi_i(\mathbf{r}) \equiv \left[ e^{-i\mathbf{k}_i \cdot \mathbf{r}} - \frac{\widehat{\psi}(\mathbf{k}_i)}{\widehat{\psi}(\mathbf{0})} \right] \psi(\mathbf{r}). \tag{18}$$

Since we will have $\widehat{\eta} \approx \eta$ with a relative accuracy $\Delta\widehat{\eta}/\eta$ of order $1/\sqrt{N}$, where $N$ is the number of galaxies in the survey, we can to a good approximation treat $\eta$ as a known constant from here on and take $\eta/\widehat{\eta} = 1$ on the last line of equation (17). Since $\widehat{\psi}_i(\mathbf{0}) = 0$, we now have $\langle \widehat{F}(\mathbf{k}_i) \rangle = 0$, so we see that we have succeeded in eliminating the above-mentioned power bias. The price for this is slightly more complicated equations. Let us now derive the expressions for the shot noise correction and normalization given in Equations (3) and (4).

Since $\widehat{\eta} \approx \eta$, we substitute the last expression of equation (17) into Equation (3) of T95, treating $\bar{n} = \eta \bar{n}_0$ as a known function, which gives

$$\langle |\widehat{F}(\mathbf{k}_i)|^2 \rangle = \frac{1}{(2\pi)^3} \int |\widehat{\psi}_i(\mathbf{k})|^2 P(k) d^3 k + \int \frac{|\psi_i(\mathbf{r})|^2}{\bar{n}(\mathbf{r})} d^3 r, \tag{19}$$

Comparing this with equation (11) and equation (2), we identify the three-dimensional window function as

$$W(\mathbf{k}) \propto |\widehat{\psi}_i(\mathbf{k})|^2 \tag{20}$$

and see that the shot noise correction is

$$\begin{aligned} \sigma_s^2(\mathbf{k}_i) &= \int \frac{|\psi_i(\mathbf{r})|^2}{\bar{n}(\mathbf{r})} d^3 r \\ &= \int \left| e^{-i\mathbf{k}_i \cdot \mathbf{r}} - \frac{\widehat{\psi}(\mathbf{k}_i)}{\widehat{\psi}(\mathbf{0})} \right|^2 \frac{\psi(\mathbf{r})^2}{\bar{n}(\mathbf{r})} d^3 r, \end{aligned} \tag{21}$$

and expanding the square completes our derivation of equation (3). Performing an angular integral of equation (20) completes the proof of equation (13). The normalization coefficient $N(\mathbf{k}_i)$ of equation (2) is determined by the requirement that the window function integrate to unity, *i.e.*,, $N(\mathbf{k}_i) = \int |\hat{\psi}_i(\mathbf{k})|^2 d^3k/(2\pi)^3$. Using Parseval's theorem, we obtain

$$\begin{aligned} N(\mathbf{k}_i) &= \int |\psi_i(\mathbf{r})|^2 d^3r \\ &= \int \left| e^{-i\mathbf{k}_i\cdot\mathbf{r}} - \frac{\hat{\psi}(\mathbf{k}_i)}{\hat{\psi}(\mathbf{0})} \right|^2 \psi(\mathbf{r})^2 d^3r, \end{aligned} \tag{22}$$

and expanding the square as above completes our derivation of equation (4).

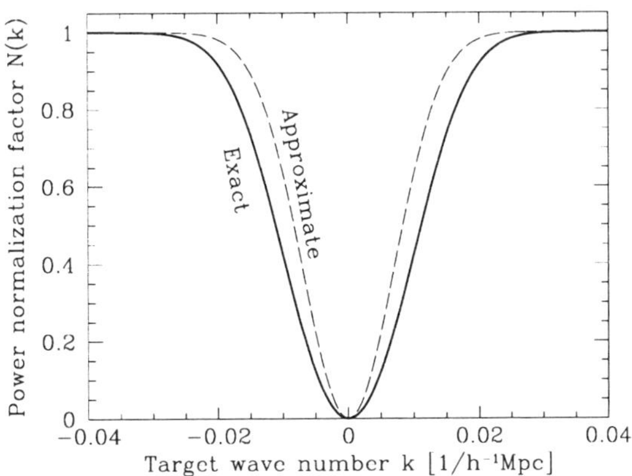


**Figure 1.**
The exact expression for $N(k)$ is plotted together with the approximation of Park *et al.* for a Gaussian weight function $\psi(\mathbf{r}) \propto \exp[-(r/R)^2/2]$, $R = 100h^{-1}$Mpc.

## 2.7. HOW IMPORTANT IS THIS CORRECTION?

Let us evaluate the integral constraint correction factor $N(\mathbf{k})$ for a couple of simple examples. We first note that for the special case of equation (7), we have $\psi(\mathbf{r})^2 \propto \psi(\mathbf{r})$. Hence $f(\mathbf{k}) \propto \hat{\psi}(\mathbf{k})$, and equation (4) reduces to

$$N(\mathbf{k}) = \left(1 - \left|\frac{\hat{\psi}(\mathbf{k})}{\hat{\psi}(\mathbf{0})}\right|^2\right) f(0), \tag{23}$$

which we recognize as the approximation of Park *et al.* (1994). For volume-limited surveys, the prescriptions given by equations (8), (9) and (10) all coincide, so we see that this approximation becomes exact for the volume-limited case with these galaxy weighting schemes. For flux-limited surveys, on the other hand, these schemes all give a decreasing weight function $\psi$, since $\bar{n}$ decreases with distance. For the simple Gaussian case $\psi(\mathbf{r}) = \exp[-(r/R)^2/2]/\pi^{1/4}R^{1/2}$, equation (4) gives

$$N(\mathbf{k}) = 1 + e^{-(Rk)^2} - 2e^{-\frac{3}{4}(Rk)^2}, \tag{24}$$

whereas the approximation (23) gives

$$N(\mathbf{k}) = 1 - e^{-(Rk)^2}. \tag{25}$$

A Taylor expansion shows that for $kR \ll 1$, the latter overestimates $N$ by a factor of two, as illustrated in Figure 1.

## 3. Finite Volume Correction for Pixelized Methods

### 3.1. PIXELIZED METHODS

Pixelized data analysis methods start by reducing the galaxy survey problem to one similar to that occurring in cosmic microwave background (CMB) experiments: estimating a power spectrum given noisy fluctuation measurements in a number of discrete "pixels". After this, the remaining steps are quite analogous to the CMB case, and involve mere linear algebra (operations such as matrix inversion, diagonalization, *etc.*). Let us define the overdensity in $N$ "pixels" $x_1, ..., x_N$ by

$$x_i \equiv \int \left[\frac{n(\mathbf{r})}{\bar{n}(\mathbf{r})} - 1\right] \psi_i(\mathbf{r}) d^3r \tag{26}$$

for some set of functions $\psi_i$. Although the specific choices of $\psi_i$ are irrelevant for our present discussion, common choices are to either make these functions fairly localized in real space (in which case the pixelization is a generalized form of counts in cells) or fairly localized in Fourier space (in which case one refers to the functions $\psi_i$ as "modes" and to $x_i$ as expansion coefficients). Let us group the pixels $x_i$ into an $N$-dimensional vector $\mathbf{x}$. All proposed pixelized methods assume that the mean and the covariance matrix of this pixel vector are

$$\langle \mathbf{x} \rangle = \mathbf{0}, \tag{27}$$

$$\langle \mathbf{x}\mathbf{x}^t \rangle = \mathbf{C}, \tag{28}$$

where $\mathbf{C}$ depends in some known way on the power spectrum. Once the problem has been cast in this form, the power spectrum can be estimated using standard machinery, with either a brute force likelihood analysis (as in *e.g.*, Bunn & White 1997), a Karhunen-Loève eigenmode analysis (Karhunen 1947; Vogeley & Szalay 1996; Tegmark, Taylor & Heavens 1997) or a direct quadratic analysis (Hamilton 1997ab; Tegmark 1997).

### 3.2. DERIVATION OF THE INTEGRAL CONSTRAINT CORRECTION

For pixelized methods of power spectrum estimation, the procedure for dealing with the integral constraint is quite analogous to that for direct

Fourier methods. However, as we will now show, it is much simpler to implement. For counts in cells, for example, one simply removes the mean from all rows and columns of the covariance matrix $\mathbf{C}$ before proceeding with the analysis. Because of this simplicity, one can, at an almost negligible numerical cost, take a more ambitious approach and allow for more than one unknown parameter in the selection function. For instance, one can impose the constraints that the radial fluctuation average equals zero for a few hundred different angular bins, thereby eliminating the sensitivity to galactic extinction variations on this scale, as well as requiring that the angular fluctuation average vanish for a number of radial bins to be insensitive to errors in estimating the precise shape of $\bar{n}$.

Let us parametrize the true selection function $\bar{n}$ as

$$\bar{n}(\mathbf{r}) = \sum_{j=1}^{M} \eta_j \bar{n}_j(\mathbf{r}), \tag{29}$$

where $\bar{n}_j$ are known functions and the "nuissance parameters" $\eta_j$, which we group into an $M$-dimensional vector $\eta$, are *a priori* unknown. Let $\bar{n}_0$ denote some *a priori* estimate of $\bar{n}$. Defining the "uncorrected" pixels as

$$x_i' \equiv \int \frac{n(\mathbf{r})}{\bar{n}_0(\mathbf{r})} \psi_i(\mathbf{r}) d^3 r, \tag{30}$$

we find that

$$\langle \mathbf{x}' \rangle = \mathbf{Z}\eta, \tag{31}$$

where the $N \times M$-dimensional matrix $\mathbf{Z}$ is defined by

$$\mathbf{Z}_{ij} \equiv \int \frac{\bar{n}_j(\mathbf{r})}{\bar{n}_0(\mathbf{r})} \psi_i(\mathbf{r}) d^3 r. \tag{32}$$

This means that in general, $\langle \mathbf{x}' \rangle \neq 0$, so the uncorrected data set does not satisfy equation (27). Instead, its statistical properties depend on the unknown nuissance parameters $\eta$. However, we can easily construct a new "corrected" data set whose mean is independent of $\eta$. Let us define

$$\mathbf{x} \equiv \mathbf{\Pi} \mathbf{x}', \tag{33}$$

where

$$\mathbf{\Pi} \equiv \mathbf{I} - \tilde{\mathbf{Z}} \tilde{\mathbf{Z}}^t, \tag{34}$$

and $\tilde{\mathbf{Z}}$ is a matrix whose rows form an orthonormal basis ($\tilde{\mathbf{Z}}^t \tilde{\mathbf{Z}} = \mathbf{I}$) for the space spanned by the rows of $\mathbf{Z}$.[2] $\mathbf{\Pi}$ is a symmetric ($\mathbf{\Pi}^t = \mathbf{\Pi}$) projection

[2]Such a matrix $\tilde{\mathbf{Z}}$ is readily constructed by orthonormalizing the rows of $\mathbf{Z}$ with a Gram-Schmidt or Cholesky procedure (as in *e.g.*, Tegmark & Bunn 1995).

matrix ($\mathbf{\Pi}^2 = \mathbf{\Pi}$) projecting onto the subspace orthogonal to the columns of $\mathbf{Z}$, *i.e.*,, $\mathbf{\Pi Z} = \mathbf{0}$. Our corrected data set $\mathbf{x}$ satisfies equation (27), since $\langle \mathbf{x} \rangle = \mathbf{\Pi Z}\eta = \mathbf{0}$. Letting $\mathbf{C}'$ denote the covariance matrix of the uncorrected data set, the corrected data will have the covariance matrix

$$\mathbf{C} \equiv \langle \mathbf{x}\mathbf{x}^t \rangle = \mathbf{\Pi C}'\mathbf{\Pi}. \tag{35}$$

Once $\mathbf{x}$ and $\mathbf{C}$ have been computed, the rest of the pixelized analysis proceeds just as if there had been no integral constraints. The only complication is that $\mathbf{C}$ is now singular, having rank $N - M$ instead of $N$. As shown in the Appendix of T97, the correct way to deal with this is to replace all occurrences of $\mathbf{C}^{-1}$ (which is of course undefined) by the "pseudo-inverse" of $\mathbf{C}$, defined as

$$\mathbf{\Pi} \left[ \mathbf{C} + \gamma \mathbf{Z}\mathbf{Z}^t \right]^{-1} \mathbf{\Pi} \tag{36}$$

for some constant $\gamma \neq 0$. The result is independent of $\gamma$, but a good choice for numerical stability is $\gamma \sim c/N$, where $c$ is the order of magnitude of a typical matrix element of $\mathbf{C}$.

The author wishes to thank Josh Frieman, Andrew Hamilton, Michael Strauss and Michael Vogeley for helpful comments on the manuscript.

## References

Baugh, C. M. & Efstathiou, G. 1994, *MNRAS*, **267**, 323.
Bunn, E. F. & White, M. 1997, *ApJ*, **480**, 6.
Efstathiou, G. P. 1994, in "Les Houches Lectures 1993", eds. Shaefer & Silk (Elsevier, Netherlands).
Feldman, H. A., Kaiser, N. & Peacock, J. A. 1994, *ApJ*, **426**, 23 ("FKP").
Fisher, K. B. *et al.* 1993, *ApJ*, **402**, 42.
Hamilton, A. J. S. 1997a, preprint astro-ph/9701008, MNRAS, in press.
Hamilton, A. J. S. 1997b, preprint astro-ph/9701009, MNRAS, in press.
Karhunen, K., *Über lineare Methoden in der Wahrscheinlichkeitsrechnung* (Kirjapaino oy. sana, Helsinki, 1947).
Park, C., Vogeley, M. S, Geller, M. J. & Huchra, J. P. 1994, *ApJ*, **431**, 569.
Peacock, J. A. & Nicholson, D. 1991, *MNRAS*, **253**, 307.
Strauss, M. A. & Willick, J. A. 1995, *Phys. Rep.*, **261**, 271.
Strauss, M. A. 1996, astro-ph/9610033, to be published in "Structure Formation in the Universe", ed. A. Dekel & J. P. Ostriker.
Tegmark, M. 1995, *ApJ*, **455**, 429 ("T95").
Tegmark, M. & Bunn, E. F. 1995, *ApJ*, **455**, 1.
Tegmark, M., Taylor, A. & Heavens, A. F. 1997, *ApJ*, **480**, 22.
Tegmark, M. 1997, astro-ph/9611174, *Phys. Rev. D*, in press.
Vogeley, M. S. 1995, in "Wide-Field Spectroscopy and the Distant Universe", eds. Maddox & Aragón-Salamanca (World Scientific, Singapore).
Vogeley, M. S. & Szalay, A. S. 1996, *ApJ*, **465**, 43.

# CONSTRAINTS ON THE LARGE-SCALE GAS DISTRIBUTION FROM CMB – X-RAY CORRELATIONS

R. KNEISSL
*Max-Planck-Institut für Astrophysik*
*Karl-Scharzschild-Str. 1, 85748 Garching, Germany*

**Abstract.** Observational constraints on the density and temperature of the intergalactic medium are discussed. The use of correlation analysis between fluctuations in the cosmic microwave and X–ray backgrounds, for the study of inhomogeneities in the IGM is demonstrated, and a practical example from results of a cross-correlation analysis between ROSAT All-Sky Survey and COBE DMR data is given.

## 1. Introduction

Important properties of the intergalactic medium (IGM), such as temperature, density, clumpiness depending on scale and ionization state, can be tested directly through observations of, *e.g.*, the cosmic microwave background (CMB), the X-ray background (XRB) or quasar absorption features. In particular, the mean temperature $T$ and density, given in units of the critical density as $\Omega_{\rm IGM}$, of the baryonic content of the IGM, are constrained by the Comptonization parameter $y$ as limited by the lack of distortions of the CMB spectrum, the intensities of the soft and hard XRB, as well as the lack of continuous absorption on the one hand (Gunn–Peterson effect), and observations of discrete absorption (mean flux decrement) on the other hand, in distant quasar spectra. Additionally the density is constrained by primordial nucleosynthesis (BBN) predictions and can be compared to the measured baryonic content of clusters of galaxies.

Present observational and theoretical constraints on the homogeneous IGM are summarized in Fig. 1. For the temperature constraints, a thermal history of the IGM has to be assumed. Given the lack of detailed knowledge a simple assumption of purely adiabatic cooling,

$$T(z) = T_0(1+z)^2, \tag{1}$$



D. Hamilton (ed.), The Evolving Universe, 371–376.

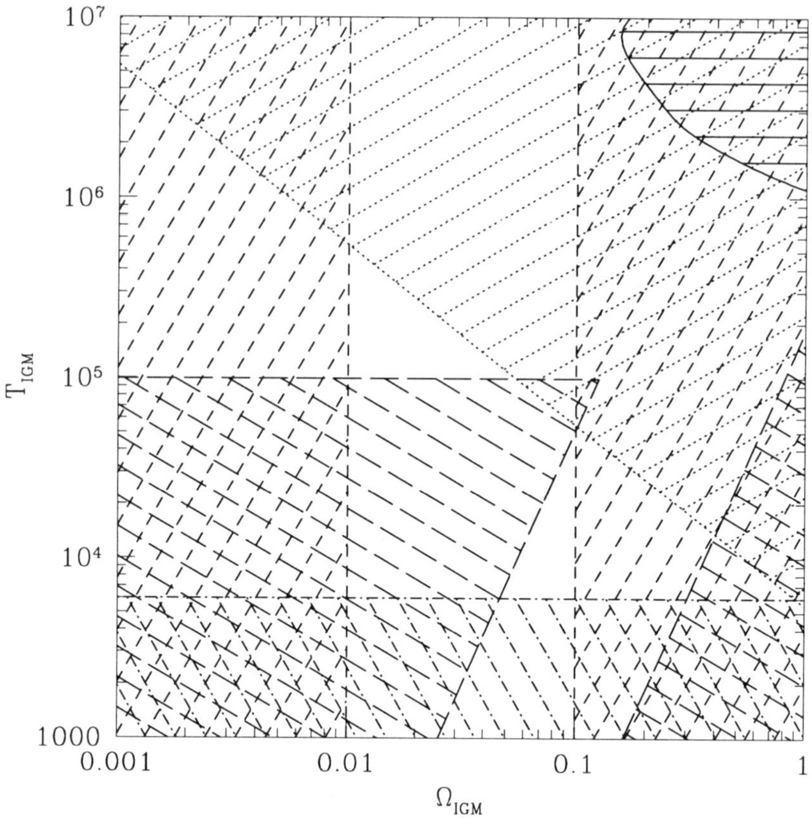


*Figure 1.* Observational constraints on the temperature and density of the homogeneous intergalactic medium set by the spectrum of the CMB (dotted line), the intensity of the soft extragalactic smooth XRB (solid line), Ly$\alpha$ absorption in quasar spectra (long dashed lines) and the effect of a photoionizing background (dotted–dashed line). In addition the constraints from BBN (short dashed lines) are shown. For details see the text.

might be employed. When heating processes, such as photoionization, are assumed to be most effective ealier than a maximum redshift ($z \sim 5$), up to which the IGM is probed by the experiments, then adiabatic cooling gives a minimum cooling rate, and conservative upper limits for the present day temperature, from observational constraints.

A hot, dense phase of the IGM is largely ruled out by measurements of the CMB spectrum. In Fig. 1 the most recent limit from the full COBE FIRAS data set of $|y| \lesssim 1.5 \times 10^{-5}$ (Fixsen *et al.* 1996) are used. For upper limits on the intensity of a truely diffuse XRB we use the analysis by

Hasinger *et al.* (1993) derived from deep PSPC observations in the direction of the Lockman hole, 2 keV $cm^{-2}$ $s^{-1}$ $sr^{-1}$ $keV^{-1}$, which is 25 % of the total extragalactic XRB intensity. It is clear from Fig. 1 that diffuse gas can contribute only a much smaller fraction to the XRB intensity, as was first demonstrated in a similar plot with the less stringent limits of the time by Barcons, Fabian and Rees (1991).

The classical BBN limits are somewhat weak; the lower one, because not all the baryons which formed initially still have to be in the IGM, and the upper BBN limit has been questioned recently due to measurements of a low deuterium–to–hydrogen ratio (Tytler *et al.* 1996). Together with the assumption of an extremely low value for $h_0$ (0.3-0.4), this is a way of reconciling the observations of a high baryon abundance in clusters of galaxies, and the implications for the baryon density (White *et al.* 1993) in a flat universe. Classical BBN is the only constraint in Fig. 1 excluding parameters around $(\Omega_{\rm IGM}, T_{\rm IGM}) = (0.2, 10^4 \text{ K})$.

The limiting line at $6 \times 10^3$ K is drawn to remind that a background photoionizing the IGM is also expected to heat it up to these temperatures, at least at redshifts of 2–3.

The limits deduced via absorption by neutral hydrogen gas in high redshift quasar spectra can be used in two ways. On one hand, the Gunn–Peterson effect sets upper limits to the HI density through the lack of continuous absorption. These can be interpreted as limits on the total hydrogen density for some ionization fraction. The temperature dependence of the ionization fraction is reflected in the limiting lines of Fig. 1. On the other hand, discrete absorption is observed in quasar spectra and the corresponding flux decrement can be employed to derive a lower limit on the baryon density causing the Ly$\alpha$ absorption lines (Weinberg *et al.* 1997). The limits shown in Fig. 1 are derived at a redshift of 3, where a flux decrement of $\bar{D} \sim 0.35$ is measured (Rauch *et al.* 1997), and a photoionization rate of $1 \times 10^{-12}$ $s^{-1}$ is assumed. This argument works around $T_{\rm IGM} \sim 10^4$ K and breaks down at higher temperatures where no significant contribution to the absorption is expected (indicated in Fig.1 by the horizontal break at $T_{\rm IGM} \sim 10^5$ K). This is suggestive of a possible 2–phase model with collapsed, shock heated, high temperature peaks of low total baryonic density on a low temperature, high total baryonic density background.

At this point the simple, homogeneous model of the IGM breaks down and needs to be replaced by more specific modeling of regions of parameter space, incorporating density perturbations by at least a simple clumping description for a specific scale. In the following the region of interest to be investigated further will be the one accessible to a combined analysis of CMB and X-ray observations.

The parameter space under consideration can be extended from the homogeneous (isotropic) case to the inhomogeneous (anisotropic) case, when spatial fluctuations in the properties of the IGM are considered. These correspond to angular fluctuations in the observed quantities, or variations along the line of sight, if they can be measured as in quasar absorption lines. The inhomogeneous case is of particular interest since considerable observational progress has been made recently in this direction. In particular, a combined fluctuation analysis of X-ray and CMB data can set improved upper limits on fluctuations in the temperature of the IGM.

## 2. Data Analysis

In a recent paper Kneissl *et al.* (1997) investigated the possibility of a correlation between the COBE DMR microwave data and the soft XRB probed by the ROSAT All-Sky Survey data. A special region of the sky ($+40° < b$, $70° < l < 250°$) and X-ray energy range (R6 band, 0.73–1.56 keV) with high sensitivity to the extragalactic soft XRB was selected. The question of a correlation with the extragalactic XRB is confused by a Galactic component of hot gas ($T \sim 2 \times 10^6$ K) of large angular scale, which aligns with structure in the DMR maps of the order of the quadrupole. This coronal Galactic feature can be removed by means of gradient fitting on the field, to give the lowest limits on an extragalactic correlation on scales between the resolution limit and the size of the field (7° - 40°).

In the best estimate for an upper limit on extragalactic correlation the authors found 4.5 $\mu$K compared to 30 $\mu$K cosmic signal in the DMR 53+90 GHz A+B and 0.09 keV cm$^{-2}$ s$^{-1}$ sr$^{-1}$ kev$^{-1}$ compared to 0.9 keV cm$^{-2}$ s$^{-1}$ sr$^{-1}$ kev$^{-1}$ fluctuation amplitude in the ROSAT R6 maps.

## 3. Constraints in $T$–$n_e$ Parameter Space for the clumped IGM

Extending the parameter space by two more parameters, the clumpiness $C = \sqrt{< n_e^2 > / < n_e >^2}$ on a fixed angular scale $\Theta$, related to physical scale by

$$d(z) = \frac{2c}{H_0} \, \Theta \, \frac{1 - (1+z)^{-\frac{1}{2}}}{1+z}, \qquad (2)$$

we are able to describe the results of the above described cross-correlation analysis of the fluctuations in the CMB and the XRB. Here we consider only density inhomogeneities over a homogeneous temperature distribution. We also simply relate an angular scale to a proper size at a certain epoch assuming one cloud per beam, and hence neglect statistical wiping out of fluctuations. Careful treatment of the problem involves the simulation of the density field either numerically, *e.g.*, Scaramella, Cen and Ostriker (1993),

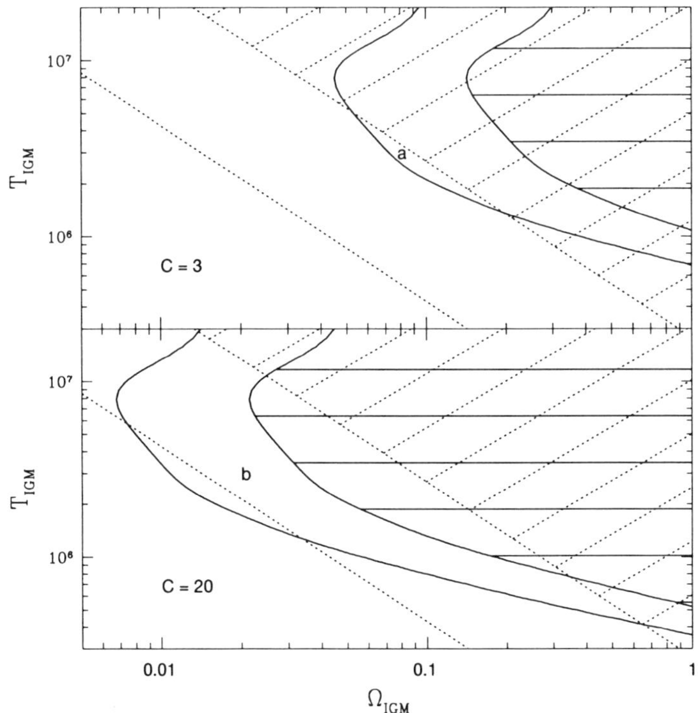


*Figure 2.* Representation of constraints on extragalactic gas parameters from a cross-correlation analysis between COBE DMR 53+90 GHz A+B (dotted lines) and ROSAT All-Sky Survey R6 data (solid lines). The shaded areas are excluded by the respective autocorrelation functions. The offset dotted and solid lines indicate the potential improvement in sensitivity by a joint analysis, but the additional constraint is represented by the overlap regions (labeled **a** for a clumping factor of 3, and **b** for $C = 20$) of the unshaded areas, between the offset lines and shaded areas. For CMB and X-ray observations the effect of the IGM out to a redshift of 1 is considered. The figure shows that significant improvement for an interesting range of clumped IGM matter can be achieved by combining the data.

or semianalytically, *e.g.*, Bond and Myers (1997), including observational results, *e.g.*, Markevitch *et al.* (1992).

The fluctuations constrained in Fig. 2 have clumping factors of 3 and 20 on scales upwards of the resolution limit of 7°, which corresponds to a physical scale of 7 $h^{-1}$ Mpc at a redshift of 0.02, and 90 $h^{-1}$ Mpc at a redshift of 0.5. The fluctuations in the CMB limit $\delta y$, the variations in the integrated pressure over a broad range of densities and temperatures, whereas the fluctuations in the XRB limit towards low densities, around the temperature at which the used X-ray detector, for given Galactic absorption, is most sensitive ($\sim 1$ keV $= 10^7$ K for the soft XRB probed by ROSAT). Combining the two measurements leads to a considerable increase in sensitivity of an order of magnitude, in a region of parameter space accessible to both measurements.

## 4. Example: Extended Gas Halo Model for Abell Clusters

Soltan *et al.* (1996) find an extended correlation out to $\sim 5°$–$7°$ between the ROSAT soft XRB and a sample of clusters of galaxies from the Abell catalog with a dependence of the extent on distance class. This correlation is attributed to extended emission around the clusters. A simple model is a halo of smoothly distributed $\sim$ 1–5 keV gas. With the measured extent of 10 $h^{-1}$ Mpc for the nearby cluster sample at an average redshift of 0.063, a gas density of $n_e = 2$–$8 \times 10^{-6} h^{\frac{1}{2}}$ cm$^{-3}$ can be estimated from the X-ray flux. Using the limits on $\delta y \lesssim 8 \times 10^{-7}$ in Kneissl *et al.* (1997) on a COBE DMR component correlated with the ROSAT template, the gas temperature of this halo can be limited to $T < 2$ keV. A more direct comparison using the Abell clusters as template is in progress.

## 5. Conclusion

We find that comparison of CMB and XRB observations probe an interesting region in the density–temperature parameter space of the clumped IGM, although properties of the homogeneous IGM at high temperatures are more strongly constrained by the measurement of the CMB spectrum.

## Acknowledgements

I thank my collaborators from the ROSAT team, in particular R. Egger for providing the detector response matrix and T. Miyaji for discussion.

## References

Barcons, X., Fabian, A.C., Rees, M.J. (1991) *Nature*, **350**, 685
Bond, J.R., Myers, S.T. (1997) *Astrophys. J. Suppl.*, **103**, 63
Fixsen, D.J., *et al.* (1996) *Astrophys. J.*, **473**, 576
Hasinger, G., *et al.* (1993) *Astr. Astrophys.*, **275**, 1
Kneissl, R., *et al.* (1997) *Astr. Astrophys.*, **320**, 685
Markevitch, M., *et al.* (1992) *Astrophys. J.*, **395**, 326
Rauch, M., *et al.* (1997) *Astrophys. J.* submitted, astro-ph/9612245
Scaramella, R., Cen, R., Ostriker, J.P. (1993) *Astrophys. J.*, **416**, 399
Soltan, A.M., *et al.* (1996) *Astr. Astrophys.*, **305**, 17
Tytler, D., Fan, X.-M., Burles, S. (1996) *Nature*, **381**, 207
Weinberg, D., *et al.* (1997) *Astrophys. J.* submitted, astro-ph/9701012
White, S.D.M., *et al.* (1993) *Nature*, **366**, 429

# ESTIMATES FOR THE LUMINOSITY FUNCTION AND SELECTION FUNCTION OF REDSHIFT SURVEYS

VOLKER SPRINGEL AND SIMON D. M. WHITE
*Max-Planck-Institut für Astrophysik*
*Karl-Schwarzschild-Straße 1*
*85740 Garching bei München, Germany*

**Abstract.** We present a new method to obtain a non-parametric maximum likelihood estimate of the luminosity function and the selection function of a flux-limited redshift survey. The method parameterizes the selection function as a series of step-wise power laws and allows possible evolution of the luminosity function.

We apply our estimators to the 1.2-Jy survey of *IRAS* galaxies. We find a far-infrared luminosity function in good agreement with previously published results.

## 1. Introduction

Flux limited redshift surveys are a major tool for studying the large-scale structure of the Universe. However, such catalogs show a strong decline of the mean number density of galaxies as a function of distance. Most of the statistics that are used to analyse these surveys require an accurate knowledge of this dependence, which is described in terms of the selection function (SF). Hence the accuracy of the adopted SF can limit the reliability of such large-scale structure studies.

Closely related to the SF is the luminosity function (LF), which describes the distribution in luminosity of the galaxy population sampled by the particular redshift survey. This quantity is of more fundamental importance, since it should be reproduced by any viable theory of galaxy formation and evolution.

Here we revisit the problem of determining the LF and SF, given only the data of a flux-limited redshift survey. Current standard methods for this task include Schmidt's (1968) $V/V_{\rm max}$ estimator, and the maximum



*D. Hamilton (ed.), The Evolving Universe*, 377–388.

likelihood techniques first introduced by Turner (1979) and Sandage *et al.* (1979). The maximum likelihood methods are generally superior to the older techniques because they allow the construction of estimators which are not systematically biased by density inhomogeneities.

Two basic procedures have been used to find maximum likelihood estimates of the LF. In the so-called parametric maximum likelihood estimate (Sandage *et al.*1979; Yahil *et al.* 1991) an analytic form for the LF (or SF) is assumed that depends on a few parameters. These are then determined by maximizing the likelihood of the observed data set.

However, because the maximum likelihood technique offers no built-in measure of goodness-of-fit, almost any functional form can be made to 'fit', although the function may provide only a poor description of the data. The parametric technique therefore requires an *a priori* knowledge of a suitable fitting form.

If this information is not available one can allow a very flexible shape of the LF by describing it by many parameters in a reasonable way. This so-called step-wise non-parametric maximum likelihood method (Nicoll & Segal 1983; Efstathiou *et al.* 1988) has been used both to find luminosity functions (Efstathiou *et al.* 1988; Loveday *et al.* 1992; Lin *et al.* 1996) and to estimate the run of radial density with distance (Saunders *et al.* 1990; Loveday *et al.* 1992). So far these applications only employed simple step functions to model the desired function.

In this work we propose a non-parametric maximum likelihood estimator that uses a new parameterization in terms of piece-wise power laws. This method provides accurate information on the shapes of the SF and the LF and has a number of computational advantages. For example, it does not require iterative solutions and it provides error-estimates easily. The method is particularly useful for justifying specific analytic fitting forms for the SF and LF.

In §2 and in §3 we present in detail our non-parametric estimators for the SF and the LF. As an illustration we apply these methods in section 4 to the 1.2-Jy survey of *IRAS* galaxies.

## 2. A non-parametric estimator for the selection function

### 2.1. DEFINITIONS

Let the field $n_z(\vec{r}, L)\,\mathrm{d}L$ describe the comoving number density of galaxies at epoch $z$, in the luminosity interval $[L, L+\mathrm{d}L]$ and at comoving spatial position $\vec{r}$. Assuming that the luminosity distribution is independent of clustering the number density field may be written as

$$n_z(\vec{r}, L) = \frac{n_z(\vec{r})}{\bar{n}_z}\Phi_z(L), \tag{1}$$

where $\Phi_z(L)$ describes the LF. Here $n_z(\vec{r}) = \int_{L_0}^{\infty} n_z(\vec{r}, L)\,\mathrm{d}L$ is the local number density and $\bar{n}_z = \langle n_z(\vec{r})\rangle$ signifies the mean number density, averaged over many realizations of the Universe. The LF is normalized as $\int_{L_0}^{\infty} \Phi_z(L)\,\mathrm{d}L = \bar{n}_z$ and the dependence on $z$ takes care of a possible time evolution, if present. The luminosity cut $L_0$ may be used to handle a possible formal divergence of the integral over $\Phi_z(L)$ at the lower end.

We define the selection function $S(z)$ as the mean comoving number density of galaxies that one expects to see in a flux-limited survey at redshift $z$. Then $S(z)$ is given by

$$S(z) = \int_{L_{\min}(z)}^{\infty} \Phi_z(L)\,\mathrm{d}L\,, \tag{2}$$

where $L_{\min}(z)$ denotes the minimum luminosity a source at redshift $z$ may have without falling below the flux limit of the catalog. In this work we neglect the peculiar velocities of galaxies and take all redshifts to be cosmological, *i.e.* we adopt a simple redshift-distance relation $z = z(|\vec{r}|)$.

## 2.2. THE LIKELIHOOD EXPRESSION

We now imagine that the catalog is drawn from an underlying parent distribution given by $n_z(\vec{r}, L)$ for $L \geq L_{\min}(z)$ and by 0 for $L < L_{\min}(z)$, where $z = z(\vec{r})$. Then the conditional probability $p(L_i|z_i)\,\mathrm{d}L$ that a source observed at redshift $z_i$ falls into the luminosity range $[L_i, L_i+\mathrm{d}L]$ takes the form

$$p(L_i|z_i)\,\mathrm{d}L = \frac{n_{z_i}(\vec{r}_i, L_i)\,\mathrm{d}L}{\int_{L_{\min}(z_i)}^{\infty} n_{z_i}(\vec{r}_i, L')\,\mathrm{d}L'}\,. \tag{3}$$

The denominator simply counts the available number of galaxies at that distance and the numerator gives the number of galaxies in the particular luminosity range. Upon insertion of equations (1) and (2) this becomes

$$p(L_i|z_i) = \frac{\Phi_{z_i}(L_i)}{S(z_i)}\,. \tag{4}$$

Note that the density fluctuations have dropped out of this expression and so the dependence on $\vec{r}$ can be dropped in equation (3). This insensitivity to density inhomogeneities makes the maximum likelihood technique used here superior compared to older methods like the ordinary $V/V_{\max}$ estimator. If one now maximizes the likelihood

$$\mathcal{L} = \prod_i p(L_i|z_i) \tag{5}$$

of the whole data set with respect to $S(z)$ one obtains an estimate of the SF that is not systematically biased by local density fluctuations.

In order to find this maximum in practice we first express $\Phi_{z_i}(L_i)$ in terms of the SF with the help of equation (2). Here a model for the evolution of the LF has to be specified. For brevity we will only treat a case with pure density evolution according to

$$\Phi_z(L) = g(z)\Phi_0(L). \tag{6}$$

However, our method can be easily generalized to include luminosity evolution as well.

We further define a maximal redshift $z_i^{\mathrm{m}}$ for each source such that $L_{\mathrm{min}}(z_i^{\mathrm{m}}) = L_i$. If then the derivative $S'(z)$ of equation (2) is evaluated at $z_i^{\mathrm{m}}$ one obtains

$$S'(z_i^{\mathrm{m}}) = -\frac{g(z_i^{\mathrm{m}})}{g(z_i)}\Phi_{z_i}(L_i)L'_{\mathrm{min}}(z_i^{\mathrm{m}}) + \frac{g'(z_i^{\mathrm{m}})}{g(z_i^{\mathrm{m}})}\, S(z_i^{\mathrm{m}}), \tag{7}$$

so that the probability of equation (4) can be expressed entirely in terms of $S(z)$ and $g(z)$. Hence one finally has to maximize

$$\begin{aligned}\Lambda = \ln \mathcal{L} \quad = \quad & \sum_i \ln\left(\frac{-S'(z_i^{\mathrm{m}})}{S(z_i)} + \frac{g'(z_i^{\mathrm{m}})}{g(z_i^{\mathrm{m}})}\,\frac{S(z_i^{\mathrm{m}})}{S(z_i)}\right) \\ & + \sum_i \ln\left(\frac{g(z_i)}{g(z_i^{\mathrm{m}})}\right) - \sum_i \ln L'_{\mathrm{min}}(z_i^{\mathrm{m}}),\end{aligned} \tag{8}$$

where the constant sum of the last term may be dropped.

The above form suggests that one might be able to maximize $\Lambda$ simultaneously for $g(z)$ and $S(z)$. However, if equation (4) is re-written for the density evolution model it becomes

$$p(L_i|z_i) = \frac{g(z_i)\Phi_0(L_i)}{\int_{L_{\mathrm{min}}(z_i)}^{\infty} g(z_i)\Phi_0(L)\,\mathrm{d}L} = \frac{\Phi_0(L_i)}{\int_{L_{\mathrm{min}}(z_i)}^{\infty} \Phi_0(L)\,\mathrm{d}L}.$$

So the function $g(z)$ drops out completely and there is no sensitivity to density evolution with this estimator. In fact, this was to be expected since estimates based on the likelihood (4) are independent of the density distribution by construction.

### 2.3. A NON-PARAMETRIC MAXIMUM LIKELIHOOD ESTIMATOR

Here we propose a new variant of the non-parametric method that models $S(z)$ as a series of continuously-linked power laws. This description seems appropriate since the SF is a smooth curve that covers a wide range of values and its local behaviour can be very well approximated by a power law.

We describe $S(z)$ by $n$ pieces. Let $S_k = S(x_k)$ be the values of $S(z)$ at a series of ascending redshifts $x_k$ where $k \in \{1, 2, \ldots, n\}$. Then bin 1 covers $0 < z \leq x_1$, bin 2 covers $x_1 < z \leq x_2$, and so forth. In each piece $k$, $S(z)$ is taken to be a power law of the form

$$S(z) = S_k \left(\frac{z}{x_k}\right)^{m_k} \qquad \text{for } x_{k-1} < z \leq x_k\,, \tag{9}$$

where $m_k$ is the logarithmic slope of the particular piece. These slopes are precisely the quantities needed to characterize the shape of the SF.

The different pieces have to join continuously, because $S(z)$ is an integral over the LF. Continuity requires the $S_k$ to be related by

$$S_k = S_1 \prod_{j=2}^{k} \left(\frac{x_j}{x_{j-1}}\right)^{m_j} \qquad \text{for } 1 < k \leq n\ . \tag{10}$$

Let us further define $a_i$ as the number of the bin to which the maximal redshift $z_i^{\mathrm{m}}$ of galaxy $i$ belongs. Similarly, let $b_i$ be the number of the interval that encloses the redshift $z_i$ of galaxy $i$. Then the likelihood (8) takes the form

$$\begin{aligned} \Lambda \;=\; & \sum_i \left[\ln\left(\frac{-m_{a_i}}{z_i^{\mathrm{m}}} + \frac{g'(z_i^{\mathrm{m}})}{g(z_i^{\mathrm{m}})}\right) + \sum_{j=b_i+1}^{a_i} m_j \ln\frac{x_j}{x_{j-1}}\right] \\ & + \sum_i m_{a_i} \ln\frac{z_i^{\mathrm{m}}}{x_{a_i}} - \sum_i m_{b_i} \ln\frac{z_i}{x_{b_i}} \\ & + \sum_i \ln\frac{g(z_i)}{g(z_i^{\mathrm{m}})} - \sum_i \ln L'_{\mathrm{min}}(z_i^{\mathrm{m}}). \end{aligned} \tag{11}$$

The best estimates for the $m_k$ can be found by solving the likelihood equations

$$\frac{\partial\Lambda}{\partial m_k} = \sum_i \frac{\delta_{k,a_i}}{m_k - z_i^{\mathrm{m}}\frac{g'(z_i^{\mathrm{m}})}{g(z_i^{\mathrm{m}})}} + T_k = 0, \tag{12}$$

where $T_k$ is defined as

$$T_k = \sum_i \sum_{j=b_i+1}^{a_i} \delta_{j,k} \ln\frac{x_j}{x_{j-1}} + \sum_i \delta_{k,a_i} \ln\frac{z_i^{\mathrm{m}}}{x_{a_i}} - \sum_i \delta_{k,b_i} \ln\frac{z_i}{x_{b_i}}. \tag{13}$$

The equations (12) are not fully linear in $m_k$, but *almost*. Since the redshift bins are narrow, replacing $z_i^{\mathrm{m}}$ by $x_{a_i}$ will give a useful first approximation $\tilde{m}_k$ to the true solution $m_k$, which might then quickly be improved by an iteration technique. This starting value can be calculated as

$$\tilde{m}_k = -\frac{n_k}{T_k} + x_k \frac{g'(x_k)}{g(x_k)}, \tag{14}$$

where $n_k$ is the number of galaxies in bin $k$. Note that in the case of no evolution the solution to equation (12) is simply given by $m_k = -n_k {T_k}^{-1}$.

The maximum likelihood method also allows an estimate of the statistical uncertainties of the derived parameters (Kendall & Stuart 1979). Asymptotically the distribution of $\mathcal{L}$ is a multivariate Gaussian around the true values of the parameters. If the information matrix $I$ is defined as

$$I_{ij} = -\frac{\partial^2 \Lambda}{\partial m_i \partial m_j}, \tag{15}$$

then the covariance matrix of the estimates is given by $\mathrm{cov}(m_i, m_j) = (I^{-1})_{ij}$ evaluated at the maximum of $\Lambda$.

Because the $T_k$ do not explicitly depend on any of the $m_k$ the information matrix is simply found to be

$$I_{lk} = \delta_{l,k} \sum_i \frac{\delta_{k,a_i}}{\left(m_k - z_i^{\mathrm{m}} \frac{g'(z_i^{\mathrm{m}})}{g(z_i^{\mathrm{m}})}\right)^2}. \tag{16}$$

One can, therefore, trivially invert this diagonal matrix and find error estimates for the $m_k$ as

$$\mathrm{var}(m_k) = \left[\sum_i \delta_{k,a_i} \left(m_k - z_i^{\mathrm{m}} \frac{g'(z_i^{\mathrm{m}})}{g(z_i^{\mathrm{m}})}\right)^{-2}\right]^{-1} \simeq \frac{m_k^2}{n_k}. \tag{17}$$

This is a surprisingly simple result. In particular, one can solve for each of the $m_k$ independent of all the others. The $m_k$ are mutually uncorrelated which shows that it is essentially the local logarithmic slope of the SF that is determined by the maximum likelihood estimator (5).

Once the non-parametric estimate of the shape of the SF is found, it can be used to find and justify an appropriate analytic fitting function for $S(z)$. For this purpose one can directly employ a minimum $\chi^2$ fit of the $m_k$ to the logarithmic slope of some fitting form for $S(z)$. Of course, once an analytic form has been selected in this way, the values of its parameters can also be determined with the parametric maximum likelihood technique by using the fitting form directly in equation (8).

### 2.4. NORMALIZATION

Because the normalization of the SF is lost with the above estimator it has to be found in a second step. For this purpose we write the SF as

$$S(z) = \psi s(z), \tag{18}$$

where $s(z)$ is the shape of the SF as determined above. Then an unbiased estimate $\tilde{\psi}$ of the factor $\psi$ is given by

$$\tilde{\psi} = \frac{\int_V m(\vec{r})w(\vec{r})\,\mathrm{d}\vec{r}}{\int_V s(\vec{r})w(\vec{r})\,\mathrm{d}\vec{r}} \tag{19}$$

for an arbitrary weight function $w(\vec{r})$. Here $m(\vec{r}) = \sum_i \delta(\vec{r} - \vec{r}_i)$ represents the observed galaxy field and we employ the shorthand notation $s(\vec{r}) = s(z(|\vec{r}|))$.

Following Davis & Huchra (1982), we choose the weight function $w(\vec{r})$ such that the expected variance of $\tilde{\psi}$ is minimized. This is to a good approximation the case for $w(\vec{r}) = [1 + J_3\tilde{\psi}s(\vec{r})]^{-1}$, where $J_3 = \int_V 4\pi r^2 \xi(r)\,\mathrm{d}r$ is the second moment of the two-point correlation function and $V$ denotes the volume used in the normalization.

## 3. A non-parametric estimator for the luminosity function

Using equation (2) it is possible to recover the underlying luminosity distribution if the SF is known, or vice versa. In particular one can readily derive a non-parametric LF estimate from the SF estimator described in the previous section.

We suppose that the SF has been determined with the estimator described above and that the function $L_{\mathrm{min}}(z)$ and its inverse $z_{\mathrm{max}}(L)$ are known. The piecewise description of the SF directly translates into a piecewise description of the LF if boundaries $L_k$ of luminosity intervals are defined by $L_k \equiv L_{\mathrm{min}}(x_k)$. Upon evaluation of the derivative of equation (2) at $z = z_{\mathrm{max}}(L)$ the present day LF results as

$$\begin{aligned} \Phi_0(L) &= \left(\frac{g'(z_{\mathrm{max}})}{g(z_{\mathrm{max}})} S(z_{\mathrm{max}}) - S'(z_{\mathrm{max}})\right) \\ &\quad \times \left(g(z_{\mathrm{max}}) L'_{\mathrm{min}}(z_{\mathrm{max}})\right)^{-1} . \end{aligned} \tag{20}$$

This translates into the piecewise description

$$\begin{aligned} \Phi_0(L) &= \left(\frac{g'(z_{\mathrm{max}})}{g(z_{\mathrm{max}})} - \frac{m_k}{z_{\mathrm{max}}}\right)\left(\frac{z_{\mathrm{max}}}{x_k}\right)^{m_k} \\ &\quad \times \frac{S_k}{g(z_{\mathrm{max}}) L'_{\mathrm{min}}(z_{\mathrm{max}})} \end{aligned} \tag{21}$$

for $L_{k-1} < L < L_k$, which is the desired non-parametric estimate.

The usual way to find a non-parametric maximum likelihood estimate of the LF utilizes a parameterization by a series of step functions (Efstathiou *et al.* 1988; Loveday *et al.* 1992). Compared to this approach the main advantage of our method is that the shape of the LF over each bin is approximately a power law, so it is able to adapt to the true shape of the LF in a flexible way. Only very small discontinuities at the boundaries of bins remain and the estimate $\Phi_0(L)$ of equation (21) traces the smooth LF quite well.

The conventional estimator on the other hand assumes constant $\Phi_0$ over each bin which leads to large discontinuous jumps in the LF at bin boundaries. Efstathiou *et al.* (1988) show that the heights $\Phi_k$ of these bins are related by $\Phi_k \approx \int_{L_k-\Delta L/2}^{L_k+\Delta L/2} \Phi \, \mathrm{d}N(L) / \int_{L_k-\Delta L/2}^{L_k+\Delta L/2} \mathrm{d}N(L)$ to the underlying luminosity distribution. It is therefore more complicated to infer the smooth underlying LF from this non-parametric estimate.

## 4. Results for the 1.2-Jy Redshift Survey

The data of the 1.2-Jy redshift survey (Strauss *et al.* 1990) of *IRAS* galaxies has been published (Fisher *et al.* 1995) and can be retrieved electronically from the Astronomical Data Center (ftp://adc.gsfc.nasa.gov). The 5321 galaxies of the survey are selected from the PSC catalog above a flux limit of 1.2 Jy in the 60 $\mu$m band. The sky coverage is 87.6 per cent, excluding only the zone of avoidance for $|b| < 5^\circ$ and a few unobserved or contaminated patches at higher latitude.

We convert the redshifts to the Local Group frame and use them subsequently to infer distances without further corrections for peculiar velocities. In order to compute the maximal redshift $z_i^{\mathrm{m}} = z_{\mathrm{max}}(L_i)$ accurate $K$- and color-corrections are necessary. Here we use the method of Fisher *et al.* (1992). We adopt an Einstein-de-Sitter universe and approximate the 60 $\mu$m band SED with a straight power law $L_\nu(\nu) \propto \nu^\alpha$ with $\alpha = -2$.

### 4.1. SELECTION FUNCTION

In our determination of the SF and LF we will assume that the LF exhibits density evolution of the form $g(z) = (1+z)^P$ with $P = 4.3$. The actual estimation of the evolutionary rate will be discussed in Springel & White (1997).

Figure 1 shows the estimated local slopes $m_k$ of the SF and a fit based on the functional form

$$S(z) = \frac{\psi}{z^\alpha \left(1 + \left(\frac{z}{z^*}\right)^\gamma\right)^{\beta/\gamma}}. \qquad (22)$$

We used 40 redshift intervals, logarithmically spaced between $x_0 = 0.003$ and $x_{39} = 0.15$, for the non-parametric estimate. A minimum $\chi^2$ fit of the measured $m_k$ to the form of equation (22) resulted in a reduced $\chi^2_\nu$ of 0.87

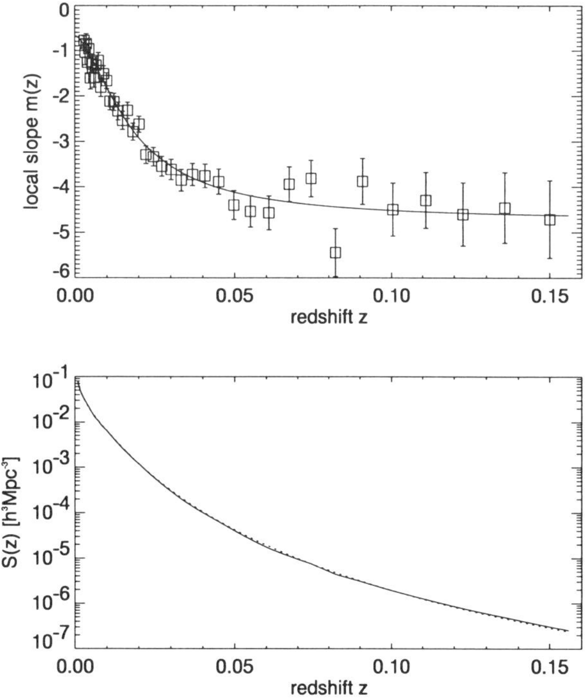


*Figure 1.* Non-parametric selection function estimate. The left panel shows the estimated slopes $m_k$ with $1\sigma$ error bars. The fit is based on equation (22). The right panel compares the actual non-parametric SF (solid) and the analytic fit (dotted) with the parameters of Table 1.

(for $\nu = 36$ degrees of freedom), which indicates a good fit. This form of the SF is slightly more general than the one used by Yahil *et al.* (1991) and Fisher *et al.* (1995) who essentially fixed $\gamma$ at the value 2. This gives a marginally worse $\chi^2_\nu$ of 0.91.

Our best estimates for the final SF parameters are listed in Table 1. They are obtained by maximizing the likelihood of the parametric form directly. Although a fit to the non-parametric estimate gives a very close result, we prefer these numbers as final estimates, because they are free of any binning effects. For each of the parameters $\alpha$, $\beta$, $\gamma$, and $z^\star$ the quoted errors give $1\sigma$ confidence intervals obtained from the bounding box of the $\Lambda_{\max} - 0.5$ likelihood contour.

In order to normalize the SF we used the volume inside $z = 0.15$. This determines the parameter $\psi$ and results in $N = (490 \pm 13)$ $\mathrm{sr}^{-1}$ expected galaxies per unit solid angle on the sky. Our final LF is also normalized to this number. The $1\sigma$ error-estimate is based on a direct computation of the variance of $\psi$ by means of equation (19). Here the uncertainty in the shape

TABLE 1. Parameters of the selection function fit.

| $\alpha$ | $\beta$ | $\gamma$ | $z^\star$ | $\psi$ $[h^3\mathrm{Mpc}^{-3}]$ |
|---|---|---|---|---|
| $0.741^{+0.128}_{-0.135}$ | $4.210^{+0.419}_{-0.344}$ | $1.582^{+0.237}_{-0.214}$ | $0.0184^{+0.00213}_{-0.00167}$ | $(486.5 \pm 13.0)10^{-6}$ |

of $s(z)$ is neglected.

## 4.2. LUMINOSITY FUNCTION

Figure 2 displays our non-parametric estimate of the far-infrared LF resulting from the method outlined in section 3. For comparison also shown is the non-parametric estimate of Saunders *et al.* (1990) which takes the form of a histogram because it uses step functions to model the LF. In order to facilitate a comparison with the literature we present the LF in terms of luminosity per decimal decade of luminosity, *viz.*,

$$\phi(L) = \Phi_0(L)\, L \ln 10, \tag{23}$$

and we define $L$ for a source with SED $L_\nu(\nu)$ as $L = \nu\, L_\nu(\nu)$ at $60\,\mu$m.

We believe that our parameterization offers an improved description of the LF, without large unphysical jumps due to binning of sparse data. In particular, the quality of a fit with an analytic function can be judged quite easily.

As a fitting form we use the function

$$\phi(L) = C \left(\frac{L}{L_\star}\right)^{1-\alpha} \left(1 + \frac{L}{L_\star \beta}\right)^{-\beta} \tag{24}$$

proposed by Lawrence *et al.* (1986) and give the best-fit parameters in Table 2. The errors are $1\sigma$ confidence intervals. The fit is based on a $\chi^2$-minimizing which takes the correlation between the non-parametric LF estimate into account. Details of this procedure will be given in Springel & White (1997).

The simple-two power law (24) does a remarkably good job in fitting the measurements as evidenced by the comparison in Figure 2 and by a reduced $\chi^2_\nu$ of 1.06. In particular we find no need to choose a functional form that shows more curvature over the full range of luminosities (Saunders *et al.* 1990). However, we must not forget that we have ignored velocity fields in this paper and that we cannot go as faint as Saunders *et al.* (1990) who used an additional set of very local galaxies. This might affect the faint end slope; we find a value not quite as shallow as Saunders *et al.* (1990), but somewhat flatter than Yahil *et al.* (1991).

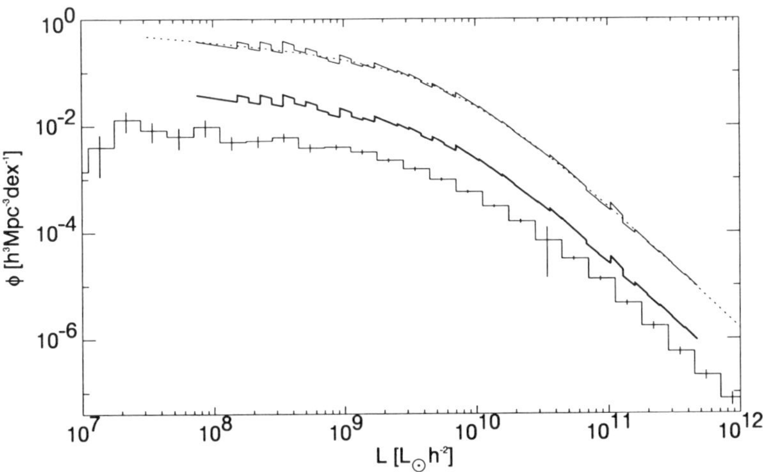


*Figure 2.* Luminosity function estimates. The top panel compares our non-parametric LF estimate (thick) with the estimate of Saunders *et al.* (1990). The latter has been rescaled to $H_0 = 100h$ km s$^{-1}$ Mpc$^{-1}$ and shifted vertically for graphical clarity. The two upper curves (also shifted) compare the non-parametric estimate with our analytic fit.

TABLE 2. Parameters of the luminosity function fit.

| $\alpha$ | $\beta$ | $L_\star$ $[h^{-2}L_\odot]$ | $C$ $[h^3\mathrm{Mpc}^{-3}]$ |
|---|---|---|---|
| $1.221^{+0.068}_{-0.072}$ | $2.116^{+0.080}_{-0.079}$ | $3.615^{+0.640}_{-0.568} \times 10^9$ | $(1.670 \pm 0.045) \times 10^{-2}$ |

## 5. Discussion

An accurate determination of the SF of a redshift survey is a prerequisite for taking full advantage of its information about the large-scale structure of the Universe. For example, statistical methods that rely on estimates of the density field in large survey volumes, like power spectrum measurements or genus statistics, depend crucially on a precise knowledge of the mean density as a function of distance.

In this work we have proposed a flexible non-parametric maximum likelihood estimator for the SF and LF. The method is independent of density inhomogeneities, gives accurate information on the shapes of SF and LF, and provides estimates of the statistical uncertainties of the derived quantities with relative computational ease. We think that the technique should be useful for upcoming redshift surveys.

## References

Davis M., Huchra J. 1982, ApJ, 254, 437
Efstathiou G., Ellis R. S., Peterson B. A. 1988, MNRAS, 232, 431
Fisher K. B., Strauss M. A., Davis M., Yahil A., Huchra J. P. 1992, ApJ, 389, 188
Fisher K. B., Huchra J. P., Strauss M. A., Davis M., Yahil A., Schlegel D. 1995, ApJS, 100, 69
Kendall M. G., Stuart A., 1979, The advanced theory of statistics. Charles Griffin, London, vol. 2
Lawrence A., Walker D., Rowan-Robinson M., Leech K. J., Penston M. V. 1986, MNRAS, 219, 687
Lin H., Kirshner R. P., Shectman S. A., Landy S. D., Oemler A., Tucker D. L., Schechter P. L. 1996, ApJ, 464, 60
Loveday J., Peterson B. A., Efstathiou G., Maddox, S. J. 1992, ApJ, 390, 338
Nicoll J. F., Segal I. E., 1983 A&A, 118, 180
Sandage A., Tammann G. A., Yahil A. 1979, ApJ, 232, 352
Saunders W., Rowan-Robinson M., Lawrence A., Efstathiou G., Kaiser N., Ellis R. S., Frenk C. S. 1990, MNRAS, 242, 318
Schmidt M. 1968, ApJ, 151, 393
Springel V., White S. D. M. 1997, in preparation
Strauss M. A., Davis M., Yahil M., Huchra J. P. 1990, ApJ, 361, 49
Turner E. L., 1979, ApJ, 231, 645
Yahil A., Strauss M. A., Davis M., Huchra J. P. 1991, ApJ, 372, 380

# THE VIRGO CONSORTIUM: THE FORMATION AND EVOLUTION OF GALAXY CLUSTERS

J. M. COLBERG AND S. D. M. WHITE
*Max-Planck-Institut für Astrophysik, Karl-Schwarzschild-Str. 1, 85740 Garching, Germany*

A. JENKINS, F. R. PEARCE AND C. S. FRENK
*Physics Dept., University of Durham, Durham DH1 3LE, UK*

P. A. THOMAS AND R. HUTCHINGS
*MAPS, University of Sussex, Brighton BN1 9QH, UK*

H. M. P. COUCHMAN
*Dept. of Astronomy, University of Western Ontario, London, Ontario N6A 3K7, Canada*

J. A. PEACOCK
*Royal Observatory, Blackford Hill, Edinburgh EH9 3HJ, UK*

G. P. EFSTATHIOU
*Dept. of Physics, Nuclear Physics Building, Keble Road, Oxford OX1 3RH, UK*

AND

A. H. NELSON
*Dept. of Physics and Astronomy, UWCC, PO Box 913, Cardiff, UK*

**Abstract.** We report on work done by the Virgo consortium, an international collaboration set up in order to study the formation and evolution of Large-Scale Structure using N–body simulations on the latest generation of parallel supercomputers. We show results of $256^3$ particle simulations of the formation of clusters in four Dark Matter models with different cosmological parameters. Normalizing the models such that one obtains the correct abundance of rich clusters yields an interesting result: The peculiar velocities of the clusters are almost independent of $\Omega$, and depend only weakly on $\Gamma$, the shape parameter of the power spectrum. Thus, it is nearly impossible to distinguish between high and low $\Omega$ models on the basis of the peculiar velocities.



*D. Hamilton (ed.), The Evolving Universe,* 389–394.

## 1. Introduction

Clusters of galaxies are the largest objects one finds today in the Universe. They are well studied from both observational and theoretical viewpoints (for a recent review *cf.*, Bahcall 1996a).

It may be possible to distinguish between certain high and low $\Omega$ Universes by using the peculiar velocities of the clusters. On this basis Bahcall *et al.* (1996b) concluded that the observed clusters velocity function "is most consistent with a low mass-density ($\Omega \sim 0.3$) CDM model". However, we show in section 4 that this is not possible if the models are normalized such that they all give the correct *abundance* of rich clusters. We present simulation results that support our theoretical considerations in section 5.

Before we discuss the details of the analysis of clusters from the simulations we give in §2 a brief description of the numerical code. In §3 we discuss the models used.

## 2. The Code

The Virgo consortium was formed in order to study the evolution of both Dark Matter and gas in the expanding Universe using the latest generation of parallel supercomputers.

The code used is called "Hydra" which was developed by Couchman *et al.* (1995) and parallelized by Pearce *et al.* (1995). Gravity is treated using an adaptive $P^3M$ technique. This places mesh refinements on the regions of strongest clustering. Large refinements are done in parallel across the processors, smaller ones are farmed out to single processors.

The simulations were run on the Cray T3D supercomputers at the computer center of the Max Planck Society in Garching and at the EPCC in Edinburgh.

## 3. The Dark Matter Simulations

We have carried out a set of very large N-body simulations of CDM universes with four different choices of parameters. The models chosen were Standard CDM (SCDM), a high $\Omega$ model with an additional radiation component ($\tau$CDM), an Open CDM model (OCDM) and a flat low density model with a cosmological constant ($\Lambda$CDM). With the exception of the Open Model, which only had $200^3$ particles, each simulation followed the evolution of $256^3$ particles in a box of $239.5\, h^{-1}$ Mpc[1] on a side.

[1] As usual we express the Hubble constant as $H_0 = 100\, h\, \mathrm{km/(Mpc\, sec)}$.

TABLE 1. The Virgo models

| Model | $\Omega$ | $\Lambda$ | $h$ | $\sigma_8$ | $\Gamma$ |
|---|---|---|---|---|---|
| SCDM | 1.0 | 0.0 | 0.5 | 0.51 | 0.50 |
| $\tau$CDM | 1.0 | 0.0 | 0.5 | 0.51 | 0.21 |
| $\Lambda$CDM | 0.3 | 0.7 | 0.7 | 0.90 | 0.21 |
| OCDM | 0.3 | 0.0 | 0.7 | 0.85 | 0.21 |

In all models, the initial fluctuation amplitude was set by requiring that the models should reproduce the observed abundance of rich clusters (for details see §4). Table 1 gives the parameters of the models.

Figure 3 shows the evolution of the same cluster[2] in the $\tau$CDM and the $\Lambda$CDM models. Structure forms earlier in the low $\Omega$ universe, but already at a redshift of 2 a very large filamentary structure can be seen in both models[3].

## 4. Peculiar Velocities of Galaxy Clusters from Linear Theory

According to linear theory, the mean-square peculiar velocity of galaxy clusters can be approximated by

$$\langle v_{3D}^2 \rangle = \frac{1}{2\pi^2} H_0^2 \, \Omega^{1.2} \int_0^\infty P(k, \Gamma) \, W^2(k, R) \, dk \, , \tag{1}$$

where $W^2(k, R) = \exp[-k^2 R^2]$ is a window function of radius $R$ corresponding to that of a sphere which initially contained the cluster mass (see, *e.g.*, Croft & Efstathiou (1994)). The power spectrum $P(k)$ is taken in the parametric form introduced by Bond & Efstathiou (1984),

$$P(k) = \frac{Bk}{\{1 + [ak + (bk)^{3/2} + (ck)^2]^\nu\}^{2/\nu}} \, , \tag{2}$$

where $a = (6.4/\Gamma) \, h^{-1}$ Mpc, $b = (3.0/\Gamma) \, h^{-1}$ Mpc, $c = (1.7/\Gamma) \, h^{-1}$ Mpc and $\nu = 1.13$. The quantity $\Gamma$ is given by

$$\Gamma = \begin{cases} \Omega_0 h & \text{for SCDM, } \Lambda\text{CDM, OCDM} \\ \Omega_0 h / [0.861 + 3.8(m_{10}^2 \tau_d)^{2/3}]^{1/2} & \text{for } \tau\text{CDM} \end{cases} \tag{3}$$

[2]The simulations were run with the same phases.

[3]The clusters were taken from an additional set of runs with the same parameters as above but box sizes of $85 \, h^{-1}$ Mpc and $141 \, h^{-1}$ Mpc for the high and low $\Omega$ models, respectively.

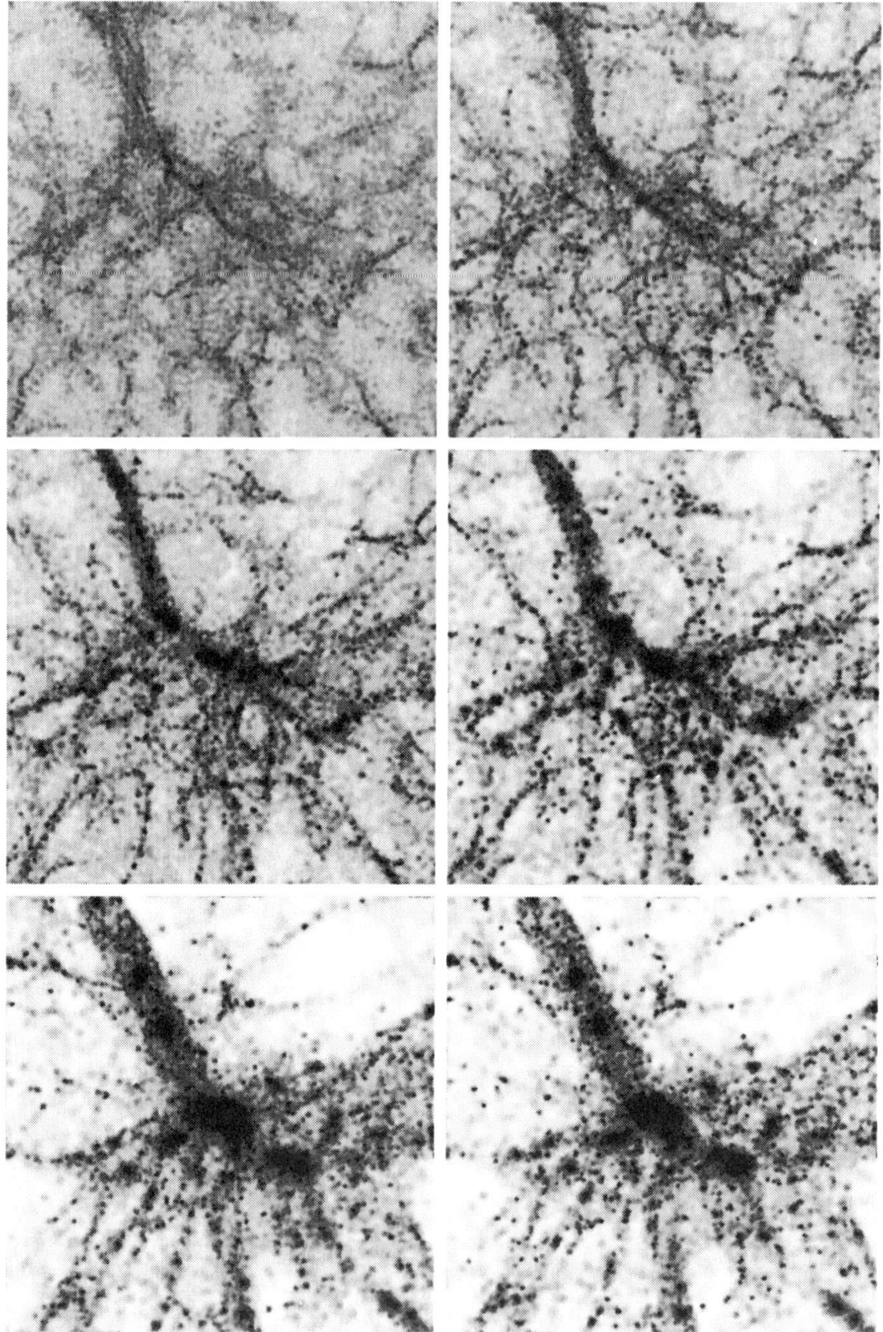

*Figure 1.* The evolution of the same cluster in the $\tau$CDM (leftmost pictures) and in the $\Lambda$CDM model (rightmost pictures), for redshifts of 2 (top), 1 (middle), and 0 (bottom). The sizes of the regions shown are $21 \times 21 \times 8\,h^{-1}\,\mathrm{Mpc}^3$ and $35 \times 35 \times 13\,h^{-1}\,\mathrm{Mpc}^3$ for $\tau$CDM and $\Lambda$CDM, respectively.

where $m_{10}$ is the neutrino mass in units of 10 keV and $\tau_d$ is its lifetime in years (White *et al.* 1995). Equation (3) is valid for any model in a spatially flat universe.

The normalisation $B$ of the power spectrum can be obtained in two different ways. Either one can use a relationship between $B$ and the COBE measurements (as shown in Efstathiou *et al.* (1992)) or one can relate $B$ to the rms linear fluctuation in the mass distribution on scales of $8\,h^{-1}$ Mpc, $\sigma_8$, which is defined by

$$\sigma_8^2 \equiv \frac{1}{(2\pi)^3} \int_0^\infty P(k,\Gamma) \left( \frac{3}{kR_8} j_1(kR_8) \right)^2 d^3k \,, \tag{4}$$

where $R_8 \equiv 8\,h^{-1}$ Mpc, and $j_1$ is a spherical Bessel function.

Values for $\sigma_8$ can be obtained by using either the mass or the X-ray temperature functions of rich clusters (White *et al.* (1993) (WEF); Viana & Liddle (1996); Eke *et al.* 1996 (ECF)). WEF obtain $\sigma_8 \approx 0.57\,\Omega_0^{-0.56}$ for cluster mass functions, while ECF get from temperature functions

$$\sigma_8 = \begin{cases} (0.52 \pm 0.04)\,\Omega_0^{-0.46+0.10\Omega_0} & \text{for } \Lambda_0 = 0 \\ (0.52 \pm 0.04)\,\Omega_0^{-0.52+0.13\Omega_0} & \text{for } \Omega_0 + \Lambda_0 = 1 \end{cases} \tag{5}$$

with the quoted statistical uncertainties obtained using bootstrap methods.

Inserting eq. (4) into eq. (1) yields

$$\langle v_{3D}^2 \rangle = 10^4 \, (\Omega^{0.6}\, \sigma_8)^2 \, f(\Gamma, R) \quad [km/sec]^2 \,, \tag{6}$$

with the abbreviation

$$f(\Gamma, R) \equiv 4\pi h^2 \frac{\int P(k) W^2(k,R) dk}{\int P(k) \left( \frac{3}{kR_8} j_1(kR_8) \right)^2 d^3k} \tag{7}$$

Comparing eqs. (5) and (6) one sees that $(\Omega^{0.6}\sigma_8)^2$ depends only very weakly on $\Omega$. Thus, the peculiar velocities of galaxy clusters are nearly a function of $\Gamma$ alone.

Taking $R = 1.5\,h^{-1}$ Mpc one obtains

$$f(\Gamma, R = 1.5\,h^{-1}\,\text{Mpc}) \approx 12.5\,\Gamma^{-1.08} + 49.4 \tag{8}$$

## 5. Peculiar Velocities of Galaxy Clusters from the Simulations

When comparing N-body simulations with the above formulae one has to find a way to select the clusters in the simulation. WEF find clusters by locating high-density regions with a friends-of-friends group finder,

TABLE 2. Peculiar velocities of galaxy clusters

| Model | $\langle v_{3D}^2 \rangle_{lt}^{1/2}$ [km/sec] | $\langle v_{3D}^2 \rangle_{Sim}^{1/2}$ [km/sec] | $N$ |
|---|---|---|---|
| SCDM | 441 | 461 | 20 |
| $\tau$CDM | 540 | 492 | 21 |
| $\Lambda$CDM | 462 | 452 | 21 |
| OCDM | 437 | 476 | 18 |

and then getting masses from the particle count within spheres of radius $r = 1.5\,h^{-1}$ Mpc. We use the same algorithm and treat objects with a mass larger than $5.5 \cdot 10^{14} h^{-1} \mathrm{M}_\odot$ as galaxy clusters.

In Table 2 we give the *rms* peculiar velocities for our four models from linear theory (first column, using eq.s (6) and (8)), from the simulations (second column), and we also give the number of clusters found in the simulations (third column). Linear theory clearly predicts the simulation results to within their uncertainties. As advertized, our simulations also all predict the same abundance of clusters by mass.

From these numbers it is obvious that the peculiar velocities are not a good way to discriminate between the models once these are normalized to reproduce the correct abundance of rich clusters. In particular, there is no difference between high and low $\Omega$ models.

## Acknowledgements

JMC would like to thank Matthias Bartelmann and Antonaldo Diaferio for numerous helpful and interesting discussions, Martin White for pointing out the correct fitting formula for the $\tau$CDM model, and Don Hamilton for his never-ending patience when reviewing a revised version of this contribution.

## References

Bahcall N.A., preprint astro-ph/9611148 (1996a)
Bahcall N.A., Oh S.P., ApJ, 462L, 49B (1996b)
Bond, J.R., Efstathiou, G., ApJ, 285, L45 (1984)
Couchman H.M.P., Thomas P.A., Pearce F.R., ApJ, 452, 797 (1995)
Croft R.A.C., Efstathiou G., MNRAS, 268, L23 (1994)
Efstathiou G., Bond J.R., White S.D.M., MNRAS, 258, 1P (1992)
Eke V.R., Cole S., Frenk C.S., MNRAS, 282, 263 (1996)
Pearce F.R., Couchman H.M.P., Jenkins A.R., Thomas P.A., 'Hydra – Resolving a parallel nightmare', in Dynamic Load Balancing on MPP systems
Viana P.T.P., Liddle A.R., MNRAS, 281, 323 (1996)
White M., Gelmini G., Silk J., Phys. Rev. D, 51, 2669 (1995)
White S.D.M., Efstathiou G., Frenk C.S., MNRAS, 262, 1023 (1993)

# TOWARD HIGH-PRECISION MEASURES OF LARGE-SCALE STRUCTURE

MICHAEL S. VOGELEY[1]
*Princeton University Observatory*
*Peyton Hall, Princeton, NJ 08544*
*vogeley@astro.princeton.edu*

**Abstract.** I review some results of estimation of the power spectrum of density fluctuations from galaxy redshift surveys and discuss advances that may be possible with the Sloan Digital Sky Survey. I then examine the realities of power spectrum estimation in the presence of Galactic extinction, photometric errors, galaxy evolution, clustering evolution, and uncertainty about the background cosmology.

## 1. INTRODUCTION

The advent of deep redshift surveys of $10^4 - 10^6$ galaxies, such as the Las Campanas Redshift Survey (LCRS; Shectman *et al.* 1996) and the AAT 2df survey (Colless 1998), and multiband photometric and spectroscopic surveys such as the Sloan Digital Sky Survey (SDSS; Gunn & Weinberg 1995; Szalay 1998), marks the beginning of a new era of investigations in large-scale structure. Rather than treat galaxies as indistinguishable tracers of mass in a static distribution, we will study the detailed dependence of clustering on galaxy type and cosmic epoch. In fact, as I will discuss, we must study the species and redshift dependence of clustering to precisely differentiate among cosmological models. In somewhat shallower wide-angle surveys (Huchra *et al.* 1983; Giovanelli & Haynes 1984; Geller & Huchra 1989; da Costa *et al.* 1994; Fisher *et al.* 1995) it has been possible to begin study of these effects; at redshift $z = 0.1$ and larger it becomes critical to do so. In addition to precisely characterizing galaxy clustering at the present epoch, we will use the anisotropy of clustering, as well as the evolution of this clustering, to constrain cosmological parameters.

[1]Hubble Fellow



*D. Hamilton (ed.), The Evolving Universe, 395–418.*

A key issue that I examine is the limiting precision with which we can hope to measure galaxy clustering at the present epoch. Although much has been made of the power of large galaxy surveys to measure cosmological parameters, most of this discussion has focussed on finding an analysis method that gives "optimal" uncertainties on model parameters, where the uncertainties are due to cosmic variance and shot noise (Vogeley & Szalay 1996; Tegmark *et al.* 1998). Much less has been said about the uncertainties caused by random and systematic errors in photometry or Galactic extinction corrections, and systematic uncertainties that arise from issues of cosmology, and evolution of galaxies and their clustering. In this paper I discuss the impact of these effects on analysis of galaxy redshift samples from the SDSS.

## 2. MEASUREMENTS OF THE POWER SPECTRUM FROM GALAXY SURVEYS

Rather than give a review of the myriad measures of galaxy clustering that we hope to apply to these new surveys, here I focus on one statistic and examine how possible random and systematic uncertainties will affect the accuracy of this measurement. Not surprisingly, I consider the power spectrum of galaxy density fluctuations, because lowest-order measures are where we first gain precision, and because uncertainties in higher-order measures often scale from the uncertainties in the variance.

### 2.1. THE DATA

Figure 1 plots the power spectrum $P(k)$ in units of $h^{-3}$ Mpc$^3$ (a power spectral density) as estimated from a number of radio (Peacock & Nicholson 1991), infrared (Fisher *et al.* 1993; Tadros & Efstathiou 1995), and optically-selected (da Costa *et al.* 1994; Lin *et al.* 1995) galaxy redshift survey samples, and the real-space power spectrum inferred from angular clustering of a photometric sample (Baugh & Efstathiou 1993). To avoid hopelessly confusing the plot, no error bars are shown (unfortunately, the various authors differ somewhat in their choice of binning and methods of computing these uncertainties, so comparison of these errors would not be as informative as one might hope). The shapes of most of these spectra roughly agree over the decade $10 - 100h^{-1}$ Mpc. However, there is large variation of their amplitudes. We usually attribute this variation to differences in the "bias parameter," which affects the amplitudes as $P_1(k)/P_2(k) = (b_1/b_2)^2$ if the bias is scale-independent. This strong dependence of clustering amplitude on method of galaxy selection implies that detailed knowledge of the selection criteria and construction of sub-samples for analysis that are homogeneous in this selection are prerequisites for precision measurement

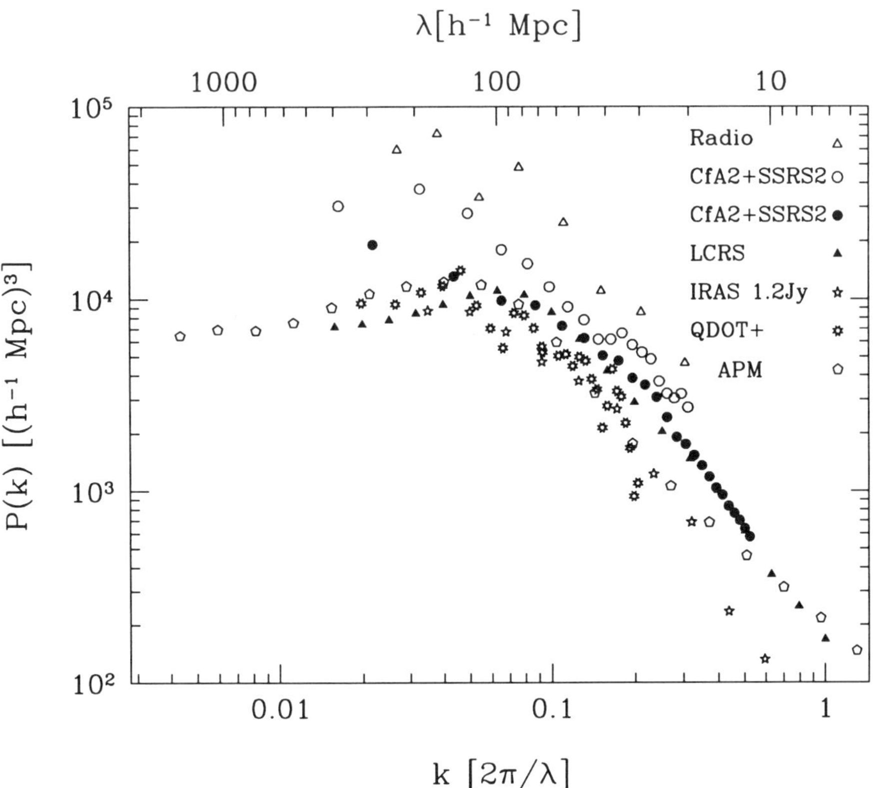


*Figure 1.* Estimates of the redshift-space power spectrum from a variety of redshift surveys, and an estimate of the real-space power spectrum inferred from angular correlations (APM).

of the power spectrum.

To clarify comparison of the shapes of the power spectra, Figure 2 plots these same data, with the amplitudes of the curves shifted so that all match $P(k)$ of the LCRS sample near wavenumber $k = 0.1$. No shift was required to obtain agreement between the LCRS and the CfA2+SSRS2-100 volume-limited sample, which is not surprising since they both include optically-selected galaxies that are roughly $M^*$ and brighter. In general, one sees excellent agreement between optical redshift samples for wavenumbers $k > 0.1$. There is some disagreement between infrared and optical samples over this same range of wavenumber, perhaps because the selected galaxies sample different physical environments. One should be careful not to

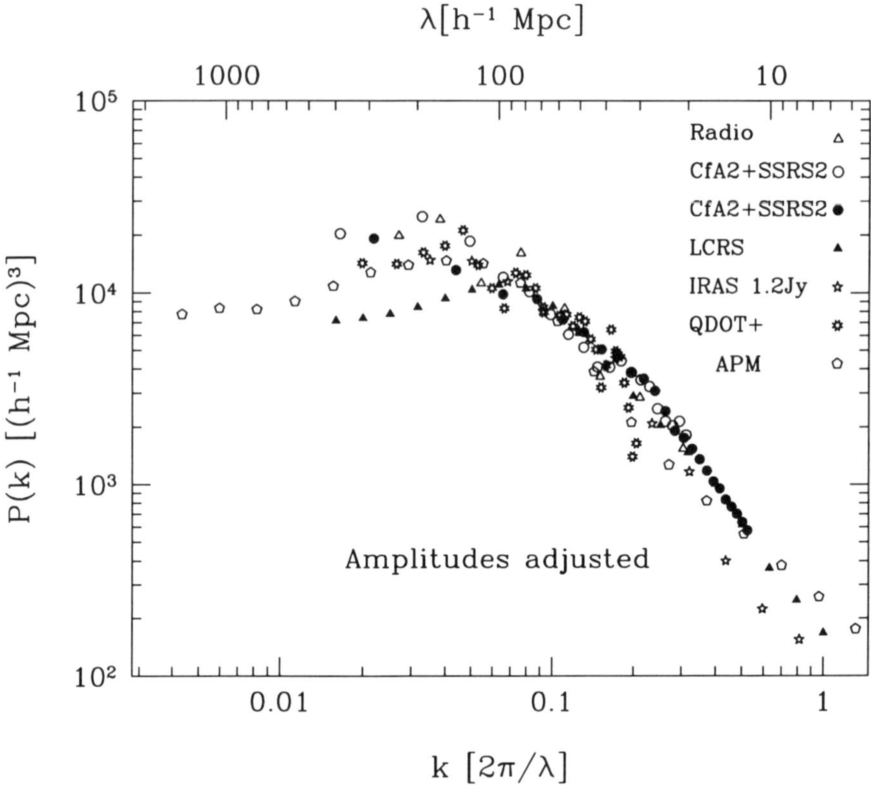


*Figure 2.* Power spectrum estimates from Figure 1 arbitrarily scaled to match at $k = 0.1$ to allow comparison of power spectrum shapes. CfA2+SSRS2-101 and LCRS were not adjusted.

over-interpret this comparison of spectral shapes. Where large shifts are necessary to match the amplitudes, this shifting procedure is questionable, because the effects of redshift distortions (both the linear boost described by Kaiser 1987 and the washing out of small-scale power due to redshift fingers of clusters) and non-linear evolution both depend on the ratio of galaxy-clustering to mass-clustering amplitude. On large scales, at wavenumber $k < 0.1$, we find significant departures in the shapes of the spectra, with a scatter of roughly a factor of three among the samples. This scatter is consistent with the uncertainty due to cosmic variance (small sample volume relative to the wavelength scales of interest).

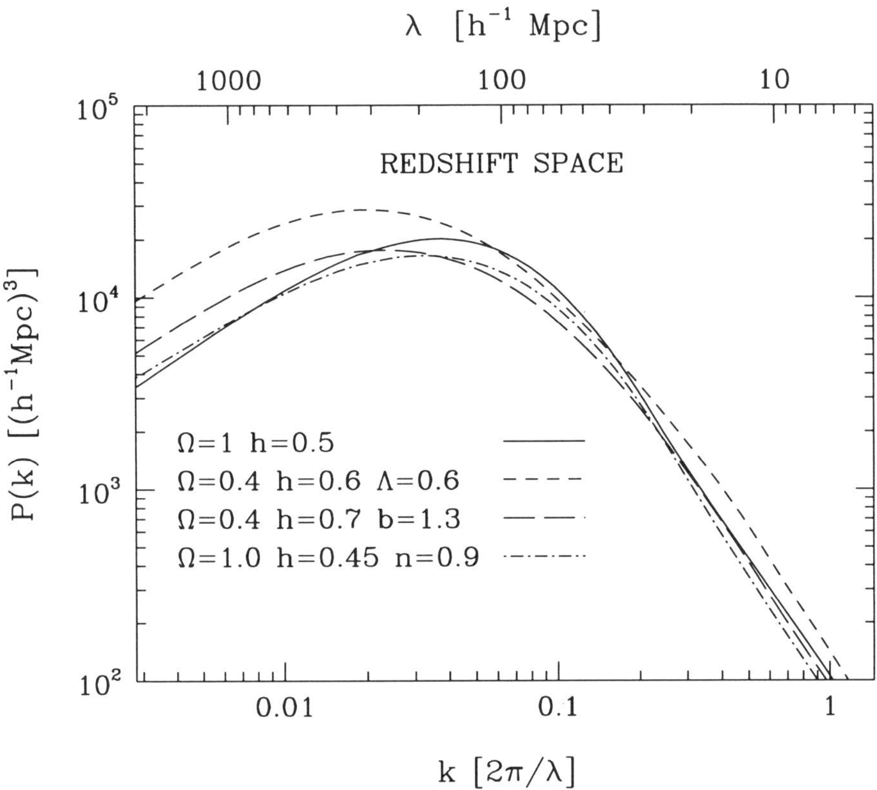


*Figure 3.* Non-linear redshift-space power spectra for a variety of COBE-normalized CDM models.

### 2.2. MODEL FITS

The shape of these spectra may be well-fitted by a linear Cold Dark Matter (CDM) power spectrum (Bardeen *et al.* 1986) with shape parameter $\Gamma = 0.25$ (where $\Gamma = \Omega h$ in the simplest models, with $h = H_0/100$). Here we use the linear CDM spectrum merely as a fitting function. It is interesting to note that, for values of $\Gamma = \Omega h \sim 0.25$, this linear curve does a reasonable job in fitting the redshift-space power spectrum of a fully non-linear evolved CDM model because the effects of non-linear evolution and redshift distortion nearly cancel out.

Does this fit to the spectral shape imply that our work is done? Hardly. This rough shape might be matched by tweaking the parameters of several

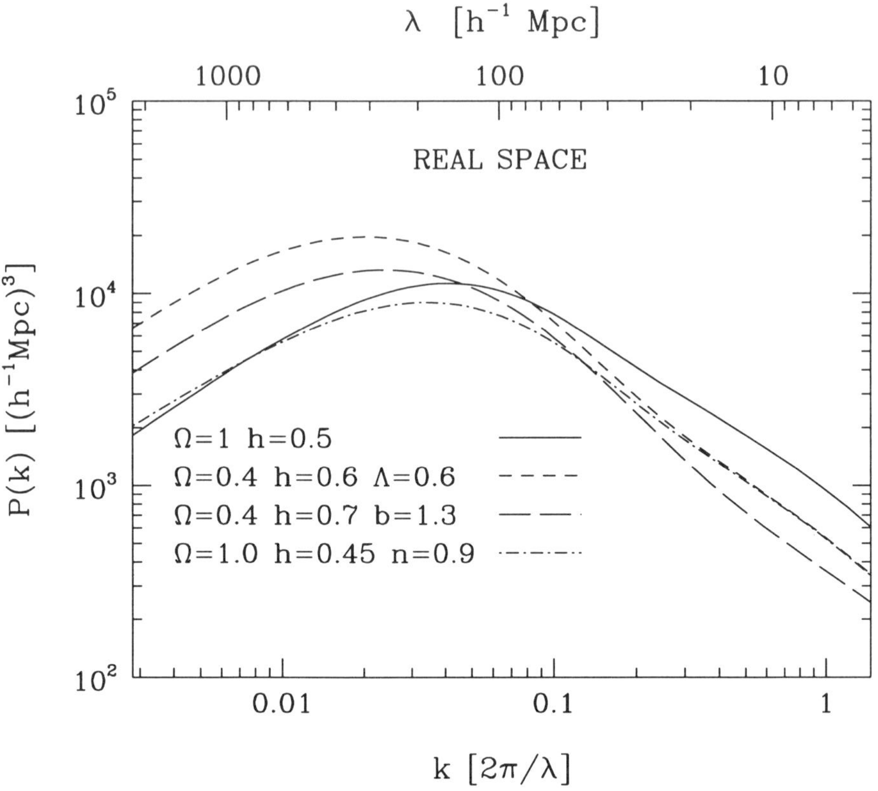


*Figure 4.* Real-space power spectra for the same models plotted in Figure 3.

competing models (*e.g.*, by adding some cosmic density in neutrinos and/or by varying the slope of the primordial power spectrum). We require higher resolution to detect features in the power spectrum that might be caused by acoustic oscillations near recombination (Eisenstein & Hu 1998) or other physical effects. Higher resolution in the Fourier domain requires a larger survey volume (see next section), which is also necessary to probe large wavelength scales. Measurements of the power spectrum to accuracies of a few percent on scales $\lambda > 100h^{-1}$Mpc would allow detailed comparison of clustering of galaxies at the present epoch with clustering of the mass at redshift $z = 10^3$, as revealed by the anisotropy of the Cosmic Microwave Background (CMB).

Figure 3 illustrates how CDM models with different parameters yield redshift-space power spectra that are similar. To compute these spectra, I use the semi-analytic formalism described by Peacock (1997) to approximate the effects of non-linear gravitational evolution and redshift distortions on the linear CDM spectrum. All of these models have a mass spectrum that is normalized to match the COBE DMR measurement of the CMB anisotropy (Wright *et al.* 1996). Even without exploring admixtures of CDM and neutrinos (*e.g.*, Primack *et al.* 1995), we see that several choices of parameters might fit the equally well (or equally poorly, depending on one's taste). Not examined here is how the small-scale power spectrum depends on the details of galaxy formation relative to the mass distribution.

### 2.3. REAL VS. REDSHIFT SPACE

One of the problems of working in redshift space is that a large amount of small-scale power creates a large small-scale velocity dispersion. In redshift space, the resulting "fingers of God" wash out much of this same small-scale power. Similarly, for fixed normalization and Hubble constant, CDM models with large $\Omega$ experience a significant boost of their redshift-space power spectrum amplitudes (1.87 if $\Omega = 1$ and $b = 1$), which makes their large-scale spectra look more like those of low $\Omega$ models.

Redshift-space power spectrum measurements with accuracy of a few percent could differentiate between the model spectra in Figure 3 but, clearly, it would be easier to break the model degeneracy that we see in redshift-space power spectra if we could access the underlying real-space (*i.e.*, configuration space) power spectrum of the galaxy distribution. Figure 4 shows the power spectra of the same models from Figure 3, now plotted in real space.

The lesson here is not that we should use the angular correlation function, and give up measuring redshifts, but rather that we should take advantage of the anisotropy of clustering in redshift space to allow measurement of both the power spectrum and $\Omega$. Andrew Hamilton's contribution to this volume examines this issue in great detail.

## 3. MINIMAL POWER SPECTRUM UNCERTAINTIES FOR THE SDSS

Power spectrum measurements with accuracy of a few percent on scales from one to a few hundred Mpc would allow us to make great strides in constraining cosmological models. Is this feasible? Here I begin to examine the accuracy with which we dare hope to measure $P(k)$ in the foreseeable future.

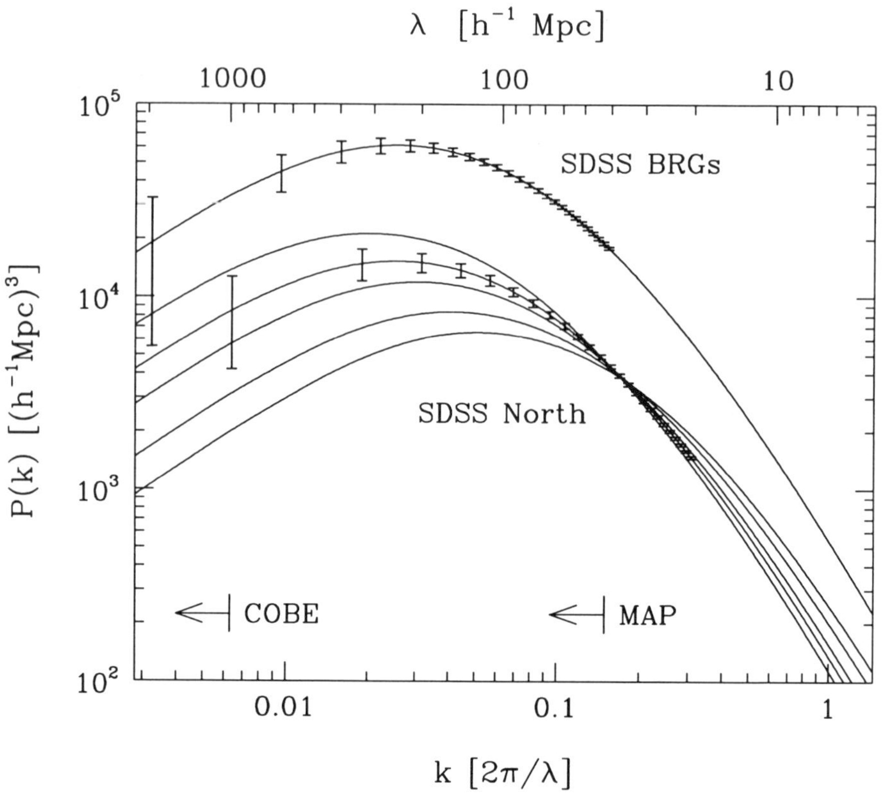


*Figure 5.* Predicted uncertainties in the power spectrum estimated from a volume-limited ($R_{max} = 500h^{-1}$ Mpc) sample of SDSS North and for the Bright Red Galaxy sample (upper set of error bars). These errors assume that the true power spectrum is that of an $\Omega h = 0.25$ CDM model and that the BRGs are more clustered than normal galaxies. Plotted for comparison to the SDSS North errors are CDM power spectra (normalized to $\sigma_8 = 1$) for a range of $\Omega h$ from 0.2 (uppermost curve) to 0.5 (lowest curve) and indications of the range of comoving scales probed by the COBE and MAP CMB anisotropy experiments.

### 3.1. APPROXIMATE UNCERTAINTY PREDICTIONS

For fixed survey strategy, the survey volume and the number density of galaxies in the redshift sample set a lower bound on the uncertainty in the estimated power spectrum. In the limit of a perfectly spherical volume-limited sample, the uncertainty in the estimated power per mode is roughly

$$\frac{\delta \tilde{P}(k)}{\tilde{P}(k)} \approx \sqrt{2\frac{V_c}{V_k}}\left[1+\frac{S(k)}{P(k)}\right] \tag{1}$$

(Feldman, Kaiser, & Peacock 1994). $P(k)$ and $\tilde{P}(k)$ are the true and estimated power spectra, respectively. $S(k) = 1/\bar{n}$ is the shot noise power for mean galaxy density $\bar{n}$. $V_c = (2\pi)^3/V_S$ is the coherence volume in the Fourier domain for a survey with volume $V_S$. It is assumed that we average the power estimates over a shell in Fourier space with volume $V_k \approx 4\pi k^2 \Delta k$. That is, we average the power over all angles, and over bins with width $\Delta k > 2\pi/R$, where $R$ is the survey depth. For a larger volume, the coherence volume in Fourier space decreases, thus we obtain a larger number of independent probes of the power spectrum. This convenient approximation breaks down when the survey volume is highly anisotropic; in this case the coherence volume in Fourier space (the "window function") becomes anisotropic and the range of true wavenumber that is probed depends on the direction of **k**. Figure 8 of Alex Szalay's contribution to this volume shows window function shapes and relative sizes for a variety of redshift surveys, including the SDSS.

A volume-limited sub-sample of the SDSS redshift survey in the North Galactic Cap region (this covers solid angle of $\pi$ steradians) that has comoving coordinate depth of $R = 500h^{-1}$ Mpc will include all galaxies with absolute magnitude roughly $0.4^m$ brighter than $M^*$, with mean number density $\bar{n} \sim 2 \times 10^{-3} h^3 \mathrm{Mpc}^{-3}$. Hereafter I will use this as a fiducial sample for computing uncertainties in the estimated power spectrum. This sample is conservative, in the sense that the SDSS will allow selection of much deeper samples, which include more volume, and because a magnitude-limited sample to the same depth would have larger average number density, hence smaller shot noise.

Figure 5 plots $1\sigma$ uncertainties for power estimates binned by $\Delta k = 2\pi/(500h^{-1}\mathrm{Mpc})$ and assuming that the true power spectrum is described by a $\Gamma = 0.25$ linear CDM spectrum. Also shown are a family of CDM spectra with different shape parameter $\Gamma$. The uncertainty in each bin of width $\Delta k$ at a wavelength scale of $\lambda = 100h^{-1}\mathrm{Mpc}$ will be $\sigma(\hat{P})/\hat{P} \approx 0.06$. Such a measurement would easily differentiate between various CDM models, as the family of CDM curves indicates. Comparison with results from the LCRS further llustrates the power of such a large survey: if a "bump" at $\lambda \sim 128h^{-1}\mathrm{Mpc}$ were found in the SDSS sample with $\delta\hat{P}/\hat{P}_{smooth} = 0.76$ in a single bin (as seen by Landy *et al.* 1996), this would be a $12\sigma$ event.

### 3.2. SAMPLES WITH LARGER VOLUME?

Could we do better than the SDSS with, for example, an all-sky survey to similar depth? This suggestion is not out of the question, if we combine all of the redshift surveys that will exist in a few years' time. The SDSS NGP survey will cover one quarter of the sky; a similar full-sky survey would have power spectrum uncertainties that are at best a factor of two smaller. More realistic at optical wavelengths would be a survey over Galactic latitude $|b| > 30$, which covers half the sky, yielding errors that are at best 40% smaller than the SDSS. In fact, the SDSS itself will cover an effective area larger than $\pi$ steradians, by including three slices in the southern Galactic hemisphere, each $3^\circ \times \sim 100^\circ$ in area. If the SDSS is combined with the AAT 2df redshift survey and the LCRS, then most of the high-latitude sky will be sampled. Galaxy selection in the infrared would allow a true full-sky survey, but no IR photometric survey exists to depth comparable to the SDSS. The IRAS PSC-Z survey (Saunders *et al.* 1994) comes closest to this goal, but has much larger shot noise (smaller galaxy density) and apparently more severe sample evolution.

One-tenth of the million galaxies targeted for spectroscopy by the SDSS will be a sample of Bright Red Galaxies (BRGs) that are volume-limited to redshift $z = 0.45$. The volume limit will be enforced by selecting the galaxies by their absolute magnitudes, estimated using the photometric redshift technique (Connolly *et al.* 1995). This sample will cover a volume eight times larger than the fiducial SDSS-500 sample described above, albeit with larger shot noise. The upper curve and error bars in Figure 5 show the expected uncertainties from cosmic variance and shot noise in the SDSS BRG sample, using equation 1 and assuming that the BRGs are biased by a factor of 2 (factor of 4 in the power spectrum) relative to the normal galaxy sample. The brightest and reddest galaxies appear to evolve least of all, so this sample will provide our best chance to measure the evolution of clustering (see section 5.3 below).

## 4. MEASUREMENT ERRORS AND EXTERNAL SYSTEMATICS

In the previous section I describe "ideal" uncertainties that are lower bounds on the total error budget. I now examine whether these expectations are realistic. What other effects limit our knowledge of clustering at the present epoch? I use the SDSS as a test case, but the following analysis is also relevant for other surveys to similar depth.

There are several possible effects about which we do not need to worry for the SDSS. Because the galaxies in the spectroscopic sample will be five magnitudes brighter than the point source detection limit, star-galaxy separation will not be a problem. Galaxies with very low central surface brightness will be excluded from the redshift sample, because we would not be able to get redshifts of these objects, but we will understand this selection bias quite well. For example, our magnitude and surface brightness cuts will be "fuzzy," such that we observe some fraction of objects beyond the nominal cuts. The redshifts that we do measure will have $20-30$ km s$^{-1}$ accuracy, so this project is overkill for simply measuring redshifts and we expect very few failures. The use of a fiber-fed spectrograph with plugplates to hold the fibers in the focal plane imposes a 55 arcsecond minimum separation for pairs of objects, except where a target lies in the overlap between two or more plugplate fields. However, during the Fall observing season, the larger ratio of spectroscopic time to photometric time (because we will image a smaller area), will allow a larger covering factor for spectroscopy, thus we will obtain spectra for at least all pairs of objects and potentially could observe all n-tuples up to $n = 7$. From these southern data, we will study the effects of excluding close pairs from the northern spectroscopy (*e.g.*, we can study the very small-scale velocity dispersion of galaxies from the southern spectroscopy).

## 4.1. EFFECTS OF EXTINCTION AND PHOTOMETRIC ERRORS

Despite our best efforts at correcting for Galactic extinction and calibrating the photometry, errors in these will add to the minimal uncertainties illustrated in Figure 5. In collaboration with Andy Connolly at Johns Hopkins, I have investigated the impact of such errors on power spectrum estimation from samples of the SDSS (Vogeley & Connolly 1998). The remainder of this section is an overview of some of our results.

Errors in either photometry or the extinction corrections affect the apparent number density of galaxies in similar fashion. Here we consider only the effects of extinction or photometric error on total magnitude selection. Although the SDSS spectroscopic samples will also be chosen on the basis of a central surface brightness cut, simulations of the SDSS indicate that surface brightness dimming from extinction will affect an insignificant number of galaxies.

In a volume-limited sample that should include all galaxies brighter than absolute magnitude $M_{lim}$, a magnitude error of $\Delta m$ over a patch on the sky causes the apparent galaxy density in that direction to differ from the true galaxy density by

$$\frac{n_{obs}}{n_{true}} = \frac{\Phi[< M - \Delta m]}{\Phi[< M]} \approx 1 - \frac{\phi(M)\Delta m}{\Phi[< M]} \quad (2)$$

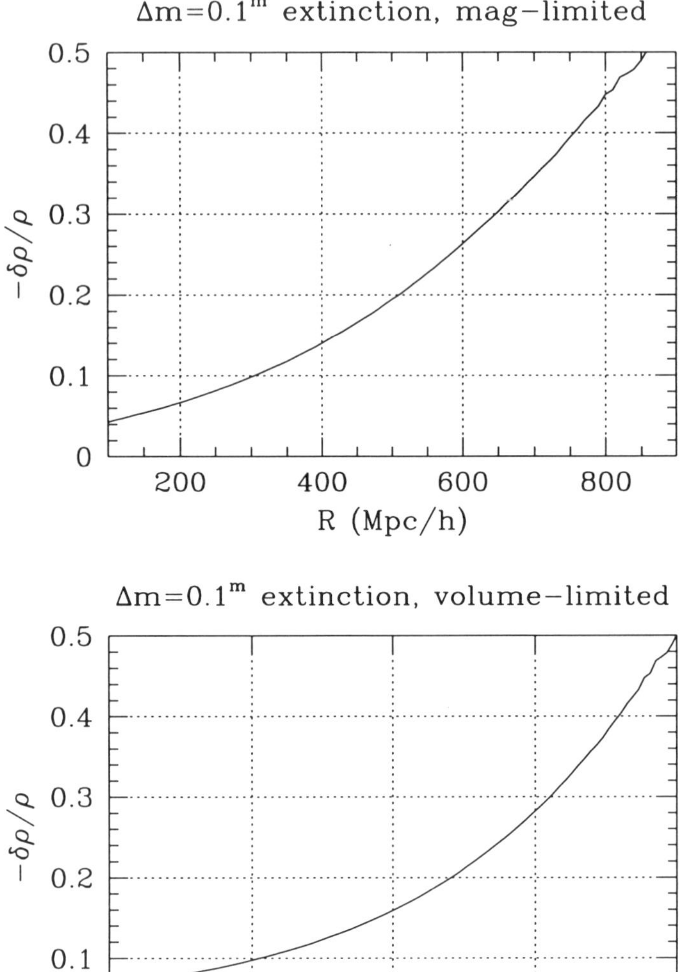


*Figure 6.* Effects on the observed galaxy density that would result from $0.1^m$ of extinction in a magnitude-limited (upper panel) or volume-limited (lower panel) redshift sample of the SDSS.

$$\sim \quad \exp\left[-\Delta m 10^{0.4(M*-M_{lim})}\right],$$

where $\phi(M)$ and $\Phi(< M)$ are the differential and integrated luminosity functions of the galaxies. The approximations are valid for a Schechter luminosity function when $M_{lim}$ is close to or brighter than $M^*$. This form clearly shows how the strength of the effect depends on the slope of the luminosity function near the magnitude limit; deeper sub-samples will be more severely affected by extinction or photometric uncertainty than shallow samples because the count slope is steeper. For a magnitude-limited sample with limit $m_{lim}$, the absolute magnitude limit varies with distance $M_{lim}(r)$ and the apparent density differs from the true galaxy density by

$$\frac{n_{obs}}{n_{true}} = \frac{\Phi[< M(r) - \Delta m]}{\Phi[< M(r)]}. \tag{3}$$

Figure 6 shows the effect of a $\Delta m = 0.1^m$ error on the apparent density of volume-limited and magnitude-limited samples of the SDSS. This modulation of the apparent density grows with distance in a magnitude-limited sample and is, on average, less severe than in an volume-limited sample that is cut off at the same depth. In the following analysis I consider volume-limited samples because (1) the effects are more severe for these, thus the analysis is conservative in the sense of giving worst-case errors and (2) we won't believe the results of magnitude-limited power spectrum analyses until we understand how the clustering amplitude differs among galaxy species (see section 2.1 above).

### 4.2. THE POWER SPECTRUM OF GALACTIC EXTINCTION

Because Galactic extinction modulates the apparent density of galaxies, it could cause erroneous clustering power to appear in our data. To leading order, the net effect is to add to the apparent clustering, $P_{obs}(k) \approx P_{true}(k) + P_{extinction}(k)$. If we make no correction at all for Galactic extinction, we expect to observe extra clustering power in the SDSS-500 sample, as shown in Figure 7. Here we predict the effect of extinction over the SDSS North region using a cosecant law, the Burstein & Heiles (1982) map, and the Stark *et al.* (1992) HI map. These maps predict similar clustering power; it is the large-scale cosecant-like variation in extinction that causes the trouble. Restricting the analysis to somewhat higher Galactic latitude ($|b| > 40°$) only slightly ameliorates the problem. Failure to notice that we live in a Galaxy that is laced with dust would cause one to infer that the power spectrum of the universe suddenly rises in power-law fashion beyond $200 - 300h^{-1}$ Mpc.

Fortunately, we are not oblivious to the effects of dust and we plan to make some correction for this effect. The SDSS will select spectroscopic targets after applying an *a priori* correction for extinction to the apparent magnitudes, thus constructing a sample that is uniform in an extragalactic sense. As of this date we plan to use the extinction maps that have

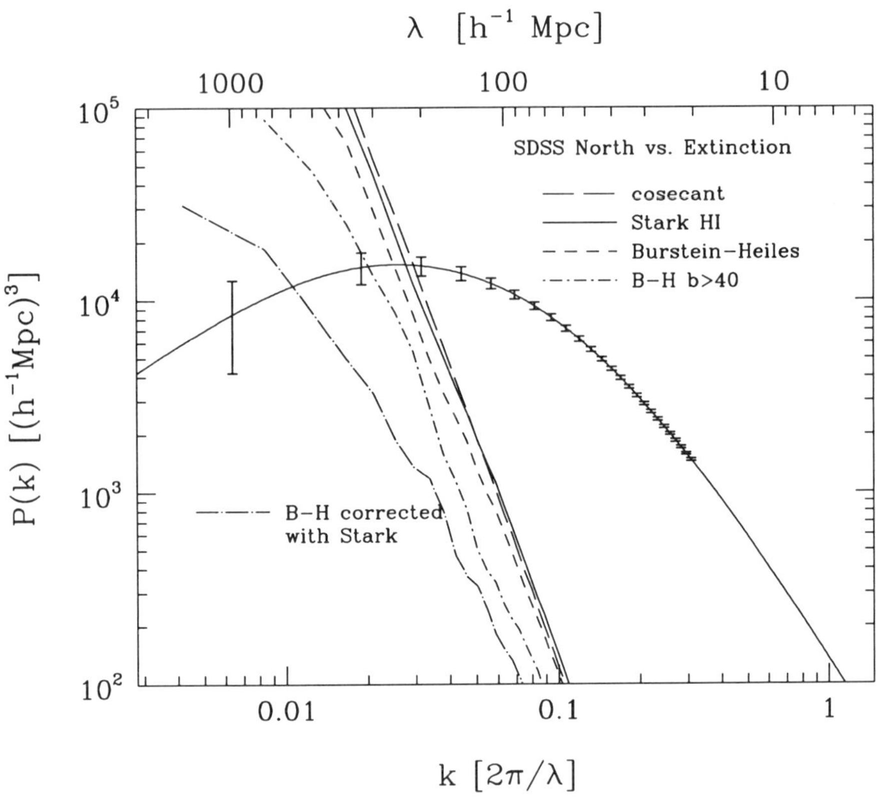


*Figure 7.* False observed clustering power that would be caused by Galactic extinction, for different assumed extinction maps. For comparison is an $\Omega h = 0.25$ CDM power spectrum and the expected uncertainties for a volume-limited sample of SDSS North. If we make no correction for extinction, then the false clustering power exceeds the true power for $\lambda \gtrsim 300h^{-1}$ Mpc. Cutting back the edges of the survey to $b > 40°$ has only a small effect. If we correct the B&H map using a crudely-calibrated map derived from the Stark *et al.* HI maps, the residual power exceeds the true power for scales $\lambda \gtrsim 700h^{-1}$ Mpc.

been constructed from the DIRBE and IRAS satellite data by Schlegel, Finkbeiner, & Davis (1998). If we or others later construct a better extinction map, we will take the residuals into account when we analyze the data. The erroneous clustering power that we really need to worry about is caused by unknown residuals between our best extinction map and the true Galactic extinction.

The lower dot-dashed line in Figure 7 shows what happens if the true

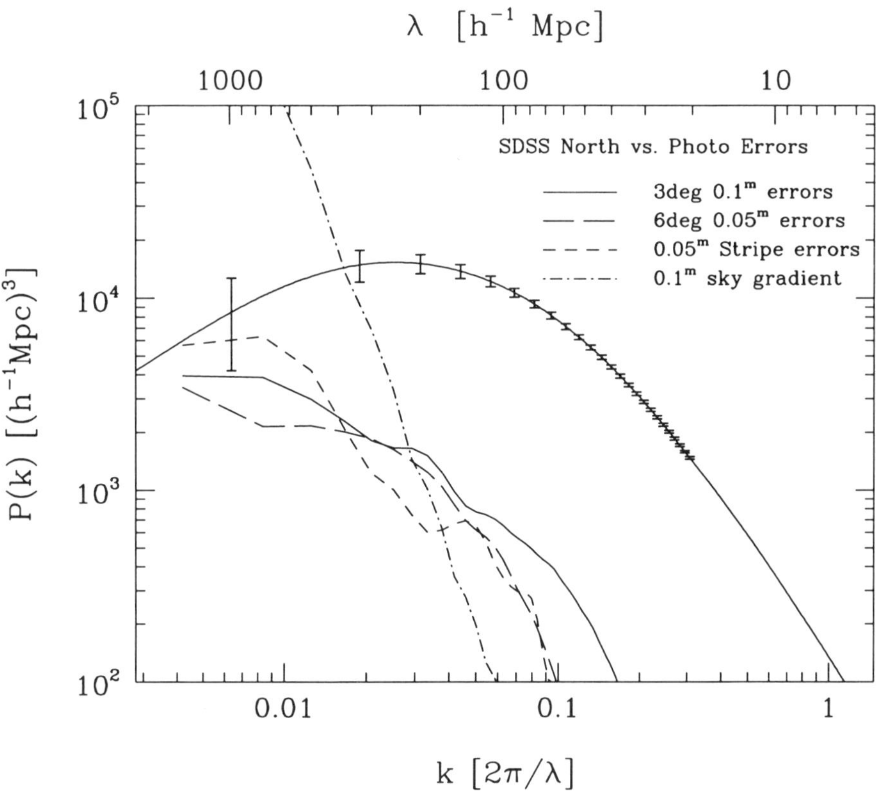


*Figure 8.* False clustering power due to fluctuations of the photometric zeropoint (or errors in the extinction map) on 3° and 6° scales, or between stripes of the SDSS. Also shown is the effect of a $0.1^m$ gradient across the sky.

extinction is described by the Burstein-Heiles map and we use a crudely-calibrated (here we applied our own *ad hoc* dust-to-gas relation) version of the Stark *et al.* HI map to estimate and correct for the extinction. Even with this roughly calibrated map, the extinction power is below the true power on scales smaller than $700h^{-1}$Mpc. A well-calibrated map would allow us to accurately probe the entire range of accessible scales.

## 4.3. THE POWER SPECTRUM OF PHOTOMETRIC ERRORS

Similar to the effects of extinction, errors in photometry that are correlated over a patch of sky will modulate the apparent density of galaxies. Random (uncorrelated between galaxies) errors will contribute to the notorious Malmquist bias; here we consider only correlated photometric errors, such as those that might arise from variations in the zero point.

Figure 8 shows the extra clustering power in the SDSS-500 sample that we would observe if the photometric zero-point varies by $0.1^m$ over patches that are 3° in diameter (the field of view of the SDSS camera), or by $0.05^m$ over patches that are 6° in diameter (the size of a Schmidt plate). These curves should match on scales much larger than the patch size, but vary somewhat because we compute these curves using a Monte Carlo realization of the errors. Even such gross errors in photometry produce erroneous clustering that is less than 10% of the true power for $\lambda < 10^3 h^{-1}$Mpc.

The relatively small amplitude of power caused by even $0.1^m$ photometry errors over the field of view of the SDSS camera is very good news for the extinction problem. This result implies that we can cure the extinction problem if we can use the SDSS data themselves to construct an extinction map that has random field-field errors that are smaller than $0.1^m$ (but, of course, without any large-scale systematic errors). Variations in faint galaxy counts and colors and colors of hot Galactic halo subdwarfs offer several means for constructing such a map, which would provide an independent method for computing extinction corrections.

The lesson of the extinction errors is that large-scale gradients, rather than small-scale random photometry errors, are the real worry. Figure 8 also shows the extra clustering power that arises if there is a zero-point gradient across the sky (pole to pole) of $0.1^m$ or if the zero-point randomly varies by $0.05^m$ between the "stripes" of the SDSS. In other words, the latter illustrates what happens if the photometry is consistent within each $\sim$ $3° \times 120°$ scan region, but the calibration zero-point varies between disjoint regions. Even such a severe error (which we would no doubt notice by comparing the overlap regions of the scans) would produce extra power that is below 10% of the true power on scales $< 10^3 h^{-1}$Mpc. To be certain, we would not be happy to have a 10% effect in our measurements. The point is that this is clearly an upper bound on the magnitude of such an effect.

It is important to note that, just as examination of the anisotropy of clustering will reveal the effects of redshift distortion on the power spectrum, comparison of clustering in the angular and radial directions will provide a test of false clustering due to extinction or photometric error. Cross-correlation of the angular clustering of samples that are selected to lie at different distance would clearly reveal the signature of correlated magnitude errors.

## 5. EVOLUTION: THE UNIVERSE, GALAXIES, AND CLUSTERING

What is the "present epoch?" For statistical analyses of clustering, a useful definition is the redshift range beyond which ignorance of, or inability to correct for, the effects of evolution of galaxies and their clustering becomes the dominant source of uncertainty. In this section I examine how the apparent power spectrum is affected by (1) the mapping between redshift and comoving coordinate distance, (2) galaxy evolution, and (3) clustering evolution. Careful analysis will allow us to learn about evolution; the question here is, when do these effects begin to dominate the apparent clustering?

### 5.1. KNOW THY COSMOLOGY

The relationship between redshift and comoving coordinate distance depends on the deceleration parameter $q_0$,

$$r(z) = \frac{c}{H_0 q_0^2 (1+z)} \left[ q_0 z + (1-q_0)\left(1 - \sqrt{1+2q_0 z}\right)\right]. \tag{4}$$

Out to redshifts of a few thousand km s$^{-1}$, it matters very little which value of $q_0$ we use. But at larger distance, this transformation begins to affect galaxy clustering in several ways. Using the wrong $q_0$ will cause fluctuations on a fixed comoving scale to appear as fluctuations on different apparent scales as a function of redshift. Over the redshift range probed by the SDSS this effect would cause only minor smearing of features in the spectrum; varying $q_0$ from 0.1 to 0.5 changes the length scale by only 4% over the redshift range $z = 0$ to $z = 0.1$.

It also follows that the comoving volume element depends on $q_0$, as

$$\frac{dV}{dz} = 2\pi \left(\frac{c}{H_0}\right)^3 \frac{\left[q_0 z + (1-q_0)(1-\sqrt{1+2q_0 z})\right]^2}{q_0^4 (1+z)^3 \sqrt{1+2q_0 z}}. \tag{5}$$

If we assume a value of $q_0$ that is too large, the volume element $dV$ will erroneously decrease with distance, and cause a radial gradient in the apparent galaxy density. Varying $q_0$ from 0.1 to 0.5 raises the comoving galaxy density by 8% at redshift $z = 0.1$.

Because construction of volume-limited samples relies on using the redshift to compute the absolute magnitudes of the galaxies, an error in $q_0$ translates into an error in the inferred absolute magnitudes,

$$\Delta M = 5 \log \left[ r(z, q_0^{true}) / r(z, q_0^{assumed}) \right]. \tag{6}$$

This error in the absolute magnitudes yields a gradient in apparent galaxy density, as described by equation 2. Using too large a value of $q_0$, we would

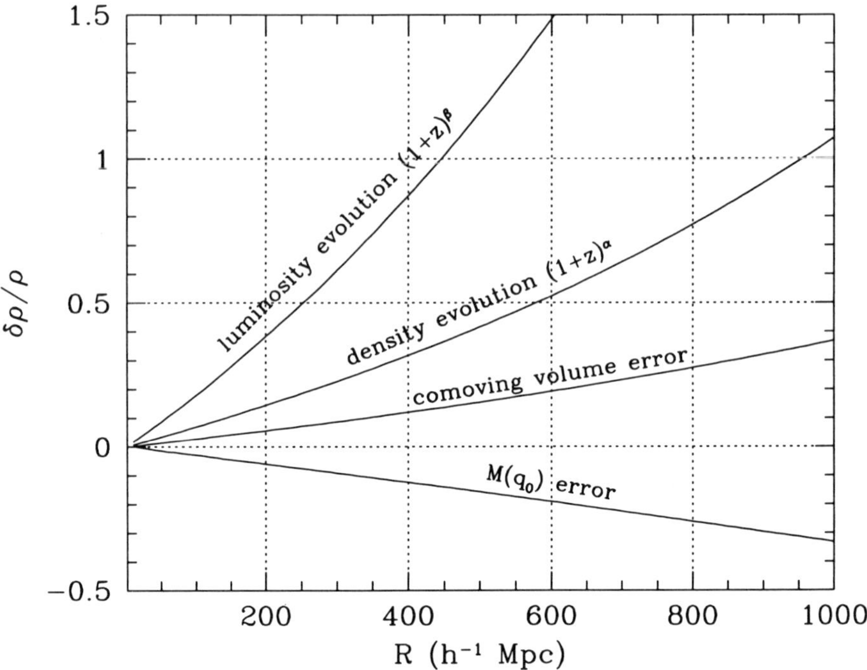


*Figure 9.* Density gradients caused by (from top to bottom) galaxy luminosity evolution with $\beta = 2.3$, density evolution with $\alpha = 2$, error in the computed comoving volume, and error in the computed absolute magnitudes in an absolute-magnitude limited sample of the SDSS. The latter two cosmological effects are shown for the case in which the true $q_0 = 0.1$ but we have used $q_0 = 0.5$ to relate redshifts to comoving coordinate distances.

infer that galaxies are fainter than their true luminosity and exclude some from our sample, thus lowering the galaxy density at large redshift. At redshift $z = 0.1$, varying $q_0$ from 0.1 to 0.5 lowers the density by 7%.

Figure 9 shows the density gradients in the SDSS-500 volume-limited sample that would result from assuming $q_0 = 0.5$ if the true value is $q_0 = 0.1$. Interestingly, the rise in apparent density that results from the change in comoving volume element nearly cancels the absolute-magnitude effect (this cancellation is not universally true because the absolute-magnitude effect depends on the slope of the integrated luminosity function at the absolute magnitude limit).

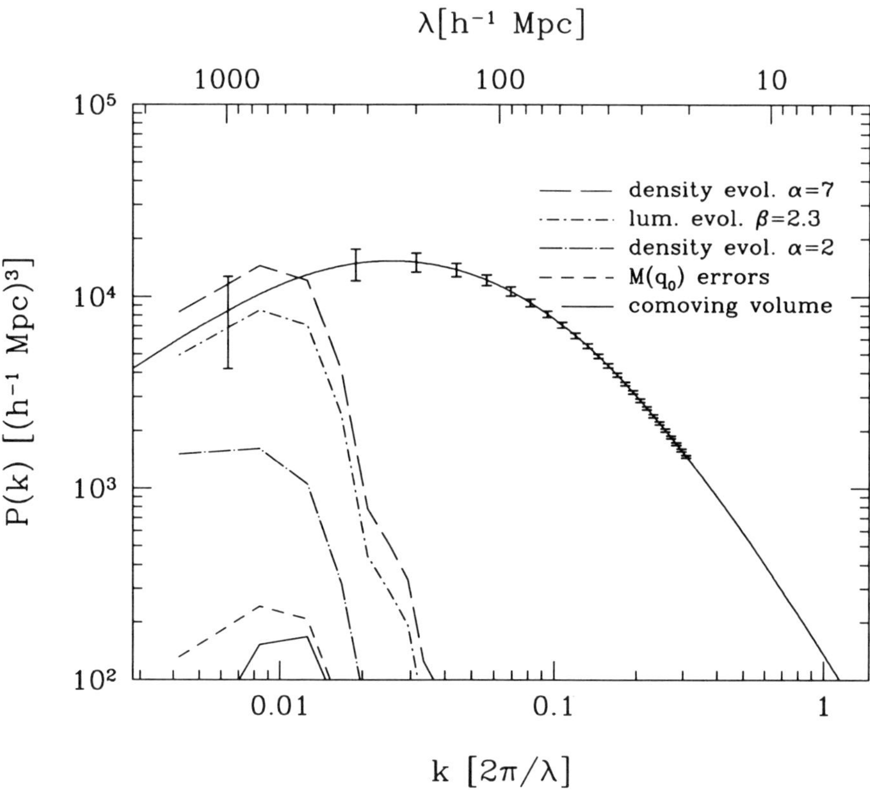


*Figure 10.* False clustering power that would be observed in the SDSS-500 volume-limited sample by (top to bottom) galaxy density evolution with $\alpha = 7$, galaxy luminosity evolution with $\beta = 2.3$, density evolution with $\alpha = 2$, and from gradients caused by using the wrong value of $q_0$. In every case, the sharp turnover at small $k$ is an artifact of our failure to correct for a numerical effect.

Figure 10 illustrates the extra clustering power that results from these effects, considered separately. Both the comoving volume effect and the absolute magnitude effect cause negligible power because the gradient is so shallow. Even if these gradients did not cancel, they would not be a cause for worry in samples that extend to $R = 500h^{-1}$ Mpc.

## 5.2. GALAXY EVOLUTION

Also shown in Figures 9 and 10 are the density gradients and extra clustering power predicted for galaxy luminosity or density evolution. Here we model density evolution as $\bar{n}(z) \propto (1+z)^\alpha$ with $\alpha = 2$ shown in Figure 9, and both $\alpha = 7$ (upper curve) and $\alpha = 2$ (middle curve) in Figure 10. If $\alpha$ is much larger than 2, this effect could cause trouble in the SDSS samples. For example, Saunders *et al.* (1990) find evidence for $\alpha = 7$ in the IRAS-selected galaxy luminosity function. We model luminosity evolution as $L(z) \propto (1+z)^\beta$ with $\beta = 2.3$, as fit by Lilly *et al.* (1995) to their $I$-band luminosity functions. Saunders *et al.* find $\beta = 3$ for IRAS galaxies. Figure 10 shows that this type of luminosity evolution would contribute a significant amount of clustering power on the largest scales probed by the SDSS-500 sample.

A related effect is the difficulty of computing the K-corrections for galaxies over a large range of redshift. Out to redshift $z = 0.2$, failing to perform the proper K-corrections in $r'$ (the selection band for SDSS spectroscopy) causes an effect on the galaxy samples that is somewhat smaller than the luminosity evolution effect.

The lesson here is that one cannot treat galaxies as indistinguishable "points" for the purpose of power spectrum analysis. We already know this from, for example, luminosity bias, whereby very bright galaxies are more strongly clustered than the average (*e.g.*, Park *et al.* 1994). As we push to larger redshift, we must first characterize the redshift dependence of the intrinsic properties of galaxies. Only then can we compute with confidence the clustering statistics of samples that are either homogeneous in their clustering properties or that have known dependence of clustering on galaxy species. Computing the requisite multivariate luminosity functions and spectral evolution of galaxies requires multiband photometry and homogeneous samples with good resolution spectra. The SDSS will provide both of these desiderata.

### 5.3. CLUSTERING EVOLUTION

Clustering of galaxies fails to be ergodic as we probe deeper into the universe for yet another reason: the pattern of clustering itself evolves with time. In linear theory, the growing-mode pertubations grow with the scale factor $a$ of the universe such that $P(k,a) \propto D^2(a)$, where $D(a)$ is a function of the cosmic matter density $\Omega_M$ and cosmological constant $\Omega_\Lambda$. Relative to the present epoch, when $D(z) \equiv 1$, the growth factor when the universe had relative scale size $a \equiv (1+z)^{-1}$ is (Carroll, Press, & Turner 1992)

$$\begin{aligned} D(a) &= \frac{5}{2}\frac{\Omega_M}{a}\left[1+\Omega_M\left(\frac{1}{a}-1\right)+\Omega_\Lambda(a^2-1)\right]^{1/2} \qquad (7)\\ &\times \int_0^a \left[1+\Omega_M\left(\frac{1}{a'}-1\right)+\Omega_\Lambda(a'^2-1)\right]^{-3/2} da'. \end{aligned}$$

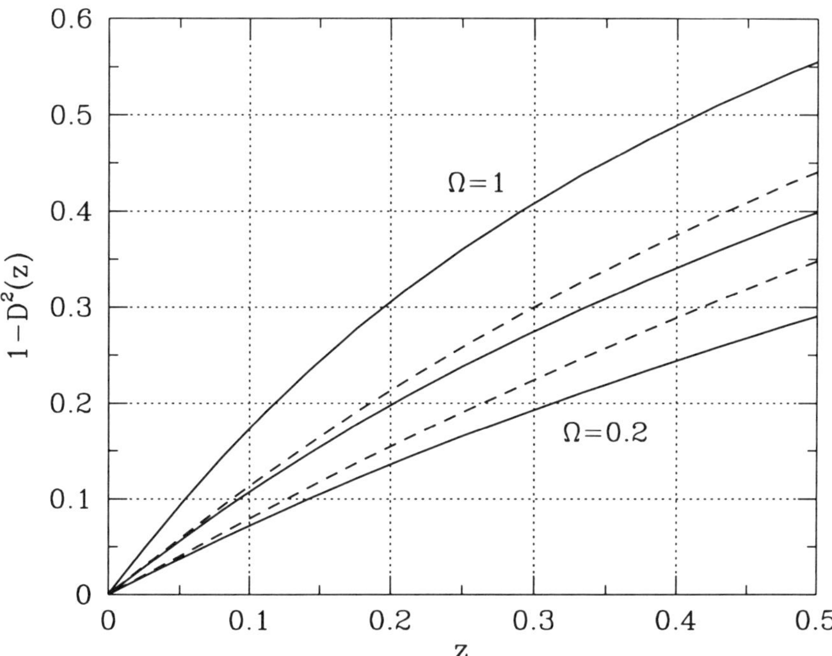


*Figure 11.* Evolution of the clustering amplitude with redshift for (top to bottom) $\{\Omega_M = 1, \Omega_\Lambda = 0\}$, $\{\Omega_M = 0.4, \Omega_\Lambda = 0.6\}$, $\{\Omega_M = 0.4, \Omega_\Lambda = 0\}$, $\{\Omega_M = 0.2, \Omega_\Lambda = 0.8\}$, and $\{\Omega_M = 0.2, \Omega_\Lambda = 0\}$. Plotted here is the difference in clustering amplitude relative to $z = 0$.

For $\Omega_M = 1$ and $\Omega_\Lambda = 0$, this relation is simply $D(z) = (1+z)^{-1}$.

Figure 11 plots the apparent decrement of clustering amplitude relative to $z = 0$, $D^2(z=1) - D^2(z) = 1 - D^2(z)$, for a few choices of spatially flat and open cosmologies. By redshift $z = 0.2$, the decrement is 30% for $\Omega_M = 1$ and 12% for an $\Omega_M = 0.2$ open universe. The difference in the decrement for these models is then larger than the statistical errors in the power spectrum on most scales for a SDSS sample to this depth.

In other words, the cosmology-dependence of the growth factor becomes the dominant source of uncertainty in estimation of the power spectrum and we reach a fundamental limit on our ability to measure the power spectrum in an model-independent fashion. This is not a problem when comparing with models that specify the cosmology, because one simply predicts the power spectrum that would be observed over the specified volume for each model. But it is not possible to compare local measures of clustering with observations at larger redshift without specifying the cosmology. For example, apparent variation in the clustering amplitude of a galaxy species that are due to galaxy evolution might be degenerate with variation of the cosmology. Thus ends the "present epoch" for the purpose of statistical large-scale structure.

On the other hand, given the large divergence of clustering amplitude for models that might otherwise match at $z = 0$, the strong cosmology-dependence of the growth factor seems like a powerful way to constrain $\Omega_M$ and $\Omega_\Lambda$. Given all of the caveats above regarding galaxy evolution, we must find a sample of objects that (1) has negligible (or, at least, very well understood) intrinsic evolution out to a redshift of a few tenths (2) are sufficiently numerous that shot noise does not dominate over the cosmic variance, and for which we (3) can easily obtain redshifts (either spectroscopic or photometric). One must use a single volume-limited sample, because the decrease of clustering amplitude with redshift might be masked by the effects of luminosity bias (brighter galaxies cluster more strongly) in a magnitude-limited sample. A good candidate for such an analysis will be the BRG sample of the SDSS, which will extend to $z = 0.45$. The brightest, reddest galaxies evolve the least and have strong spectral features that ease redshift determination for these objects. Dividing this redshift sample into broad bins of redshift and averaging the power estimates over broad bands should allow determination of the clustering amplitude to within a few percent at each redshift.

## 6. CONCLUSIONS

The large sample volume and dense sampling of the galaxy distribution in the SDSS redshift survey will set lower bounds to the uncertainties that could, in principle, allow detection of small deviations from smooth spectra and easily differentiate between similar cosmological models. Clever use of the anisotropy of clustering would allow simultaneous estimation of both $\Omega_M$ and the real-space power spectrum. Further, we could use the growth rate of clustering to constrain $\Omega_M$ and $\Omega_\Lambda$ in a different way. However, all of these dreams hinge on our ability to cleanly separate true galaxy clustering from contamination by extinction and photometric error, and to differentiate between clustering evolution and evolution of the galaxies themselves.

Thus, the future of large-scale structure lies in studying evolution, both of galaxies and of the clustering pattern itself. This evolution is strongly species-dependent and will require estimation of multivariate luminosity functions to interpret. In a few years' time, the SDSS, together with the 2MASS infrared survey and FIRST radio surveys, which will also cover (really, have already begun to cover) the same region of sky, will provide a database of optical, near-IR, and radio information about millions of objects and will be ideal for just this sort of investigation.

The author acknowledges the collaborative effort of many of the participants in the SDSS, in particular Andy Connolly and Alex Szalay, that contributed to the results presented in this volume. Support for this work was provided by NASA through grant HF-01078.01-94A from the Space Telescope Science Institute, which is operated by AURA, Inc. under NASA contract NAS5-26555,

## References

Bardeen, J.M., Bond, J.R., Kaiser, N., & Szalay, A.S. 1986, ApJ, 304, 15.
Baugh, C.M., & Efstathiou, G. 1993, MNRAS 265, 145.
Burstein, D., & Heiles, C. 1982, AJ, 87, 1165.
Carroll, S.M., Press, W.H., & Turner, E.L. 1992, ARAA, 30, 499.
Colless, M. 1998, Phil. Trans. R. Soc. Lond. A, submitted (preprint astro-ph/9804079).
Connolly, A.J., Csabai, I., Szalay, A., Koo, D.C., Kron, R.G., & Munn, J.A. 1995, AJ, 110, 2655.
da Costa, L.N., *et al.* 1994, ApJ, 424, L1.
da Costa, L.N., Vogeley, M.S., Geller, M.J., Huchra, J.P., & Park, C. 1994, ApJL 437, 1.
Eisenstein, D.J., & Hu, W. 1998, ApJ, 496, 605.
Feldman, H., Kaiser, N., & Peacock, J. 1994, ApJ 426, 23.
Fisher, K.B., Davis, M., Strauss, M.A., Yahil, A., & Huchra, J.P. 1993, ApJ 402, 42.
Fisher, K.B., Huchra, J.P., Davis, M., Strauss, M.A., Yahil, A., & Schlegel, D. 1995, ApJSuppl, 100, 69.
Geller, M.J., & Huchra, J.P. 1989, Science 246, 897.
Giovanelli, R., & Haynes, M.P. 1984, AJ, 89, 1.
Gunn, J.E. & Weinberg, D.H. 1995, in Wide Field Spectroscopy and the Distant Universe, proc. of 35th Herstmonceux Conf., eds. S.J. Maddox and A. Aragón-Salamanca, 3.
Hamilton, A.J.S. 1998, this volume, pp 185-274.
Kaiser, N. 1987, MNRAS, 227, 1.
Landy, D.S., Shectman, S.A., Lin, H., Kirshner, R.P., Oemler, A.A., & Tucker, D. 1996, ApJL 456, 1.
Lilly, S.J., Tresse, L., Hammer, F., Crampton, D. & Le Fevre, O., 1995, ApJ, 455, 108.
Lin, H., Kirshner, R.P., Shectman, S.A., Landy, S.D., Oemler, A., Tucker, D.L., & Schechter, P.L. 1995, ApJ, 471, 617.
Park, C., Vogeley, M.S., Geller, M.J., & Huchra, J.P. 1994, ApJ 431, 569.
Peacock, J.A. 1997, MNRAS, 284, 885.
Peacock, J.A., & Nicholson, D. 1991, MNRAS, 253, 307.
Primack, J.R., Holtzman, J., Klypin, A., & Caldwell, D.O. 1995, Phys. Rev. Lett., 74, 2160
Saunders, W., Rowan-Robinson, M., Lawrence, A., Efstathiou, G., Kaiser, N., Ellis, R.S., & Frenk, C.S. 1990, MNRAS, 242, 318.
Saunders, W. *et al.* 1994, in Wide Field Spectroscopy and the Distant Universe, proc. of 35th Herstmonceux Conference, eds. S.J. Maddox and A. Aragón-Salamanca, 88.
Schlegel, D.J., Finkbeiner, D.P., & David, M. 1998, ApJ, 499, in press.
Shectman, S.A., Landy, S.D., Oemler, A., Tucker, D.L., Lin, H., Kirshner, R.P., & Schechter, P.L. 1996, ApJ 470, 172.

Stark, A.A., Gammie, C.F., Wilson, R.W., Bally, J., Linke, R.A., Heiles, C., & Hurwitz, M. 1992, ApJS, 79, 77.
Szalay, A.S. 1998, this volume, pp 277-289.
Tadros, H., & Efstathiou, G. 1995, MNRAS, 276, L45.
Tegmark, M., Hamilton, A.J.S., Strauss, M.A., Vogeley, M.S., & Szalay, A.S. 1998, ApJ, 499, 555.
Vogeley, M.S., & Connolly, A.J. 1998, in preparation.
Vogeley, M.S., & Szalay, A.S. 1996, ApJ, 465, 34.
Wright, E.L., Bennett, C.L., Gorski, K., Hinshaw, G., & Smoot, G.R. 1996, ApJL, 464, L21.

# SEARCH FOR PRIMEVAL GALAXIES WITH THE CALAR ALTO DEEP IMAGING SURVEY

E.THOMMES, K.MEISENHEIMER,
R. FOCKENBROCK, H. HIPPELEIN, H.-J. RÖSER
*Max-Planck-Institut für Astronomie,*
*Königstuhl 17, D-69117 Heidelberg, Germany*

The **C**alar **A**lto **D**eep **I**maging **S**urvey (CADIS) is a very deep emission line survey (using a Fabry-Pérot (FP)), combined with deep broad- and medium-band photometry. This survey project is specifically designed to detect primeval galaxies (PGs), but it will in addition produce a large data base for the investigations of the faint end of the luminosity function and three dimensional correlation function of faint galaxies at intermediate redshifts ($0.2 < z < 1.2$). Its multifilter technique also allows the classification and redshift determination of early type galaxies, of faint QSOs beyond z=3, and of faint M stars.

The central point of the strategy is the search for faint emission line objects ($S_{lim}(5\sigma) \approx 3 \times 10^{-20}\mathrm{W/m^2}$) with an imaging Fabry-Pérot in the wavelength intervals [694,706nm], [814,826nm] and [910,926nm]. These wavelength intervals are located in prominent windows of the night sky emission, corresponding to Ly-$\alpha$ redshifts of z=4.75, 5.75 and 6.55. The very deep flux limit for line detection allows one to detect star formation rates (SFRs) of 10 to 50 $M_{\odot}$/yr in PGs at $z \geq 5$. A completely new feature of the CADIS strategy are very deep images in a set of narrow band filters ($R = \lambda/\Delta\lambda \approx 50$, typical exposure times: 10 h at a 2.2 m telescope), which are selected in such a way, that for every prominent emission line of a foreground object falling into the Fabry-Pérot scan (e.g. $H\alpha$) a second or third line (e.g. [OII]372.7nnm or [OIII]500.7nm ) should show up in them (fig. 1). We call these narrow band filters veto-filters since a signal in one of them excludes that the line detected in the Fabry-Pérot is Ly-$\alpha$. Combining the veto-filter information with the accurate spectroscopy through the Fabry-Pérot, we expect to determine the redshift of the majority of emission line galaxies ($0.25 < z < 1.0$) to an accuracy of 100 ... 200 km/s (depending on the S/N ratio). Since the detection of emission line objects requires several "off-line" continuum exposures anyway, we decided to add in a complete set



*D. Hamilton (ed.), The Evolving Universe,* 419–422.

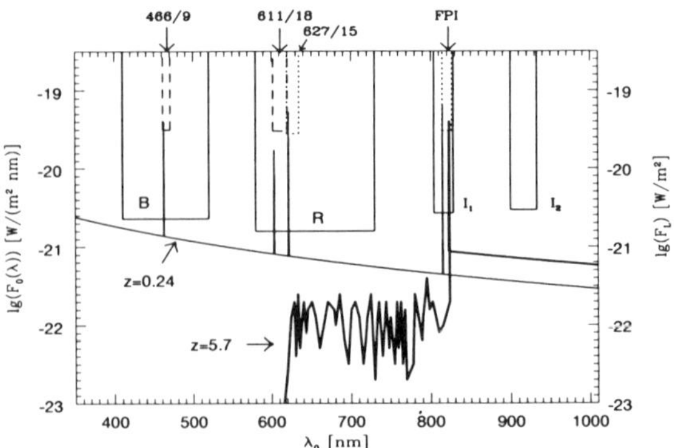


*Figure 1.* *Demonstration of the veto-filter strategy. Solid boxes indicate the band width and $5\sigma$ limit of the deepest broad and medium band filters (B, R, $I_1$, $I_2$ refer to the scale on the left). The dotted box represents the $5\sigma$ limit for line detection in the Fabry-Pérot scan, and dashed boxes refer to $3\sigma$ limits in the veto-filters 466/9, 611/18 and 627/15 (scale on the right). Note that a typical Ly-$\alpha$ galaxy at $z = 5.7$ will only be detected in the Fabry-Pérot, the I-bands and perhaps the R-band, whereas a foreground galaxy at $z = 0.24$ with very strong emission lines should show up in both 466/9, 627/15 and perhaps 611/18.*

of broad band and medium band images, which are optimized both for the continuum determination and the identification of one of the most severe contaminations - faint M-stars in our Galaxy. The optical multi-color survey is supplemented by a K' ($5\sigma$ limit 21.0 mag) survey with the new OMEGA camera at the prime focus of the Calar Alto 3.5 m-telescope ($6.6 \times 6.6$ □′ field). This gives us a global view of the spectral energy distribution (SED) of every object in the field. The narrow band and medium band images will allow us to discriminate very effectively between foreground emission line objects and good candidates for Ly-$\alpha$ emitting PGs at high redshifts (Fig. 1). CADIS will survey 10 fields (each 120 □′) distributed over the northern sky $\delta \geq -5°$, which were selected for their absence of bright stars ($R \leq 16.0$) and an extra-ordinary low flux on the IRAS 100$\mu$m maps ($\leq$ 2 MJy/sr). Thus the total survey area will be at least 0.3 □°. CADIS will use the 2.2m- and the 3.5m-telescopes at Calar Alto (Spain). The details of the CADIS survey are given in Meisenheimer *et al.* 1997 .

Here we present some first results which we got from the first data recorded with the CADIS strategy. These data were taken with the 2.2m telescope at Calar Alto in the CADIS field 9H (with slightly different filters than the standard CADIS filter). We employed a 1k×1k CCD (field of view 8'× 8'). We got four FP settings in the wavelength region 814nm to 818.5nm (resolution=1.8nm). Every setting consists of 7 individual exposures of 1500 s integration. We reached a $5\sigma$ detection limit of $S_{lim}(5\sigma) \approx 5\times10^{-20}\mathrm{W/m^2}$. To get an estimate of the continuum near the emission lines, we did exposures with a filter $\lambda/\Delta\lambda$=812/17 nm ($F_{lim}(5\sigma) \approx 6 \times 10^{-21}$W/(m$^2$nm)). The FP exposures were supplemented by broad band exposures with the filters $BV$ (centered at 500 nm, $5\sigma$ limit $\approx 25.^m3$), $R_c$ ($5\sigma$ limit $\approx 24.^m8$) and $I$ ($5\sigma$ limit $\approx 22.^m5$). Further narrow band exposures with the filters 466/9 ($F_{lim}(5\sigma) \approx 13 \times 10^{-21}$W/(m$^2$nm)), 612/10 ($F_{lim}(5\sigma) \approx 11 \times 10^{-21}$W/(m$^2$nm)) and 614/28 ($F_{lim}(5\sigma) \approx 7 \times 10^{-21}$W/(m$^2$nm)) enable to detect further emission lines of foreground objects.

In this exploratory CADIS data set we already found 147 emission line galaxies, which were selected by the requirement, that they have a 5 $\sigma$ detection in at least one FP wavelength setting and that the line flux exceeds the continuum by more than $3.5\sigma$. 104 of these show at least a marginal detection in the BV band and are therefore classified as faint blue galaxies in the foreground. 74 of these emission line objects with a blue detection in addition show signals in the additional narrow band exposures. These allowed us to identify them as being either at $z \approx 0.24$ or $z \approx 0.63/0.68$. Of the 43 emission line objects without detection in the blue band, 35 show at least a marginal detection in one of the additional narrow band filters and therefore also could be identified as galaxies at $z \approx 0.24$ or $z \approx 0.63/0.68$. From the remaining 8 galaxies **7** are promising candidates for **Ly-$\alpha$ emitting PGs at** $z \approx 5.7$. Three of these 7 candidates have no continuum flux shortward of the emission line at $\lambda \approx 816$ nm and 4 candidates showed a marginal detection in our red filter ($R \leq 26.0$). Five of the candidates are clearly resolved ($\geq 2''$), while two objects may be unresolved (see Fig. 2).

About 30 % of the foreground emission line objects have no detectable continuum in the BV and R band. These abundant galaxies constitute a new class of objects which has been overlooked by broad band selected redshift surveys. We derived lower limits to the blue band luminosity function from our data for the emission line objects in the redshift bins 0.24 and 0.64. Although the area covered by the present data ( $\approx 60\ \square'$ ) is yet too small to draw definite conclusions our results are in good agreement with the results of the Autofib redshift survey ( see Colles, 1997 and Thommes *et. al.*, 1997 ). We expect that our sample of 7 candidates for primeval Ly-$\alpha$ emitting galaxies at $z \approx 5.7$ is still contaminated by several types of foreground objects. Statistical considerations indicate, that artifacts and emission line galaxies at $z = 0.24$ and $z = 0.63$ could hardly account for more than three of the candidates. The unknown fraction of galaxies at $z = 1.2$ with strong [OII] line but undetectable blue continuum could however well make up for half of the candidates. Therefore, high S/N ratio slit-spectroscopy with a 10m-class telescope at medium resolution ($\Delta\lambda \approx 0.5$ nm) is required to identify the true Ly-$\alpha$ galaxies among our candidates. We are pretty optimistic that at least some of the objects will, in fact, turn out to be at $z \approx 5.7$.

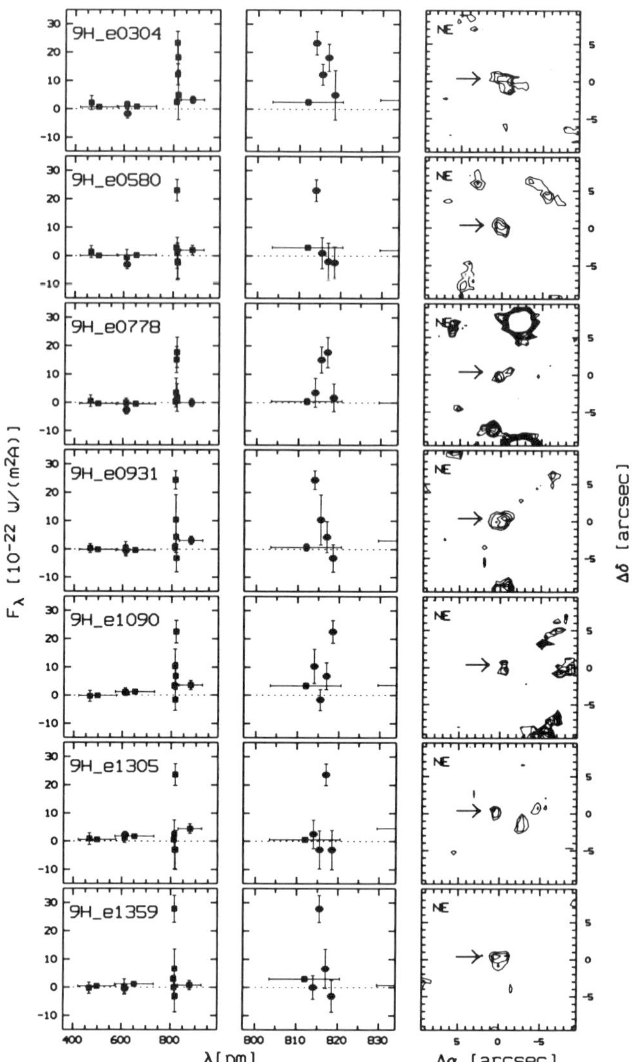


*Figure 2.* *Photometric spectra (left panel) and contour plots (right panel) of the PG candidates. The magnified part (panel in the middle) shows the wavelength region of the 4 Fabry-Pérot settings at 814.0, 815.5, 817.0 and 818.5 nm.*

## References


Meisenheimer, K., Beckwith, S., Fockenbrock, R., Fried, J., Hippelein, H., Hopp, U., Leinert, C., Röser, H.-J., Thommes, E. and Wolf, C., 1997, in *The Early Universe with the VLT*, Proceedings of the ESO Workshop Held at Garching, Germany, 1-4 April 1996, ed. Bergeron, J., Springer, Berlin Heidelberg New York, 165

Thommes, E., Fockenbrock, R., Hippelein, H., Meisenheimer, K., Röser,H.-J., 1997, in *The Early Universe with the VLT*, ed. Bergeron, Springer, 1997, 173

Colles, M., 1997, in *The Early Universe with the VLT*, ed. Bergeron, Springer, 1997, 87

INDEX

**Telescopes & Instruments**